Vector Integration and Stochastic Integration in Banach Spaces

PURE AND APPLIED MATHEMATICS

A Wiley-Interscience Series of Texts, Monographs, and Tracts

Founded by RICHARD COURANT
Editors Emeriti: PETER HILTON and HARRY HOCHSTADT
Editors: MYRON B. ALLEN III, DAVID A. COX, PETER LAX,
 JOHN TOLAND

A complete list of the titles in this series appears at the end of this volume.

Vector Integration and Stochastic Integration in Banach Spaces

Nicolae Dinculeanu

A Wiley-Interscience Publication
JOHN WILEY & SONS, INC.
New York • Chichester • Weinheim • Brisbane • Singapore • Toronto

This text is printed on acid-free paper. ∞

Copyright © 2000 by John Wiley & Sons, Inc.

All rights reserved. Published simultaneously in Canada.

No part of this publication may be reproduced, stored in a retrieval system or transmitted in any form or by any means, electronic, mechanical, photocopying, recording, scanning or otherwise, except as permitted under Section 107 or 108 of the 1976 United States Copyright Act, without either the prior written permission of the Publisher, or authorization through payment of the appropriate per-copy fee to the Copyright Clearance Center, 222 Rosewood Drive, Danvers, MA 01923, (978) 750-8400, fax (978) 750-4744. Requests to the Publisher for permission should be addressed to the Permissions Department, John Wiley & Sons, Inc., 605 Third Avenue, New York, NY 10158-0012, (212) 850-6011, fax (212) 850-6008, E-Mail: PERMREQ @ WILEY.COM.

For ordering and customer service, call 1-800-CALL-WILEY.

Library of Congress Cataloging in Publication Data:

Dinculeanu, N. (Nicolae)
   Vector integration and stochastic integration / Nicolae Dinculeanu.
      p. cm — (Pure and applied mathematics series)
   Includes bibliographical references and index.
   ISBN 0-471-37738-4 (cloth : alk. paper)
     1. Stochastic integrals.   2. Banach spaces.   3. Vector spaces.   I. Title.   II. Pure and applied mathematics (John Wiley & Sons : Unnumbered)

QA274.22.D56   2000
519.2—dc21
                                                                          99-054734

Printed in the United States of America

10 9 8 7 6 5 4 3 2 1

Preface

This book consists of two parts. The first part, Chapter 1, is devoted to vector integration. The rest of the book, chapters 2–10, is devoted to stochastic integration in Banach spaces.

Vector integration of various kinds has been presented in other books. We mention especially the books by N. Dunford and J. Schwartz [D–S], N. Dinculeanu [D.1], J. Diestel and J. J. Uhl Jr. [D–U] and A. U. Kussmaul [Kus.1]. In the text, we refer the reader to these books for the proof of some important theorems that we do not want to repeat.

The core of Chapter 1 is §5, devoted to integration of vector-valued functions with respect to vector measures with finite *semivariation*. This kind of integration is not contained in any other book and was presented first in [B–D.2]. Kussmaul [Kus.1] considers a similar kind of integration but only for real-valued functions.

Among the applications of the integral of §5 we quote: the Riesz representation theorem, the integral representation of continuous linear operators on L^p-spaces, the Stieltjes integral with respect to vector-valued functions with *finite semivariation* (which was not considered before) and, especially, the stochastic integration in Banach spaces.

The reader interested in integration theory only, could use only chapter 1 and the paragraphs 19, 21, 29 and 31.

For the part devoted to stochastic integration we assume familiarity with the definitions and the results of the general theory of stochastic processes, as presented, for example, in the book by C. Dellacherie and P. A. Meyer [D–M] and often we refer the reader to this book.

The classical theory of stochastic integration for real-valued processes, reduces, essentially, to integration with respect to a square integrable martingale. This is done by defining the stochastic integral first for simple processes and then extending it to a larger class of processes, by means of an isometry between certain L^2-type spaces of processes. This method has been also used by Kunita [Ku] for processes with values in Hilbert spaces, by using the existence of the inner product to prove the above-mentioned isometry. But this approach cannot be used for Banach spaces, which lack an inner product. A new approach is needed for Banach-valued processes.

On the other hand, the classical stochastic integral as described above is not a genuine integral, in the sense that it is not an integral with respect to a measure.

It is desirable, as in classical Measure Theory, to have a space of "integrable" processes with a norm on it, for which it is a Banach space, and to have an integral for integrable processes, which would be the stochastic integral. Also desirable would be to have Vitali- and Lebesgue-type convergence theorems. Such a goal is legitimate and many attempts have been made to fulfill it.

Any measure-theoretic approach to stochastic integration has to use an integration theory with respect to a vector measure. Pellaumail [P] was the first to attempt such an approach, but due to the lack of a satisfactory integration theory, this goal was not achieved. Kussmaul [Kus.1] used the idea of Pellaumail and was able to define a consistent, measure theoretic stochastic integral, but only for real-valued processes. He used in [Kus.2] the same method for Hilbert-valued processes, but the goal was only partially fulfilled, again due to the lack of a satisfactory general integration theory. The integration theory presented here in §5 seems to be tailor-made for application to the stochastic integral. It was presented for the first time in [B–D.2].

In order to apply the integration theory to define a stochastic integral with respect to a Banach-valued process X, we associate to it a measure I_X on the ring $\mathcal{R}$ generated by the predictable rectangular sets. The process X is called *summable*, if I_X can be extended to a σ-additive measure with finite semivariation on the σ-algebra $\mathcal{P}$ of predictable sets. Roughly speaking, the stochastic integral $H \cdot X$ is the process $(\int_{[0,t]} H dI_X)_{t \geq 0}$ of integrals with respect to I_X.

The summable processes play, in this theory, the role played by the square integrable martingales in the classical theory. It turns out that every Hilbert-valued, square integrable martingale is summable and the processes with integrable variation are also summable. In addition, a new class of summable processes emerges: the processes with integrable *semivariation*. Moreover, the stochastic integral with respect to such a process can be computed pathwise, as a Stieltjes integral (itself a Stieltjes integral with respect to a function of finite *semivariation*, rather than finite variation). This new class of summable processes could not be made evident in the classical case of scalar processes, since for these processes, the variation and the semivariation are equal. It is only for processes with values in an infinite dimensional Banach space that the semivariation is different from the variation.

Our space of integrable processes with respect to a summable process is a Lebesgue-type space, endowed with a seminorm, for which it is complete and in which the Vitali and the Lebesgue convergence theorems are valid. The legitimate goal mentioned above is thus fulfilled.

It is worth mentioning the following summability criterion: X is summable iff I_X is bounded and has finite semivariation on the ring $\mathcal{R}$. It is quite unexpected that the mere boundedness of I_X on $\mathcal{R}$ implies not only that I_X is σ-additive on $\mathcal{R}$, but that I_X can be extended to a σ-additive measure on $\mathcal{P}$.

Using the same measure-theoretic approach, we extend the theory of Stochastic integration for vector valued, two-parameter processes, in Chapters 7–10.

The same measure-theoretic approach can be used to extend the theory of stochastic integration for process measures and martingale measures in Banach spaces ([D.9], [D.10], [Di–Mu]). This extension, which is not included in this book, has applications in the theory of stochastic partial differential equations [W.3].

Each chapter is divided into paragraphs, numbered in continuation from one chapter to the last one, from 1 to 32. Each paragraph is divided into several sections indicated anew, in each paragraph, by capital letters. The numbering of definitions and theorems starts anew in each paragraph. The quotation, in the text, of definitions and theorems, is done in the following way: if we refer in a paragraph to a theorem from the same paragraph, then we quote it by its number in that paragraph; if we refer to a theorem from a different paragraph, then we quote it by a pair of numbers, in the form Theorem $a.b$, the first number a indicating the paragraph, and the second number b indicating the number of the theorem in paragraph a.

Gainesville, Florida   N. Dinculeanu
February 26, 1999

Contents

	Preface		*v*
Chapter 1	**Vector Integration**		**1**
§1.	*Preliminaries*		*1*
	A.	*Banach spaces*	*1*
	B.	*Classes of sets*	*2*
	C.	*Measurable functions*	*3*
	D.	*Simple measurability of operator-valued functions*	*7*
	E.	*Weak measurability*	*8*
	F.	*Integral of step functions*	*9*
	G.	*Totally measurable functions and the immediate integral*	*10*
	H.	*The Riesz representation theorem*	*11*
	I.	*The classical integral*	*13*
	J.	*The Bochner integral*	*15*
	K.	*Convergence theorems*	*17*
§2.	*Measures with finite variation*		*20*
	A.	*The variation of vector measures*	*20*

	B.	Boundedness of σ-additive measures	23
	C.	Variation of real-valued measures	24
	D.	Integration with respect to vector measures with finite variation	26
	E.	The indefinite integral	28
	F.	Integration with respect to gm	32
	G.	The Radon–Nikodym theorem	36
	H.	Conditional expectations	43
§3.	σ-additive measures		49
	A.	σ-additive measures on σ-rings	49
	B.	Uniform σ-additivity	50
	C.	Uniform absolute continuity and uniform σ-additivity	52
	D.	Weak σ-additivity	57
	E.	Uniform σ-additivity of indefinite integrals	57
	F.	Weakly compact sets in $L^1_F(\mu)$	60
§4.	Measures with finite semivariation		63
	A.	The semivariation	63
	B.	Semivariation and norming spaces	66
	C.	The semivariation of σ-additive measures	69
	D.	The family $m_{F,Z}$ of measures	71
§5.	Integration with respect to a measure with finite semivariation		77
	A.	Measurability with respect to a vector measure	77
	B.	The seminorm $\tilde{m}_{F,G}(f)$	79
	C.	The space of integrable functions	82
	D.	The integral	85
	E.	Convergence theorems	87
	F.	Properties of the space $\mathcal{F}_D(\mathcal{B},\tilde{m}_{F,G})$	90
	G.	Relationship between the spaces $\mathcal{F}_D(m)$	92
	H.	The indefinite integral of measures with finite semivariation	95
§6.	Strong additivity		102
§7.	Extension of measures		109
§8.	Applications		117
	A.	The Riesz representation theorem	117
	B.	Integral representation of continuous linear operations on L^p-spaces	117

	C. Random Gaussian measures	120

Chapter 2   The Stochastic Integral 123

§9. Summable processes 124
    A. Notations 124
    B. The measure I_X 125
    C. Summable processes 126
    D. Computation of I_X for predictable rectangles 127
    E. Computation of I_X for stochastic intervals 129

§10. The stochastic integral 133
    A. The space $\mathcal{F}_D(\tilde{I}_{F,L_G^p})$ 133
    B. The integral $\int H dI_X$ 134
    C. A convergence theorem 134
    D. The stochastic integral $H \cdot X$ 136

§11. The stochastic integral and stopping times 140
    A. Stochastic integral of elementary processes 140
    B. Stopping the stochastic integral 146
    C. Summabilty of stopped processes 148
    D. The jumps of the stochastic integral 151

§12. Convergence theorems 153
    A. The completeness of the space $L^1_{F,G}(X)$ 153
    B. The Uniform Convergence Theorem 156
    C. The Vitali and the Lebesgue Convergence Theorems 156
    D. The stochastic integral of σ-elementary and of caglad processes as a pathwise Stieltjes integral 157

§13. Summability of the stochastic integral 161

§14. Summability criterion 164
    A. Quasimartingales and the Doléans measure 164
    B. The summability criterion 168

§15. Local summability and local integrability 171
    A. Definitions 171
    B. Basic properties 172
    C. Convergence theorems 175

		D.	Additional properties	179

Chapter 3 Martingales 181

§16.	Stochastic integral of martingales		181
§17.	Square integrable martingales		185
	A.	Extension of the measure I_M	185
	B.	Summability of square integrable martingales	188
	C.	Properties of the space $\mathcal{F}_{F,G}(M)$	190
	D.	Isometrical isomorphism of $L^1_{F,G}(M)$ and $L^2_F(\mu_{\langle M\rangle})$	193

Chapter 4 Processes with Finite Variation 199

§18.	Functions with finite variation and their Stieltjes integral		200
	A.	Functions with finite variation	200
	B.	The variation function $|g|$	201
	C.	The measure associated to a function	203
	D.	The Stieltjes integral	208
§19.	Processes with finite variation		211
	A.	Definition and properties	211
	B.	Optional and predictable measures	216
	C.	The measure μ_X	217
	D.	Summability of processes with integrable variation	233
	E.	The stochastic integral as a Stieltjes integral	236
	F.	The pathwise stochastic integral	239
	G.	Semilocally summable processes	242

Chapter 5 Processes with Finite Semivariation 243

§20.	Functions with finite semivariation and their Stieltjes integral		244
	A.	Functions with finite semivariation	244

	B.	Semivariation and norming spaces	245
	C.	The measure associated to a function	247
	D.	The Stieltjes integral with respect to a function with finite semivariation	250
§21.	Processes with finite semivariation	253
	A.	The semivariation process	253
	B.	The measure μ_X	258
	C.	Summability of processes with integrable semivariation	262
	D.	The pathwise stochastic integral	267
§22.	Dual projections	272
	A.	Dual projection of measures	272
	B.	Dual projections of processes	274
	C.	Existence of dual projections	278
	D.	Processes with locally integrable variation or semivariation	279
	E.	Examples of processes with locally integrable variation or semivariation	281
	F.	Decomposition of local martingales	285

Chapter 6   The Itô Formula   289

§23.	The Itô formula	289
	A.	Preliminary results	289
	B.	The vector quadratic variation	296
	C.	The quadratic variation	298
	D.	The process of jumps	304
	E.	Itô's formula	312

Chapter 7   Stochastic Integration in the Plane   321

§24.	Preliminaries	321
	A.	Order relation in $\mathbb{R}^2$	321
	B.	The increment $\Delta_{zz'} g$	322
	C.	Right continuity	322
	D.	The filtration	323
	E.	The predictable σ-algebra	323
	F.	Stopping times	323

		G.	Stochastic processes	324
		H.	Extension of processes from $\mathbb{R}_+^2 \times \Omega$ to $\mathbb{R}^2 \times \Omega$	325
	§25.		Summable processes	327
		A.	The measure I_X	327
		B.	Summable processes	327
		C.	The seminorm $\tilde{I}_X$ and the space $\mathcal{F}_{F,G}(X)$	328
		D.	The integral $\int H dI_X$	328
		E.	The stochastic integral $H \cdot X$	329
	§26.		Properties of the stochastic integral	331
		A.	Convergence theorems	331
		B.	Extension of I_X to $\mathcal{P}(\infty)$	332
		C.	Existence of left limits of X in L_E^p	332
		D.	Some properties of the integral $\int H dI_X$	333
		E.	Summability of stopped processes	338
		F.	Summability of the stochastic integral	340

Chapter 8   Two-Parameter Martingales   343

	§27.		Martingales	343
	§28.		Square integrable martingales	349
		A.	A decomposition theorem	349
		B.	The measures $\|I_M(\cdot)\|_{L_E^2}^2$ and $\mu_{\langle M \rangle}$	349
		C.	Summability of the square integrable martingales in Hilbert spaces	353
		D.	The space $\mathcal{F}_{F,G}(I_M)$	355
		E.	Isometric isomorphism of $L_{F,G}^1(M)$ and $L_F^2(\mu_{\langle M \rangle})$	357

Chapter 9   Two-Parameter Processes with Finite Variation   363

	§29.		Functions with finite variation in the plane	363
		A.	Monotone functions	363
		B.	Partitions	364
		C.	Variation corresponding to a partition	366
		D.	Variation of a function on a rectangle	366
		E.	Limits of the variation	369

	F.	The variation function $|g|$	374
	G.	Functions with finite variation	375
	H.	Functions vanishing outside a quadrant	376
	I.	Variation of real-valued functions	378
	J.	Lateral limits	379
	K.	Measures associated to functions	385
	L.	σ-additivity of the measure m_g	387
	M.	The Stieltjes integral	390
§30.		Processes with finite variation	391
	A.	Processes with integrable variation	391
	B.	The measure μ_X	391
	C.	Summability of processes with integrable variation	400
	D.	The stochastic integral as a Stieltjes integral	400

Chapter 10 Two-Parameter Processes with Finite Semivariation   403

§31.		Functions with finite semivariation in the plane	403
	A.	Functions with finite semivariation	403
	B.	Semivariation and norming spaces	405
	C.	The measure associated to a function	406
	D.	The Stieltjes integral for functions with finite semivariation in $\mathbb{R}^2$	408
§32.		Processes with finite semivariation in the plane	410
	A.	Processes with finite semivariation	410
	B.	The measure μ_X	411
	C.	Summability of processes with integrable semivariation	412

References   *413*

Chapter 1
Vector Integration

§1. PRELIMINARIES

In this paragraph we establish the notations used in the book and present the immediate integral, the classical integral, and the Bochner integral. Special attention is given to measurability of vector-valued functions.

A. Banach spaces

1. Throughout the book, E, F, G, D will denote Banach spaces. All Banach spaces are over the real field $\mathbb{R}$.

Numbers $\alpha \geq 0$ are called *positive* (rather than nonnegative). A sequence (α_n) of numbers such that $\alpha_n \leq \alpha_{n+1}$ for every n is called *increasing* (rather than nondecreasing).

For any Banach space M, the norm of an element $x \in M$ is denoted by $|x|$ or $|x|_M$; the dual of M is denoted by M^* and the unit ball of M by M_1. The duality between M and M^* is denoted by $\langle x, x^* \rangle$ or x^*x, $\langle x^*, x \rangle$ or even xx^*.

If M is a Hilbert space, the inner product of two elements $x, y \in M$ is denoted by $\langle x, y \rangle$, or $\langle x, y \rangle_M$ or even xy.

The space of bounded linear operators from F into G is denoted by $L(F, G)$. We write $E \subset L(F, G)$ to mean that E is continuously embedded into $L(F, G)$, that is, $|xy| \leq |x|\,|y|$, for $x \in E$ and $y \in F$. Special mention will be made in case the embedding is an isometry.

Examples of such isometries are: $E = L(\mathbb{R}, E)$; $E \subset L(E^*, \mathbb{R}) = E^{**}$; $L \subset L(F, E\hat{\otimes}_\pi F)$; $E \subset L(F, F\hat{\otimes}_\pi E)$; if E is a Hilbert space, $E = L(E, \mathbb{R})$;

if E and F are Hilbert spaces, $E \subset L(F, E \hat{\otimes}_{HS} F)$, where HS denotes the Hilbert–Schmidt norm.

We write $c_0 \not\subset M$ to mean that M does not contain a copy of c_0, that is, M does not contain a subspace which is isomorphic to the Banach space c_0.

If M is a Banach space, a subspace $Z \subset M^*$ is said to be a *norming* space for M, if for every $x \in M$ we have

$$|x| = \sup\{|\langle x, z \rangle| : z \in Z_1\}.$$

Obviously, M^* is norming for M and if we consider $M \subset M^{**}$, then M is norming for M^*.

A useful example of a norming space, which will be used in the sequel, is the following one:

Let $(\Omega, \mathcal{F}, P)$ be a probability space and $1 \leq p \leq +\infty$. Denote $L_E^p = L_E^p(P)$, the space of Bochner-integrable functions $f : \Omega \to E$ with respect to P (see Section J on the Bochner integral). If $\frac{1}{p} + \frac{1}{q} = 1$, then $L_{E^*}^q$ is isometrically contained in $(L_E^p)^*$. If E^* has the Radon–Nikodym Property (RNP) and if $p < \infty$, then $(L_E^p)^* = L_{E^*}^q$. But even if E^* does not have the RNP and even if $p = \infty$, $L_{E^*}^q$ is a norming space for L_E^p. Moreover, if $\mathcal{R}$ is a ring generating the σ-algebra $\mathcal{F}$, the subspace $\mathcal{S}_{E^*}(\mathcal{R})$ of E^*-valued, $\mathcal{R}$-step functions is a norming space for L_E^p.

B. Classes of sets

2. Throughout the first chapter, S is a set and $\mathcal{P}, \mathcal{R}, \mathcal{A}, \mathcal{D}, \mathcal{S}, \Sigma$ are respectively a semiring, a ring, an algebra, a δ-ring, a σ-ring and a σ-algebra of subsets of S.

A *semiring* $\mathcal{P}$ is a class of subsets of S, closed under intersection $A \cap B$ and satisfying the following condition: for any pair (A, B) of sets from $\mathcal{P}$ such that $A \subset B$, there is a finite family $(C_i)_{0 \leq i \leq n}$ of sets from P with

$$A = C_0 \subset C_1 \subset \ldots \subset C_n = B$$

and

$$C_i - C_{i-1} \in \mathcal{P}, \text{ for } i = 1, 2, \ldots, n.$$

An important example of semiring is the class of the intervals of the form $(a, b]$.

A *ring* is a class of subsets of S closed under union $A \cup B$ and difference $A - B$.

Any ring is a semiring.

An *algebra* is a ring containing S.

A *δ-ring* is a ring closed under countable intersections.

A *σ-ring* is a ring closed under countable unions.

A *σ-algebra* is a σ-ring containing S.

For any class $\mathcal{C}$ of subsets of S we denote by $r(\mathcal{C})$, $a(\mathcal{C})$, $\delta r(\mathcal{C})$, $\sigma r(\mathcal{C})$, $\sigma a(\mathcal{C})$ respectively the ring, the algebra, the δ-ring, the σ-ring, and the σ-algebra generated by $\mathcal{C}$.

If $\mathcal{P}$ is a semiring, the ring $r(\mathcal{P})$ generated by $\mathcal{P}$ consists of all the finite, disjoint unions of sets from $\mathcal{P}$.

The σ-ring $\sigma r(\mathcal{D})$ generated by a δ-ring $\mathcal{D}$ consists of all countable unions of disjoint sets from $\mathcal{D}$.

For any class $\mathcal{C}$ of subsets of S we denote by $\mathcal{C}_{\text{loc}}$ the class of sets $A \subset S$ that are "locally" in $\mathcal{C}$, that is, such that $A \cap B \in \mathcal{C}$ for every $B \in \mathcal{C}$.

If $\mathcal{R}$ is a ring, then $\mathcal{R}_{\text{loc}}$ is an algebra; if $\mathcal{D}$ is a δ-ring and $\mathcal{S}$ is a σ-ring, then $\mathcal{D}_{\text{loc}}$ and $\mathcal{S}_{\text{loc}}$ are σ-algebras.

For any class $\mathcal{C}$ of subsets of S and any set $A \subset S$ we denote

$$\mathcal{C} \cap A = \{B \cap A : B \in \mathcal{C}\}.$$

If $\mathcal{C}$ is a ring, δ-ring, σ-ring, then so is $\mathcal{C} \cap A$.

The characteristic function of a set $A \subset S$ is denoted by φ_A, 1_A or I_A.

If $\mathcal{R}$ is a ring (or any other class), we denote by $\mathcal{S}_F(\mathcal{R})$, the set of $\mathcal{R}$-step functions (or $\mathcal{R}$-simple functions) $f : S \to F$ of the form

$$f = \sum_{1 \leq i \leq n} \varphi_{A_i} x_i,$$

with $A_i \in \mathcal{R}$ and $x_i \in F$. If $\mathcal{R}$ is a ring, the sets A_i can be taken mutually disjoint. In this case

$$|f| = \sum_{1 \leq i \leq n} \varphi_{A_i} |x_i|.$$

If $\mathcal{P}$ is a semiring and $\mathcal{R} = r(\mathcal{P})$, then $\mathcal{S}_F(\mathcal{P}) = \mathcal{S}_F(\mathcal{R})$.

C. Measurable functions

Measurability will be defined with respect to a σ-algebra.

Let Σ be a σ-algebra of subsets of S.

We start with the usual definition of measurability of numerical functions.

3. Definition. *A function $f : S \to \overline{\mathbb{R}}$ is Σ-measurable if $f^{-1}(B) \in \Sigma$ for every Borel set $B \subset \overline{\mathbb{R}}$.*

The Σ-step functions are Σ-measurable.

We state the following characterization of measurability in terms of step functions:

4. Theorem. a) *A function $f : S \to \overline{\mathbb{R}}$ is Σ-measurable iff there is a sequence (f_n) of finite, real-valued, Σ-step functions such that $f_n \to f$ pointwise.*

b) *If $f : S \to \overline{\mathbb{R}}$ is Σ-measurable there is a sequence (f_n) of finite, real-valued, Σ-step functions such that $f_n \to f$ pointwise and $|f_n| \leq |f|$, for each n.*

If $f \geq 0$, the sequence (f_n) can be chosen to be increasing and $f_n \geq 0$.
If f is bounded, one can choose the sequence (f_n) to converge uniformly to f.

For vector-valued functions we take the statement in Theorem 4 a) as a definition of measurability:

5. Definition. *A function $f : S \to F$ is said to be Σ-measurable if there is a sequence (f_n) of F-valued, Σ-step functions such that $f_n \to f$ pointwise.*

In particular, the Σ-step functions are Σ-measurable. It follows that if $f : S \to F$ is Σ-measurable, then $|f|$ is also Σ-measurable.

Since the range of a step function is finite, it follows that the range of a Σ-measurable function is separable.

Assertion b) in Theorem 4 remains valid for vector-valued, measurable functions:

6. Theorem. *If $f : S \to F$ is Σ-measurable, then there is a sequence (f_n) of F-valued, Σ-step functions, such that $f_n \to f$ pointwise and $|f_n| \leq |f|$ for every n.*

Proof. Let (g_n) be a sequence of F-valued, Σ-step functions such that $g_n \to f$ pointwise. Then $|g_n| \to |f|$ pointwise. Since $|g_n|$ are positive Σ-step functions, by Theorem 4 a), the function $|f|$ is Σ-measurable. By Theorem 4 b), there is an increasing sequence (h_n) of positive, finite, Σ-step functions such that $h_n \to |f|$ pointwise. Then $|g_n| - h_n \to 0$ pointwise.

For each n, we can represent g_n and h_n using the same sets of Σ:

$$g_n = \sum_{i \in I(n)} \varphi_{A_i} x_i \text{ and } h_n = \sum_{i \in I(n)} \varphi_{A_i} \alpha_i$$

with $(A_i)_{i \in I(n)}$ a finite family of mutually disjoint sets from Σ, $x_i \in F$ and $\alpha_i \geq 0$. For each n we define

$$f_n = \sum_{i \in I(n)} \varphi_{A_i} x_i |x_i|^{-1} \alpha_i,$$

where we set $x_i |x_i|^{-1} \alpha_i = 0$ if $x_i = 0$. Then

$$|f_n| \leq \sum_{i \in I(n)} \varphi_{A_i} \alpha_i = h_n \leq |f|$$

and

$$|f_n - g_n| = \sum_{i \in I(n)} \varphi_{A_i} \big| x_i |x_i|^{-1} \alpha_i - x_i \big| \leq$$
$$\leq \sum_{i \in I(n)} \varphi_{A_i} \big| \alpha_i - |x_i| \big| = \big| h_n - |g_n| \big| \to 0.$$

As $g_n \to f$ pointwise, we deduce that $f_n \to f$ pointwise. ∎

The property used in Definition 3 is preserved under pointwise limits for real-valued functions, as well as for vector-valued functions.

7. Proposition. *Let (f_n) be a sequence of functions $f_n : S \to F$ (or $\overline{\mathbb{R}}$) converging pointwise to a function $f : S \to F$ (or $\overline{\mathbb{R}}$).*
Assume that for each n and each Borel set $B \subset F$ (or $\overline{\mathbb{R}}$) we have $f_n^{-1}(B) \in \Sigma$. Then $f^{-1}(B) \in \Sigma$ for each Borel set $B \subset F$ (or $\overline{\mathbb{R}}$).

Proof. Let $G \subset F$ be an open set and for each $k \in \mathbb{N}$ let G_k be the set of all points $x \in F$ with distance $d(x, G^c) > \frac{1}{k}$. Then G_k is open, $\overline{G}_k \subset G$ and $\bigcup_{k \in \mathbb{N}} G_k = G$ and we have

$$f^{-1}(G) = \bigcup_{k \geq 1} \bigcup_{n \geq 1} \bigcap_{p \geq 1} f_{n+p}^{-1}(G_k) \in \Sigma.$$

It follows that $f^{-1}(B) \in \Sigma$ for every Borel set $B \subset F$.

The above proof remains valid for functions with values in $\overline{\mathbb{R}}$, if we take a distance d on $\overline{\mathbb{R}}$ compatible with its topology. ∎

The property in Definition 3 can now be used to characterize Σ-measurability of vector-valued functions (cf. [N.1], p. 101):

8. Theorem. *A function $f : S \to F$ is Σ-measurable iff it has separable range and $f^{-1}(B) \in \Sigma$ for every Borel set $B \subset F$.*

Proof. Assume first that f is Σ-measurable and let (f_n) be a sequence of F-valued, Σ-step functions such that $f_n \to f$ pointwise.

For each step function f_n we have $f_n^{-1}(B) \in \Sigma$ for every Borel set $B \subset F$. By Proposition 7, we have also $f^{-1}(B) \in \Sigma$ for every Borel set $B \subset F$. We already mentioned above that a Σ-measurable function has separable range. The first implication is proved.

To prove the converse implication, assume that f has separable range and $f^{-1}(B) \in \Sigma$ for every Borel set $B \subset F$. Let F_0 be a separable subspace of F containing the range of f.

Let $(y_n)_{n \geq 0}$ be a sequence dense in F_0 with $y_0 = 0$. For each $n \in \mathbb{N}$ define $\varphi_n : F_0 \to \{y_0, y_1, \ldots, y_n\}$ for each $x \in F_0$ as the first y_k with $0 \leq k \leq n$ for which the minimum $\min_{0 \leq m \leq n} |x - y_m|$ is attained; that is, for $k \leq n$ we have $\varphi(x) = y_k$ if $|x - y_k| < |x - y_m|$, for $m = 0, 1, \ldots, k-1$ and $|x - y_k| \leq |x - y_m|$, for $m = k+1, \ldots, n$. Then $\varphi_n : F_0 \to F$ is a Borel function, since the mapping $x \to |x - y_m|$ from F_0 into $\mathbb{R}_+$ is continuous and for each $i \leq n$ we have

$$\varphi_n^{-1}\{y_i\} = \{x \in F_0; \varphi_n(x) = y_i\}$$
$$= \bigcap_{m < i} \{x \in F_0 : |x - y_m| < |x - y_i|\} \cap \bigcap_{m > i} \{x \in F_0 : |x - y_m| \leq |x - y_i|\},$$

which is a finite intersection of open or closed sets.

On the other hand, $\lim_n \varphi_n(x) = x$ for $x \in F_0$, since

$$|x - \varphi_n(x)| = \min_{0 \leq m \leq n} |x - y_m| \downarrow 0 \text{ as } n \to \infty, \text{ for each } x \in F_0,$$

because (y_n) is dense in F_0. For each n set $f_n = \varphi_n \circ f : S \to F_0$. For each $i \leq n$, the set $\varphi_n^{-1}\{y_i\}$ is a Borel set, hence $f_n^{-1}(y_i) = f^{-1}(\varphi_n^{-1}(y_i)) \in \Sigma$, hence the function f_n is a Σ-step function and we have $f_n \to f$ pointwise. ∎

For functions with values in a separable space, the property in Definition 3 is a complete characterization of measurability.

9. Corollary. *If F is separable, then a function $f : S \to F$ is Σ-measurable iff $f^{-1}(B) \in \Sigma$ for every Borel set $B \subset F$.*

From Proposition 7 and Theorem 8 we deduce that Σ-measurability is preserved under pointwise convergence:

10. Theorem. *If (f_n) is a sequence of F (or $\overline{\mathbb{R}}$)-valued, Σ-measurable functions, converging pointwise to a function f, then the limit f is also Σ-measurable.*

As stated in Theorem 4 b), a real-valued, bounded, Σ-measurable function is the uniform limit of a sequence of Σ-step functions. This is still true for vector-valued functions $f : S \to F$ with *relatively compact range*, but not necessarily for bounded functions. However, an *arbitrary* Σ-measurable function $f : S \to F$ is the *uniform limit* of a sequence of Σ-measurable functions with *countable range*.

A Σ-measurable function $g : S \to F$ with countable range is called a Σ-measurable, σ-step function, or simply, a σ-step function, if the σ-algebra Σ is understood. It is of the form

$$g = \sum_{1 \leq n < \infty} \varphi_{A_n} x_n$$

with $A_n \in \Sigma$ mutually disjoint and $x_n \in F$.

11. Proposition. *If $f : S \to F$ is Σ-measurable, then there is a sequence (f_n) of F-valued, Σ-measurable, σ-step functions, converging to f uniformly on S.*

Proof. Let $\varepsilon > 0$ and let (x_k) be a sequence dense in the range of f. For each k let $B_k = f^{-1}(S_\varepsilon(x_k)) \in \Sigma$, where $S_\varepsilon(x_k)$ is the ball centered at x_k with radius ε. We have $\bigcup_{1 \leq k < \infty} B_k = S$. Denote $A_1 = B_1$ and $A_k = B_k - \bigcup_{i < k} B_i$ for $k > 1$. The sets A_k are mutually disjoint, belong to Σ and their union is S. Define the function $g_\varepsilon : S \to F$ by

$$g_\varepsilon = \sum_{1 \leq k < \infty} \varphi_{A_k} x_k.$$

Then g_ε is a Σ-measurable, σ-step function and
$$|f(x) - g_\varepsilon(s)| \le \varepsilon, \text{ for } s \in S.$$
Taking $\varepsilon = \frac{1}{n}$ and $f_n = g_{\frac{1}{n}}$, we obtain the desired sequence (f_n). ∎

D. Simple measurability of operator-valued functions

12. Definition. *A function $U : S \to E \subset L(F, G)$ is said to be simply Σ-measurable, if for every $x \in F$, the function $Ux : S \to G$ is Σ-measurable.*

It is clear that if U is measurable, then it is simply measurable. We state below a few useful properties of simply measurable functions.

13. Proposition. *If $U : S \to E \subset L(F, G)$ is simply Σ-measurable and $f : S \to F$ is Σ-measurable, then the function $Uf : S \to G$ is Σ-measurable.*

Proof. The proposition is true first, for Σ-step functions f and then, by taking limits, for any Σ-measurable function f. ∎

A simply measurable function U is not necessarily measurable; and even $|U|$ is not necessarily measurable. We give below sufficient condition for $|U|$ or U to be measurable.

14. Proposition. *If $U : S \to E \subset L(F, G)$ is such that $|Ux|$ is Σ-measurable for every $x \in F$ and if F is separable, then the function $|U|$ is Σ-measurable.*

Proof. Let (x_n) be a sequence dense in F with $x_n \ne 0$. For each n, the functions $|Ux_n|$ and $\frac{|Ux_n|}{|x_n|}$ are Σ-measurable. Since
$$|U(s)| = \sup_n \frac{|U(s)x_n|}{|x_n|}, \text{ for } s \in S,$$
we deduce that $|U|$ is Σ-measurable. ∎

15. Proposition. *Assume the embedding $E \subset L(F, G)$ is an isometry. If $U : S \to E \subset L(F, G)$ is simply Σ-measurable and separably valued, then U is Σ-measurable.*

Proof. Assume U is simply Σ-measurable with separable range. Let B be a closed sphere in E with center a and radius r and show that $U^{-1}(B) \in \Sigma$.

Let (a_n) be a sequence dense in the range of U, with $a_1 = a$. Since $E \subset L(F, G)$ isometrically, for each n there is an $x_{nm} \in F$ with $|x_{nm}| = 1$ and $|a_n x_{nm}| > |a_n| - \frac{1}{m}$. Then
$$|a_n| = \sup_m |a_n x_{nm}|, \text{ for each } n.$$

Let V be the closed vector space in E generated by the sequence (a_n). For each $v \in V$ we have
$$|v| = \sup_{n,m} |v x_{nm}|.$$

For each n and m, the function $Ux_{nm} - ax_{nm}$ is Σ-measurable, hence $|Ux_{nm} - ax_{nm}|$ is Σ-measurable; therefore

$$A_{nm} = \{s : |U(s)x_{nm} - ax_{nm}| \leq r\} \in \Sigma.$$

Since

$$|U(s) - a| = \sup_{nm} |(U(s) - a)x_{nm}|, \text{ for } s \in S,$$

we deduce that

$$U^{-1}(B) = \bigcap_{nm} A_{nm} \in \Sigma,$$

hence U is Σ-measurable. ∎

E. Weak measurability

Particular cases of simple measurability are weak measurability and weak star measurability.

16. Definition. *We say that a function $f : S \to F$ is weakly Σ-measurable, if for every $x^* \in F^*$, the real function $\langle f, x^* \rangle$ is Σ-measurable.*

A function $g : S \to F^*$ is said to be weak star Σ-measurable, if for every $x \in F$, the real function $\langle x, g \rangle$ is Σ-measurable.

Weak star measurable functions are also called weak * measurable.

If we want to emphasize the difference between different kinds of measurability, the functions that are Σ-measurable in the usual sense are called *strongly* Σ-measurable.

There is a more general weak measurability, with respect to a space $Z \subset F^*$, norming for F.

17. Definition. *Let $Z \subset F^*$ be a space norming for F. We say that a function $f : S \to F$ is Z-weakly Σ-measurable, if for every $z \in Z$, the real function $\langle f, z \rangle$ is Σ-measurable.*

Taking $Z = F^*$, the F^*-weak measurability is the weak measurability of Definition 16. If $g : S \to F^*$ is a function, considering $F \subset (F^*)^*$ and taking $Z = F$, the F-weak measurability is the weak star measurability of Definition 16.

Z-weak measurability is itself a particular case of simple measurability if we consider the isometric embedding $F \subset L(Z, \mathbb{R})$.

From the properties of simple measurability we deduce then the properties of Z-weak measurability, where $Z \subset F^*$ is a space norming for F.

18. Proposition. *If $f : S \to F$ is Z-weakly Σ-measurable and $g : S \to Z$ is Σ-measurable, then the real function $\langle f, g \rangle$ is Σ-measurable.*

19. Proposition. *If $f : S \to F$ is Z-weakly Σ-measurable and Z is separable, then $|f|$ is Σ-measurable.*

20. Proposition. *If $f : S \to F$ is Z-weakly Σ-measurable and has separable range, then f is Σ-measurable.*

In particular, the above properties are valid for weakly measurable and for weak* measurable functions.

21. Proposition. *If $f : S \to F$ is weakly Σ-measurable and has separable range, then f is strongly Σ-measurable.*
If F is separable, then for functions $f : S \to F$ weak measurability and strong measurability are equivalent.

22. Proposition. *If $g : S \to F^*$ is weak star Σ-measurable and has separable range, then g is strongly Σ-measurable.*
If F^* is separable, then for functions $g : S \to F^*$, weak measurability, weak star measurability, and strong measurability are equivalent.

F. Integral of step functions

23. A set function $m : \mathcal{R} \to E$ defined on a *ring* $\mathcal{R}$ is called an *additive measure*, if, for every pair (A, B) of disjoint sets from $\mathcal{R}$ we have

$$m(A \cup B) = m(A) + m(B).$$

An additive measure is *finitely additive*, that is,

$$m(\bigcup_{1 \leq i \leq n} A_i) = \sum_{1 \leq i \leq n} m(A_i)$$

for any finite family $(A_i)_{1 \leq i \leq n}$ of mutually disjoint sets from $\mathcal{R}$.
A set function $m : \mathcal{R} \to E$ is called a *σ-additive measure*, if, for any sequence (A_n) of mutually disjoint sets from $\mathcal{R}$ with union in $\mathcal{R}$, we have

$$m(\bigcup_n A_n) = \sum_n m(A_n).$$

If $m : \mathcal{R} \to E \subset L(F, G)$ is an additive measure and $f = \sum_{i \in I} \varphi_{A_i} x_i$ is an $\mathcal{R}$-step function from $\mathcal{S}_F(\mathcal{R})$, the integral $\int f dm$ is an element of G defined by the equality

$$\int f dm = \sum_{i \in I} m(A_i) x_i.$$

Since m is additive, the integral $\int f dm$ is independent of the particular representation of f as an $\mathcal{R}$-step function.
If we want to define the integral for a larger class of functions, we should impose additional conditions on $\mathcal{R}$ and m.

G. Totally measurable functions and the immediate integral

An immediate extension of the integral $\int f dm$ is for totally measurable functions f, provided that the measure m has *finite semivariation*. This immediate integral is used in the Riesz representation theorem.

Let $\mathcal{R}$ be a ring of subsets of S.

24. Definition. *A function $f : S \to F$ is said to be totally $\mathcal{R}$-measurable, if it vanishes outside a set $A \in \mathcal{R}$ and if it is the uniform limit of a sequence (f_n) of F-valued, $\mathcal{R}$-step functions.*

The set of totally $\mathcal{R}$-measurable functions $f : S \to E$ is denoted by $TM_F(\mathcal{R})$.

If $F = \mathbb{R}$ we write $TM(\mathcal{R})$ instead of $TM_\mathbb{R}(\mathcal{R})$. Any totally measurable function is bounded. We consider on $TM_F(\mathcal{R})$ the topology of uniform convergence, defined by the sup norm:

$$\|f\| = \sup_{s \in S} |f(s)|.$$

Then the set $\mathcal{S}_F(\mathcal{R})$ is dense in $TM_F(\mathcal{R})$.

According to Theorem 4, if Σ is a σ-algebra, then a *real-valued*, measurable function f is totally Σ-measurable iff f is bounded. But a vector-valued, bounded, Σ-measurable function need not be totally measurable.

We remark also that a totally Σ-measurable function is Σ-measurable.

If S is a locally compact, Hausdorff space and $\mathcal{B}$ is the δ-ring of the relatively compact, Borel subsets of S, then $TM_F(\mathcal{B})$ contains the space $\mathcal{K}_F(S)$ of continuous functions $f : S \to F$ with compact support.

25. Definition. *Let $m : \mathcal{R} \to E \subset L(F, G)$ be an additive measure defined on a ring $\mathcal{R}$. The semivariation of m on a set $A \in \mathcal{R}$, relative to the pair (F, G), is a number denoted by $\tilde{m}_{F,G}(A)$ and defined by the following equality:*

$$\tilde{m}_{F,G}(A) = \sup\{|\int f dm| : f \in \mathcal{S}_F(\mathcal{R}), |f| \le \varphi_A\}.$$

We say m has finite semivariation relative to (F, G) if $\tilde{m}_{F,G}(A) < \infty$ for every $A \in \mathcal{R}$.

The semivariation will be studied in detail in §4.

26. Proposition. *Let $m : \mathcal{R} \to E \subset L(F, G)$ be an additive measure with finite semivariation $\tilde{m}_{F,G}$. Then for every $\mathcal{R}$ simple function $f \in \mathcal{E}_F(\mathcal{R})$ with support $A \in \mathcal{R}$ we have*

$$|\int f dm| \le \|f\| \tilde{m}_{F,G}(A).$$

Proof. The inequality is evident, if $\|f\| = 0$. If $\|f\| \ne 0$, then $\left|\frac{1}{\|f\|} f\right| \le \varphi_A$ and the inequality follows from the definition of the integral of step functions. ∎

27. It follows that the linear mapping $f \to \int f dm$ is continuous on $\mathcal{S}_F(\mathcal{R})$ for the sup norm. Since $\mathcal{S}_F(\mathcal{R})$ is dense in $TM_F(\mathcal{R})$, this linear mapping can be extended by continuity to the whole space $TM_F(\mathcal{R})$. The value of the extension for a function $f \in TM_F(\mathcal{R})$ is still denoted by $\int f dm$ and is called the *immediate integral* of f with respect to m. We still have

$$|\int f dm| \leq \|f\| \tilde{m}_{F,G}(A),$$

for $f \in TM_F(\mathcal{R})$ with support in A.

H. The Riesz representation theorem

The immediate integral is easily defined, but it does not have too many properties. For example, the Lebesgue convergence theorem cannot be proved in this context. But it is good enough to represent continuous linear operators:

28. Theorem. *Let $\mathcal{A}$ be an algebra of subsets of S and $U: TM_F(\mathcal{A}) \to G$ a continuous linear operator. Then there is an additive measure $m: \mathcal{A} \to L(F,G)$ with finite semivariation $\tilde{m}_{F,G}(S)$ such that*

$$U(f) = \int f dm, \text{ for } f \in TM_F(\mathcal{A})$$

and

$$\|U\| = \tilde{m}_{F,G}(S).$$

The proof of this theorem is immediate.

The measure m is defined by $m(A) = U(\varphi_A)$, for $A \in \mathcal{A}$. For more details see [D.1].

But not so immediate is the proof of the following Riesz-type representation theorem.

29. Theorem. *Let K be a compact Hausdorff space and $C_F(K)$ the space of continuous functions $f: K \to F$, endowed with the sup norm.*

Let $U: C_F(K) \to G$ be a continuous linear operation.

Then there is an additive Borel measure $m: \mathcal{B}(K) \to L(F, G^{**})$ with finite semivariation $\tilde{m}_{F,G^{**}}$ such that

$$U(f) = \int f dm, \text{ for } f \in C_F(K)$$

and

$$\|U\| = \tilde{m}_{F,G^{**}}(K).$$

Moreover, for each $z \in G^*$, the measure $m_z: \mathcal{B}(K) \to F^*$ defined by

$$\langle x, m_z(A) \rangle = \langle m(A)x, z \rangle, \text{ for } x \in F \text{ and } A \in \mathcal{B}(K),$$

is regular, σ-additive and with finite variation.

For the proof see [D.1] and [D–U].

30. An additive measure $m : \mathcal{B}(K) \to L(F, G^{**})$ with finite semivariation $\tilde{m}_{F,G}$ and such that m_z is regular, σ-additive, and with finite variation for every $z \in G^*$, is sometimes called a *representing measure*.

31. There are cases when the measure m in the above theorem is neither σ-additive, nor regular, nor with values in $L(F, G)$.

It is an open problem to give a characterization of the operations U for which the corresponding measure m has one or more of the above-mentioned properties.

There are partial answers to this problem:
a) *If $U : C_{\mathbb{R}}(K) \to G$ is weakly compact, then the corresponding measure m is σ-additive, regular, and has values in G.*

For a complete presentation of this case see [D–U].

b) An operation $U : C_F(K) \to G$ is said to be *dominated* if there is a positive, regular Borel measure μ such that

$$|U(f)| \leq \int |f| d\mu, \text{ for } f \in C_F(K).$$

If $U : C_F(K) \to G$ is dominated, then the corresponding measure m is σ-additive, regular, with values in $L(F, G)$, and with finite variation $|m|$.

For a complete presentation of this case see [D.1].

A continuous linear functional $U : C_F(K) \to \mathbb{R}$ is dominated; therefore the corresponding measure $m : \mathcal{B}(K) \to F^*$ is σ-additive, regular, and with finite variation.

We give one more case which answers the above problem.

c) *If $G = D^*$ is a dual of a Banach space and $U : C_F(K) \to D^*$ is a continuous linear operation, then the corresponding measure m has values in $L(F, D^*)$, is σ-additive, and regular.*

32. If an additive measure $m : \mathcal{R} \to E \subset L(F, G)$ defined on a ring $\mathcal{R}$ has finite semivariation $\tilde{m}_{F,G}$, it might not be possible to extend the integral $\int f dm$ beyond the space $TM_F(\mathcal{R})$ of totally measurable functions.

In order to define the integral for a larger class of functions we need an additive measure $m : \mathcal{D} \to E \subset L(F, G)$ defined on a δ-ring, with finite semivariation $\tilde{m}_{F,G}$ such that the measures $m_z : \mathcal{D} \to F^*$ are σ-additive for every z in a subspace $Z \subset G^*$ norming for G. This integral will be presented in §5 and is an extension of the immediate integral. It follows that this integral can be used in the Riesz representation theorem instead of the immediate integral.

33. There are 4 stages in the development of the integral $\int f dm$:
I) The classical integral, with $m \geq 0$ and f *real-valued* ;
II) The Bochner integral, with $m \geq 0$ and f *vector-valued* ;

III) The integral $\int f dm$, where m is a *vector measure with finite variation* and f is *vector-valued*;
IV) The integral $\int f dm$, where m is a *vector measure with finite semivariation* and f is *vector-valued*.

The most important stage is the first one. The other stages can be performed to the extent that they can be reduced to the first stage. In particular, integration with respect to a vector-valued measure can be performed if it can be reduced to integration with respect to positive measures.

We shall review succinctly the first three stages, in order to fix the notations and for further reference.

The rest of this chapter will be devoted to integration with respect to measures with *finite semivariation*.

I. The classical integral

34. We assume that the reader is familiar with the content of this section. We present only a few facts for further reference.

The framework for this section is a measure space (S, Σ, μ), consisting of a set S, a σ-algebra Σ of subsets of S, and a positive, σ-additive measure μ on Σ, with finite or infinite values.

We shall always assume that μ has the *finite measure property* (FMP), i.e., for every set $A \in \Sigma$ we have

$$\mu(A) = \sup\{\mu(B) : B \in \Sigma, B \subset A, \mu(B) < \infty\}.$$

If μ does not have the FMP, we replace it with the measure $\mu' : \Sigma \to \overline{\mathbb{R}}_+$ defined by

$$\mu'(A) = \sup\{\mu(B) : B \in \Sigma, B \subset A, \mu(B) < \infty\}$$

.

If $\mu(S) < \infty$, that is, if μ has only finite values, we say that (S, Σ, μ) is a *finite measure* space. We say the measure μ is σ-finite if there is a sequence (S_n) from Σ with $\mu(S_n) < \infty$ for each n and $S = \bigcup_n S_n$. Evidently, a σ-finite measure has the FMP.

A set $A \subset S$ is μ-negligible if there is a set $B \in \Sigma$ with $A \subset B$ and $\mu(B) = 0$. A property $P(s)$ defined for every $s \in S$ is said to be true μ-almost everywhere (μ-a.e.) if the set of the points $s \in S$ for which $P(s)$ is false is μ-negligible. A function $f : S \to F$ or $\overline{\mathbb{R}}$ is μ-negligible if $f = 0$, μ-a.e.

A function $f : S \to F$ or $\overline{\mathbb{R}}$ is said to be μ-measurable if it is equal μ-a.e. to a Σ-measurable function.

An F or $\overline{\mathbb{R}}$-valued function f defined μ-a.e. on S is said to be μ-measurable, if it has a μ-measurable extension on the whole space S. Then any extension of f on S is μ-measurable.

Functions which are equal μ-a.e. will be often identified.

We denote by $\Sigma_f(\mu)$ the δ-ring of sets $A \in \Sigma$ with $\mu(A) < \infty$.

$L^1(\mu)$ is the vector space of (equivalence classes of) real-valued functions $f: S \to \mathbb{R}$ which are μ-integrable, with integral $\int f d\mu$, and is equipped with the seminorm $\|f\|_1 = \int |f| d\mu$. We shall write also $L^1(\Sigma, \mu)$ instead of $L^1(\mu)$.

$L^1(\mu)$ is a complete vector space for the seminorm $\|f\|_1$ and the mapping $f \mapsto \int f d\mu$ is a continuous linear functional on $L^1(\mu)$.

The topology defined on $L^1(\mu)$ by the seminorm $\|f\|_1$ is called the *topology of convergence in the mean*.

In the construction of the integral $\int f d\mu$ of μ-integrable functions, only the restriction of the measure μ to the δ-ring $\Sigma_f(\mu)$ is involved. For integration theory it is enough to assume that the measure μ is *finite* and defined on a δ-ring (rather than to be finite or infinite and defined on a σ-algebra).

For $1 \leq p < \infty$, $L^p(\mu)$ is the vector space of (equivalence classes of) real-valued, μ-measurable functions $f: S \to \mathbb{R}$ with $|f|^p \in L^1(\mu)$ and is equipped with the seminorm $\|f\|_p = (\int |f|^p d\mu)^{1/p}$. For this seminorm, $L^p(\mu)$ is complete and the $\Sigma_f(\mu)$-step functions are dense in $L^p(\mu)$. Moreover, if $\mathcal{R} \subset \Sigma_f(\mu)$ is a ring generating the δ-ring $\Sigma_f(\mu)$, then the $\mathcal{R}$-step functions are dense in $L^p(\mu)$.

The Fatou's Lemma, the Monotone Convergence Theorem, the Vitali Convergence Theorem, and the Lebesgue Dominated Convergence Theorem are valid in $L^p(\mu)$.

$L^\infty(\mu)$ is the space of (equivalence classes of) μ-measurable functions $f: S \to \mathbb{R}$ with $\|f\|_\infty < \infty$, where

$$\|f\|_\infty = \inf\{\alpha : 0 \leq \alpha \leq \infty, \ |f(s)| \leq \alpha, \ \mu-\text{a.e.}\}.$$

The set $\mathcal{S}_\mathbb{R}(\Sigma)$ of Σ-step functions is dense in $L^\infty(\mu)$.

Let $1 \leq p < \infty$ and $1 < q \leq +\infty$ with $\frac{1}{p} + \frac{1}{q} = 1$. In case $p = 1$, assume μ is σ-finite. Then the dual $(L^p(\mu))^*$ is isometrically isomorphic to $L^q(\mu)$. More precisely, the correspondence $L \mapsto g$ between $L \in (L^p(\mu))^*$ and $g \in L^q(\mu)$ is given by

$$L(f) = \int f g d\mu, \text{ for } f \in L^p(\mu),$$

and we have $\|L\| = \|g\|_q$.

It is convenient to define the integral for positive, μ-measurable functions, with finite or infinite values, defined μ-a.e.

If f is a positive, μ-measurable function, defined μ-a.e. on S, we define the integral $\int f d\mu$ by the equality

$$\int f d\mu = \sup\{\int s d\mu : s \in \mathcal{S}(\Sigma), \ 0 \leq s \leq f, \mu\text{-a.e.}\} \leq +\infty.$$

If $\int f d\mu < \infty$, we still say that f is μ-integrable. If f is μ-integrable, then f is finite μ-a.e. If $\int f d\mu = 0$, then $f = 0$, μ-a.e.

The Fatou's Lemma and the Monotone Convergence Theorem are still valid for positive, μ-measurable functions, integrable or not.

A function $f: S \to \mathbb{R}$ is μ-integrable iff f is μ-measurable and $\int |f| d\mu < \infty$.

J. The Bochner integral

Let (X, Σ, μ) be a measure space and F a Banach space. We shall define in this section the Bochner integral $\int f d\mu$ for vector-valued functions $f : S \to F$.

We denote, as before, with $\Sigma_f(\mu)$, the δ-ring of the sets $A \in \Sigma$ with $\mu(A) < \infty$. In the definition of the Bochner integral $\int f d\mu$, only the restriction of the measure μ to the δ-ring $\Sigma_f(\mu)$ is involved.

35. Definition. *A function $f : S \to F$ is said to be Bochner–integrable with respect to μ, or Bochner μ-integrable, if f is μ-measurable and if $|f|$ is μ-integrable.*

The space of Bochner μ-integrable functions $f : S \to F$ is denoted by $L_F^1(\mu)$ or $L_F^1(\Sigma, \mu)$.

For $f \in L_F^1(\mu)$ we define the seminorm

$$\|f\|_1 = \int |f| d\mu = \| \, |f| \, \|_1.$$

Then $L_F^1(\mu)$ is complete for this seminorm. If $\mathcal{R} \subset \Sigma_f(\mu)$ is a ring generating the δ-ring $\Sigma_f(\mu)$, then the set of $\mathcal{R}$-step function is dense in $L_F^1(\mu)$.

For a $\Sigma_f(\mu)$-step function $f = \sum \varphi_{A_i} x_i$ with $A_i \in \Sigma_f(\mu)$ mutually disjoint and $x_i \in F$, we have $|f| = \sum \varphi_{A_i} |x_i|$. Then

$$\int f d\mu = \sum \mu(A_i) x_i \in F$$

and

$$|\int f d\mu| = |\sum \mu(A_i) x_i| \le \sum \mu(A_i) |x_i| = \int |f| d\mu = \|f\|_1.$$

The mapping $f \mapsto \int f d\mu$ from the subspace $\mathcal{S}_F(\Sigma_f(\mu))$ of the F-valued, $\Sigma_f(\mu)$-step functions, into the space F, is continuous for the seminorm $\|f\|_1$; therefore it can be extended uniquely to a linear, continuous mapping from the whole space $L_F^1(\mu)$ into F. The value of the extension for a function $f \in L_F^1(\mu)$ is denoted by $\int f d\mu$ and is called the Bochner integral of f with respect to μ. We still have

$$|\int f d\mu| \le \int |f| d\mu = \| \, |f| \, \|_1, \text{ for } f \in L_F^1(\mu).$$

If $f \in L_F^1(\mu)$ and $A \in \Sigma$, then $\varphi_A f \in L_F^1(\mu)$. We denote

$$\int_A f d\mu = \int \varphi_A f d\mu.$$

The mapping $A \mapsto \int_A f d\mu$ from Σ into F is a σ-additive measure.

For $1 \le p < \infty$, we define the space $L_F^p(\mu)$ to be the set of (equivalence classes of) μ-measurable functions $f : S \to F$ with $|f| \in L^p(\mu)$. We define on

$L_F^p(\mu)$ the seminorm

$$\|f\|_p = \| \, |f| \, \|_p = (\int |f|^p d\mu)^{\frac{1}{p}}, \text{ if } 1 \leq p < \infty.$$

Then $L_F^p(\mu)$ is complete for the seminorm $\|f\|_p$. If $1 \leq p < \infty$ and if $\mathcal{R} \subset \Sigma_f(\mu)$ is a ring generating the δ-ring Σ_f, then the set $\mathcal{S}_F(\mathcal{R})$ of $\mathcal{R}$-step function is dense in $L_F^p(\mu)$.

For $p = \infty$, we define $L_F^\infty(\mu)$ to be the set of (equivalence classes of) μ-measurable functions $f : S \to F$ with $|f| \in L^\infty(\mu)$ with seminorm $\|f\|_\infty = \| \, |f| \, \|_\infty$. With this seminorm, $L_F^\infty(\mu)$ is complete.

If F is infinite dimensional, the Σ-step functions are no longer dense in $L_F^\infty(\mu)$.

If $1 \leq p \leq \infty$ and $\frac{1}{p} + \frac{1}{q} = 1$, we have a linear isometry $g \mapsto L$ of $L_{F^*}^q(\mu)$ into $(L_F^p(\mu))^*$, given by

$$L(f) = \int \langle f(s), g(s) \rangle d\mu, \text{ for } f \in L_F^p(\mu),$$

and we have $\|L\| = \|g\|_q$.

In general, this mapping is not surjective. However, if $1 \leq p < \infty$, if μ is σ-finite in case $p = 1$ and if F^* has the Radon–Nikodym property (RNP), then $(L_F^p(\mu))^*$ is isometrically isomorphic to $L_{E^*}^q(\mu)$.

Examples of Banach spaces with the RNP are the reflexive Banach spaces and the separable duals of Banach spaces.

We mention the following theorems which will be used in the sequel:

36. Theorem. *Let $T : F \to D$ be a continuous linear operator. If $f \in L_F^1(\mu)$, then $Tf \in L_D^1(\mu)$ and we have*

$$\int Tf d\mu = T \int f d\mu, \text{ for } f \in L_F^1(\mu).$$

The theorem is proved first for Σ-step functions f of $L_F^1(\mu)$ and then for any function $f \in L_F^1(\mu)$, which is the limit in $L_F^1(\mu)$ of a sequence of Σ-step functions.

37. Corollary. *Assume $E \subset L(F, G)$. If $f \in L_E^1(\mu)$ and $x \in F$, then $fx \in L_G^1(\mu)$ and we have*

$$(\int f d\mu) x = \int fx d\mu.$$

In particular, if $f \in L_E^1(\mu)$ and $x^* \in E^*$, then $\langle f, x^* \rangle \in L_\mathbb{R}^1(\mu)$ and we have

$$\langle \int f d\mu, x^* \rangle = \int \langle f, x^* \rangle d\mu.$$

We apply Theorem 36 to the continuous linear operator $T: E \to G$ defined by $Te = ex$, for $e \in E$, and to the continuous linear operator $T: E \to \mathbb{R}$ defined by $Te = \langle e, x^* \rangle$, for $e \in E$.

The computation of the seminorm $\|f\|_p$ is given in the following theorem:

38. Theorem. *Assume $E \subset L(F, G)$ isometrically and let $\mathcal{R}$ be a ring generating the δ-ring Σ_f. Let $1 \leq p, q \leq +\infty$ with $1/p + 1/q = 1$ and let $f \in L_E^p(\mu)$. Then:*
a)
$$\|f\|_p = \sup \int |fg|d\mu = \sup \int |f|\,|g|d\mu,$$
the supremum being taken for $g \in \mathcal{S}_F(\mathcal{R})$ with $\|g\|_q \leq 1$.
b) *If $G = \mathbb{R}$ (for example, if $E \subset F^*$ or if $F \subset E^*$), then*
$$\|f\|_p = \sup |\int fg d\mu|, \text{ for } g \in \mathcal{S}_F(\mathcal{R}) \text{ with } \|g\|_q \leq 1.$$

For the proof the reader is referred to [D.1], Theorem 5, p. 228.

The following theorem gives a necessary and sufficient condition for a function f to belong to L_E^p.

39. Theorem. *Assume $E \subset L(F, G)$ isometrically. Let $1 \leq p, q \leq +\infty$ with $1/p + 1/q = 1$ and let $f: S \to E$ be a μ-measurable function. Then $f \in L_E^p(\mu)$ iff $fg \in L_G^1(\mu)$ for every $g \in L_F^q(\mu)$.*

For the proof see [D.1], Theorem 7, p. 233.

40. Theorem. *Let $1 \leq p < \infty$. If $c_0 \not\subset E$ then $c_0 \not\subset L_E^p(\mu)$.*

See [Kw].

K. Convergence theorems

We state first some properties of σ-additivity and absolute continuity of the indefinite integral.

41. Theorem. *Let $f \in L_F^1(\mu)$. Then*
a) *The mapping $A \mapsto \int_A |f| d\mu$ from Σ into F is σ-additive;*
b) $\lim_{\mu(A) \to 0} \int_A |f| d\mu = 0$;
c) *for every $\varepsilon > 0$, there is a set $S_\varepsilon \in \Sigma$ with $\mu(S_\varepsilon) < \infty$ such that*
$$\int_{S-S_\varepsilon} |f| d\mu < \varepsilon.$$

Let (f_n) be a Cauchy sequence in $L_F^1(\mu)$. Then
a') *The measures $A \mapsto \int_A |f_n| d\mu$ are uniformly σ-additive, for $n \in \mathbb{N}$;*
b') $\lim_{\mu(A) \to 0} \int_A |f_n| d\mu = 0$, *uniformly for $n \in \mathbb{N}$;*

c') for every $\varepsilon > 0$, there is a set $S_\varepsilon \in \Sigma$ with $\mu(S_\varepsilon) < \infty$ such that

$$\int_{S-S_\varepsilon} |f| d\mu < \varepsilon, \text{ for all } n \in \mathbb{N}.$$

Then we state the Egorov theorem:

42. Theorem. (Egorov) *Let $f_n, f : S \to F$, $n = 1, 2, \ldots$ be μ-measurable functions such that $f_n \to f$, μ-a.e. Then for every set $A \in \Sigma$ with $\mu(A) < \infty$ and for every $\varepsilon > 0$, there is a set $B \in \Sigma$ with $B \subset A$ and $\mu(A - B) < \varepsilon$, such that $f_n \to f$ uniformly on B.*

43. Corollary. *Assume $\mu(S) < \infty$ and let $f_n, f : S \to F$, $n = 1, 2, \ldots$, be μ-measurable functions. If $f_n \to f$, μ-a.e., then $f_n \to f$ in μ-measure, i.e.,*

$$\lim_n \mu\{|f_n - f| > \varepsilon\} = 0, \text{ for every } \varepsilon > 0.$$

Uniform convergence implies convergence in $L_F^p(\mu)$.

44. Theorem. *Let $1 \leq p < \infty$ and (f_n) be a sequence of functions from $L_F^p(\mu)$, vanishing μ-a.e. outside a set $A \in \Sigma$ with $\mu(A) < \infty$. If $f_n \to f$ uniformly on A, then $f\varphi_A \in L_F^p(\mu)$ and $f_n\varphi_A \to f\varphi_A$ in $L_F^p(\mu)$.*

Next we state the Vitali convergence theorem and the Lebesgue convergence theorem, for nets of functions converging in μ-measure.

45. Theorem. (Vitali) *Let $1 \leq p < \infty$, $(f_i)_{i \in I}$ a net of functions from $L_F^p(\mu)$ and $f : S \to F$ a μ-measurable function.*
Then $f \in L_F^p(\mu)$ and $\lim_{i \in I} f_i = f$ in $L_F^p(\mu)$ iff the following conditions are satisfied:
a) $\lim_{i \in I} f_i = f$, in μ-measure;
b) $\lim_{\mu(A) \to 0} \int_A |f_i|^p d\mu = 0$, uniformly for $i \in I$;
c) for every $\varepsilon > 0$ there is a set $S_\varepsilon \in \Sigma$ with $\mu(S_\varepsilon) < \infty$ such that

$$\int_{S-S_\varepsilon} |f_i|^p d\mu < \varepsilon, \text{ for all } i \in I.$$

For the proof see [D-S], Theorems III.3.6 and III.9.5.

46. Theorem. (Lebesgue) *Let $1 \leq p < \infty$, $(f_i)_{i \in I}$ a net of funcitons from $L_F^p(\mu)$, $f : S \to F$ a μ-measurable function, and $g \in L_+^p(\mu)$, such that $|f_i| \leq g$, μ-a.e. for each $i \in I$.*
Then $f \in L_F^p(\mu)$ and $\lim_{i \in I} f_i = f$ in $L_F^p(\mu)$ iff $\lim_{i \in I} f_i = f$ in μ-measure.

For the proof see [D–S], Theorem III.3.7.
Finally, we state Vitali's and Lebesgue's convergence theorems for sequences of functions converging pointwise.

47. Theorem. (Vitali) Let $1 \leq p < \infty$, (f_n) a sequence from $L_F^p(\mu)$ and $f : S \to F$ a μ-measurable function such that $f_n \to f$, μ-a.e.

Then $f \in L_F^p(\mu)$ and $f_n \to f$ in $L_F^p(\mu)$ iff the following conditions are satisfied:

b') $\lim\limits_{\mu(A) \to 0} \int_A |f_n|^p d\mu = 0$, uniformly for $n \in \mathbb{N}$;

c') for every $\varepsilon > 0$, there is a set $S_\varepsilon \in \Sigma$ with $\mu(S_\varepsilon) < \infty$, such that

$$\int_{S-S_\varepsilon} |f_n|^p d\mu < \varepsilon, \text{ for all } n \in \mathbb{N}.$$

For the proof see [D–S], Theorem III.6.15.

48. Theorem. (Lebesgue) Let $1 \leq p < \infty$, (f_n) a sequence from $L_F^p(\mu)$, $f : S \to F$ a μ-measurable function, and $g \in L_+^p(\mu)$ such that $f_n \to f$ μ-a.e. and $|f_n| \leq g$, μ-a.e., for each n.

Then $f \in L_F^p(\mu)$ and $f_n \to f$ in $L_F^p(\mu)$.

See [D–S], Theorem III.6.16.

Remark. If $\mu(S) < \infty$, conditions c) and c') in Vitali's Convergence Theorems 45 and 47 are superfluous.

§2. MEASURES WITH FINITE VARIATION

In this paragraph we present the third stage in the development of the integral, namely, the integral with respect to a vector measure with finite variation. For a detailed presentation of this stage the reader is referred to [D.1].

A. The variation of vector measures

The framework for this section is a ring $\mathcal{R}$ of subsets of S, a Banach space E and an additive measure $m : \mathcal{R} \to E$.

1. Definition. *For every set $A \subset S$ (not necessarily from $\mathcal{R}$ or $\mathcal{R}_{loc}$) the variation of m on A is a number, finite or $+\infty$, denoted by $var(m, A)$ or $\overline{m}(A)$ and defined by the following equality:*

$$var(m, A) = \sup \sum |m(A_i)|,$$

the supremum being taken for all finite families $(A_i)_{i \in I}$ of mutually disjoint sets from $\mathcal{R}$ contained in A.

The above definition can be stated for any set function m (not necessarily additive) defined on any class $\mathcal{C}$ containing ϕ (not necessarily a ring), such that $m(\phi) = 0$.

For $A \in \mathcal{R}_{loc}$ we usually denote $|m|(A) = var(m, A)$.

If $A \in \mathcal{R}$, in the definition of $|m|(A)$ we can take only families $(A_i)_{i \in I}$ of disjoint sets from $\mathcal{R}$ with union *equal* to A.

We say that m has *finite* (respectively *bounded*) variation, if $|m|(A) < \infty$ for every $A \in \mathcal{R}$ (respectively if $|m|(S) < \infty$).

We state now some properties of the variation.

2. *If $\mathcal{P}$ is a semiring generating the ring $\mathcal{R}$, then in the definition of $var(m, A)$ we can take the supremum only for finite families $(A_i)_{i \in I}$ of mutually disjoint sets A_i from $\mathcal{P}$ contained in A.*

In fact, every set of $\mathcal{R}$ is a finite union of disjoint sets from $\mathcal{P}$.

If $m' : \mathcal{P} \to E$ is the restriction of m to $\mathcal{P}$, the above property means that

$$var(m', A) = var(m, A), \text{ for any } A \subset S.$$

3. $|m(A)| \leq |m|(A)$, for $A \in \mathcal{R}$.

4. If $A \subset B$, then $var(m, A) \leq var(m, B)$.

5. $var(m, A) = 0$ iff $m(B) = 0$ for every $B \in \mathcal{R}$ with $B \subset A$.

6. $\overline{m}$ is superadditive: for an arbitrary family $(A_i)_{i \in I}$ of disjoint subsets of S we have

$$\overline{m}(\bigcup_{i \in I} A_i) \geq \sum_{i \in I} \overline{m}(A_i).$$

Proof. We consider first two disjoint sets A, B in S and an arbitrary number $\theta < \overline{m}(A) + \overline{m}(B)$. There are two numbers α, β such that $\alpha < \overline{m}(A)$, $\beta < \overline{m}(B)$ and $\alpha + \beta = \theta$.

We find a family $(A_i)_{1 \leq i \leq n}$ of disjoint sets from $\mathcal{R}$, contained in A such that
$$\alpha < \sum_{1 \leq i \leq n} |m(A_i)|$$
and a family $(B_j)_{1 \leq j \leq m}$ of disjoint sets from $\mathcal{R}$, contained in B such that
$$\beta < \sum_{1 \leq j \leq m} |m(B_j)|.$$

The sets $A_1, \ldots, A_n, B_1, \ldots, B_m$ are mutually disjoint, belong to $\mathcal{R}$, and are contained in $A \cup B$, hence
$$\theta = \alpha + \beta \leq \sum_{1 \leq i \leq n} |m(A_i)| + \sum_{1 \leq j \leq m} |m(B_j)| \leq \overline{m}(A \cup B).$$

It follows that $\overline{m}(A) + \overline{m}(B) \leq \overline{m}(A \cup B)$.

By induction, we deduce that for every finite family $(A_i)_{i \in J}$ of disjoint subsets of S we have
$$\sum_{j \in J} \overline{m}(A_i) \leq \overline{m}(\bigcup_{j \in J} A_i).$$

If $(A_i)_{i \in I}$ is an arbitrary family of disjoint subsets of S, then, for every finite subset $J \subset I$ we have
$$\sum_{i \in J} \overline{m}(A_i) \leq \overline{m}(\bigcup_{i \in J} A_i) \leq \overline{m}(\bigcup_{i \in I} A_i),$$
therefore,
$$\sum_{i \in I} \overline{m}(A_i) \leq \overline{m}(\bigcup_{i \in I} A_i).$$

■

7. $\overline{m}$ *is the smallest of all positive, increasing, and superadditive set functions ν defined for all subsets of S and satisfying*
$$|m(A)| \leq \nu(A), \text{ for } A \in \mathcal{R}.$$

8. *If $m, n : \mathcal{R} \to E$ are additive measures and α is a number, then, for any set $A \subset S$ we have*
$$var(m + n, A) \leq var(m, A) + var(n, A)$$

and
$$var(\alpha m, A) = |\alpha| var(m, A).$$

9. *If $\mathcal{R}'$ is a subring of $\mathcal{R}$ and if m' is the restriction of m to $\mathcal{R}'$, then, for any set $A \subset S$, we have*
$$var(m', A) \leq var(m, A).$$

In some cases we have equality in Property 9 (see Property 2 and Propositions 11 and 12 below).

10. Proposition. *The variation $|m| : \mathcal{R}_{loc} \to [0, \infty]$ is additive. If m is σ-additive on $\mathcal{R}$, then $|m|$ is σ-additive on $\mathcal{R}_{loc}$.*

Proof. Let $(A_i)_{i \in I}$ be a family of disjoint sets from $\mathcal{R}_{loc}$ with union $A \in \mathcal{R}_{loc}$. We assume that this family is finite if m is additive and countable if m is σ-additive and we have to prove that
$$|m|(A) = \sum_{i \in I} |m|(A_i).$$

Let $(B_j)_{1 \leq j \leq m}$ be a family of disjoint sets from $\mathcal{R}$ contained in A. For each $i \in I$, the family $(A_i \cap B_j)_{1 \leq j \leq m}$ consists of disjoint sets from $\mathcal{R}$ contained in A_i; for each j, the family $(A_i \cap B_j)_{i \in I}$ consists of disjoint sets from $\mathcal{R}$ with union B_j. Then
$$\sum_{1 \leq j \leq m} |m(B_j)| = \sum_{1 \leq j \leq m} |m(\bigcup_{i \in I} A_i \cap B_j)|$$
$$= \sum_{1 \leq j \leq m} |\sum_{i \in I} m(A_i \cap B_j)| \leq \sum_{1 \leq j \leq m} \sum_{i \in I} |m(A_i \cap B_j)|$$
$$= \sum_{i \in I} \sum_{1 \leq j \leq m} |m(A_i \cap B_j)| \leq \sum_{i \in I} |m|(A_i).$$

It follows that
$$|m|(A) \leq \sum_{i \in I} |m|(A_i).$$
Since, by Property 6, $|m|$ is superadditive, we deduce that
$$|m|(A) = \sum_{i \in I} |m|(A_i).$$
∎

The following proposition is the analog of Property 2 and states that the variation is the same whether the measure is defined on a σ-ring or on a δ-ring generating it.

11. Proposition. *Let $m : S \to E$ be a σ-additive measure defined on a σ-ring S and $\mathcal{D}$ a δ-ring generating S. If $m' : \mathcal{D} \to E$ is the restriction of m to $\mathcal{D}$, then for every set $A \subset S$ we have*
$$\mathrm{var}(m', A) = \mathrm{var}(m, A).$$

Proof. Let $A \subset S$ and $(A_i)_{i \in I}$ be a finite family of disjoint sets from S contained in A. Each set A_i is a countable union, $A_i = \bigcup_{j \in \mathbb{N}} A_{ij}$ of mutually disjoint sets A_{ij} from $\mathcal{D}$. Then

$$\sum_i |m(A_i)| = \sum_i |m(\bigcup_j A_{ij})| = \sum_i |\sum_j m'(A_{ij})|$$
$$\leq \sum_{ij} |m'(A_{ij})| = \sup_n \sum_{1 \leq j \leq n} \sum_{i \in I} |m'(A_{ij})|$$
$$\leq \mathrm{var}(m', A),$$

therefore $\mathrm{var}(m, A) \leq \mathrm{var}(m', A)$. The converse inequality follows from Property 9. ■

We mention here the following proposition, which will follow from Theorem 7.4 on extension of measures.

12. Proposition. *Let $m : \mathcal{D} \to E$ be a σ-additive measure with finite variation $|m|$, defined on a δ-ring $\mathcal{D}$ and let $\mathcal{R} \subset \mathcal{D}$ be a ring generating the δ-ring $\mathcal{D}$. If $m' : \mathcal{R} \to E$ is the restriction of m to $\mathcal{R}$, then*

$$|m'|(A) = |m|(A), \text{ for } A \in \mathcal{R}.$$

In particular, $\mathcal{D}$ can be a σ-ring or a σ-algebra.

B. Boundedness of σ-additive measures

As we shall see in the next section, for real-valued measures there is a close relationship between finite variation and boundedness.

13. Definition. *An additive measure $m : \mathcal{R} \to E$ on a ring $\mathcal{R}$ is said to be bounded if*
$$\sup\{|m(B)| : B \in \mathcal{R}\} < \infty.$$

The measure m is said to be locally bounded if for every set $A \in \mathcal{R}$ we have
$$\sup\{|m(B)| : B \in \mathcal{R} \cap A\} < \infty.$$

A σ-additive measure on a ring is not necessarily bounded or locally bounded; and a finitely additive measure on a σ-ring or on a δ-ring is not necessarily bounded or locally bounded either.

But a σ-additive measure on a σ-ring is bounded, as the following theorem shows.

24 Ch.1 VECTOR INTEGRATION

14. Theorem. a) *Any σ-additive measure $m : \mathcal{S} \to E$ on a σ-ring is bounded.*
b) *Any σ-additive measure $m : \mathcal{D} \to E$ on a δ-ring is locally bounded.*

Proof. Assume $m : \mathcal{S} \to E$ is a σ-additive measure on a σ-ring $\mathcal{S}$ but m is not bounded and show that this leads to a contradiction. We say that a set $A \in \mathcal{S}$ is unbounded if $\sup\{|m(B)| : B \in \mathcal{S} \cap A\} = \infty$.

α) For any set $A \in \mathcal{S}$ on which m is unbounded and for every number $N > |m(A)|$, there is a set $A' \in \mathcal{S} \cap A$ on which m is unbounded, with $|m(A')| > N$.

In fact, there is a set $B \in \mathcal{S} \cap A$ with $|m(B)| > 2N$. Then

$$|m(A - B)| = |m(A) - m(B)| \geq |m(B)| - |m(A)| \geq 2N - N = N.$$

Since m is unbounded on $\mathcal{S} \cap A$, it is unbounded either on $\mathcal{S} \cap B$, or on $\mathcal{S} \cap (A - B)$. We take A' one of the sets B or $A - B$, on which m is unbounded.

β) There are sets $A \in \mathcal{S}$ on which m is unbounded. In fact, since m is unbounded, there is a sequence (C_n) from $\mathcal{S}$ such that $|m(C_n)| > n$ for each n. Then $A = \bigcup C_n \in \mathcal{S}$ and m is unbounded on A.

γ) Starting with a set $A_1 \in \mathcal{S}$ on which m is unbounded and with a number $N_1 > \max(1, |m(A_1)|)$ and using α), we can find a decreasing sequence (A_n) from $\mathcal{S}$ and a sequence (N_n) of numbers with $N_n > \max(n, |m(A_n)|)$ and $|m(A_n)| > N_{n-1} > n - 1$. Then $|m(\bigcap A_n)| = \lim_n |m(A_n)| = \infty$ and we reached a contradiction. It follows that m is bounded on $\mathcal{S}$.

If $m : \mathcal{D} \to E$ is σ-additive on a δ-ring $\mathcal{D}$, then, for each set $A \in \mathcal{D}$ we consider the restriction of m to the σ-ring $\mathcal{D} \cap A$; by the first part of the proof, m is bounded on $\mathcal{D} \cap A$. Hence m is locally bounded on $\mathcal{D}$. ∎

C. Variation of real-valued measures

We consider first the variation of positive measures.

15. Proposition. *If $\mu : \mathcal{R} \to \mathbb{R}_+$ is a positive, finite, additive measure, then*

$$var(\mu, A) = \mu(A), \text{ for } A \in \mathcal{R}$$

and

$$var(\mu, A) = \sup\{\mu(B) : B \in \mathcal{R}, B \subset A\}, \text{ for } A \subset S.$$

Proof. Let $A \subset S$ and $(A_i)_{i \in I}$ be a finite family of disjoint sets from $\mathcal{R}$ contained in A and denote $B = \bigcup_{i \in I} A_i$. We have $B \in \mathcal{R}$, $B \subset A$ and

$$\sum_{i \in I} |\mu(A_i)| = \sum_{i \in I} \mu(A_i) = \mu(B) \leq \sup\{\mu(B) : B \in \mathcal{R}, B \subset A\},$$

hence

$$\overline{\mu}(A) \leq \sup\{\mu(B) : B \in \mathcal{R}, B \subset A\}$$
$$\leq \sup\{\overline{\mu}(B) : B \in \mathcal{R}, B \subset A\} \leq \overline{\mu}(A)$$

and the second equality in the statement follows. If $A \in \mathcal{R}$ then
$$\mu(A) \leq \overline{\mu}(A) = \sup\{\mu(B) : B \in \mathcal{R}, \ B \subset A\} \leq \mu(A)$$
and the first equality in the statement follows. ∎

For real-valued measures, finite variation is equivalent to boundedness.

16. Proposition. *Let $\mu : \mathcal{R} \to \mathbb{R}$ be a real-valued, additive measure on a ring $\mathcal{R}$. Then:*
a) *for every set $A \subset S$ we have*
$$\sup\{|\mu(B)| : B \in \mathcal{R}, B \subset A\} \leq \overline{\mu}(A) \leq 2\sup\{|\mu(B)| : B \in \mathcal{R}, B \subset A\};$$
b) *μ is locally bounded (respectively bounded) on $\mathcal{R}$ iff μ has finite (respectively bounded) variation.*

Proof. Let $A \subset S$. Then for every set $B \in \mathcal{R}$ with $B \subset A$ we have
$$|\mu(B)| \leq \overline{\mu}(B) \leq \overline{\mu}(A)$$
and the first inequality in assertion a) follows.

Let $(A_i)_{i \in I}$ be a finite family of disjoint sets from $\mathcal{R}$ contained in A. Let I' be the set of indices $i \in I$ with $\mu(A_i) > 0$ and I'' the set of indices $i \in I$ with $\mu(A_i) \leq 0$ and set
$$B' = \bigcup_{i \in I'} A_i \text{ and } B'' = \bigcup_{i \in I''} A_i.$$
Then the sets B', B'' belong to $\mathcal{R}$, are disjoint and contained in A, and we have
$$\sum_{i \in I} |\mu(A_i)| = \sum_{i \in I'} \mu(A_i) - \sum_{i \in I''} \mu(A_i) = \mu(B') - \mu(B'')$$
$$= |\mu(B')| + |\mu(B'')| \leq 2\sup\{|\mu(B)| : B \in \mathcal{R}, B \subset A\},$$
hence
$$\overline{\mu}(A) \leq 2\sup\{|\mu(B)| : B \in \mathcal{R}, \ B \subset A\},$$
which is the second inequality in assertion a).

Assertion b) follows from assertion a). ∎

Remark. If $\mu : \mathcal{R} \to \mathbb{C}$ is a complex-valued, additive measure, for every $A \subset S$ we have
$$\overline{\mu}(A) \leq 4\sup\{|\mu(B)| : B \in \mathcal{R}, B \subset A\}.$$
In fact, $\mu = \mu_1 + i\mu_2$ with μ_1, μ_2 real-valued, additive measures.

From Theorem 14 and Proposition 16 we deduce the following corollary:

17. Corollary. *Any real-valued, σ-additive measure $\mu : \mathcal{S} \to \mathbb{R}$ on a σ-ring $\mathcal{S}$ has bounded variaton.*

Any real-valued, σ-additive measure $\mu : \mathcal{D} \to \mathbb{R}$ on a δ-ring $\mathcal{D}$ has finite variation.

D. Integration with respect to vector measures with finite variation

In this section we present the third stage in the development of the integral.

The framework for this section consists of a δ-ring $\mathcal{D}$ of subsets of S, three Banach spaces E, F, G with $E \subset L(F, G)$, and a σ-additive measure $m : \mathcal{D} \to E$ with finite variation $|m|$.

By Proposition 10, the variation $|m|$ is σ-additive on the σ-algebra $\mathcal{D}_{\text{loc}}$. This σ-algebra contains $\mathcal{D}$, hence it contains the σ-algebra $\Sigma = \sigma a(\mathcal{D})$ generated by $\mathcal{D}$. Therefore the semivariation $\|m\|$ is σ-additive on Σ, possibly taking on infinite values.

We shall reduce integrability of vector-valued functions $f : S \to E$ with respect to m, to the Bochner integrability of f with respect to the variation $|m|$.

18. Definition. *We say that a set $A \subset S$ is m-negligible (resp. m-measurable) if it is $|m|$-negligible (resp. $|m|$-measurable).*

We say a function $f : S \to F$ is m-negligible, m-measurable, m-integrable if it has the same property with respect to the variation $|m|$, as defined in Sections 1.I and 1.J on the classical integral and on the Bochner integral.

It follows that a set $A \in \Sigma$ is m-negligible iff $m(B) = 0$ for every set $B \in \mathcal{D} \cap A$, iff $|m|(A) = 0$. A set $A \subset S$ is m-negligible iff there is a set $B \in \Sigma$ with $A \subset B$ and $|m|(B) = 0$.

For $1 \leq p \leq +\infty$, the space $L_F^p(|m|)$ is defined in §1.J on the Bochner integral.

19. Definition. *For $1 \leq p \leq +\infty$ we denote*

$$L_F^p(m) := L_F^p(|m|)$$

and endow $L_F^p(m)$ with the seminorm of $L_F^p(|m|)$:

$$\|f\|_p = \| \, |f| \, \|_p = (\int |f|^p d|m|)^{\frac{1}{p}}, \text{ if } 1 \leq p < \infty$$

and

$$\|f\|_\infty = \| \, |f| \, \|_\infty.$$

We shall write also $L_F^p(\Sigma, m)$ instead of $L_F^p(m)$. If $1 \leq p < \infty$, then $L_F^p(m)$ contains all the characteristic functions of the sets $A \in \Sigma$ with $|m|(A) < \infty$.

We have the following immediate properties:

20. $L_F^p(m)$ is complete.

21. If $1 \leq p < \infty$ and if $\mathcal{R}$ is a ring generating the δ-ring $\mathcal{D}$, then the $\mathcal{R}$-step functions $f : S \to F$ are dense in $L_F^p(m)$.

In particular, the $\mathcal{D}$-step functions are dense in $L_F^p(m)$.

22. If $1 \leq p < \infty$, the Vitali and the Lebesgue convergence theorems are valid in $L_F^p(m)$.

23. For an m-measurable function $f : S \to F$ the following assertions are equivalent:

a) f is m-integrable;

b) f is $|m|$-integrable;

c) $|f|$ is $|m|$-integrable.

For a $\mathcal{D}$-step function $f = \sum \varphi_{A_i} x_i$ with $A_i \in \mathcal{D}$ and $x_i \in F$, we defined the integral
$$\int f dm = \sum m(A_i) x_i \in G.$$
If we take the sets A_i mutually disjoint, then $|f| = \sum \varphi_{A_i} |x_i|$ and
$$\left| \int f dm \right| = \left| \sum m(A_i) x_i \right| \leq \sum |m(A_i)| \, |x_i|$$
$$\leq \sum |m|(A_i) |x_i| = \int |f| d|m| = \|f\|_1,$$
therefore, the mapping $f \mapsto \int f dm$ from the set $\mathcal{S}_F(\mathcal{D})$ of F-valued, $\mathcal{D}$-step functions into G is linear and continuous for the seminorm $\|f\|_1$.

Since $\mathcal{S}_F(\mathcal{D})$ is dense in $L_F^1(m)$, we can extend uniquely the mapping $f \mapsto \int f dm$ to a linear continuous mapping on the whole space $L_F^1(m)$ with values in G.

The value of this extension for a function $f \in L_F^1(m)$ is denoted $\int f dm$ and is called the integral of f with respect to m.

We still have
$$\left| \int f dm \right| \leq \int |f| d|m| = \|f\|_1, \text{ for } f \in L_F^1(m).$$

We can extend m to the δ-ring $\Sigma_f = \{A \in \Sigma : |m|(A) < \infty\}$ by $m(A) = \int \varphi_A dm$, for $A \in \Sigma_f$.

If $f \in L_F^1(m)$ and $A \in \Sigma$, then $\varphi_A f \in L_F^1(m)$. We denote
$$\int_A f dm = \int \varphi_A f dm.$$

We have the following properties of the integral:

24. If $f_n \to f$ in $L_F^1(m)$, then $\int f_n dm \to \int f dm$ in G.

25. If $f \in L_F^1(m)$, then the mapping $A \mapsto \int_A f dm$ from Σ into G is σ-additive and
$$\lim_{|m|(A) \to 0} \int_A |f| d|m| = 0.$$

26. Theorem. Let $m : \mathcal{D} \to E \subset L(F,G)$ be a σ-additive measure with finite variation $|m|$, and let $x \in X$ and $z \in G^*$. Consider the measure $m_z : \mathcal{D} \to F^*$ defined by
$$\langle y, m_z(A) \rangle = \langle m(A)y, z \rangle, \text{ for } A \in \mathcal{D} \text{ and } y \in F.$$
Then:

a) The measures $m(\cdot)x : \mathcal{D} \to G$, $\langle m(\cdot)x, z \rangle : \mathcal{D} \to \mathbb{R}$ and $m_z : \mathcal{D} \to F^*$ are σ-additive and have finite variation.

b) If $\phi \in L^1_\mathbb{R}(m)$, then $\phi \in L^1_\mathbb{R}(m(\cdot)x)$, $\phi \in L^1_\mathbb{R}(\langle m(\cdot)x, z\rangle)$ and $\phi x \in L^1_F(m)$ and we have
$$\left(\int \phi dm\right)x = \int \phi dm(\cdot)x = \int \phi x dm$$
and
$$\left\langle \left(\int \phi dm\right)x, z \right\rangle = \int \phi d\langle m(\cdot)x, z\rangle = \left\langle \int \phi x dm, z \right\rangle$$
$$= \int \phi x dm_z.$$

c) If $f \in L^1_F(m)$, then $f \in L^1_F(m_z)$ and we have
$$\left\langle \int f dm, z \right\rangle = \int f dm_z.$$

Proof. The fact that the measures mentioned above are σ-additive and have finite variation follows from the inequalities $|m(A)x| \leq |m(A)||x|$, $|\langle m(A)x, z\rangle| \leq |m(A)|\,|x|\,|z|$ and $|m_z(A)| \leq |m(A)|\,|z|$, for $A \in \mathcal{D}$

The equalities in b) and c) are true for $\mathcal{D}$-step functions ϕ and f. Since the $\mathcal{D}$-step functions are dense in $L^1_\mathbb{R}(m)$ and $L^1_F(m)$, respectively, then, by the continuity of the integrals, these equalities remain valid for $\phi \in L^1_\mathbb{R}(m)$ and $f \in L^1_F(m)$. ∎

Remark. Unfortunately, most of the interesting measures do not have finite variation. But they may have finite semivariation instead. Paragraph 5 will be devoted to the construction and properties of the integral with respect to a measure with finite semivariation. This will be done by reducing the problem to a *family* of positive measures, rather than one single measure $|m|$, as it was in the case of measures with finite variation presented in this section.

E. The indefinite integral

Let $m : \mathcal{D} \to E \subset L(F,G)$ be a σ-additive measure with finite variation $|m|$ on a δ-ring $\mathcal{D}$ and let Σ be the σ-algebra generated by $\mathcal{D}$.

27. Definition. Let $g \in L^1_F(m)$. We denote by $gm : \Sigma \to G$ the σ-additive measure defined by
$$(gm)(A) = \int_A g dm, \text{ for } A \in \Sigma$$

and we call gm the indefinite integral of g with respect to m, or the measure with density g and base m, or even the product of g and m.

If $g \in L^1_F(m)$, then $|g| \in L^1_F(|m|)$ and we can consider the positive, indefinite integral $|g|\,|m| : \Sigma \to \mathbb{R}_+$:

$$(|g|\,|m|)(A) = \int_A |g| d|m|, \text{ for } A \in \Sigma.$$

We have the following property:

28. Proposition. *If $g \in L^1_F(m)$, then the measure gm has finite variation $|gm|$ satisfying*

$$|gm| \leq |g|\,|m|.$$

In fact, the inequality

$$\left|\int_A gdm\right| \leq \int_A |g|d|m|, \text{ for } A \in \Sigma$$

means

$$|(gm)(A)| \leq (|g|\,|m|)(A), \text{ for } A \in \Sigma$$

therefore $|gm| \leq |g|\,|m|$.

If either m or g is real-valued, we have the equality $|gm| = |g|\,|m|$, as the following theorem states.

29. Theorem. *Let $g \in L^1_F(m)$ and let*

$$m'(A) = \int_A gdm, \text{ for } A \in \Sigma.$$

a) *If either m or g is real-valued, then*

$$|m'|(A) = \int_A |g|d|m|, \text{ for } A \in \Sigma,$$

that is,

$$|gm| = |g|\,|m|.$$

b) *If $|gm| = |g||m|$ and if*

$$\int_A gdm = 0, \text{ for every } A \in \Sigma,$$

then $g = 0, m$-a.e..

Proof. We assume first that g is a $\mathcal{D}$-step function,

$$g = \sum_{i \in I} \varphi_{A_i} x_i$$

with $A_i \in \mathcal{D}$ disjoint and $x_i \in F$ if m is real-valued, or $x_i \in \mathbb{R}$ if m is E-valued. Then, for every $A \in \Sigma$ we have

$$m'(A) = \sum_{i \in I} m(A \cap A_i) x_i.$$

Let $M > \sum_{i \in I} |x_i|$. Let $A \in \Sigma$ and $\varepsilon > 0$. For each $i \in I$, there is a finite family $(B_{ij})_{j \in J(i)}$ of disjoint sets from $\mathcal{D}$ with union $A \cap A_i$, such that

$$\sum_{j \in J(i)} |m(B_{ij})| > |m|(A \cap A_i) - \varepsilon/M.$$

Since either m or g is real-valued, we have, for $j \in J(i)$,

$$|m'(B_{ij})| = |\sum_{k \in I} m(A_k \cap B_{ij}) x_k|$$
$$= |m(B_{ij}) x_i| = |m(B_{ij})| \, |x_i|.$$

Then

$$|m'|(A) \geq \sum_{i \in I} \sum_{j \in J(i)} |m'(B_{ij})| = \sum_{i \in I} |x_i| \sum_{j \in J(i)} |m(B_{ij})|$$
$$> \sum_{i \in I} |x_i| \, |m|(A \cap A_i) - \varepsilon = \int_A |g| d|m| - \varepsilon,$$

therefore, ε being arbitrary,

$$|m'|(A) \geq \int_A |g| d|m|.$$

The converse inequality being always true, we have the equality

$$|m'|(A) = \int_A |g| d|m|.$$

Assume now g is m-integrable. Since the $\mathcal{D}$-step functions are dense in $L_F^1(m)$, there is a sequence (g_n) of $\mathcal{D}$-step functions such that $g_n \to g$ in $L_F^1(m)$. Then, for every $A \in \Sigma$ we have $\varphi_A |g_n| \to \varphi_A |g|$ in $L^1(m)$; therefore

$$\int_A |g_n| d|m| \to \int_A |g| d|m|.$$

We shall prove now that

$$\lim_n |m'_n|(A) = |m'|(A), \text{ for } A \in \Sigma,$$

where $m'_n = g_n m$.

Let $A \in \Sigma$ and $\varepsilon > 0$. There is an N such that for every $n \geq N$ we have
$$\int_A |g_n - g|d|m| < \varepsilon.$$
Fix $n \geq N$. Let $(B_j)_{j \in J}$ be a finite family of disjoint set from Σ with union A such that
$$|m'|(A) - \sum_{j \in J} |\int_{B_j} g \, dm| < \varepsilon$$
and let $(C_k)_{k \in K}$ be a finite family of disjoint sets from Σ with union A such that
$$|m'_n|(A) - \sum_{k \in K} |\int_{C_k} g_n \, dm| < \varepsilon.$$
Let $I = J \times K$ and for each $i = (j, k) \in I$ denote $A_i = B_j \cap C_k$. The sets A_i are mutually disjoint, their union is A, and we have
$$|m'|(A) \geq \sum_{i \in I} |\int_{A_i} g \, dm| = \sum_{j \in J} \sum_{k \in K} |\int_{B_j \cap C_k} g \, dm|$$
$$\geq \sum_{j \in J} |\sum_{k \in K} \int_{B_j \cap C_k} g \, dm| = \sum_{j \in J} |\int_{B_j} g \, dm| \geq |m'|(A) - \varepsilon,$$
hence
$$|m'|(A) - \sum_{i \in I} |\int_{A_i} g \, dm| < \varepsilon.$$
Similarly,
$$|m'_n|(A) - \sum_{i \in I} |\int_{A_i} g_n \, dm| < \varepsilon.$$
We have also
$$|\sum_{i \in I} |\int_{A_i} g \, dm| - \sum_{i \in I} |\int_{A_i} g_n \, dm| |$$
$$\leq \sum_{i \in I} |\int_{A_i} (g - g_n) dm| \leq \int_A |g - g_n| d|m| < \varepsilon.$$
It follows that
$$| |m'|(A) - |m'_n|(A) | < 3\varepsilon.$$
We deduce then that
$$\lim_n |m'_n|(A) = |m'|(A).$$
Since (g_n) are $\mathcal{D}$-step functions, by the first part of of the proof we have
$$|m'_n|(A) = \int_A |g_n| d|m| \to \int_A |g| d|m|.$$

It follows that
$$|m'|(A) = \int_A |g|d|m|.$$
This proves assertion a).

To prove assertion b), assume $|gm| = |g||m|$. If $\int_A gdm = 0$ for every $A \in \Sigma$, then $gm = 0$ on Σ. Then $|gm| = 0$; therefore $|g||m| = 0$. It follows that $\int_S |g|d|m| = 0$; therefore $|g| = 0$, $|m|$-a.e.; consequently, $g = 0$, m-a.e.. ∎

Remarks. a) The fact that either m or g is real-valued was used in the equality
$$|m(B_{ij})x_i| = |m(B_{ij})| \, |x_i|.$$
The theorem remains valid as long as we have $|xy| = |x| \, |y|$ for $x \in E$ and $y \in F$. This is the case, for example, if $E \subset L(F, E \hat{\otimes}_\pi F)$.

b) Assertion b) in Theorem 29 can be stated as follows: if $|gm| = |g||m|$ and if $gm = 0$, then $g = 0$, m-a.e.

F. Integration with respect to gm

We prove first a theorem of integration with respect to the measure gm in case the measure m is real-valued.

30. Theorem. *Let $\mu : \mathcal{D} \to \mathbb{R}$ be a real-valued, σ-additive measure on a δ-ring $\mathcal{D}$. Let $|\mu|$ be its variation.*

Assume $E \subset L(F, G)$. Let $g \in L_E^1(\mu)$ and $f : S \to F$ a Σ-measurable function. Then

a) *f is $|g| \, |\mu|$-integrable iff f is $g\mu$-integrable.*

b) *If f is $g\mu$-integrable, then fg is μ-integrable, and we have*
$$\int f d(g\mu) = \int fg d\mu$$
and the associativity formula
$$f(g\mu) = (fg)\mu.$$

c) *Assume both μ and g are real-valued. Then f is $g\mu$-integrable iff fg is μ-integrable.*

Proof. Assertion a) is evident, in view of the equality $|g\mu| = |g| \, |\mu|$ (Theorem 29).

We shall first prove assertion b) for a Σ-step function f,
$$f = \sum_{1 \le i \le n} \varphi_{A_i} x_i, \text{ with } A_i \in \Sigma \text{ disjoint and } x_i \in F.$$

The function
$$fg = \sum_i \varphi_{A_i} g x_i$$

is Bochner μ-integrable (Corollary 1.37) and we have

$$\int f d(g\mu) = \sum_i (g\mu)(A_i) x_i = \sum_i (\int_{A_i} g d\mu) x_i$$
$$= \int (\sum_i \varphi_{A_i} x_i) g d\mu = \int f g d\mu.$$

We have also
$$|f| = \sum_i \varphi_{A_i} |x_i|$$

and, as above,
$$\int |f| d(|g| \, |\mu|) = \int |f| \, |g| d|\mu|.$$

Assume now f is $g\mu$-integrable. There is a sequence (f_n) of Σ-step function $f_n : S \to F$ with $f_n \to f$ and $|f_n| \leq |f|$.

By Lebesgue's Theorem we have $f_n \to f$ in $L_F^1(g\mu)$; therefore

$$\int f_n d(g\mu) \to \int f d(g\mu).$$

Since $|g\mu| = |g| \, |\mu|$, f is $|g| \, |\mu|$-integrable, and by Lebesgue's theorem we have $f_n \to f$ in $L_F^1(|g| \, |\mu|)$, hence (f_n) is a Cauchy sequence in $L_F^1(|g| \, |\mu|)$:

$$\lim_{mn} \int |f_n - f_m| d(|g| \, |\mu|) = 0.$$

Since $f_n - f_m$ are step functions, by the first part of the proof we have

$$\int |f_n - f_m| d(|g| \, |\mu|) = \int |f_n - f_m| \, |g| d|\mu|$$

and

$$\int |f_n g - f_m g| d|\mu| \leq \int |f_n - f_m| \, |g| d|\mu|.$$

It follows that
$$\lim_{mn} \int |f_n g - f_m g| d|\mu| = 0.$$

Since $f_n g \to fg$ pointwise and since $(f_n g)$ is a Cauchy sequence in $L_G^1(|\mu|) = L_G^1(\mu)$, it follows that fg is $|\mu|$-integrable, that is, μ-integrable and

$$\int f_n g d\mu \to \int f g d\mu.$$

For each n we have, by the first part of the proof,

$$\int f_n d(g\mu) = \int f_n g d\mu.$$

Passing to the limit we get

$$\int f d(g\mu) = \int fg d\mu.$$

Replacing f with $\varphi_A f$ for every $A \in \Sigma$ we deduce that

$$\int_A f d(g\mu) = \int_A fg d\mu, \text{ for } A \in \Sigma,$$

that is, the associativity formula

$$f(g\mu) = (fg)\mu.$$

This proves assertion b).

To prove assertion c), assume g is also real-valued and fg is μ-integrable and prove that f is $g\mu$-integrable.

Let (f_n) be a sequence of F-valued, Σ-step functions such that $f_n \to f$ and $|f_n| \le |f|$. Then $f_n g \to fg$ and $|f_n g| = |f_n|\,|g| \le |f|\,|g| = |fg|$. By Lebesgue's theorem, $f_n g \to fg$ in $L^1_F(\mu)$, hence

$$\lim_{m,n} \int |f_n g - f_m g| d|\mu| = 0.$$

Then

$$\int |f_n - f_m| d|g\mu| = \int |f_n - f_m| d(|g|\,|\mu|)$$
$$= \int |f_n - f_m|\,|g| d|\mu| = \int |f_n g - f_m g| d|\mu|,$$

therefore (f_n) is a Cauchy sequence in $L^1_F(g\mu)$; consequently f is $g\mu$-integrable. This proves assertion c). ∎

In the following theorem the measure m is vector-valued, but one of the functions f and g is real-valued.

31. Theorem. *Assume $E \subset L(F,G)$ and let $m : \mathcal{D} \to E \subset L(F,G)$ be a σ-additive measure with finite variation $|m|$ on a δ-ring $\mathcal{D}$.*

Assume that either

(i) *$g \in L^1(m)$ is a real-valued function and $f : S \to F$ is Σ-measurable,*

or

(ii) *$g \in L^1_F(m)$ is a vector-valued function and $f : S \to \mathbb{R}$ is a real-valued, Σ-measurable function.*

Then:

a) *f is $|g|\,|m|$-integrable iff fg is m-integrable.*
b) *If fg is m-integrable, then f is gm-integrable and we have*

$$\int f d(gm) = \int fg dm$$

and the associativity formula
$$f(gm) = (fg)m.$$

c) *If g is real-valued, then f is gm-integrable iff fg is m-integrable.*

Proof. In both cases (i) and (ii), assertion a) follows from the equivalence of the following statements:
f is $|g|\ |m|$-integrable;
$f|g|$ is $|m|$-integrable (by Theorem 30 c);
$|f|g||$ is $|m|$-integrable;
$|fg|$ is $|m|$-integrable (since $|fg| = |f|\ |g| = |f|g||$);
fg is m-integrable.

To prove assertion b), assume fg is m-integrable. Then, by assertion a), f is $|g|\ |m|$-integrable and from the inequality $|gm| \leq |g|\ |m|$ we deduce that f is $|gm|$-integrable, that is, f is gm-integrable. To prove the equalities in assertion b), let (f_n) be a sequence of Σ-step functions (with values in the same space as f) such that $f_n \to f$ and $|f_n| \leq |f|$. Since f is gm-integrable, by Lebesgue's theorem we deduce that $f_n \to f$ in the mean with respect to the measure gm; therefore
$$\int f_n d(gm) \to \int f d(gm).$$

Since either f or g is real-valued, we have
$$|f_n g| = |f_n|\ |g| \leq |f|\ |g| = |fg|.$$

Since $f_n g \to fg$ and $|fg|$ is m-integrable, by Lebesgue's theorem we deduce that $f_n g \to fg$ in the mean with respect to the measure m; therefore
$$\int f_n g\, dm \to \int fg\, dm.$$

Since for each n we have
$$\int f_n d(gm) = \int f_n g\, dm$$

it follows that
$$\int f d(gm) = \int fg\, dm.$$

Replacing f with $\varphi_A f$ with $A \in \Sigma$ we obtain
$$f(gm) = (fg)m$$

and assertion b) is proved.

To prove assertion c), assume g is real-valued and f is gm-integrable. Then f is $|gm|$-integrable. By Theorem 29, we have $|gm| = |g|\ |m|$, hence f is $|g|\ |m|$-integrable; by assertion a), fg is m-integrable and assertion c) is proved. ∎

G. The Radon–Nikodym theorem

In this section we shall prove a Radon–Nikodym-type theorem for vector measures, which will be used in Chapter 4 to prove the equality $|\mu_X| = \mu_{|X|}$ for the measure associated to a process X with integrable variation $|X|$.

Let $\mu : \mathcal{R} \to \overline{\mathbb{R}}_+$ be a positive, additive measure defined on a ring $\mathcal{R}$ and $m : \mathcal{R} \to E$ an additive measure. We say that m is *absolutely continuous with respect to* μ, or that m is μ-*absolutely continuous* and we write $m \ll \mu$, if

$$\lim_{\mu(A) \to 0} m(A) = 0.$$

We state first, without proof, the classical Radon–Nikodym theorem for real-valued measures on a σ-algebra Σ.

32. Theorem. *Let* $\mu : \Sigma \to \overline{\mathbb{R}}_+$ *be a positive, σ-finite, σ-additive measure and* $m : \Sigma \to \mathbb{R}$ *a real valued, σ-additive measure, such that* $m \ll \mu$.

Then there is a Σ-measurable function $g \in L^1(\mu)$ *such that* $m = g\mu$, *that is,*

$$m(A) = \int_A g\, d\mu, \text{ for } A \in \Sigma.$$

Then

$$|m|(A) = \int_A |g|\, d\mu, \text{ for } A \in \Sigma.$$

33. Remarks. a) The theorem is valid under more general conditions, for measures with the direct sum property, in particular for regular measures on a locally compact space (See [D.1]).
b) The theorem remains valid for vector-valued measures $m : \Sigma \to E$ with finite variation $|m|$, provided that E has the Radon–Nikodym Property.

A Banach space E has the Radon–Nikodym Property (RNP) if for every finite measure space (S, Σ, μ) and every σ-additive measure $m : \Sigma \to E$ with finite variation $|m|$ such that $m \ll \mu$, there is a function $g \in L^1_E(\mu)$ such that $m = g\mu$, that is,

$$m(A) = \int_A g\, d\mu, \text{ for } A \in \Sigma.$$

Then, by Theorem 29, we have

$$|m|(A) = \int_A |g|\, d\mu, \text{ for } A \in \Sigma.$$

Examples of Banach spaces with the RNP are: reflexive spaces and separable duals of Banach spaces.

For a detailed study of the Radon–Nikodym Theorem for vector measures, the reader is referred to [D–U].

We shall state and prove now a more general and very useful version of the Radon–Nikodym Theorem, without imposing the Banach space involved to have the RNP; but the density g is no longer measurable.

34. Theorem. Let $m : \Sigma \to E \subset L(F, G)$ be a σ-additive measure with bounded variation $|m|$ and $\mu : \Sigma \to \overline{\mathbb{R}}_+$ a positive, σ-finite, σ-additive measure such that $m \ll \mu$.

Let $Z \subset G^*$ be a space norming for G. Assume F and Z are separable. Then there is a function $V_m : S \to L(F, Z^*)$ having the following properties:
a) For every $x \in F$ and $z \in Z$, the function $\langle V_m x, z \rangle$ is μ-integrable, Σ-measurable and

$$\langle m(A)x, z \rangle = \int_A \langle V_m x, z \rangle d\mu, \text{ for } A \in \Sigma.$$

b) $|V_m|$ is μ-integrable, Σ-measurable, and $|m| = |V_m|\mu$, that is,

$$|m|(A) = \int_A |V_m| d\mu, \text{ for } A \in \Sigma.$$

c) If $|m| = \mu$, then $|V_m| \equiv 1$.
d) For every function $f \in L_F^1(m)$ and every $z \in Z$, the function $\langle V_m f, z \rangle$ is μ-integrable and

$$\left\langle \int f dm, z \right\rangle = \int \langle V_m f, z \rangle \mu.$$

Remark. The theorem is true without assuming F and Z separable; but the proof involves the lifting theory ([D.1] p. 260 and [IT]).

Proof. Assume first that $|m| = \mu$. If $m = 0$, we take $V_m = 0$ and the theorem is proved. Assume $m \neq 0$. For every $x \in F$ and $z \in Z$, the real-valued measure $m_{x,z} : \Sigma \to \mathbb{R}$ defined by

$$m_{x,z}(A) = \langle m(A)x, z \rangle, \text{ for } A \in \Sigma$$

is σ-additive and has bounded variation $|m_{x,z}|$ satisfying

$$|m_{xz}| \leq |x| |z| |m|.$$

By the Radon–Nikodym Theorem 32, there is a Σ-measurable, $|m|$-integrable, real-valued function $g_{x,z}$ such that

$$m_{x,z} = g_{x,z}|m|.$$

By Theorem 29 we have

$$|m_{x,z}| = |g_{x,z}| |m|.$$

It follows that

$$|g_{x,z}| |m| \leq |x| |z| |m|,$$

hence

(*) $\quad |g_{x,z}| \leq |x| |z|, \; |m|$-a.e.

Consequently $g_{x,z} \in L^\infty(|m|)$. Let $N(x,z)$ be an $|m|$-negligible set of Σ such that
$$|g_{x,z}| \leq |x|\,|z|, \text{ for } s \notin N(x,z).$$
For $x, x' \in F$, $z, z' \in Z$ and $\alpha, \alpha', \beta, \beta' \in \mathbb{R}$ we have
$$m_{\alpha x + \alpha' x', z} = \alpha m_{x,z} + \alpha' m_{x',z}$$
and
$$m_{x, \beta z + \beta' z'} = \beta m_{x,z} + \beta' m_{x,z'},$$
therefore

(**) $$g_{\alpha x + \alpha' x', z} = \alpha g_{x,z} + \alpha' g_{x',z}, \; |m|\text{-a.e.}$$

and

(***) $$g_{x, \beta z + \beta' z'} = \beta g_{x,z} + \beta' g_{x,z'}, \; |m|\text{-a.e.}$$

Let $N(x, x', z, z', \alpha, \alpha', \beta, \beta')$ be an $|m|$-negligible set such that the equalities (**) and (***) are true everywhere outside this set.

Let (x_n) be a sequence dense in F and F_0 the set of linear combinations of elements of the sequence (x_n) with rational coefficients. Then F_0 is a countable linear space over the rational field and F_0 is dense in F.

Similarly, let (z_n) be a sequence dense in Z and Z_0 the linear space over the rational field, generated by (z_n). Then Z_0 is countable and dense in Z.

Let N be the $|m|$-negligible set, union of all the sets $N(x,z)$ and $N(x, x', z, z', \alpha, \alpha', \beta, \beta')$, with $x, x' \in F_0$, $z, z' \in Z_0$, and $\alpha, \alpha', \beta, \beta'$ rationals.

For $s \notin N$ redefine $g_{x,z}(s) = 0$. Then the equalities (*), (**) and (***) are valid for every $s \in S$, $x, x' \in F_0$, $z, z' \in Z_0$, and $\alpha, \alpha', \beta, \beta'$ rational.

For each $s \in S$ and $x \in F_0$, the mapping $g_x(s) : z \mapsto g_{x,z}(s)$ is a continuous functional on Z_0, linear with respect to the rational field; therefore it can be extended to a continuous linear functional on Z, still denoted by $g_x(s)$ and we have $g_x(s) \in Z^*$,
$$\langle g_x(s), z \rangle = g_{x,z}(s), \text{ for } z \in Z$$
and
$$|g_x(s)| \leq |x|.$$
For fixed $s \in S$, the mapping $U_m(s) : x \mapsto g_x(s)$ of F_0 into Z^* is continuous and linear with respect to the rational field; therefore it can be extended to a continuous linear mapping of F into Z^*, still denoted by $U_m(s)$, and we have $U_m(s) \in L(F, Z^*)$,
$$\langle U_m(s) x, z \rangle = g_{x,z}(s), \text{ for } x \in F \text{ and } z \in Z,$$
and
$$|U_m| \leq 1.$$

Then, for $x \in F$ and $z \in Z$, the function $\langle U_m x, z \rangle$ is $|m|$-integrable, Σ-measurable, and

$$\langle m(A)x, z \rangle = \int_A \langle U_m(s)x, z \rangle d|m|, \text{ for } A \in \Sigma.$$

Taking $V_m = U_m$, assertion a) is proved.

We shall prove now assertion d). From the above equality we deduce that for every Σ-step function $f : S \to F$ and $z \in Z$ we have

$$\langle \int f dm, z \rangle = \int \langle U_m f, z \rangle d|m|.$$

Let now $f \in L^1_F(|m|)$ be a Σ-measurable function and let (f_n) be a sequence of F-valued, Σ-step functions such that $f_n \to f$ pointwise and $|f_n| \leq |f|$. Then

$$\langle U_m f_n, z \rangle \to \langle U_m f, z \rangle$$

and

$$|\langle U_m f_n, z \rangle| \leq |f_n| \, |z| \leq |f| \, |z|.$$

By Lebesgue's Theorem we deduce that

$$\int f_n dm \to \int f dm$$

and

$$\int \langle U_m f_n, z \rangle d|m| \to \int \langle U_m f, z \rangle d|m|.$$

Since for each n we have

$$\langle \int f_n dm, z \rangle = \int \langle U_m f_n, z \rangle d|m|,$$

taking the limit as $n \to \infty$ we deduce that

$$\langle \int f dm, z \rangle = \int \langle U_m f, z \rangle d|m|$$

Taking $V_m = U_m$, assertion d) is proved.

We prove now assertion c), that is, $|U_m(s)| = 1$, $|m|$-a.e.

We prove first that $|U_m|$ is Σ-measurable. In fact, for each $x \in F$, the mapping $U_m x = g_x : S \to Z^*$ is Z-weakly Σ-measurable. Since Z is separable, by Proposition 1.19, the function $|U_m x|$ is Σ-measurable. Then, by Proposition 1.14, $|U_m|$ is Σ-measurable.

From the inequality $|U_m| \leq 1$ proved above, we deduce that $|U|$ is $|m|$-integrable.

For $A \in \Sigma, x \in F$, and $z \in Z$ we have

$$|\langle m(A)x, z \rangle| \leq \int_A |x| \, |z| \, |U_m(s)| d|m|,$$

therefore
$$|m(A)| \leq \int_A |U_m(s)| d|m| = (|U_m| \, |m|)(A).$$

It follows that
$$|m|(A) \leq (|U_m| \, |m|)(A), \text{ for } A \in \Sigma;$$

therefore $|U_m| \geq 1$, $|m|$-a.e.; consequently $|U_m| = 1$, $|m|$-a.e. Taking $V_m = U_m$, assertion c) is proved. Then assertion b) follows immediately.

Consider now the general case. Since $m \ll \mu$ we have $|m| \ll \mu$. By the Radon–Nikodym Theorem 32, there is a μ-integrable, Σ-measurable function $g \geq 0$ such that $|m| = g\mu$.

Set $V_m = U_m g$. Then $|V_m| = |U_m|g = g$, hence $|m| = |V_m|\mu$ and assertion b) is proved.

For $x \in F$ and $z \in Z$ we have
$$\langle V_m x, z \rangle = \langle U_m, x, z \rangle g.$$

Since $\langle U_m x, z \rangle$ is $|m| = g\mu$-integrable, it follows that $\langle U_m x, z \rangle g$ is μ-integrable, that is, $\langle V_m x, z \rangle$ is μ-integrable and we have (Theorem 30)
$$\langle m(A)x, z \rangle = \int_A \langle U_m x, z \rangle d|m|$$
$$= \int_A \langle V_m x, z \rangle d\mu, \text{ for } A \in \Sigma$$

and assertion a) is proved.

Finally, to prove assertion d), let $f \in L^1_F(|m|)$ and $z \in Z$. The function $\langle U_m f, z \rangle$ is $|m| = g\mu$-integrable and
$$\langle \int f dm, z \rangle = \int \langle U_m f, z \rangle d|m|.$$

By Theorem 30 again, the function
$$\langle V_m f, z \rangle = \langle U_m f, z \rangle g$$

is μ-integrable and
$$\langle \int f dm, z \rangle = \int \langle U_m f, z \rangle g d\mu = \int \langle V_m f, z \rangle d\mu$$

and assertion d) is proved. ■

The following theorem is, in a certain sense, the converse of the preceding one. It will be used in §22 to prove the existence of dual predictable projections of processes with integrable variation.

35. Theorem. *Let $\mu : \Sigma \to \overline{\mathbb{R}}_+$ be a positive, σ-finite, σ-additive measure, $Z \subset G^*$ norming for G and $U : S \to E \subset L(F, G)$ a function having the* *following properties:*

(i) *The function $\langle Ux, z\rangle$ is μ-measurable for every $x \in E$ and $z \in Z$;*
(ii) *The function $|U|$ is μ-integrable.*
Then there is a σ-additive measure $m : \Sigma \to L(F, Z^*)$ with finite variation $|m|$ satisfying the following conditions:
a) *For every $A \in \Sigma$, $x \in F$, and $z \in Z$, the function $\varphi_A \langle Ux, z\rangle$ is μ-integrable and*
$$\langle m(A)x, z\rangle = \int_A \langle Ux, z\rangle d\mu;$$
b) $|m|(A) \leq \int_A |U| d\mu$, *for $A \in \Sigma$, that is, $|m| \leq |U|\mu$.*
b') *If F and Z are separable, then we have the equality $|m| = |U|\mu$.*
c) *For $f \in L^1_F(|U|\mu)$ and $z \in Z$, the function $\langle Uf, z\rangle$ is μ-integrable and we have*
$$\langle \int f dm, z\rangle = \int \langle Uf, z\rangle d\mu.$$
d) *The measure m has values in $L(F, G)$ if either G is separable, or G is the dual of a Banach space H and we take $Z = H$.*

Proof. For every $x \in F$ and $z \in Z$, the function $\langle Ux, z\rangle$ is μ-integrable since it is μ-measurable, $|\langle Ux, z\rangle| \leq |U| |x| |z|$ and $|U|$ is μ-integrable. Set
$$m_{x,z}(A) = \int_A \langle Ux, z\rangle d\mu, \text{ for } A \in \Sigma, \ x \in F \text{ and } z \in Z.$$

For fixed $A \in \Sigma$ and $x \in F$, the mapping $m_x(A) : z \mapsto m_{x,z}(A)$ is a continuous linear functional on Z:
$$|m_{x,z}(A)| \leq |x| |z| \int_A |U| d\mu;$$
therefore $m_x(A) \in Z^*$ and
$$|m_x(A)| \leq |x| \int_A |U| d\mu.$$

For fixed $A \in \Sigma$, the mapping $m(A) : x \to m_x(A)$ of F into Z^* is linear and continuous; therefore $m(A) \in L(F, Z^*)$ and
$$|m(A)| \leq \int_A |U| d\mu.$$

The measure $\lambda = |U|\mu$ is σ-additive, finite and we have $|m(A)| \leq \lambda(A)$ for $A \in \Sigma$. We have also
$$\langle m(A)x, z\rangle = m_{x,z}(A) = \int_A \langle Ux, z\rangle d\mu,$$
for $A \in \Sigma$, $x \in F$ and $z \in Z$.

From this last equality we deduce that m is finitely additive. From the inequality $|m(A)| \leq \lambda(A)$ for $A \in \Sigma$ we deduce that m is σ-additive and has finite variation $|m|$ satisfying $|m| \leq \lambda$. This proves assertion a) and the inequality in assertion b).

For an F-valued, Σ-step function $f = \sum \varphi_{A_i} x_i$ with $A_i \in \Sigma$ and $x_i \in F$ and for $z \in Z$ we have

$$\langle \int f\, dm, z \rangle = \sum \langle m(A_i) x_i, z \rangle = \sum \int_{A_i} \langle U x_i, z \rangle d\mu$$
$$= \int \langle U f, z \rangle d\mu.$$

Let $f \in L^1_F(\lambda)$ be a Σ-measurable function and let (f_n) be a sequence of F-valued, Σ-step functions such that $f_n \to f$ and $|f_n| \leq |f|$. Let $z \in Z$. Then

$$\langle U f_n, z \rangle \to \langle U f, z \rangle.$$

From $|f_n| \leq |f|$ we deduce that $|U| |f_n| \leq |U| |f|$, hence

$$\langle U f_n, z \rangle| \leq |U| |f| |z|.$$

Since $|f|$ is λ-integrable, the function $|U| |f|$ is μ-integrable. We can apply Lebesgue's Theorem in the space $L^1(\mu)$ an deduce that $\langle U f, z \rangle$ is μ-integrable and

$$\int \langle U f_n, z \rangle d\mu \to \int \langle U f, z \rangle d\mu.$$

On the other hand, from the inequality $|m| \leq \lambda$ and the fact that $f_n \to f$ in $L^1_F(\lambda)$, we deduce that $f_n \to f$ in $L^1_F(m)$, hence

$$\langle \int f_n\, dm, z \rangle \to \langle \int f\, dm, z \rangle.$$

Since for each n we have

$$\langle \int f_n\, dm, z \rangle = \int \langle U f_n, z \rangle d\mu,$$

it follows that

$$\langle \int f\, dm, z \rangle = \int \langle U f, z \rangle d\mu$$

and assertion c) is proved.

To prove the equality $|m| = |U|\mu$ of assertion b'), assume F and Z are separable. Since $m \ll \mu$, by Theorem 34 there is a function $V : S \to L(F, Z^*)$ such that $|m| = |V|\mu$ and

$$\langle \int f\, dm, z \rangle = \int \langle V(s) f(s), z \rangle d\mu, \text{ for } f \in L^1_F(m) \text{ and } z \in Z.$$

We deduce that for $f \in L_F^1(m)$ and $z \in Z$ we have

$$\int \langle V(s)f(s), z\rangle d\mu = \int \langle U(s)f(s), z\rangle d\mu.$$

In particular, for $f = \varphi_A x$ with $A \in \Sigma$ and $x \in F$ we get

$$\int_A \langle V(s)x, z\rangle d\mu = \int_A \langle U(s)x, z\rangle d\mu.$$

It follows that
$$\langle V(s)x, z\rangle = \langle U(s)x, z\rangle, \mu\text{-a.e.}$$

Let $N(x, z)$ be the μ-negligible set outside of which we have the above equality. Let (z_n) be a sequence dense in Z and $N(x) = \bigcup_n N(x, z_n)$. Then $N(x)$ is μ-negligible and for $s \notin N(x)$ we have

$$V(s)x = U(s)x.$$

Let (x_n) be a sequence dense in F and $N = \bigcup_n N(x_n)$. Then N is μ-negligible and for $s \notin N$ we have $V(s) = U(s)$. Then

$$|m| = |V|\mu = |U|\mu$$

and assertion b') is proved.

To prove assertion d), assume G is separable. Then for every $x \in F$, the function $Ux : S \to G$ has separable range and is Z-weakly Σ-measurable. By Proposition 1.20, Ux is Σ-measurable and from $|Ux| \leq |U|\,|x|$ we deduce that Ux is μ-integrable. Then, for every $A \in \Sigma$ we have

$$\langle m(A)x, z\rangle = \int_A \langle Ux, z\rangle d\mu = \langle \int_A Ux d\mu, z\rangle,$$

hence
$$m(A)x = \int_A Ux d\mu \in G.$$

It follows that $m : \Sigma \to L(F, G)$.

If G is the dual of H and we take $Z = H$, then $Z^* = H^* = G$ and the measure m takes on values in $L(F, G)$. ∎

H. Conditional expectations

In this section we define the conditional expectation for vector-valued functions. The definition and the existence of the conditional expectation involve the Bochner integral and could be placed in §1. However, in the proof of the uniqueness of the conditional expectation we use the properties of vector-valued measures with finite variation.

Let (S, Σ, μ) be a finite measure space and $\mathcal{F} \subset \Sigma$ a sub σ-algebra.

Every $\mathcal{F}$-measurable function $f : S \to F$ or $\overline{\mathbb{R}}$ is also Σ-measurable.

Consider the restriction $\mu_\mathcal{F}$ of μ to $\mathcal{F}$. Then $(S, \mathcal{F}, \mu_\mathcal{F})$ is a measure space. The following proposition states the relationship between the integrability with respect μ and $\mu_\mathcal{F}$.

36. Proposition. *We have* $L^1_F(\mathcal{F}, \mu_\mathcal{F}) \subset L^1_F(\Sigma, \mu)$ *and*

$$\int f d\mu_\mathcal{F} = \int f d\mu, \text{ for } f \in L^1_F(\mathcal{F}, \mu_\mathcal{F}).$$

Proof. If f is an $\mathcal{F}$-step function, then the equality in the statement of the proposition is immediate.

Let now $f \in L^1_F(\mathcal{F}, \mu_\mathcal{F})$ and let (f_n) be a sequence of F-valued, $\mathcal{F}$- step functions, such that $f_n \to f$, $\mu_\mathcal{F}$-a.e. (hence μ-a.e.) and in $L^1_F(\mathcal{F}, \mu_\mathcal{F})$; therefore $\int f_n d\mu_\mathcal{F} \to \int f d\mu_\mathcal{F}$. The sequence (f_n) is also Cauchy in $L^1_F(\Sigma, \mu)$:

$$\int |f_n - f_m| d\mu = \int |f_n - f_m| d\mu_\mathcal{F} \to 0, \text{ as } n, m \to \infty.$$

It follows that $f \in L^1_F(\Sigma, \mu)$ and $f_n \to f$ in $L^1_F(\Sigma, \mu)$; therefore $\int f_n d\mu \to \int f d\mu$.

Since for each n we have $\int f_n d\mu_\mathcal{F} = \int f_n d\mu$, it follows that $\int f d\mu_\mathcal{F} = \int f d\mu$. ∎

In view of the above proposition we shall write $L^1_F(\mathcal{F}, \mu)$ instead of $L^1_F(\mathcal{F}, \mu_\mathcal{F})$ and $\int f d\mu$ instead of $\int f d\mu_\mathcal{F}$, for functions $f \in L^1_F(\mathcal{F}, \mu)$.

The uniqueness μ-a.e. of the conditional expectation will follow from the following proposition:

37. Proposition. a) *Let* $g \in L^1_F(\mathcal{F}, \mu)$ *be a function satisfying*

$$\int_A g d\mu = 0, \text{ for every } A \in \mathcal{F}.$$

Then $g = 0, \mu$-*a.e.*

b) *Let* $g, g' \in L^1_F(\mathcal{F}, \mu)$ *be two functions satisfying*

$$\int_A g d\mu = \int_A g' d\mu = 0, \text{ for every } A \in \mathcal{F}.$$

Then $g = g', \mu$-*a.e.*

Proof. To prove assertion a), let $m : \mathcal{F} \to F$ be the measure defined by

$$m(A) = \int_A g d\mu_\mathcal{F}, \text{ for } A \in \mathcal{F}.$$

Then $m = g\mu_\mathcal{F}$ is a σ-additive measure with finite variation $|m| = |g|\mu_\mathcal{F}$ (Theorem 29 a)). Since, by hypothesis, we have $m = 0$, we deduce that $|m| = 0$, that is, $|g|\mu_{\mathcal{F}} = 0$. By Theorem 29 b), it follows that $g = 0$, $\mu_{\mathcal{F}}$-a.e.,that is, μ-a.e..

Assertion b) follows from assertion a). ∎

38. Definition. *Let $f \in L_F^1(\Sigma, \mu)$. The conditional expectation of f is any $\mathcal{F}$-measurable and μ-integrable function, denoted by $E(f|\mathcal{F})$, satisfying*

$$\int_A E(f|\mathcal{F})d\mu = \int_A f d\mu, \text{ for every } A \in \mathcal{F}.$$

By Proposition 37, the conditional expectation $E(f|\mathcal{F})$ is unique, μ-a.e.

The existence of the conditional expectation $E(f|\mathcal{F})$ for any function $f \in L_F^1(\Sigma, \mu)$ is proved in Theorems 39 and 50 below.

We prove first the existence of the conditional expectation for real-valued functions. We shall write again $L_F^1(\mu)$ instead of $L_F^1(\Sigma, \mu)$.

39. Theorem. *For any function $f \in L^1(\mu)$, the conditional expectation $E(f|\mathcal{F})$ exists.*

Proof. Consider the measure $\nu : \mathcal{F} \to \mathbb{R}$ defined by

$$\nu(A) = \int_A f d\mu, \text{ for } A \in \mathcal{F}.$$

Then ν is σ-additive and absolutely continuous with respect to the restriction $\mu_{\mathcal{F}}$ of μ to $\mathcal{F}$. By the classical Radon–Nikodym Theorem, there is an $\mathcal{F}$-measurable function $g \in L^1(\mu_{\mathcal{F}})$ such that

$$\nu(A) = \int_A g d\mu_{\mathcal{F}} = \int_A g d\mu, \text{ for } A \in \mathcal{F}.$$

It follows that

$$\int_A g d\mu = \int_A f d\mu \text{ for } A \in \mathcal{F}.$$

We take $E(f|\mathcal{F}) = g$. ∎

Before considering vector-valued functions, we list some properties of the conditional expectation for real-valued functions $f \in L^1(\mu)$.

40. If $\mathcal{F}_1 \subset \mathcal{F}_2 \subset \Sigma$ are σ-algebras and $f \in L^1(\mu)$, then

$$E(E(f|\mathcal{F}_1)|\mathcal{F}_2) = E(f|\mathcal{F}_1), \mu\text{-a.e.}$$

The iterated conditional expectation is written

$$E(f|\mathcal{F}_1|\mathcal{F}_2).$$

41. If $f_1, f_2 \in L^1(\mu)$ and $c_1, c_2 \in \mathbb{R}$, then

$$E(c_1 f_1 + c_2 f_2 | \mathcal{F}) = c_1 E(f_1|\mathcal{F}) + c_2 E(f_2|\mathcal{F}), \mu\text{-a.e.}$$

42. If $f \geq 0$, then $E(f|\mathcal{F}) \geq 0, \mu$-a.e.

43. If $f \leq g$, then $E(f|\mathcal{F}) \leq E(g|\mathcal{F}), \mu$-a.e.

44. $|E(f|\mathcal{F})| \leq E(|f||\mathcal{F}), \mu$-a.e.

More generally:

45. Jensen's inequality. If $c : \mathbb{R} \to \mathbb{R}$ is a convex function then

$$c \circ E(f|\mathcal{F}) \leq E(c \circ f|\mathcal{F}), \mu\text{-a.e..}$$

In particular, taking $c(x) = |x|^p$ with $1 \leq p < \infty$ we get:

46. If $f \in L^p(\mu)$ with $1 \leq p < \infty$, then

$$|E(f|\mathcal{F})|^p \leq E(|f|^p|\mathcal{F}), \mu\text{-a.e.}$$

47. If $f \in L^p(\mu)$ with $1 \leq p \leq \infty$, then $E(f|\mathcal{F}) \in L^p(\mu)$ and

$$\|E(f|\mathcal{F})\|_p \leq \|f\|_p.$$

48. If $f \in L^1(\mu)$ and $g : S \to \mathbb{R}$ is $\mathcal{F}$-measurable and if $fg \in L^1(\mu)$, then

$$E(fg|\mathcal{F}) = gE(f|\mathcal{F}), \mu\text{-a.e.}$$

49. If $f, g \in L^1(\mu)$ and if $E(f|\mathcal{F})g$ and $E(g|\mathcal{F})f$ are μ-integrable, then

$$\int E(f|\mathcal{F})g d\mu = \int E(g|\mathcal{F})f\mu.$$

We prove now the existence of the conditional expectation for vector-valued functions.

50. Theorem. *For any function $f \in L^1_F(\mu)$, the conditional expectation $E(f|\mathcal{F})$ exists.*

Proof. If F has the RNP, the proof of Theorem 39 can be applied in this case. We assume that F does not have the RNP. Let $f = \sum_{i \in I} \varphi_{A_i} x_i$ be a Σ-step function with $A_i \in \Sigma$ and $x_i \in F$.

Let $f = \sum_{j \in J} \varphi_{B_j} y_j$ with $B_j \in \Sigma$ and $y_j \in F$ be another representation of f and prove that

$$\sum_i E(\varphi_{A_i}|\mathcal{F})x_i = \sum_j E(\varphi_{B_j}|\mathcal{F})y_j, \mu\text{-a.e.}$$

In fact, for every $A \in \mathcal{F}$ we have

$$\int_A \sum_i E(\varphi_{A_i}|\mathcal{F})x_i d\mu = \sum_i x_i \int_A E(\varphi_{A_i}|\mathcal{F})d\mu$$

$$= \sum_i x_i \int_A \varphi_{A_i} d\mu = \int_A f d\mu$$

and similarly
$$\int_A \sum_j E(\varphi_{B_j}|\mathcal{F})y_j d\mu = \int_A f d\mu.$$

It follows that
$$\int_A \sum_i E(\varphi_{A_i}|\mathcal{F})x_i d\mu = \int_A \sum_j E(\varphi_{B_j}|\mathcal{F})y_j d\mu, \text{ for } A \in \mathcal{F}.$$

Since the functions under the integral sign are $\mathcal{F}$-measurable, by Proposition 37 we have
$$\sum_i E(\varphi_{A_i}|\mathcal{F})x_i = \sum_j E(\varphi_{B_j}|\mathcal{F})y_j, \mu\text{-a.e.}$$

We can now define, unambiguously, μ-a.e.,
$$E(f|\mathcal{F}) = \sum_i E(\varphi_{A_i}|\mathcal{F})x_i.$$

Then $E(f|\mathcal{F})$ is indeed the conditional expectation of f with respect to $\mathcal{F}$:
$$\int_A E(f|\mathcal{F})d\mu = \int_A f d\mu, \text{ for } A \in \mathcal{F}.$$

It is clear that the linearity Property 41 remains valid for Σ-step functions of $L_F^1(\mu)$. Property 44 remains valid in this case too. In fact, if we choose the sets A_i mutually disjoint, we have $|f| = \sum_i \varphi_{A_i}|x_i|$, hence
$$|E(f|\mathcal{F})| = |\sum_i E(\varphi_{A_i}|\mathcal{F})x_i|$$
$$\leq \sum_i E(\varphi_{A_i}|\mathcal{F})|x_i|$$
$$= E(\sum_i \varphi_{A_i}|x_i||\mathcal{F}) = E(|f||\mathcal{F}).$$

Then property 47 remains valid in this case:
$$\|E(f|\mathcal{F})\|_1 \leq \|E(|f||\mathcal{F})\|_1 \leq \|f\|_1.$$

Assume now $f \in L_F^1(\mu)$ and let (f_n) be a sequence of Σ-step functions from $L_F^1(\mu)$ converging in the mean to f. Then $\int_A f_n d\mu \to \int_A f d\mu$, for $A \in \mathcal{F}$.

The sequence $(E(f_n|\mathcal{F}))$ of conditional expectations is Cauchy in $L_F^1(\mu)$:
$$\|E(f_n|\mathcal{F}) - E(f_m|\mathcal{F})\|_1 = \|E(f_n - f_m|\mathcal{F})\|_1 \leq \|f_n - f_m\|_1 \to 0, \text{ as } n, m \to \infty.$$

Let $g \in L_F^1(\mu)$ be such that $E(f_n|\mathcal{F}) \to g$ in $L_F^1(\mu)$. We can take g to be $\mathcal{F}$-measurable and we have
$$\int_A E(f_n|\mathcal{F})d\mu \to \int_A g d\mu, \text{ for } A \in \mathcal{F}.$$

Since for each n we have

$$\int_A E(f_n|\mathcal{F})d\mu = \int_A f_n d\mu, \text{ for } A \in \mathcal{F},$$

we deduce that

$$\int_A g d\mu = \int_A f d\mu, \text{ for } A \in \mathcal{F}.$$

We take then $E(f|\mathcal{F}) = g$. ∎

Properties 40–49 remain valid for the conditional expectations of vector valued functions, as long as the terms involved are defined.

§3. σ-ADDITIVE MEASURES

In this paragraph we state a series of important theorems about σ-additive measures on δ-rings or σ-algebras, about uniform σ-additivity and uniform absolute continuity, and about the existence of control measures. For some of the proofs, the reader will be referred to [D–S].

A. σ-additive measures on σ-rings

We already proved in Theorem 2.14 that a σ-additive measure on a σ-ring is bounded and a σ-additive measure on a δ-ring is locally bounded.

The following theorem shows that a σ-additive measure on a σ-ring can be replaced with a σ-additive measure on a σ-algebra.

It follows that the natural domain for a σ-additive measure is either a δ-ring or a σ-algebra.

1. Theorem. (Kluvanek) *If $m : \mathcal{S} \to E$ is a σ-additive measure defined on a σ-ring $\mathcal{S}$, then there exists a set $S_0 \in \mathcal{S}$ such that $m(B) = 0$ for every set $B \in \mathcal{S}$ with $B \cap S_0 = \phi$.*

Proof. a) Let $(A_i)_{i \in I}$ be a family of mutually disjoint sets from $\mathcal{S}$. The set $\{i \in I; m(A_i) \neq 0\}$ is at most countable.

In fact, let $\varepsilon > 0$. Then the set $I_\varepsilon = \{i \in I : |m(A_i)| > \varepsilon\}$ is finite. If not, there is a sequence $(A_{i_n})_{n \in \mathbb{N}}$ from the family, with $|m(A_{i_n})| > \varepsilon$ for each n. Since $\mathcal{S}$ is a σ-ring, the union $A = \bigcup_{n \in \mathbb{N}} A_{i_n}$ belongs to $\mathcal{S}$; since m is σ-additive, we have

$$\sum_{n \in \mathbb{N}} m(A_{i_n}) = m(A).$$

It follows that $\lim_n m(A_{i_n}) = 0$, which contradicts $|m(A_{i_n})| > \varepsilon$ for every n.

Taking $\varepsilon = \frac{1}{n}$, the set $I_\infty = \bigcup_{n \in \mathbb{N}} I_{\frac{1}{n}} = \{i \in I : m(A_i) \neq 0\}$ is at most countable.

b) Let $\mathcal{F}$ be the set of all families $\mathcal{A} = (A_i)_{i \in I}$ of mutually disjoint sets from $\mathcal{S}$ with $m(A_i) \neq 0$ for every $i \in I$. We order $\mathcal{F}$ by inclusion. Then $\mathcal{F}$ is inductively ordered. By Zorn's Lemma, there is a maximal family $\mathcal{A}_0 = (A_i)_{i \in I_0}$ of $\mathcal{F}$. By the first part of the proof, $\mathcal{A}_0$ is at most countable, hence, the union $S_0 = \bigcup_{i \in I_0} A_i$ belongs to $\mathcal{S}$.

If $B \in \mathcal{S}$ is disjoint from S_0, then by the maximality of the family $\mathcal{A}_0$, we have $m(B) = 0$ and the theorem is proved. ∎

2. Remarks. a) The theorem is true, in particular, for a positive, finite, σ-additive measure $\mu : \mathcal{S} \to \mathbb{R}_+$ on a σ-ring $\mathcal{S}$.

If $\mathcal{S}$ is a δ-ring, the theorem is no longer true.

b) If $m : \mathcal{S} \to E$ is a σ-additive measure on a σ-ring, we have two alternatives to reduce it to a measure on a σ-algebra.

The first alternative is to extend m to the σ-algebra $\Sigma = \sigma a\,(\mathcal{S})$ generated by $\mathcal{S}$, by defining $m(A) = m(A \cap S_0)$, for $A \in \Sigma$.

The second alternative is to replace S with S_0 and $\mathcal{S}$ with the σ-algebra $\mathcal{S} \cap S_0$ of subsets of S_0 and to consider the restriction of m to $\mathcal{S} \cap S_0$.

It follows that we can restrict ourselves only to σ-additive measures defined on a δ-ring or on a σ-algebra.

B. Uniform σ-additivity

In this section we define the uniform σ-additivity and prove some simple properties.

3. Definition. *A set K of σ-additive measures $m : \mathcal{R} \to E$ on a ring $\mathcal{R}$ is said to be uniformly σ-additive, if for every sequence (A_n) of disjoint sets from $\mathcal{R}$ with union $A \in \mathcal{R}$ we have $m(A) = \sum_n m(A_n)$, the series being uniformly convergent with respect to $m \in K$.*

It is easy to see that K is uniformly σ-additive iff for every increasing (resp. decreasing) sequence (A_n) from $\mathcal{R}$ with limit $A \in \mathcal{R}$ we have $\lim_n m(A_n) = m(A)$, uniformly for $m \in K$.

If for every decreasing sequence $A_n \downarrow \phi$ from $\mathcal{R}$, we have $\lim_n m(A_n) = 0$, uniformly for $m \in K$, then K is uniformly σ-additive.

Another characterization of uniform σ-additivity is given by the following lemma.

4. Lemma. *A set K of σ-additive measures $m : \Sigma \to E$ on a σ-algebra Σ is uniformly σ-additive, iff for every sequence (A_n) of disjoint sets from Σ we have $\lim_n m(A_n) = 0$, uniformly for $m \in K$.*

Proof. If K is uniformly σ-additive, then for any sequence (A_n) of disjoint sets from Σ with union $A \in \Sigma$, the series $\Sigma_n m(A_n)$ is uniformly convergent for $m \in K$, hence $\lim_n m(A_n) = 0$ uniformly for $m \in K$.

Conversely, assume that for every sequence (A_n) of disjoint sets from Σ, we have $\lim_n m(A_n) = 0$ uniformly for $m \in K$ but that K is not uniformly σ-additive. Then there is a sequence (B_n) of disjoint sets from Σ and an $\varepsilon_0 > 0$, such that for every $k \in \mathbb{N}$, there are numbers $m_k > n_k > k$ and a measure $m_k \in K$ such that

$$\left| \sum_{n_k < n \leq m_k} m_k(B_n) \right| > \varepsilon_0.$$

By induction, we can take $n_k < m_k < n_{k+1}$ for each k. Denote $A_k = \bigcup_{n_k < n \leq n_k} B_n$. Then the sets (A_k) are mutually disjoint and for each k we have

$$|m_k(A_k)| = |\Sigma_{n_k < n \leq m_k} m_k(B_n)| > \varepsilon_0,$$

which contradicts the hypothesis that $m(A_k) \to 0$ uniformly for $m \in K$. ■

The preceding lemma is used in the following proposition. If K is a set of σ-additive measures $m : \Sigma \to E$ with finite variation $|m|$, we denote $|K| = \{|m| : m \in K\}$.

5. Proposition. *Let K be a set of real-valued, σ-additive measures $m : \Sigma \to \mathbb{R}$ on a σ-algebra Σ. Then K is uniformly σ-additive iff $|K|$ is uniformly σ-additive.*

Proof. The σ-additive measures $m : \Sigma \to \mathbb{R}$ have finite variation $|m|$ (Corollary 2.17). If $|K|$ is uniformly σ-additive, then, evidently, K is uniformly σ-additive.

Conversely, assume K is uniformly σ-additive but $|K|$ is not uniformly σ-additive. There is a sequence (A_n) of disjoint sets from Σ, an $\varepsilon_0 > 0$ and a sequence (m_n) from K such that $|m_n|(A_n) > 2\varepsilon_0$, for each n.

By Proposition 2.16, for each n there is a set $B_n \in \Sigma$ with $B_n \subset A_n$ such that $|m_n|(A_n) \leq 2|m_n(B_n)|$. It follows that $|m_n(B_n)| > \varepsilon_0$ for each n. Since the sets B_n are mutually disjoint and K is uniformly σ-additive, by Lemma 4 we have $\lim_n m(B_n) = 0$ uniformly for $m \in K$, which contradicts the above inequalities $|m_n(B_n)| > \varepsilon_0$ for each n. ∎

The following lemma will be used in Theorem 11, to prove the existence of a positive measure λ such that $K \ll \lambda$ uniformly.

6. Lemma. *Let K be a set of uniformly σ-additive measures $m : \Sigma \to E$ with finite variation $|m|$ on a σ-algebra Σ.*

Then, for every $\varepsilon > 0$, there is a $\delta > 0$ and a finite subset $J \subset K$ such that for every set $A \in \Sigma$ with $|m|(A) < \delta$ for $m \in J$ we have $|m(A)| < \varepsilon$ for all $m \in K$.

Proof. Deny the conclusion: there is an $\varepsilon_0 > 0$ such that for every $\delta > 0$ and for every finite set $J \subset K$, there is a set $E_J \in \Sigma$ and a measure $m_J \in K$ such that $|m|(E_J) < \delta$ for all $m \in J$ and $|m_J(E_J)| > \varepsilon_0$.

Take $\delta = \varepsilon_0/2$ and an arbitrary $m_1 \in K$ and $J = \{m_1\}$: there is a set $E_1 \in \Sigma$ and a measure $m_2 \in K$ such that $|m_1|(E_1) < \varepsilon_0/2$ and $|m_2(E_1)| > \varepsilon_0$.

Then take $\delta = \varepsilon_0/2^2$ and $J = \{m_1, m_2\}$: there is a set $E_2 \in \Sigma$ and a measure $m_3 \in K$ such that

$$|m_1|(E_2) < \varepsilon_0/2^2, |m_2|(E_2) < \varepsilon_0/2^2 \text{ and } |m_3(E_2)| > \varepsilon_0.$$

By induction, we can find a sequence (m_n) of measures from K and a sequence (E_n) of sets from Σ such that, for each n we have

$$|m_1|(E_n) < \varepsilon_0/2^n, \ldots, |m_n|(E_n) < \varepsilon_0/2^n \text{ and } |m_{n+1}(E_n)| > \varepsilon_0.$$

Set $F_n = \bigcup_{i \geq n} E_i$. Then, for $j \leq n \leq i$ we have

$$|m_j|(F_n) \leq \sum_{i \geq n} |m_j|(E_i) \leq \sum_{i \geq n} \varepsilon_0/2^i = \varepsilon_0/2^{n-1}$$

and
$$|m_{n+1}(F_n)| = |m_{n+1}(E_n) - m_{n+1}(F_n - E_n)|$$
$$\geq |m_{n+1}(E_n)| - |m_{n+1}(F_n - E_n)| \geq \varepsilon_0 - |m_{n+1}|(F_n - E_n)$$
$$\geq \varepsilon_0 - |m_{n+1}|(F_{n+1}) \geq \varepsilon_0 - \varepsilon_0/2^n > \varepsilon_0/2.$$

Set $F = \bigcap_n F_n$. Then, for $n > j$ we have
$$|m_j|(F) \leq |m_j|(F_n) \leq \varepsilon_0/2^{n-1} \to 0, \text{ as } n \to \infty;$$

therefore $|m_j|(F) = 0$ for each j. Then $m_j(F) = 0$ for each j; therefore, for each n we have
$$|m_{n+1}(F_n - F)| = |m_{n+1}(F_n) - m_{n+1}(F)|$$
$$= |m_{n+1}(F_n)| > \varepsilon_0/2.$$

Since $F_n - F \downarrow \phi$ we have $m(F_n - F) \to 0$, as $n \to \infty$, uniformly for $m \in K$; so, for $\varepsilon_0/2$, there is an n_0 such that for every $n \geq n_0$ we have $|m(F_n - F)| < \varepsilon_0/3$ for every $m \in K$. We have reached a contradiction and this proves the lemma. ∎

C. Uniform absolute continuity and uniform σ-additivity

We start with the absolute continuity and local absolute continuity of one measure. Absolute continuity has already been defined in §2.G on the Radon–Nikodym theorem.

7. Definition. *Let $m : \mathcal{R} \to E$ be an additive measure defined on a ring $\mathcal{R}$ and $\mu : \mathcal{R} \to \overline{\mathbb{R}}_+$ a positive, additive measure. We say that m is absolutely continuous with respect to μ, or that m is μ-absolutely continuous and we write $m \ll \mu$, if*
$$\lim_{\mu(A) \to 0} m(A) = 0,$$

that is, if for every $\varepsilon > 0$, there is a $\delta > 0$, such that for every $A \in \mathcal{R}$ with $\mu(A) < \delta$ we have $|m(A)| < \varepsilon$.

We say that m is locally μ-absolutely continuous and we write $m \ll \mu$ locally, if for every set $A \in \mathcal{R}$ we have
$$\lim_{\substack{\mu(B) \to 0 \\ B \subset A}} m(B) = 0.$$

We state first, without proof, the following two well-known properties.

8. Proposition. *Let $m : \Sigma \to E$ be a σ-additive measure with finite variation $|m|$ on a σ-algebra Σ and $\mu : \Sigma \to \overline{\mathbb{R}}_+$ a positive, σ-additive measure on Σ. Then*
a) *$m \ll \mu$ iff for every set $A \in \Sigma$ with $\mu(A) = 0$ we have $m(A) = 0$.*
b) *$m \ll \mu$ iff $|m| \ll \mu$.*

9. Proposition. *Let $m : \mathcal{D} \to E$ be a σ-additive measure with finite variation $|m|$ on a δ-ring $\mathcal{D}$ and $\mu : \mathcal{D} \to \overline{\mathbb{R}}_+$ a positive σ-additive measure on $\mathcal{D}$. Then*
a') *$m \ll \mu$ locally iff for every set $A \in \mathcal{D}$ with $\mu(A) = 0$ we have $m(A) = 0$.*
b') *$m \ll \mu$ locally iff $|m| \ll \mu$ locally.*

For a family of measures, one of the main problems is whether or not the measures are uniformly absolutely continuous.

10. Definition. *Let K be a set of σ-additive measures $m : \mathcal{R} \to E$ on a ring $\mathcal{R}$ and $\mu : \mathcal{R} \to \overline{\mathbb{R}}_+$ a positive measure.*

We say that $K \ll \mu$ uniformly (or that $m \ll \mu$ uniformly for $m \in K$), if for every $\varepsilon > 0$, there is a $\delta > 0$ such that for every set $A \in \mathcal{R}$ with $\mu(A) < \delta$ and for every measure $m \in K$ we have $|m(A)| < \varepsilon$.

We say that $K \ll \mu$ locally uniformly (or that $m \ll \mu$ locally, uniformly for $m \in K$), if for every set $A \in \mathcal{R}$ we have $K \ll \mu$ uniformly on $\mathcal{R} \cap A$, that is, for every $A \in \mathcal{R}$ and for every $\varepsilon > 0$, there is a $\delta > 0$ such that for every $B \in \mathcal{R}$ with $B \subset A$ and $\mu(B) < \delta$ and for every $m \in K$ we have $|m(B)| < \varepsilon$.

It is possible to have $K \ll \mu$ locally uniformly, without having $K \ll \mu$ uniformly on $\mathcal{R}$.

If each measure $m \in K$ is μ-absolutely continuous, we write $K \ll \mu$.

The existence of a positive measure λ such that $K \ll \lambda$ uniformly is proved in the following theorem.

11. Theorem. *Let K be a set of σ-additive measures $m : \Sigma \to E$ defined on a σ-algebra Σ.*

If K is uniformly σ-additive, then there exists a positive, finite, σ-additive measure $\lambda : \Sigma \to \mathbb{R}_+$ such that $K \ll \lambda$ uniformly and

$$\lambda(A) \le \sup_{m \in K} \sup_{B \subset A} |m(B)|, \text{ for } A \in \Sigma.$$

If K consists of real-valued measures, then λ can be taken of the form

$$\lambda = \sum_n c_n |m_n|$$

with $c_n \ge 0$, $\sum_n c_n \le 1$ and (m_n) a sequence of measures from K.

Proof. Assume first that K consists of real-valued measures. Then each measure $m \in K$ has finite variation $|m|$, by Corollary 2.17. We can apply Lemma 6: for each $n \in \mathbb{N}$, take $\epsilon = \frac{1}{2^n}$; there is a $\delta_n = \delta(\frac{1}{2^n})$ and measures $m_1, m_2, ..., m_{k(n)}$ from K, such that, for every set $A \in \Sigma$ with $|m_i|(A) < \delta_n$ for $1 \le i \le k(n)$ we have $|m(A)| < \frac{1}{2^n}$ for all $m \in K$. Set

$$\lambda_n = \sum_{1 \le i \le k(n)} (\frac{1}{2^i}) |m_i|$$

and
$$\lambda = \frac{1}{4}\sum_n (\frac{1}{2^n})\frac{\lambda_n}{1+\lambda_n(X)}.$$

Then λ_n and λ are positive, finite, σ-additive measures. We shall prove that $K \ll \lambda$ uniformly. In fact, let $\epsilon > 0$ and take $n \in \mathbb{N}$ such that $\frac{1}{2^n} < \epsilon$. Then take $\delta(\epsilon) = \frac{1}{4}\delta_n \frac{1}{2^{n+k(n)}} \frac{1}{1+\lambda_n(X)}$ and let $A \in \Sigma$ with $\lambda(A) < \delta(\epsilon)$. Then

$$\frac{1}{4}\frac{1}{2^n}\frac{\lambda_n(A)}{1+\lambda_n(X)} \leq \lambda(A) < \delta(\epsilon) = \frac{1}{4}\delta_n \frac{1}{2^{n+k(n)}} \frac{1}{1+\lambda_n(X)},$$

hence
$$\lambda_n(A) < \delta_n \frac{1}{2^{k(n)}},$$

consequently
$$(\frac{1}{2^i})|m_i|(A) < \delta_n \frac{1}{2^{k(n)}}, \text{ for } 1 \leq i \leq k(n);$$

therefore
$$|m_i|(A) < \delta_n, \text{ for } 1 \leq i \leq k(n).$$

It follows that $|m(A)| < \epsilon$ for all $m \in K$; consequently $K \ll \lambda$ uniformly. By Proposition 5 we have $|K| \ll \lambda$ uniformly. Moreover, for every $A \in \Sigma$ we have

$$\lambda_n(A) \leq \sup_{m \in K}|m|(A) \sum_{1 \leq i \leq k(n)} \frac{1}{2^i}$$
$$\leq \sup_{m \in K}|m|(A) \leq 4 \sup_{m \in K}\sup_{B \subset A}|m(B)|,$$

hence
$$\lambda(A) \leq \frac{1}{4}\sup \lambda_n(A) \leq \sup_{m \in K}\sup_{B \subset A}|m(B)|.$$

Assume now K consists of E-valued, uniformly σ-additive measures. Then the set $K^* = \{x^*m : m \in K, x^* \in E_1^*\}$ consists of real-valued, uniformly σ-additive measures. By the first part of the proof, there is a positive, finite, σ-additive measure λ on Σ such that $K^* \leq \lambda$ uniformly and

$$\lambda(A) \leq \sup_{x^*m \in K^*}\sup_{B \subset A}|x^*m(B)|$$
$$= \sup_{m \in K}\sup_{B \subset A}|m(B)|.$$

From $K^* \leq \lambda$ uniformly, it follows that $K \ll \lambda$ uniformly. ∎

The above theorem can be extended for measures on a δ-ring.

12. Theorem. *Let K be a set of σ-additive measures $m : \mathcal{D} \to E$ defined on a δ-ring $\mathcal{D}$. Assume that $S = \bigcup_n S_n$ with $S_n \in \mathcal{D}$. If K is uniformly*

σ-additive, then there is a positive, finite, σ-additive measure λ on $\mathcal{D}$ such that $K \ll \lambda$ locally uniformly and

$$\lambda(A) \leq \sup_{m \in K} \sup_{B \subset A} |m(B)|, \text{ for } A \in \mathcal{D}.$$

Proof. For each set $A \in \mathcal{D}$ we apply Theorem 11 to the set K_A of the restrictions of the measures $m \in K$ to the σ-algebra $\mathcal{D} \cap A$ of subsets of A and deduce the existence of a positive, finite, σ-additive measure $\lambda_A : \mathcal{D} \cap A \to \mathbb{R}_+$ such that $m_A \ll \lambda_A$ on $\mathcal{D} \cap A$ uniformly for $m_A \in K_A$ and

$$\lambda_A(B) \leq \sup_{m \in K} \sup_{C \subset B} |m(C)|, \text{ for } B \in \mathcal{D} \cap A.$$

It follows that if $B \in \mathcal{D} \cap A$, then $\lambda_A(B) = 0$ iff $m(C) = 0$ for every $m \in K$ and every $C \in \mathcal{D} \cap B$.

We deduce that if $A, A' \in \mathcal{D}$, then the measures λ_A and $\lambda_{A'}$ are equivalent on $\mathcal{D} \cap A \cap A'$.

We take now $A = S_n$ and denote $\lambda_n = \lambda_{S_n}$. Without loss of generality we can assume that the sets S_n are mutually disjoint. We define

$$\lambda(A) = \sum_{1 \leq n < \infty} 2^{-n} \frac{\lambda_n(A \cap S_n)}{1 + \lambda_n(S_n)}, \text{ for } A \in \mathcal{D}.$$

Then λ is positive, finite and σ-additive on $\mathcal{D}$. For $A \in \mathcal{D}$ the restriction of λ to $\mathcal{D} \cap A$ is equivalent to λ_A. In fact, if $B \in \mathcal{D} \cap A$ and $\lambda(B) = 0$, then $\lambda_n(B \cap S_n) = 0$ for every n; since λ_A and λ_n are equivalent on $\mathcal{D} \cap (A \cap S_n)$, we deduce that $\lambda_A(B \cap S_n) = 0$; therefore

$$\lambda_A(B) = \sum_{1 \leq n < \infty} \lambda_A(B \cap S_n) = 0.$$

Conversely, if $\lambda_A(B) = 0$, then $\lambda_A(B \cap S_n) = 0$ for every n, hence $\lambda_n(B \cap S_n) = 0$ for every n; consequently $\lambda(B) = 0$.

Since $m_A \ll \lambda_A$ on $\mathcal{D} \cap A$, uniformly for $m \in K$, it follows that $m \ll \lambda$ on $\mathcal{D} \cap A$, uniformly for $m \in K$ and the theorem is proved. ∎

If the set K in Theorems 11 and 12 consists of one measure, we have the following theorem about the existence of a control measure.

13. Theorem. *If $m : \Sigma \to E$ is a σ-additive measure on a σ-algebra Σ, then there exists a positive, finite, σ-additive measure λ on Σ such that $m \ll \lambda$ and*

$$\lambda(A) \leq \sup_{B \subset A} |m(B)|, \text{ for } A \in \Sigma.$$

The measure λ is called a control measure for m.

Remark. We shall prove later (Theorem 4.20) that we have also $\tilde{m}_{\mathbb{R},E} \ll \lambda$, where $\tilde{m}_{\mathbb{R},E}$ is the semivariation of m.

14. Theorem. *If $m : \mathcal{D} \to E$ is a σ-additive measure defined on a δ-ring $\mathcal{D}$ and if $S = \bigcup S_n$ with $S_n \in \mathcal{D}$, then there is a local control measure λ of m, that is, a positive, finite, σ-additive measure $\lambda : \mathcal{D} \to \mathbb{R}_+$ such that $m \ll \lambda$ locally and*

$$\lambda(A) \leq \sup_{B \subset A} |m(B)|, \text{ for } A \in \mathcal{D}.$$

Remark. We shall see later (Theorem 4.21) that we have also $\tilde{m}_{\mathbb{R},E} \ll \lambda$ locally.

As a consequence of Theorem 13, we have the following theorem stating that the values of a σ-additive measure on a σ-algebra can be approximated with its values on a ring generating the σ-algebra.

15. Theorem. *Let $m : \Sigma \to E$ be a σ-additive measure on a σ-algebra Σ and $\mathcal{R}$ a ring generating Σ. Then, for every set $A \in \Sigma$ and every $\varepsilon > 0$, there is a set $B \in \mathcal{R}$ such that $|m(A) - m(B)| < \varepsilon$.*

Proof. By Theorem 13, there is a control measure λ for m. Let $A \in \Sigma$ and $\varepsilon > 0$. Since $m \ll \lambda$, there is a $\delta > 0$ such that for every set $C \in \Sigma$ with $\lambda(C) < \delta$ we have $m(C) < \varepsilon/2$.

On the other hand, for the above $\delta > 0$, there is a set $B \in \mathcal{R}$ such that $\lambda(A \triangle B) < \delta$, hence $\lambda(A - B) < \delta$ and $\lambda(B - A) < \delta$; consequently $|m(A - B)| < \varepsilon/2$ and $|m(B - A)| < \varepsilon/2$. Then

$$|m(A) - m(B)| = |m(A - B) - m(B - A)|$$
$$\leq |m(A - B)| + |m(B - A)| < \varepsilon.$$

∎

The following theorem shows that under certain conditions, $K \ll \mu$ implies $K \ll \mu$ uniformly.

16. Theorem. (Vitali–Hans–Saks) *Let (m_n) be a sequence of σ-additive measures $m_n : \Sigma \to E$ on a σ-algebra Σ and $\lambda : \Sigma \to \mathbb{R}_+$ be a positive, finite, σ-additive measure.*

If $m(A) = \lim_n m_n(A)$ exists for every $A \in \Sigma$ and if $m_n \ll \lambda$ for every $n \in \mathbb{N}$, then $m_n \ll \lambda$ uniformly for $n \in \mathbb{N}$ and the limit $m : \Sigma \to E$ is σ-additive.

For the proof, the reader is referred to [D–S], III.7.2 and III.7.3.

Using the above theorem we can prove, under certain conditions, the uniform σ-additivity of a sequence of measures.

17. Theorem. (Nikodym) *Let (m_n) be a sequence of σ-additive measures $m_n : \mathcal{D} \to E$ on a δ-ring $\mathcal{D}$. If $m(A) = \lim_n m_n(A)$ exists for each set $A \in \mathcal{D}$, then $m : \mathcal{D} \to E$ is σ-additive and the sequence (m_n) is uniformly σ-additive.*

Proof. Assume first that $\mathcal{D}$ is a σ-algebra. By Theorem 13, for each n there is a positive, finite, σ-additive measure λ_n on $\mathcal{D}$ such that $m_n \ll \lambda_n$. The measure $\lambda : \mathcal{D} \to \mathbb{R}_+$ defined by

$$\lambda(A) = \sum_{1 \leq n < \infty} \frac{1}{2^n} \frac{\lambda_n(A)}{1 + \lambda_n(S)}, \text{ for } A \in \mathcal{D},$$

is positive, finite and σ-additive and we have $\lambda_n \ll \lambda$ for each n. It follows that $m_n \ll \lambda$ for each n. By the Vitali–Hans–Saks Theorem 16, the measure m is σ-additive and $m_n \ll \lambda$ uniformly. It follows that the sequence (m_n) is uniformly σ-additive.

Assume now $\mathcal{D}$ is a δ-ring. Let (A_k) be a sequence of disjoint sets from $\mathcal{D}$ with union $A \in \mathcal{D}$. We apply the first part of the proof to the restrictions of the measures m_n to the σ-algebra $\mathcal{D} \cap A$ of subsets of A and deduce that the restriction of m to $\mathcal{D} \cap A$ is σ-additive and that (m_n) are uniformly σ-additive on $\mathcal{D} \cap A$. It follows that m is σ-additive on $\mathcal{D}$ and that (m_n) are uniformly σ-additive on $\mathcal{D}$. ∎

D. Weak σ-additivity

18. Definition. *We say that an additive measure $m : \mathcal{R} \to E$ defined on a ring $\mathcal{R}$ is weakly σ-additive if for every $x^* \in E^*$, the real measure x^*m is σ-additive.*

By contrast, the σ-additive measures are called strongly σ-additive.

Every strongly σ-additive measure is, evidently, weakly σ-additive. The converse is also true if the domain of the measure is a δ-ring.

19. Theorem. (Pettis) *If $m : \mathcal{D} \to E$ is a weakly σ-additive measure on a δ-ring $\mathcal{D}$, then m is strongly σ-additive.*

Proof. The theorem is true if $\mathcal{D}$ is a σ-algebra ([D–S], IV.10.1).

Assume now $\mathcal{D}$ is a δ-ring and m is weakly σ-additive. Let (A_n) be a sequence of disjoint sets from $\mathcal{D}$ with union $A \in \mathcal{D}$. The restriction of m to the σ-algebra $\mathcal{D} \cap A$ of subsets of A is weakly σ-additive, hence, by the first part of the proof, it is strongly σ-additive on $\mathcal{D} \cap A$; therefore

$$m(A) = \sum_n m(A_n).$$

It follows that m is strongly σ-additive on $\mathcal{D}$. ∎

E. Uniform σ-additivity of indefinite integrals

20. Let (S, Σ, μ) be a measure space and consider the space $L^1_F(\mu)$. For each function $f \in L^1_F(\mu)$ we can consider the σ-additive measure $f\mu : \Sigma \to F$

defined by the indefinite integral of f:

$$(f\mu)(A) = \int_A f d\mu, \text{ for } A \in \Sigma.$$

The measure $f\mu$ has finite variation $|f\mu|$ (Theorems 2.28, 2.29) satisfying

$$|f\mu|(A) = \int_A |f| d\mu, \text{ for } A \in \Sigma.$$

For any set $K \subset L^1_F(\mu)$ we denote $|K| = \{|f| : f \in K\}$ and $K\mu = \{f\mu : f \in K\}$. Then $|K|\mu = \{|f|\mu : f \in K\}$.

For each $f \in K$ we have $f\mu \ll \mu$; that is, $K\mu \ll \mu$.

21. Definition. Let $K \subset L^1_F(\mu)$. We say K is uniformly σ-additive, if $K\mu$ is uniformly σ-additive.

We say K is uniformly absolutely continuous with respect to μ if $K\mu \ll \mu$ uniformly, i.e., if

$$\lim_{\mu(A) \to 0} \int_A f d\mu = 0, \text{ uniformly for } f \in K.$$

We write $K \ll \mu$ uniformly.

The following theorem gives a characterization of uniform σ-additivity of $|K|$.

22. Theorem. Let $K \subset L^1_F(\mu)$. The following assertions are equivalent:
a) $|K|$ is uniformly σ-additive;
b) For each $g \in L^\infty_{F^*}(\mu)$, the set $\langle K, g \rangle \mu = \{\int_{(\cdot)} \langle f, g \rangle d\mu; f \in K\}$ is uniformly σ-additive;
c) For each $g \in L^\infty_{F^*}(\mu)$, the set $|\langle K, g \rangle|\mu = \{\int_{(\cdot)} |\langle f, g \rangle| d\mu; f \in K\}$ is uniformly σ-additive.

Proof. The implications a) $\Longrightarrow$ c) and c) $\Longrightarrow$ b) follow from the inequalities

$$\left| \int_A \langle f, g \rangle d\mu \right| \leq \int_A |\langle f, g \rangle| d\mu \leq \|g\|_\infty \int_A |f| d\mu,$$

for $A \in \Sigma$, $f \in K$ and $g \in L^\infty_{F^*}(\mu)$.

To prove b) $\Longrightarrow$ a) deny $|K|$ is uniformly σ-additive: there is a $\delta > 0$, a sequence (f_n) from K and a sequence (A_n) of disjoint sets from Σ such that

$$\int_{A_n} |f_n| d\mu > \epsilon, \text{ for each } n.$$

By Theorem 1.38 b), for each n there is a Σ-step function $g_n : X \to F^*$ with $|g_n| \leq 1$ such that

$$\int_{A_n} \langle f_n, g_n \rangle d\mu > \epsilon.$$

If we set $g = \sum_n \varphi_{A_n} g_n$, then $|g| \leq 1$, hence $g \in L^\infty_{F^*}(\mu)$ and

$$\int_{A_n} \langle f_n, g \rangle d\mu = \int_{A_n} \langle f_n, g_n \rangle d\mu > \epsilon, \text{ for each } n,$$

hence $\langle K, g \rangle \mu$ is not uniformly σ-additive. ∎

The following theorem states the relationship between uniform σ-additivity and uniform absolute continuity of sets of $L^1_F(\mu)$:

23. Theorem. *Let $K \subset L^1_F(\mu)$. Then*
a) *$|K|$ is uniformly σ-additive iff:*
i) *$|K| \ll \mu$ uniformly;*
and
ii) *for every $\epsilon > 0$ there is a set $S_\epsilon \in \Sigma$ with $\mu(S_\epsilon) < \infty$ such that*

$$\int_{S - S_\epsilon} |f| d\mu < \epsilon, \text{ for all } f \in K.$$

b) *Assume $\mu(S) < \infty$. Then $|K|$ is uniformly σ-additive iff $|K| \ll \mu$ uniformly.*

Proof. Assume $|K|$ is uniformly σ-additive and deny i): there is an $\epsilon_0 > 0$ such that for every $\delta > 0$, there is a set $E_\delta \in \Sigma$ with $\mu(E_\delta) < \delta$ and a function $f_\delta \in K$ such that

$$\int_{E_\delta} |f_\delta| d\mu > \epsilon_0,$$

hence

$$\sup_{f \in K} \int_{E_\delta} |f| d\mu > \epsilon_0.$$

Taking $\delta = \frac{1}{2^n}$ and denoting $E_n = E_\delta$, we have $\mu(E_n) < \frac{1}{2^n}$ and

$$\sup_{f \in K} \int_{E_n} |f| d\mu > \epsilon_0, \text{ for each } n.$$

If we set $F_n = \bigcup_{k > n} E_k$, then the sequence (F_n) is decreasing and $\mu(F_n) \leq \sum_{k > n} \mu(E_k) \leq \frac{1}{2^n}$ for each n, hence the set $F = \bigcap F_n$ is μ-negligible. Since $|K|$ is uniformly σ-additive, we have

$$\limsup_n \sup_{f \in K} \int_{F_n} |f| d\mu = 0,$$

which contradicts the above inequality

$$\sup_{f \in K} \int_{F_n} |f| d\mu \geq \sup_{f \in K} \int_{E_n} |f| d\mu > \epsilon_0, \text{ for each } n.$$

It follows that if $|K|$ is uniformly σ-additive then $|K| \ll \mu$ uniformly.

Assume now $|K|$ is uniformly σ-additive and prove ii). Let $\epsilon > 0$. By Lemma 6, there is a $\delta > 0$ and a finite family $(f_i)_{i \in J}$ from K such that for every set $A \in \Sigma$ with $\int_A |f_i| d\mu < \delta$ for $i \in J$ we have $\int_A |f| d\mu < \varepsilon$ for all $f \in K$.

For each $i \in J$ there is a set $S_i \in \Sigma$ with $\mu(S_i) < \infty$ such that $\int_{S-S_i} |f_i| d\mu < \delta$. If we set $S_\epsilon = \bigcup_{i \in J} S_i$, then $\mu(S_\epsilon) < \infty$ and

$$\int_{S-S_\epsilon} |f_i| d\mu \leq \int_{S-S_i} |f_i| d\mu < \delta, \text{ for } i \in J.$$

It follows that $\int_{S-S_\epsilon} |f| d\mu < \epsilon$ for all $f \in K$ and assertion ii) is proved.

Conversely assume i) and ii) and prove $|K|$ is uniformly σ-additive. Let $A_n \downarrow \phi$ with $A_n \in \Sigma$ and prove $\int_{A_n} |f_i| d\mu \to 0$, uniformly for $f \in K$. Let $\epsilon > 0$. By assumption ii), there is a set $S_\epsilon \in \Sigma$ with $\mu(S_\epsilon) < \infty$ such that

$$\int_{S-S_\epsilon} |f| d\mu < \frac{\epsilon}{2}, \text{ for all } f \in K.$$

Then $A_n \cap S_\epsilon \downarrow \phi$, hence $\mu(A_n \cap S_\epsilon) \to 0$. By assumtion i), there is a $\delta > 0$ such that for every set $A \in \Sigma$ with $\mu(A) < \delta$ we have $\int_A |f| d\mu < \frac{\epsilon}{2}$ for all $f \in K$. Let N be such that for $n \geq N$ we have $\mu(A_n \cap S_\epsilon) < \delta$. Then for $n \geq N$ we have

$$\int_{A_n \cap S_\epsilon} |f| d\mu < \frac{\epsilon}{2}, \text{ for all } f \in K,$$

hence

$$\int_{A_n} |f| d\mu = \int_{A_n \cap S_\epsilon} |f| d\mu + \int_{A_n - S_\epsilon} |f| d\mu$$
$$\leq \int_{A_n \cap S_\epsilon} |f| d\mu + \int_{S-S_\epsilon} |f| d\mu < \epsilon$$

for all $f \in K$, that is, $|K|$ is uniformly σ-additive.

Assertion b) follows from a) since if $\mu(X) < \infty$, then condition ii) is automatically satisfied, with $S_\epsilon = S$. ∎

F. Weakly compact sets in $L_F^1(\mu)$

Let (S, Σ, μ) be a measure space. Consider the space $L_F^1(\mu)$ with the topology $\sigma' = \sigma(L_F^1(\mu), L_{F^*}^\infty(\mu))$.

If F^* has the Radon–Nikodym property and if μ is σ-finite, then $L_{F^*}^\infty(\mu)$ is the dual of $L_F^1(\mu)$ and σ' is the weak topology of $L_F^1(\mu)$.

In particular, if μ is σ-finite, then $L^\infty(\mu)$ is the dual of $L^1(\mu)$ and σ' is the weak topology of $L^1(\mu)$.

24. Definition. *We say that a set $K \subset L_F^1(\mu)$ is relatively σ'-compact if the closure of K in the σ'-topology is compact. We say that K is conditionally*

σ'-compact if every sequence from K contains a subsequence which is Cauchy for the σ'-topology.

To say that a sequence (f_n) from $L^1_F(\mu)$ is Cauchy for the σ'-topology means that $\lim_n \int \langle f_n, g \rangle \, d\mu$ exists for each $g \in L^\infty_{F^*}(\mu)$.

If σ' is the weak topology of $L^1_F(\mu)$ and if K is relatively σ'-compact (resp. conditionally σ'-compact), we say that K is relatively weakly compact (resp. conditionally weakly compact).

Remark. More generally, if E is a locally convex space, a set $K \subset E$ is said to be relatively weakly compact if its closure in the weak topology of E is compact. We say that K is conditionally weakly compact if every sequence (x_n) from K contains a weak Cauchy subsequence (x_{n_k}), that is, such that $\lim_k \langle x_{n_k}, x^* \rangle$ exists for every $x^* \in E^*$.

The following theorem gives a characterization of conditionally σ'-compact sets in terms of uniform σ-additivity.

25. Theorem. *Let $K \subset L^1_F(\mu)$ be conditionally σ'-compact. Then:*
a) *K is bounded in $L^1_F(\mu)$;*
b) *$|K|$ is uniformly σ-additive;*
c) *For each set $A \in \Sigma$ with $\mu(A) < \infty$, the set*

$$K(A) = \left\{ \int_A f \, d\mu : f \in K \right\}$$

is conditionally weakly compact in F.

Proof. Assume K is conditionally σ'-compact. We prove first that K is bounded in $L^1_F(\mu)$. Deny it: there is a sequence (f_n) from K such that $\int |f_n| \, d\mu > n$ for each n. By extracting a subsequence, if necessary, we can assume that (f_n) is Cauchy for the σ'-topology, i.e., $\lim_n \int \langle f_n, g \rangle \, d\mu$ exists for each $g \in L^\infty_{F^*}(\mu)$. For each n consider the continuous linear functional $T_n : L^\infty_{F^*}(\mu) \to \mathbb{R}$ defined by

$$T_n(g) = \int \langle f_n, g \rangle \, d\mu, \qquad \text{for } g \in L^\infty_{F^*}(\mu).$$

By Theorem 1.38 b) we have $\|T_n\| = \|f_n\|_1$ for each n.

Since $\lim_n T_n(g)$ exists for each $g \in L^\infty_{F^*}(\mu)$, it follows that $(T_n(g))$ is bounded for each n. By the uniform boundedness theorem ([D-S], II.1.11) we have $\lim_{\|g\|_\infty \to 0} T_n(g) = 0$, uniformly for $n \in \mathbb{N}$. Let $\epsilon = 1$; there is a $\delta > 0$ such that if $\|g\|_\infty < \delta$ then $|T_n(g)| \leq 1$ for all n. Then for $\|g\|_\infty \leq 1$ we obtain $|T_n(g)| \leq 1/\delta$ for all n, hence $\sup_n \|T_n\| \leq 1/\delta$. It follows that $\sup_n \|f_n\|_1 \leq 1/\delta$, which contradicts $\|f_n\|_1 > n$ for each n. We deduce that K is bounded in $L^1_F(\mu)$.

We prove now that $|K|$ is uniformly σ-additive. Deny it: there is a sequence (A_n) of disjoint sets from Σ, an $\epsilon > 0$ and a sequence (f_n) from K such that $\int_{A_n} |f_n| \, d\mu > \epsilon$ for all n (Lemma 4).

Since K is conditionally σ'-compact, (f_n) contains a Cauchy subsequence for the σ'-topology. We can assume that (f_n) is itself a Cauchy sequence for σ'.

Since $\int_{A_n} |f_n|\, d\mu > \epsilon$, by Theorem 1.38 b), for each n there is a function $g_n \in L^\infty_{F^*}(\mu)$ with $|g_n| \leq 1$ and $\int_{A_n} \langle f_n, g\rangle\, d\mu > \epsilon$.

If we set $g = \sum_n \varphi_{A_n} g_n$, then $|g| \leq 1$, hence $g \in L^\infty_{F^*}(\mu)$ and

$$\int_{A_n} \langle f_n, g\rangle\, d\mu = \int_{A_n} \langle f_n, g_n\rangle\, d\mu > \epsilon, \qquad \text{for each } n.$$

Since (f_n) is a Cauchy sequence for the σ'-topology and since $\varphi_A g \in L^\infty_{F^*}(\mu)$ for each $A \in \Sigma$, the limit

$$\sigma(A) = \lim_n \int_A \langle f_n, g\rangle\, d\mu$$

exists for each $A \in \Sigma$. By the Nikodym Theorem 17, the indefinite integrals $\int_{(\cdot)} \langle f_n, g\rangle\, d\mu$ are uniformly σ-additive. This contradicts the above inequalities $\int_{A_n} \langle f_n, g\rangle\, d\mu > \epsilon$, for all $n \in \mathbb{N}$. It follows that $|K|$ is uniformly σ-additive.

To prove c), let $A \in \Sigma$ with $\mu(A) < \infty$ and let (f_n) be a sequence from K. Taking a subsequence, if necessary, we can assume that (f_n) is Cauchy for the σ'-topology, i.e., $\lim_n \int \langle f_n, g\rangle\, d\mu$ exists for each $g \in L^\infty_{F^*}(\mu)$.

Let $x^* \in E^*$ be arbitrary. Then $\varphi_A x^* \in L^\infty_{F^*}(\mu)$ and

$$\left\langle \int_A f_n\, d\mu, x^* \right\rangle = \int_A \langle f_n, x^*\rangle\, d\mu = \int_A \langle f_n, \varphi_A x^*\rangle\, d\mu.$$

Taking $g = \varphi_A x^*$, we deduce that $\lim_n \left\langle \int_A f_n\, d\mu, x^*\right\rangle$ exists, hence $\left(\int_A f_n\, d\mu\right)$ is weak Cauchy in F and this proves condition c). ∎

26. Remark. It can be proved that conditions a), b) and c) of Theorem 25 are sufficient for K to be conditionally σ'-compact (see [D·2], Theorem 1).

Using Theorems 23 and 25 we can deduce the following corollary:

27. Corollary. *Let $K \subset L^1_F(\mu)$ be conditionally σ'-compact. Then:*
a) *K is bounded in $L^1_F(\mu)$;*
b) *$|K| << \mu$, uniformly;*
c) *$K(A) = \{\int_A f\, d\mu : f \in K\}$ is conditionally weakly compact in F, for each set $A \in \Sigma$ with $\mu(A) < \infty$;*
d) *For every $\epsilon > 0$, there is a set $S_\epsilon \in \Sigma$ with $\mu(S_\epsilon) < \infty$, such that $\int_{S - S_\epsilon} |f|\, d\mu < \epsilon$, for all $f \in K$.*

§4. MEASURES WITH FINITE SEMIVARIATION

The semivariation is the basis for the fourth stage in the development of the integral, which is the object of paragraph 5.

A. The semivariation

The framework for this section is a ring $\mathcal{R}$ of subsets of S, three Banach spaces E, F, G such that $E \subset L(F, G)$ continuously, and an additive measure $m : \mathcal{R} \to E \subset L(F, G)$.

1. Definition. *For every set $A \subset S$ (not necessarily in $\mathcal{R}$ or $\mathcal{R}_{loc}$) the semivariation of m on A relative to the embedding $E \subset L(F,G)$ (or relative to the pair (F,G)) is a number, finite or $+\infty$, denoted $svar_{F,G}(m, A)$ or $\tilde{m}_{F,G}(A)$ and defined by the equality*

$$\tilde{m}_{F,G}(A) = \sup | \sum_{i \in I} m(A_i) x_i |,$$

where the supremum is taken for all finite families $(A_i)_{i \in I}$ of disjoint sets from $\mathcal{R}$ contained in A and all families $(x_i)_{i \in I}$ of elements from F_1.

The above definition can be stated for any set function m defined on any class $\mathcal{C}$ containing ϕ, such that $m(\phi) = 0$.

If $A \in \mathcal{R}$, in the definition of $\tilde{m}_{F,G}(A)$ we can restrict ourselves to the families $(A_i)_{i \in I}$ of disjoint sets from $\mathcal{R}$ with *union equal to A*.

The semivariation $\tilde{m}_{F,G}$ depends only on the norms of the spaces F and G, but not on the norm of E.

For the embedding $E = L(\mathbb{R}, E)$ we sometimes write $svar\, m$ or $\tilde{m}$ or $\|m\|$ for $\tilde{m}_{\mathbb{R},E}$. We agree also to write $\tilde{m}$ instead of $\tilde{m}_{F,G}$, if the pair (F, G) is understood.

We say that m has *finite* (respectively *bounded*) *semivariation* relative to (F, G) if $\tilde{m}_{F,G}(A) < \infty$ for every $A \in \mathcal{R}$ (respectively $\tilde{m}_{F,G}(S) < \infty$).

Remark. From Definition 1 we deduce that for any set $A \subset S$, the semivariation $\tilde{m}_{F,G}(A)$ can also be defined by the equality

$$\tilde{m}_{F,G}(A) = \sup | \int s\, dm |,$$

where the supremum is taken for all $\mathcal{R}$-step functions $s : S \to F$ with $|s| \le \varphi_A$.

We state now some properties of the semivariation similar to those of the variation.

2. *If $\mathcal{P}$ is a semiring generating the ring $\mathcal{R}$, then in the definition of the semivariation $\tilde{m}_{F,G}(A)$ we can take only families (A_i) of disjoint sets from $\mathcal{P}$ contained in A.*

If $m' : \mathcal{P} \to E$ is the restriction of m to $\mathcal{P}$, this means that

$$svar_{F,G}(m', A) = svar_{F,G}(m, A).$$

3. If the embedding $E \subset L(F,G)$ is an isometry then

$$|m(A)| \leq \tilde{m}_{F,G}(A), \text{ for } A \in \mathcal{R}.$$

4. If $A \subset B$, then $\tilde{m}_{F,G}(A) \leq \tilde{m}_{F,G}(B)$.

5. $\tilde{m}_{F,G}(A) = 0$ iff $m(B) = 0$ for every set $B \in \mathcal{R}$ with $B \subset A$.

6. $\tilde{m}_{F,G}(A) \leq \overline{m}(A)$, for every $A \subset S$.

7. If $m, n : \mathcal{R} \to E$ are additive measures and $\alpha \in \mathbb{R}$, then

$$(m+n)\tilde{}_{F,G} \leq \tilde{m}_{F,G} + \tilde{n}_{F,G}$$

and

$$(\alpha m)\tilde{}_{F,G} = |\alpha|\tilde{m}_{F,G}.$$

8. If $\mathcal{R}'$ is a subring of $\mathcal{R}$ and if m' is the restriction of m to $\mathcal{R}'$, then for every set $A \subset S$ we have

$$svar_{F,G}(m', A) \leq svar_{F,G}(m, A)$$

9. Proposition. *The set function $\tilde{m}_{F,G}$ is finitely subadditive on $\mathcal{R}_{loc}$.*

If m is σ-additive on $\mathcal{R}$, then for any sequence (A_n) of disjoint sets from $\mathcal{R}_{loc}$ with union A not necessarily in $\mathcal{R}_{loc}$ we have $\tilde{m}_{F,G}(A) \leq \sum_n \tilde{m}_{F,G}(A_n)$.

Proof. Let (A_n) be a sequence of sets from $\mathcal{R}_{loc}$ with union A (not necessarily in $\mathcal{R}_{loc}$). In case m is finitely additive, we assume $m(A_n) = 0$, except for a finite set of indices. Set $A'_1 = A_1$ and $A'_n = A_n - \bigcup_{i<n} A_i$, for $n > 1$. The sets A'_n are mutually disjoint, belong to $\mathcal{R}_{loc}$, and their union is A.

Let $(B_j)_{1 \leq j \leq m}$ be a family of disjoint sets from $\mathcal{R}$ contained in A and $(x_j)_{1 \leq j \leq m}$ a family of elements of F such that $|x_j| \leq 1$ for each j.

For each i, the family $(A'_i \cap B_j)_{1 \leq j \leq m}$ consists of disjoint sets of $\mathcal{R}$, contained in A'_i; for each j, the sequence $(A'_i \cap B_j)_{1 \leq i < \infty}$ consists of sets of $\mathcal{R}$, with union B_j. Then

$$\left|\sum_{1 \leq j \leq m} m(B_j)x_j\right| = \left|\sum_{1 \leq j \leq m} m\left(\bigcup_{1 \leq i < \infty}(A'_i \cap B_j)\right)x_j\right|$$

$$= \left|\sum_{1 \leq j \leq m}\sum_{1 \leq i < \infty} m(A'_i \cap B_j)x_j\right| = \left|\sum_{1 \leq i < \infty}\sum_{1 \leq j \leq m} m(A'_i \cap B_j)x_j\right|$$

$$\leq \sum_{1 \leq i < \infty}\left|\sum_{1 \leq j \leq m} m(A'_i \cap B_j)x_j\right| \leq \sum_{1 \leq i < \infty} \tilde{m}(A'_i) \leq \sum_{1 \leq i < \infty} \tilde{m}(A_i),$$

hence
$$\tilde{m}(A) \leq \sum_{1 \leq i < \infty} \tilde{m}(A_i).$$

∎

If $G = \mathbb{R}$, the semivariation and the variation are equal.

10. Proposition. *Assume $E \subset L(F, \mathbb{R})$ isometrically. If $m : \mathcal{R} \to E \subset L(F, \mathbb{R})$ is an additive measure, then $\tilde{m}_{F,\mathbb{R}} = \overline{m}$.*

Proof. Let $A \subset S$ and $\varepsilon > 0$. Let $(A_i)_{1 \leq i \leq n}$ be a finite family of disjoint sets from $\mathcal{R}$ contained in A. For each $i \leq n$, there is an element $x_i \in F$ with $|x_i| = 1$ such that $\langle x_i, m(A_i) \rangle \geq 0$ and
$$|m(A_i)| \leq \langle x_i, m(A_i) \rangle + \frac{\varepsilon}{n}.$$
Then
$$\sum_{1 \leq i \leq n} |m(A_i)| \leq \sum_{1 \leq i \leq n} \langle x_i, m(A_i) \rangle + \varepsilon$$

$$= |\sum_{1 \leq i \leq n} m(A_i) x_i| + \varepsilon \leq \tilde{m}_{F,\mathbb{R}}(A) + \varepsilon.$$

It follows that $\overline{m}(A) \leq \tilde{m}_{F,R}(A) + \varepsilon$. Since ε was arbitrary, we deduce that $\overline{m}(A) \leq \tilde{m}_{F,\mathbb{R}}(A)$. The converse inequality is stated in Property 6. ∎

11. Corollary. *If $\mu : \mathcal{R} \to \mathbb{R}$ is a real valued, additive measure and if we consider $\mathbb{R} = L(\mathbb{R}, \mathbb{R})$, then the variation $\overline{\mu}$ and the semivariation $\tilde{\mu}_{\mathbb{R},\mathbb{R}}$ are equal.*

12. Proposition. *If $m : \mathcal{R} \to E \subset L(F, G)$ is an additive measure, then*
$$\tilde{m}_{F,G} \leq \tilde{m}_{E^*,\mathbb{R}} = \overline{m}.$$
If the embedding $E \subset L(F, G)$ is an isometry, then
$$\tilde{m}_{\mathbb{R},E} \leq \tilde{m}_{F,G}.$$

Proof. Let $A \subset S$, $(A_i)_{i \in I}$ a finite family of disjoint sets from $\mathcal{R}$ contained in A and $(x_i)_{i \in I}$ a family of elements from F with $|x_i| \leq 1$. Let $z \in F^*$ with $|z| \leq 1$. For each $i \in I$ consider the element $y_i \in E^*$ defined by $\langle u, y_i \rangle = \langle u(x_i), z \rangle$, for $u \in E$. Then $|y_i| \leq |x_i| \, |z| \leq 1$ and we have
$$|\sum_{i \in I} \langle m(A_i) x_i, z \rangle| = |\sum_{i \in I} \langle m(A_i), y_i \rangle| \leq \tilde{m}_{E^*,\mathbb{R}}(A).$$

Taking the supremum for $|z| \leq 1$ we obtain

$$|\sum_{i \in I} m(A_i)x_i| \leq \tilde{m}_{E^*,\mathbb{R}}(A);$$

therefore

$$\tilde{m}_{F,G}(A) \leq \tilde{m}_{E^*,\mathbb{R}}(A).$$

The equality $\tilde{m}_{E^*,\mathbb{R}} = \overline{m}$ follows from Proposition 10, since $E \subset L(E^*, R)$ isometrically.

Assume now that $E \subset L(F, G)$ isometrically. Then, with A and $(A_i)_{i \in I}$ as above and $(\alpha_i)_{i \in I}$ a family of real numbers with $|\alpha_i| \leq 1$, we have, for every $x \in F$ with $|x| \leq 1$,

$$|\sum_{i \in I} m(A_i)\alpha_i x| \leq \tilde{m}_{F,G}(A).$$

Taking the supremum for $|x| \leq 1$ we get

$$|\sum_{i \in I} m(A_i)\alpha_i| \leq \tilde{m}_{F,G}(A),$$

therefore

$$\tilde{m}_{\mathbb{R},E}(A) \leq \tilde{m}_{F,G}(A).$$

∎

Remark. If the embedding $E \subset L(F, G)$ is continuous but not an isometry, then, in general, there is no relationship between the semivariations $\tilde{m}_{\mathbb{R},E}$ and $\tilde{m}_{F,G}$.

B. Semivariation and norming spaces

One of the most important properties of the semivariation is that it can be computed by means of a family of positive measures obtained in connection with a norming space for G. This property is very important for the definition of the integral and for the proof of its properties and reduces integration with respect to a vector measure to integration with respect to a family of positive measures.

We assume $m : \mathcal{R} \to E \subset L(F, G)$ is an additive measure on a ring $\mathcal{R}$ and $Z \subset G^*$ is a space norming for G. In particular, we can take $Z = G^*$.

For each $z \in Z$ we define the set function $m_z : \mathcal{R} \to F^*$ by the equality

$$\langle x, m_z(A) \rangle = \langle m(A)x, z \rangle, \text{ for } x \in F \text{ and } A \in \mathcal{R}.$$

Then m_z is an additive measure. If we consider the embedding $F^* = L(F, \mathbb{R})$, by Proposition 10 we have $(\tilde{m}_z)_{F,\mathbb{R}} = \overline{m}_z$.

As usual, we denote by $|m_z|$ the restriction of the variation $\overline{m}_z$ to $\mathcal{R}_{\text{loc}}$.

The semivariation $\tilde{m}_{F,G}$ can be computed by means of the variations $\overline{m}_z$.

13. Proposition. *For any space $Z \subset G^*$ norming for G we have*
$$\tilde{m}_{F,G} = \sup_{z \in Z_1} \overline{m}_z.$$

Proof. Let $A \subset S$, $(A_i)_{i \in I}$ a family of disjoint sets from $\mathcal{R}$ contained in A, $(x_i)_{i \in I}$ a family of elements from F_1 and $z \in Z_1$. Then
$$|\langle \sum_{i \in I} m(A_i) x_i, z \rangle| = |\sum_{i \in I} \langle m(A_i) x_i, z \rangle|$$
$$= |\sum_{i \in I} m_z(A_i) x_i| \leq (\tilde{m}_z)_{F, \mathbb{R}}(A) = \overline{m}_z(A),$$

therefore
$$|\sum m(A_i) x_i| \leq \sup_{z \in Z_1} \overline{m}_z(A),$$

consequently
$$\tilde{m}_{F,G}(A) \leq \sup_{s \in Z_1} \overline{m}_z(A).$$

On the other hand,
$$|\sum_{i \in I} m_z(A_i) x_i| = |\langle \sum_{i \in I} m(A_i) x_i, z \rangle|$$
$$\leq |\sum_{i \in I} m(A_i) x_i| \leq \tilde{m}_{F,G}(A),$$

hence,
$$\overline{m}_z(A) = (\tilde{m}_z)_{F, \mathbb{R}}(A) \leq \tilde{m}_{F,G}(A),$$

whence,
$$\sup_{z \in Z_1} \overline{m}_z(A) \leq \tilde{m}_{F,G}(A),$$

and the equality of the statement follows. ∎

Remark. Proposition 13 is valid for any norming set Z (not necessarily a vector space).

If we consider the embedding $E = L(\mathbb{R}, E)$ and if $Z \subset E^*$ is a space norming for E, then, for each $z \in E^*$ the measure $m_z = zm$ is real-valued, $m_z : \mathcal{R} \to \mathbb{R}^* = \mathbb{R}$. We can use the inequality in Proposition 2.16 to prove a similar inequality for $\tilde{m}_{\mathbb{R}, E}$.

14. Proposition. *If $m : \mathcal{R} \to E = L(\mathbb{R}, E)$ is an additive measure on a ring $\mathcal{R}$, then:*
a) *for every set $A \subset S$ we have*
$$\tilde{m}_{\mathbb{R}, E}(A) \leq 2 \sup\{|m(B)| : B \in \mathcal{R}, \ B \subset A\}.$$

b) *m is locally bounded (respectively bounded) on $\mathcal{R}$ iff $\tilde{m}_{\mathbb{R}, E}(A) < \infty$ for every $A \in \mathcal{R}$ (respectively $\tilde{m}_{\mathbb{R}, E}(S) < \infty$).*

Proof. Let $A \subset S$. Let $Z \subset E^*$ be a norming space for E and $z \in Z_1$. Then

$$m_z(B) = \langle m(B), z \rangle, \text{ for } B \in \mathcal{R}.$$

Since m_z is a real-valued measure, by Proposition 2.16, we have

$$\overline{m}_z(A) \leq 2\sup\{|m_z(B)| : B \in \mathcal{R}, B \subset A\}$$
$$= 2\sup\{|\langle m(B), z \rangle| : B \in \mathcal{R}, B \subset A\}.$$

Taking the supremum for $z \in Z_1$ we obtain, by Proposition 13,

$$\tilde{m}(A) = \sup_{z \in Z_1} \overline{m}_z(A) \leq 2\sup\{|m(B)| : B \in \mathcal{R}, B \subset A\}$$

and this proves assertion a). Assertion b) follows from assertion a). ∎

From Proposition 13 we deduce that if m has finite semivariation $\tilde{m}_{F,G}(A)$ for some set $A \subset S$, then all measures m_z have finite variation $\overline{m}_z(A)$.

The converse is also true, provided that the norming space Z is closed in G^*.

15. Proposition. *Let $m : \mathcal{R} \to E \subset L(F, G)$ be an additive measure defined on a ring $\mathcal{R}$ and let Z be a closed subspace of G^*, norming for G. For any set $A \subset S$ we have*

$$\tilde{m}_{F,G}(A) < \infty \text{ iff } \overline{m}_z(A) < \infty, \text{ for every } z \in Z.$$

Proof. Let $A \subset S$. Let M be the set of all $\mathcal{R}$-step functions $s : S \to F$ with $|s| \leq \varphi_A$. For each $s \in M$ define $T_s \in Z^*$ by

$$T_s(z) = \int s \, dm_z, \text{ for } z \in Z.$$

We have $|T_s| = |\int s \, dm|$. For each $z \in Z$ we have then

$$\sup_{s \in M} |T_s(z)| = \sup_{s \in M} |\int s \, dm_z|$$
$$= (\tilde{m}_z)_{F,\mathbb{R}}(A) = \overline{m}_z(A).$$

Assume $\overline{m}_z(A) < \infty$ for every $z \in Z$. Then $\sup_{s \in M} |T_s(z)| < \infty$, for $z \in Z$, hence, by the Banach–Steinhauss Theorem, we obtain

$$\tilde{m}_{F,G}(A) = \sup_{s \in M} |\int s \, dm| = \sup_{s \in M} |T_s| < \infty.$$

The converse implication follows from the equality of Proposition 13. ∎

C. The semivariation of σ-additive measures

The semivariation of σ-additive measures has additional properties.

16. Proposition. *Let $m : \mathcal{R} \to E \subset L(F, G)$ be a σ-additive measure on a ring $\mathcal{R}$.*

Then, for each $z \in G^*$, the measure $m_z : \mathcal{R} \to F^*$ is σ-additive and its variation $|m_z|$ is σ-additive on the algebra $\mathcal{R}_{loc}$.

Proof. Let $A \in \mathcal{R}$, $x \in F$ and $z \in Z$. Then

$$|\langle x, m_z(A) \rangle| = |\langle m(A)x, z \rangle| \le |m(A)| \, |x| \, |z|,$$

hence

$$|m_z(A)| \le |m(A)| \, |z|.$$

Since m is σ-additive, from this inequality it follows that m_z is σ-additive. Then, by Proposition 2.10, the variation $|m_z|$ is σ-additive on $\mathcal{R}_{loc}$. ∎

17. Corollary. *Let $m : \mathcal{D} \to E \subset L(F, G)$ be a σ-additive measure on a δ-ring $\mathcal{D}$.*

Then for any $z \in G^*$, the measure m_z is σ-additive on $\mathcal{D}$ and the variation $|m_z|$ is σ-additive on the σ-algebra $\Sigma = \sigma a(\mathcal{D})$ generated by $\mathcal{D}$.

In fact, Σ is contained in the σ-algebra $\mathcal{D}_{loc}$.

Remark. It is possible that $|m_z|$ is σ-additive for every $z \in G^*$ but the measure m is not σ-additive.

The following proposition is an analog of Property 2. It states that the semivariation is the same whether the measure is defined on a σ-ring or on a δ-ring generating it.

18. Proposition. *Let $m : \mathcal{S} \to E \subset L(F, G)$ be an additive measure on a σ-ring $\mathcal{S}$; let $\mathcal{D} \subset \mathcal{S}$ be a δ-ring generating $\mathcal{S}$, and $m' : \mathcal{D} \to E$ the restriction of m to $\mathcal{D}$. If $|m_z|$ is σ-additive for every z in a space $Z \subset G^*$ norming for G, then, for every set $A \subset S$ we have*

$$svar_{F,G}(m', A) = svar_{F,G}(m, A).$$

Proof. By Proposition 2.11, for any set $A \subset S$ and for any $z \in G^*$ we have

$$var(m'_z, A) = var(m_z, A).$$

Taking the supremum for $z \in G_1^*$ and using Proposition 13 we get the equality in the statement of the proposition. ∎

A similar property is valid for a δ-ring and a ring generating it. Its proof requires a result on extension of measures, in §7.

19. Proposition. *Let $m : \mathcal{D} \to E \subset L(F,G)$ be an additive measure on a δ-ring (resp. a σ-algebra) $\mathcal{D}$, $\mathcal{R}$ a ring generating the δ-ring (resp. the σ-algebra) $\mathcal{D}$, and $m' : \mathcal{R} \to E$ the restriction of m to $\mathcal{R}$.*
If $|m_z|$ is σ-additive for every z in a set $Z \subset G^*$ norming for G, then

$$svar_{F,G}(m', A) = svar_{F,G}(m, A), \text{ for } A \in \mathcal{R}.$$

Proof. Let $z \in G^*$. Then the restriction to $\mathcal{R}$ of the measure $m_z : \mathcal{D} \to F^*$ is equal to $m'_z = (m')_z$. By Proposition 7.4 infra, we have

$$|m_z|(A) = |m'_z|(A), \text{ for } A \in \mathcal{R}.$$

Using Proposition 13 we get the equality in the statement of the proposition. ∎

Remark. We shall prove (Corollary 7.6) that if m' has finite semivariation on $\mathcal{R}$ relative to (F, G), then m has finite semivariation on $\mathcal{D}$ relative to (F, G).

20. Theorem. *Let $m : \Sigma \to E$ be a σ-additive measure on a σ-algebra Σ. Then:*
a) *m has bounded semivariation $\tilde{m}_{\mathbb{R},E}$,*
b) *There is a positive, finite, σ-additive measure $\lambda : \Sigma \to \mathbb{R}_+$ such that*

$$\tilde{m}_{\mathbb{R},E} \ll \lambda \text{ and } \lambda \leq \tilde{m}_{\mathbb{R},E}.$$

Proof. By Theorem 2.14, m is bounded. Then by Proposition 14 b), we have $\tilde{m}_{\mathbb{R},E}(S) < \infty$ and this proves assertion a). From the inequality

$$|x^*m(A)| \leq |m(A)| \, |x^*|, \text{ for } A \in \Sigma \text{ and } x^* \in E^*,$$

we deduce that the set $K = \{x^*m : x^* \in E_1^*\}$ consists of σ-additive, real-valued measures on Σ, which are uniformly σ-additive. By Theorem 3.11, there is a positive, finite, σ-additive measure λ on Σ such that $K \ll \lambda$, uniformly and

$$\lambda(A) \leq \sup_{x^* \in E_1^*} \sup_{B \subset A} |x^*m(B)|, \text{ for } A \in \Sigma.$$

Since $x^*m \ll \lambda$ uniformly for $x^* \in E_1^*$, it follows that $|x^*m| \ll \lambda$ uniformly for $x^* \in E_1^*$ (Proposition 3.8); therefore $\sup_{x^* \in E_1^*} |x^*m| \ll \lambda$, that is, using Proposition 13, $\tilde{m}_{\mathbb{R},E} \ll \lambda$. At the same time, for $A \in \Sigma$ we have, by the above,

$$\lambda(A) \leq \sup_{x^* \in E_1^*} \sup_{B \subset A} |x^*m(B)| \leq \sup_{x^* \in E_1^*} \sup_{B \subset A} |x^*m|(B)$$
$$= \sup_{B \subset A} \tilde{m}_{\mathbb{R},E}(B) \leq \tilde{m}_{\mathbb{R},E}(A),$$

and this proves assertion b). ∎

From the above proof it follows also that $m \ll \lambda$ and that λ and m have the same negligible sets.

We have a similar theorem for measures on a δ-ring.

21. Theorem. *Let $m : \mathcal{D} \to E$ be a σ-additive measure on a δ-ring $\mathcal{D}$. Then:*
a) *m has finite semivariation $\tilde{m}_{\mathbb{R},E}$;*
b) *If $S = \cup S_n$ with $S_n \in \mathcal{D}$, then there is a positive, finite, σ-additive measure λ on $\mathcal{D}$ such that*
$$\tilde{m}_{\mathbb{R},E} \ll \lambda \text{ locally and } \lambda \leq \tilde{m}_{\mathbb{R},E}.$$

Proof. By Theorem 2.14, m is locally bounded. Then, by Proposition 14 b), we have $\tilde{m}_{\mathbb{R},E}(A) < \infty$ for every $A \in \mathcal{D}$ and assertion a) is proved. From the inequality
$$|x^*m(A)| \leq |m(A)|\, |x^*|, \text{ for } A \in \mathcal{D} \text{ and } x^* \in E^*,$$
it follows that the set $K = \{x^*m : x^* \in E_1^*\}$ consists of real-valued measures on $\mathcal{D}$, which are uniformly σ-additive. By Theorem 3.12, there is a positive, finite, σ-additive measure λ on $\mathcal{D}$ such that $K \ll \lambda$ locally uniformly and
$$\lambda(A) \leq \sup_{x^* \in E_1^*} \sup_{B \subset A} |x^*m(B)|, \text{ for } A \in \mathcal{D}.$$

We continue as in the proof of Theorem 20 to deduce that $\tilde{m}_{\mathbb{R},E} \ll \lambda$ locally and $\lambda \leq \tilde{m}_{\mathbb{R},E}$. ■

D. The family $m_{F,Z}$ of measures

Let $m : \mathcal{R} \to E \subset L(F,G)$ be an additive measure with finite semivariation $\tilde{m}_{F,G}$ on a ring $\mathcal{R}$ and $Z \subset G^*$ a space norming for G.

22. Definition. *We denote by $m_{F,Z}$ the set of positive, finite measures $\{|m_z| : z \in Z_1\}$ defined on $\mathcal{R}$.*

By Proposition 16, if m is σ-additive, then $|m_z|$ is σ-additive for every $z \in G^*$. But the converse is not true, in general. It is possible that m_z is σ-additive for every $z \in Z$, without m being σ-additive. In particular, it is possible that x^*m is σ-additive for x^* in a subspace $Z \subset E^*$, norming for E, without m being σ-additive.

However, the converse of Proposition 16 is true, provided that $m_{F,Z}$ is uniformly σ-additive, as Theorem 23 below shows. It shows also that the uniform σ-additivity is independent of the norming space Z. For another instance when the converse of Proposition 16 is true, see Theorem 7.7 infra.

23. Theorem. *Let $m : \mathcal{R} \to E \subset L(F,G)$ be an additive measure with finite semivariation $\tilde{m}_{F,G}$ on a ring $\mathcal{R}$ and $Z \subset G^*$ a norming space for G. The following 5 assertions are equivalent:*
a) *m_{F,G^*} is uniformly σ-additive on $\mathcal{R}$;*

72   Ch.1 VECTOR INTEGRATION

b) $m_{F,Z}$ is uniformly σ-additive on $\mathcal{R}$.
c) $\tilde{m}_{F,G}(A_n) \to 0$ if $A_n \downarrow \phi$ in $\mathcal{R}$.
d) $\tilde{m}_{F,G}(A_n) \to \tilde{m}_{F,G}(A)$ for every decreasing sequence (A_n) of sets from $\mathcal{R}$ with intersection $A = \bigcap A_n \in \mathcal{R}$.
e) $\tilde{m}_{F,G}(A_n) \to 0$ for every sequence (A_n) of disjoint sets from $\mathcal{R}$ with union $A = \bigcup_n A_n \in \mathcal{R}$.

Each of these assertions implies that m is σ-additive.

Proof. The implication a) $\Longrightarrow$ b) is evident. The equivalence b) $\Longleftrightarrow$ c) follows from the equality

$$\tilde{m}_{F,G}(A) = \sup_{|z| \leq 1} |m_z|(A), \text{ for } A \in \mathcal{R},$$

proved in Propositon 13.

To prove the implication c) $\Longrightarrow$ d), let $A_n \downarrow A$ with $A_n, A \in \mathcal{R}$. Then $A_n - A \downarrow \phi$, hence by assertion c), we have $\tilde{m}_{F,G}(A_n - A) \to 0$. Then

$$0 \leq \tilde{m}_{F,G}(A_n) - \tilde{m}_{F,G}(A) \leq \tilde{m}_{F,G}(A_n - A)$$

hence $\tilde{m}_{F,G}(A_n) \to \tilde{m}_{F,G}(A)$, which proves d).

The implication d) $\Longrightarrow$ c) is evident, since $\tilde{m}_{F,G}(\phi) = 0$.

Assume now b) and prove e). Let (A_n) be a sequence of disjoint sets from $\mathcal{R}$ with union $A \in \mathcal{R}$. Then

$$\lim_n \sum_{1 \leq i \leq n} |m_z|(A_i) = |m_z|(A), \text{ uniformly for } z \in Z_1,$$

hence

$$\lim_n |m_z|(A_n) = 0, \text{ uniformly for } z \in Z_1,$$

consequently

$$\lim_n \tilde{m}_{F,G}(A_n) = \lim_n \sup_{z \in Z_1} |m_z|(A_n) = 0,$$

which is assertion e).

To prove the implication e) $\Longrightarrow$ a), assume e) and deny a): there is a sequence (A_n) of disjoint sets from $\mathcal{R}$ with union $A \in \mathcal{R}$ and an $\varepsilon > 0$ such that for every n we have

$$\sup_{z \in G_1^*} \sum_{i \geq n} |m_z|(A_i) > 2\varepsilon.$$

There is an increasing sequence (n_j) from $\mathbb{N}$ such that

$$\sup_{z \in G_1^*} \sum_{n_j < i \leq n_{j+1}} |m_z|(A_i) > \varepsilon,$$

that is,
$$\sup_{z \in G_1^*} |m_z|(\bigcup_{n_j < i \leq n_{j+1}} A_i) > \varepsilon.$$

If we set $B_j = \bigcup_{n_j < i \leq n_{j+1}} A_i$, the sets B_j belong to $\mathcal{R}$, are mutually disjoint and their union is A. The above inequality can be written

$$\tilde{m}_{F,G}(B_j) > \varepsilon, \text{ for each } j,$$

which contradicts the assumption e). It follows that e) $\Longrightarrow$ a). ∎

The following proposition states the relationship between the absolute continuity of $\tilde{m}_{F,G}$ and the uniform absolute continuity of $m_{F,Z}$.

24. Proposition. *Let λ be a positive, finite, additive measure on $\mathcal{R}$. Then*
a) *$\tilde{m}_{F,G} \ll \lambda$ iff $m_{F,Z} \ll \lambda$ uniformly.*
b) *$\tilde{m}_{F,G} \ll \lambda$ locally iff $m_{F,Z} \ll \lambda$ locally uniformly.*

The proof follows from the equality proved in Proposition 13.

Remark. If the embedding $E \subset L(F,G)$ is an isometry, from Property 3 we deduce that if $\tilde{m}_{F,G} \ll \lambda$ (resp. $\tilde{m}_{F,G} \ll \lambda$ locally), then $m \ll \lambda$ (resp. $m \ll \lambda$ locally). But if $m \ll \lambda$, it does not follow that $\tilde{m}_{F,G} \ll \lambda$.

The following two theorems relate the uniform σ-additivity to the uniform absolute continuity of m_{F,Z^*}.

25. Theorem. *Let $m : \Sigma \to E \subset L(F,G)$ be a σ-additive measure with finite semivariation $\tilde{m}_{F,G}$ on a σ-algebra Σ. Then m_{F,G^*} is uniformly σ-additive iff there is a positive, finite, σ-additive measure λ on Σ of the form $\lambda = \sum_n c_n \mu_n$ with $c_n \geq 0$, $\sum_n c_n \leq 1$ and $\mu_n \in m_{F,G^*}$ such that $\tilde{m}_{F,G} \ll \lambda$ and $\lambda \leq \tilde{m}_{F,G}$.*

Proof. Assume first that there is a positive measure λ on Σ such that $\tilde{m}_{F,G} \ll \lambda$. If $A_n \downarrow \phi$ in Σ, then $\lambda(A_n) \to 0$, hence $\tilde{m}_{F,G}(A_n) \to 0$; from the inequality $|m_z| \leq \tilde{m}_{F,G}$ for $z \in G^*$, we deduce that $|m_z|(A_n) \to 0$, uniformly for $z \in G_1^*$, that is, m_{F,G^*} is uniformly σ-additive.

Conversely, assume that m_{F,G^*} is uniformly σ-additive. Then, by Theorem 3.11, there is a positive, finite, σ-additive measure λ on Σ such that $m_{F,G^*} \ll \lambda$ uniformly and
$$\lambda(A) \leq \sup_{|z| \leq 1} \sup_{B \subset A} |m_z|(B), \text{ for } A \in \Sigma.$$

Moreover, λ can be taken of the form $\lambda = \sum c_n \mu_n$ with $c_n \geq 0$, $\sum c_n \leq 1$ and $\mu_n \in m_{F,G^*}$. From the above inequality we deduce that
$$\lambda(A) \leq \tilde{m}_{F,G}(A), \text{ for } A \in \Sigma.$$

From $|m_z| \ll \lambda$ uniformly for $z \in G_1^*$ and from Proposition 13 we deduce that $\tilde{m}_{F,G} \ll \lambda$. ∎

26. Theorem. *Let $m : \mathcal{D} \to E \subset L(F,G)$ be a σ-additive measure on a δ-ring $\mathcal{D}$ with finite semivariation $\tilde{m}_{F,G}$ and assume $S = \bigcup S_n$ with $S_n \in \mathcal{D}$.*

Then m_{F,G^*} is uniformly σ-additive iff there is a positive, finite, σ-additive measure λ on $\mathcal{D}$ such that

$$\tilde{m}_{F,G} \ll \lambda, \text{ locally on } \mathcal{D}$$

and

$$\lambda \leq \tilde{m}_{F,G}.$$

Proof. The proof is similar to that of Theorem 25, replacing Σ with $\mathcal{D}$ and using Theorem 3.12. ∎

Next we state some theorems that give sufficient conditions for m_{F,G^*} to be uniformly σ-additive.

For a σ-additive measure m on a δ-ring, the set $m_{\mathbb{R},E^*}$ is always uniformly σ-additive:

27. Theorem. *Let $m : \mathcal{D} \to E$ be a σ-additive measure on a δ-ring $\mathcal{D}$. Then the semivariation $\tilde{m}_{\mathbb{R},E}$ is finite and the set of measures $m_{\mathbb{R},E^*}$ is uniformly σ-additive.*

Proof. The fact that the semivariation $\tilde{m}_{\mathbb{R},E}$ is finite was proved in Theorem 21. We want now to prove assertion e) of Theorem 23. Deny it: there is a sequence (A_n) of disjoint sets from $\mathcal{D}$ with union $A \in \mathcal{D}$ and an $\varepsilon > 0$ such that

$$\tilde{m}_{\mathbb{R},E}(A_n) > \varepsilon, \text{ for every } n.$$

By Proposition 14, for each n there is a set $B_n \in \mathcal{D}$ with $B_n \subset A_n$ and

$$\tilde{m}_{\mathbb{R},E}(A_n) < 3|m(B_n)|,$$

hence

$$|m(B_n)| > \varepsilon/3, \text{ for every } n.$$

The sets B_n are mutually disjoint and contained in the set $A \in \mathcal{D}$; therefore their union $B = \bigcup_n B_n$ belongs to $\mathcal{D}$. Since m is σ-additive on $\mathcal{D}$, we have $m(B_n) \to 0$, which contradicts $|m(B_n)| > \varepsilon/3$ for every n. We deduce that assertion e) of Theorem 23 is true; then, by the same Theorem 23, $m_{\mathbb{R},E^*}$ is uniformly σ-additive. ∎

If $c_0 \not\subset G$, for a σ-additive measure m on a δ-ring, the set m_{F,G^*} is uniformly σ-additive. This is true, in particular, if G is reflexive, or weakly sequentially complete.

28. Theorem. *Let $m : \mathcal{D} \to E \subset L(F,G)$ be a σ-additive measure with finite semivariation $\tilde{m}_{F,G}$ on a δ-ring $\mathcal{D}$. If $c_0 \not\subset G$, then m_{F,G^*} is uniformly σ-additive.*

§4. MEASURES WITH FINITE SEMIVARIATION

Proof. Assume $c_0 \not\subset G$ and prove assertion e) of Theorem 23. Deny it: there is a sequence (A_n) of disjoint sets from $\mathcal{D}$ with union $A \in \mathcal{D}$ and an $\varepsilon > 0$, such that for every n we have

$$\tilde{m}_{F,G}(A_n) > \varepsilon.$$

By Proposition 13, for each n there is a $z_n \in G_1^*$ such that

$$|m_{z_n}|(A_n) > \varepsilon.$$

Then, for each n we can find a finite partition $(B_{nj})_{j \in J(n)}$ of A_n into sets $B_{nj} \in \mathcal{D}$ such that

$$\sum_{j \in J(n)} |m_{z_n}(B_{nj})| > \varepsilon.$$

For each n and j we can find an $x_{nj} \in F_1$ such that

$$\langle x_{nj}, m_{z_n}(B_{nj}) \rangle \geq 0$$

and

$$\sum_{j \in J(n)} \langle x_{nj}, m_{z_n}(B_{nj}) \rangle > \varepsilon,$$

that is,

$$\sum_{j \in J(n)} \langle m(B_{nj}) x_{nj}, z_n \rangle > \varepsilon.$$

Then

$$\left| \sum_{j \in J(n)} m(B_{nj}) x_{nj} \right| \geq \left| \langle \sum_{j \in J(n)} m(B_{nj}) x_{nj}, z_n \rangle \right|$$

$$= \sum_{j \in J(n)} \langle m(B_{nj}) x_{nj}, z_n \rangle > \varepsilon.$$

For each $n \in \mathbb{N}$ set

$$y_n = \sum_{j \in J(n)} m(B_{nj}) x_{nj}.$$

Then $y_n \in G$ and $|y_n| > \varepsilon$ for each n. For each $z \in G^*$ we have

$$\sum_{1 \leq i \leq n} |\langle y_i, z \rangle| = \sum_{1 \leq i \leq n} \left| \sum_{j \in J(i)} m_z(B_{ij}) x_{ij} \right|$$

$$\leq \sum_{1 \leq i \leq n} \sum_{j \in J(i)} |m_z|(B_{ij}) = |m_z|\left(\bigcup_{1 \leq i \leq n} A_i \right)$$

$$\leq |m_z|(A) < \infty,$$

therefore

$$\sum_{1 \leq i < \infty} |\langle y_i, z \rangle| \leq |m_z|(A) < \infty.$$

It follows that the series $\sum y_n$ is weakly unconditionally convergent. By the Bessaga–Pelczynski theorem, the series $\sum y_n$ is convergent, which contradicts $|y_n| > \varepsilon$ for every n.

It follows that assertion e) of Theorem 23 is true; therefore, by the same theorem, m_{F,G^*} is uniformly σ-additive. ∎

Finally, for a σ-additive measure m with finite variation $|m|$ on a ring, the set m_{F,G^*} is uniformly σ-additive.

29. Theorem. *If $m : \mathcal{R} \to E$ is a σ-additive measure with finite variation $|m|$ on a ring $\mathcal{R}$, then m_{F,G^*} is uniformly σ-additive for any embedding $E \subset L(F,G)$.*

Proof. From the inequality $\tilde{m}_{F,G}(A) \leq |m|(A)$ for $A \in \mathcal{R}$, it follows that $\tilde{m}_{F,G}(A_n) \to 0$ as $A_n \downarrow \phi$ in $\mathcal{R}$; then by Theorem 23 c), m_{F,G^*} is uniformly σ-additive. ∎

§5. INTEGRATION WITH RESPECT TO A MEASURE WITH FINITE SEMIVARIATION

The framework for this paragraph is a δ-ring $\mathcal{D}$ of subsets of S, an *additive* measure $m : \mathcal{D} \to E \subset L(F,G)$ with *finite semivariation* $\tilde{m}_{F,G}$, and a space $Z \subset G^*$ norming for G such that for each $z \in Z$, the measure $m_z : \mathcal{D} \to F^*$ is σ-*additive*.

$\Sigma = \sigma a(\mathcal{D})$ is the σ-algebra generated by $\mathcal{D}$.

We do not assume m to be σ-additive on $\mathcal{D}$. If m is σ-additive, then $|m_z|$ is σ-additive for every $z \in G^*$ (Proposition 4.16); in this case we can take $Z = G^*$. Conversely, if the family of measures $m_{F,Z}$ is *uniformly* σ-*additive*, then m is σ-additive (Theorem 4.23) and we can take $Z = G^*$.

In this paragraph we develop an integration theory with respect to m, for functions $f : S \to F$. For this purpose we define a seminorm $\tilde{m}_{F,G}(f)$ for such functions, then the space $\mathcal{F}_{F,G}(m)$ of measurable functions f with $\tilde{m}_{F,G}(f) < \infty$, and then an integral $\int f dm \in Z^*$ for functions $f \in \mathcal{F}_{F,G}(m)$.

Some of the results are valid under additional conditions, such as:

a) $\tilde{m}_{F,G}(S) < \infty$,

or

b) $S = \bigcup S_n$, with $S_n \in \mathcal{D}$,

or

c) m is σ-additive.

Conditions a) and b) are satisfied, evidently, if $S \in \mathcal{D}$, i.e., if $\mathcal{D}$ is a σ-algebra.

But most results are valid without imposing these conditions and can be used in a wider range of applications, such as the integral representation of Gaussian measures or the Riesz representation theorem stated in §8.

If the reader is not interested in this generality, he or she can assume from the very beginning that $\mathcal{D}$ is a σ-algebra and m is σ-additive.

A. Measurability with respect to a vector measure

Measurability with respect to a vector measure is defined the same way it was defined with respect to a positive measure.

1. a) *A set $A \in \Sigma$ is said to be m-negligible, if $m(B) = 0$ for every set $B \in \mathcal{D}$ with $B \subset A$.*

It follows that a set $A \in \Sigma$ is m-negligible iff $|m|(A) = 0$, iff $\tilde{m}_{F,G}(A) = 0$ for any embedding $E \subset L(F,G)$; in particular, $A \in \Sigma$ is m-negligible iff $\tilde{m}_{\mathbb{R},E}(A) = 0$.

A set $A \in \Sigma$ is m-negligible iff A is m_z-negligible for every $z \in Z$.

Since the measures m_z are σ-additive, it follows that the union of a countable family of m-negligible sets from Σ is again m-negligible.

b) *A set $A \subset S$ is said to be m-negligible if it is contained in an m-negligible set $B \in \Sigma$.*

78   Ch.1 VECTOR INTEGRATION

It follows that the union of a sequence of m-negligible sets is again m-negligible.

If A is m-negligible, then A is m_z-negligible for every $z \in Z$. The converse is not necessarily true. The converse is true if Z is separable, or if $m_{F,Z}$ is uniformly σ-additive and $S = \bigcup S_n$ with $S_n \in \mathcal{D}$ (see Proposition 5 infra).

2. As usual, *a property valid outside an m-negligible set is said to be valid m-almost everywhere (m-a.e.)*.

3. *A function $f : S \to D$ (or $\overline{\mathbb{R}}_+$) is said to be m-negligible if the set $\{f \neq 0\}$ is m-negligible.*

It follows that a Σ-measurable function f is m-negligible iff f is m_z-negligible for every $z \in Z$.

If a (not necessarily Σ-measurable) function f is m-negligible, then it is m_z-negligible for every $z \in Z$; the converse is true if Z is separable or if $m_{F,Z}$ is uniformly σ-additive and $S = \bigcup S_n$ with $S_n \in \mathcal{D}$.

4. *A function $f : S \to D$ (or $\overline{\mathbb{R}}_+$) is said to be m-measurable if it is equal m-a.e. to a Σ-measurable function, i.e., if there is a sequence (f_n) of D (or $\overline{\mathbb{R}}_+$)-valued, Σ-step functions converging to f, m-a.e.*

By Theorem 1.6, we can take the functions f_n such that $|f_n| \leq |f|$, m-a.e., for each n.

In case $S = \bigcup S_n$ with $S_n \in \mathcal{D}$, a function f is m-measurable iff there is a sequence (f_n) of $\mathcal{D}$-step functions such that $f_n \to f$, m-a.e.

It follows that if f is m-measurable, then f is m_z-measurable for every $z \in Z$. The converse is not necessarily true. However, if Z is separable, then the converse is true. The converse is also true if $m_{F,Z}$ is uniformly σ-additive and $S = \bigcup S_n$ with $S_n \in \mathcal{D}$, as the following proposition states.

5. Proposition. *Assume $m_{F,Z}$ is uniformly σ-additive and $S = \bigcup_{1 \leq n < \infty} S_n$ with $S_n \in \mathcal{D}$.*
a) *A set $A \subset S$ is m-negligible iff A is m_z-negligible for every $z \in Z$.*
b) *A function $f : S \to D$ (or $\overline{\mathbb{R}}_+$) is m-negligible iff f is m_z-negligible for every $z \in Z$.*
c) *A function $f : S \to D$ (or $\overline{\mathbb{R}}_+$) is m-measurable iff f is m_z-measurable for every $z \in Z$.*

Proof. Assume first that $\mathcal{D}$ is a σ-algebra, i.e., $\mathcal{D} = \Sigma$. Since the semivariation $\tilde{m}_{F,G}$ is finite and $m_{F,Z}$ is uniformly σ-additive, by Theorems 4.23 and 4.25 there is a positive, finite, σ-additive measure λ on Σ, of the form

$$\lambda = \sum_{1 \leq n < \infty} c_n \mu_n$$

for some $c_n \geq 0$ with $\sum_n c_n \leq 1$ and some measures $\mu_n \in m_{F,G^*}$, such that $\tilde{m}_{F,G} \ll \lambda$ and $\lambda \leq \tilde{m}_{F,G}$. Let $A \subset S$ be an $|m_z|$-negligible set for every $z \in Z$. Then A is μ_n-negligible for every n, hence there is a set $B_n \in \Sigma$ with $A \subset B_n$

§5. INTEGRATION WITH RESPECT TO A MEASURE WITH FINITE SEMIVARIATION   79

and $\mu_n(B_n) = 0$. The set $B = \bigcap_{1 \leq n < \infty} B_n$ belongs to Σ, we have $A \subset B$ and B is μ_n-negligible for each n, hence $\lambda(B) = 0$. Since $\tilde{m}_{F,G} \ll \lambda$, we deduce that $\tilde{m}_{F,G}(B) = 0$; therefore A is m-negligible. This proves assertion a). This also proves assertion b).

Let now $f : S \to D$ (or $\overline{\mathbb{R}}_+$) be an $|m_z|$-measurable function for every $z \in Z$. Then f is μ_n-measurable for every n.

Since f is μ_1-measurable, there is a sequence (f_{1n}) of D (or $\mathbb{R}_+$)-valued, Σ-step functions, converging to f on $S - A_1$, for some set $A_1 \in \Sigma$ with $\mu_1(A_1) = 0$; we can take $f_{1n} = 0$ on A_1.

Since f is μ_2-measurable, there is a sequence (f_{2n}) of D (or $\mathbb{R}_+$)-valued, Σ-step functions converging to f on $A_1 - A_2$ for some set $A_2 \in \Sigma$ with $\mu_2(A_2) = 0$; we can take $f_{2n} = 0$ on $A_2 \cup A_1^c$.

By induction, for each i there is a sequence $(f_{in})_{n \in \mathbb{N}}$ of D (or $\mathbb{R}_+$)-valued, Σ-step functions, converging to f on $A_{i-1} - A_i$, for some $A_i \in \Sigma$ with $\mu_i(A_i) = 0$; we can take $f_{in} = 0$ on $A_i \cup A_{i-1}^c$.

If we set $A_0 = \bigcap_{1 \leq i < \infty} A_i$, then $\lambda(A_0) = 0$, hence A_0 is m-negligible. The sets $S - A_1$, $A_1 - A_2, \ldots$ are mutually disjoint and their union is $S - A_0$.

For each n, set $f_n = \sum_{1 \leq i \leq n} f_{in}$. Then f_n is a Σ-step function and $f_n \to f$ outside A_0, that is, m-a.e. It follows that f is m-measurable and this proves assertion c).

Assume now $\mathcal{D}$ is a δ-ring and $S = \bigcup_n S_n$ with $S_n \in \mathcal{D}$. We can assume the sets S_n are mutually disjoint. Considering the restriction of m to the σ-algebra $\mathcal{D} \cap S_n$ of subsets of S_n, we can deduce, as above, that f is equal m-a.e. on S_n to a $\mathcal{D} \cap S_n$-measurable function g_n which can be taken 0 outside S_n. Then the functions $f_n = \sum_{1 \leq i \leq n} g_i$ are Σ-measurable and $f_n \to f$, m-a.e., hence f is m-measurable. ∎

6. Corollary. *Assume $m : \mathcal{D} \to E$ is σ-additive and $S = \bigcup_{1 \leq n < \infty} S_n$ with $S_n \in \mathcal{D}$. Then:*
a) *A set $A \subset S$ is m-negligible iff A is $|x^*m|$-negligible for every $x^* \in E^*$.*
b) *A function $f : S \to D$ (or $\overline{\mathbb{R}}_+$) is m-measurable (resp. m-negligible) iff f is $|x^*m|$-measurable (resp. $|x^*m|$-negligible) for every $x^* \in E^*$.*

In fact, by Theorem 4.27, $m_{\mathbb{R},E^*}$ is uniformly σ-additive and we can apply Proposition 5.

B. The seminorm $\tilde{m}_{F,G}(f)$

We shall extend the definition of the semivariation $\tilde{m}_{F,G}$ to functions.

7. Definition. *For every function $f : S \to D$ (or $\overline{\mathbb{R}}_+$), (measurable or not), we define*

$$\tilde{m}_{F,G}(f) = \tilde{m}_{F,G}(|f|) = \sup \left| \int s\, dm \right|$$

where the supremum is taken for all $\mathcal{D}$-step functions $s : S \to F$ with $|s| \leq |f|$.

If f is m-measurable we can define $\tilde{m}_{F,G}(f)$ using the norming space Z.

8. Proposition. *If $f : S \to D$ (or $\overline{\mathbb{R}}_+$) is m-measurable, then*

$$\tilde{m}_{F,G}(f) = \sup\{\int |f|d|m_z| : z \in Z_1\}.$$

Proof. If $s : S \to F$ is a $\mathcal{D}$-step function such that $|s| < |f|$ and if $z \in Z_1$, then $|\langle \int sdm, z\rangle| \leq \int |s|d|m_z|$.

Taking the supremum for $z \in Z_1$ and for $|s| \leq |f|$ we get

$$\tilde{m}_{F,G}(f) \leq \sup\{\int |f|d|m_z| : z \in Z_1\}.$$

Conversely, let $\varepsilon > 0$ and $a < \sup\{\int |f|d|m_z| : z \in Z_1\}$. There is a $z \in Z_1$ such that $a < \int |f|d|m_z|$. Then there is a positive, finite, $\mathcal{D}$-step function $\varphi \leq |f|$ such that $a < \int \varphi d|m_z|$. Let

$$\varphi = \sum_{1 \leq i \leq n} \varphi_{A_i} \alpha_i.$$

with $A_i \in \mathcal{D}$ mutually disjoint and $\alpha_i > 0$. For each i there is a finite family $(B_{ij})_j$ of disjoint sets from $\mathcal{D}$ with union A_i, such that

$$|m_z|(A_i) < \sum_j |m_z(B_{ij})| + \varepsilon/(2n\alpha_i).$$

We can choose elements $x_{ij} \in F_1$ such that

$$\sum_j |m_z(B_{ij})| < \sum_j \langle x_{ij}, m_z(B_{ij})\rangle + \varepsilon/(2n\alpha_i).$$

If we denote

$$s = \sum_{ij} \alpha_i x_{ij} \varphi_{B_{ij}},$$

then s is a $\mathcal{D}$-step function with $|s| \leq |f|$ and

$$a < \int \varphi d|m_z| < \langle \int sdm, z\rangle + \varepsilon$$
$$\leq |\int sdm| + \varepsilon \leq \tilde{m}_{F,G}(f) + \varepsilon.$$

Since $\varepsilon > 0$ and a are arbitrary, we deduce that

$$\sup\{\int |f|d|m_z| : z \in Z_1\} \leq \tilde{m}_{F,G}(f)$$

§5. INTEGRATION WITH RESPECT TO A MEASURE WITH FINITE SEMIVARIATION

and the equality follows. ∎

We now list some properties of $\tilde{m}_{F,G}(f)$. All functions are assumed to be m-measurable. For simplicity we shall write $\tilde{m}$ instead of $\tilde{m}_{F,G}$ if the spaces F, G are understood.

9. $\tilde{m}(f) = \tilde{m}(|f|)$.

10. $\tilde{m}(f + g) \leq \tilde{m}(f) + \tilde{m}(g)$; $\tilde{m}(\alpha f) = (\alpha)\tilde{m}(f)$.

11. $\tilde{m}(f) \leq \tilde{m}(g)$ if $|f| \leq |g|$.

12. $\tilde{m}(\sup_n f_n) = \sup_n \tilde{m}(f_n)$, for every increasing sequence (f_n) of positive, m-measurable functions.

We use Proposition 8 and the Monotone Convergence Theorem for each $|m_z|$.

In particular:

13. $\tilde{m}(f) = \sup_n \tilde{m}(f \varphi_{\{|f| \leq n\}})$.

14. $\tilde{m}(\sum_n f_n) \leq \sum_n \tilde{m}(f_n)$, for every sequence (f_n) of positive, m-measurable functions.

15. $\tilde{m}(\liminf f_n) \leq \liminf \tilde{m}(f_n)$, for every sequence (f_n) of positive, m-measurable functions.

In fact,
$$\tilde{m}(\inf_{p \geq n} f_p) \leq \tilde{m}(f_p), \text{ for } p \geq n,$$

hence
$$\tilde{m}(\inf_{p \geq n} f_p) \leq \inf_{p \geq n} \tilde{m}(f_p).$$

The sequence $(\inf_{p \geq n} f_p)_{n \in \mathbb{N}}$ is increasing and we apply Property 12.

16. If $f : S \to D$ is m-measurable and $c > 0$, then
$$\tilde{m}(\{|f| > c\}) \leq \frac{1}{c}\tilde{m}(f) \leq +\infty.$$

In fact, for each $z \in Z_1$ we have
$$|m_z|(\{|f| > c\}) \leq \frac{1}{c}\int |f|d|m_z| \leq \frac{1}{c}\tilde{m}(f).$$

17. Let $f : S \to \overline{\mathbb{R}}_+$ be an m-measurable function with $\tilde{m}(f) < \infty$. Then
a) $f < \infty$, m-a.e.
b) The set $\{f \neq 0\}$ is contained in the union of a sequence (A_n) from Σ with $\tilde{m}(A_n) < \infty$.

Proof. Let $g : S \to \overline{\mathbb{R}}_+$ be a Σ-measurable function such that $f = g$, m-a.e. Then $\tilde{m}(g) < \infty$, hence, for every $z \in Z$ we have $\int g d|m_z| < \infty$; therefore the set $\{g = \infty\}$ belongs to Σ and is $|m_z|$-negligible. It follows that

$\tilde{m}(\{g = \infty\}) = 0$, that is, $g < \infty$, m-a.e.; consequently, $f < \infty$, m-a.e. and this proves assertion a).

To prove assertion b) we remark that

$$\{f \neq 0\} = \{f \neq g\} \cup \{g \neq 0\}$$

The set $\{f \neq g\}$ is m-negligible; therefore it is contained in an m-negligible set $A_1 \in \Sigma$. Then we take $A_n = \{|g| > \frac{1}{n}\} \in \Sigma$ for $n \geq 2$. By Property 16 we have $\tilde{m}(A_n) < \infty$. Since $\{g \neq 0\} = \bigcup_{n \geq 2} A_n$, assertion b) is proved. ∎

18. Definition. *Let $f_n, f : S \to D$ be m-measurable functions, for $n \in \mathbb{N}$.*

We say that the sequence (f_n) converges to f in m-measure, if for every $\varepsilon > 0$ we have

$$\lim_n \tilde{m}(\{|f_n - f| > \varepsilon\}) = 0.$$

19. Proposition. *Let $f_n, f : S \to D$, $n \in \mathbb{N}$, be m-measurable functions. If $\tilde{m}(f_n - f) \to 0$, then $f_n \to f$ in m-measure.*

The proof follows from Property 16.

C. The space of integrable functions

We can define now the m-integrable functions and prove their properties.

20. Definition. *We denote by $\mathcal{F}_D(\tilde{m}_{F,G})$ or by $\mathcal{F}_D(m_{F,G})$ the set of all m-measurable functions $f : S \to D$ with $\tilde{m}_{F,G}(f) < \infty$.*

We call $\mathcal{F}_D(\tilde{m}_{F,G})$ the space of D-valued, $\tilde{m}_{F,G}$-integrable functions. However, the reader can choose any subspace of $\mathcal{F}_D(\tilde{m}_{F,G})$ as the space of integrable functions, according to the requirements of the specific problem to which this integration theory is applied. In order to cover all possible applications, we shall study below the properties of the whole space $\mathcal{F}_D(\tilde{m}_{F,G})$.

If the spaces F, G are understood we shall write $\mathcal{F}_D(\tilde{m})$ or $\mathcal{F}_D(m)$ instead of $\mathcal{F}_D(\tilde{m}_{F,G})$. From the properties of $\tilde{m}(f)$ we deduce that $\mathcal{F}_D(\tilde{m})$ is a vector space and that $f \mapsto \tilde{m}(f)$ is a seminorm on this space.

21. Since for each function $f \in \mathcal{F}_D(\tilde{m})$ we have $\int |f| d|m_z| \leq \tilde{m}(f)$ for $z \in Z_1$, we deduce that

$$\mathcal{F}_D(\tilde{m}) \subset L_D^1(|m_z|), \text{ for every } z \in Z,$$

hence

$$\mathcal{F}_D(\tilde{m}) \subset \bigcap_{z \in Z} L_D^1(|m_z|).$$

If the norming space Z is closed in G^*, the converse inclusion is also true. This will follows from the next proposition, which is an extension of Proposition 4.15.

§5. INTEGRATION WITH RESPECT TO A MEASURE WITH FINITE SEMIVARIATION

22. Proposition. *Assume Z is closed in G^*. Let $f : S \to D$ be an m-measurable function. Then*

$$\tilde{m}(f) < \infty \text{ iff } \int |f|d|m_z| < \infty, \text{ for every } z \in Z.$$

The proof is the same as that of Proposition 4.15, taking M the set of D-valued, $\mathcal{D}$-step functions s such that $|s| \leq |f|$.

23. Corollary. *If Z is closed in G^*, then*

$$\mathcal{F}_D(\tilde{m}) = \bigcap_{z \in Z} L_D^1(|m_z|).$$

The following theorem proves that the space $\mathcal{F}_D(\tilde{m})$ is complete.

24. Theorem. *If (f_n) is a Cauchy sequence in $\mathcal{F}_D(\tilde{m})$, then there is a function $f \in \mathcal{F}_D(\tilde{m})$ and a subsequence (f_{n_k}) such that:*
a) $f_n \to f$ in $\mathcal{F}_D(\tilde{m})$;
b) $f_{n_k} \to f$, m-a.e., as $k \to \infty$.

Proof. Let (f_n) be a Cauchy sequence in $\mathcal{F}_D(\tilde{m})$.
By induction, we can find a subsequence (f_{n_k}) such that

$$\tilde{m}(f_{n_k} - f_{n_{k-1}}) < \frac{1}{2^k}, \text{ for } k \geq 2.$$

By Property 14 we have

$$\tilde{m}(\sum_{k \geq 2} |f_{n_k} - f_{n_{k-1}}|) \leq \sum_{k \geq 2} \tilde{m}(f_{n_k} - f_{n_{k-1}}) < \infty.$$

Then, by Property 17 we have

$$\sum_{k \geq 2} |f_{n_k} - f_{n_{k-1}}| < \infty, \ m\text{-a.e.},$$

therefore the series

$$\sum_{k \geq 2} (f_{n_k} - f_{n_{k-1}})$$

is convergent in D, m-a.e. Define $g : S \to D$ by

$$g(s) = \sum_{k \geq 2} (f_{n_k}(s) - f_{n_{k-1}}(s))$$

if the series is convergent and $g(s) = 0$ otherwise. Then g is m-measurable and

$$\tilde{m}(g) \leq \sum_{k \geq 2} \tilde{m}(f_{n_k} - f_{n_{k-1}}) < \infty,$$

therefore $g \in \mathcal{F}_D(\tilde{m})$. The partial sums of the series are

$$g_k = \sum_{2 \leq i \leq k} (f_{n_i} - f_{n_{i-1}}) = f_{n_k} - f_{n_1},$$

hence $\lim_k (f_{n_k} - f_{n_1}) = \lim_k g_k = g$, m-a.e. If we set $f = g + f_{n_1}$, then $f \in \mathcal{F}_D(\tilde{m})$ and

$$\lim_k f_{n_k} = f, \ m - \text{a.e.}$$

and this proves assertion b). For each $k \geq 2$ we have

$$\tilde{m}(f_{n_k} - f) = \tilde{m}(f_{n_k} - f_{n_1} - g)$$
$$= \tilde{m}\Big(\sum_{2 \leq i \leq k}(f_{n_i} - f_{n_{i-1}}) - g\Big)$$
$$= \tilde{m}\Big(|\sum_{i > k}(f_{n_i} - f_{n_{i-1}})|\Big) \leq \tilde{m}\Big(\sum_{i > k}|f_{n_i} - f_{n_{i-1}}|\Big)$$
$$\leq \sum_{i > k} \tilde{m}(f_{n_i} - f_{n_{i-1}}) \leq \sum_{i > k} \frac{1}{2^i} = \frac{1}{2^k} \to 0, \text{ as } k \to \infty,$$

hence $\lim_k f_{n_k} = f$ in $\mathcal{F}_D(\tilde{m})$. Since (f_n) is a Cauchy sequence, we have also $f_n \to f$ in $\mathcal{F}_D(\tilde{m})$ and assertion a) is also proved. ∎

25. Corollary. *The space $\mathcal{F}_D(\tilde{m})$ is complete.*

26. Corollary. *If (f_n) is a Cauchy sequence in $\mathcal{F}_D(\tilde{m})$ and if $f_n \to f$, m-a.e., then $f \in \mathcal{F}_D(\tilde{m})$ and $f_n \to f$ in $\mathcal{F}_D(\tilde{m})$.*

27. Corollary. *If (f_n) is a sequence from $\mathcal{F}_D(\tilde{m})$ and if $f_n \to f$ in $\mathcal{F}_D(\tilde{m})$ and $f_n \to g$, m-a.e., then $f = g$, m-a.e.*

28. *The space $\mathcal{F}_D(\tilde{m})$ contains the set $\mathcal{S}_D(\mathcal{D})$ of $\mathcal{D}$-step functions $f : S \to D$.*

However, unlike the classical integration theory, the set $\mathcal{S}_D(\mathcal{D})$ is not necessarily dense in $\mathcal{F}_D(\tilde{m})$.

We denote by $\mathcal{F}_D(\mathcal{B}, \tilde{m})$ or by $\mathcal{F}_D(\mathcal{B}, m)$ the closure in $\mathcal{F}_D(\tilde{m})$ of the set $\mathcal{B}_D(\mathcal{D})$ of *bounded* functions of $\mathcal{F}_D(\tilde{m})$.

In general, for any set $\mathcal{C} \subset \mathcal{F}_D(\tilde{m})$, we denote by $\mathcal{F}_D(\mathcal{C}, \tilde{m})$ the closure of $\mathcal{C}$ in $\mathcal{F}_D(\tilde{m})$.

If $\tilde{m}_{F,G}(S) < \infty$, then the set $\mathcal{S}_D(\Sigma)$ of Σ-step functions $f : S \to D$ is contained in $\mathcal{F}_D(\mathcal{B}, \tilde{m})$, but it is not necessarily dense. We shall see later (Proposition 43) that if $\tilde{m}_{F,G}(S) < \infty$, then $\mathcal{S}_\mathbb{R}(\Sigma)$ is dense in $\mathcal{F}_\mathbb{R}(\mathcal{B}, \tilde{m}_{F,G})$ and that if $\tilde{m}_{F,G}(S) < \infty$ and m_{F,G^*} is uniformly σ-additive, then $\mathcal{S}_D(\Sigma)$ is dense in $\mathcal{F}_D(\mathcal{B}, \tilde{m}_{F,G})$.

We shall see also (Proposition 44) that if $S \in \mathcal{D}$ and m_{F,G^*} is uniformly σ-additive and if $\mathcal{R}$ is a ring generating the σ-algebra $\mathcal{D}$, then $\mathcal{S}_D(\mathcal{R})$ is dense in $\mathcal{F}_D(\mathcal{B}, \tilde{m}_{F,G})$.

§5. INTEGRATION WITH RESPECT TO A MEASURE WITH FINITE SEMIVARIATION   85

There are special cases (Theorem 8.3) when $\mathcal{F}_D(\mathcal{B}, \tilde{m}_{F,G}) = \mathcal{F}_D(\tilde{m}_{F,G})$.

The functions of $\mathcal{F}_D(\mathcal{B}, \tilde{m}_{F,G})$ have some properties similar to those of the Bochner-integrable functions. We state one of these properties in the following proposition, which will be used in the proof of the Vitali Convergence Theorem 36.

Other properties will be stated in Section F on properties of the space $\mathcal{F}_D(\mathcal{B}, \tilde{m})$.

29. Proposition. *Let $f \in \mathcal{F}_D(\mathcal{B}, \tilde{m})$. Then*

$$\lim_{\tilde{m}(A) \to 0} \tilde{m}(f\varphi_A) = 0.$$

Proof. Let $\varepsilon > 0$. There is a bounded function $g \in \mathcal{F}_D(\tilde{m})$ such that $\tilde{m}(f - g) < \varepsilon/2$. Denote

$$a = \sup_{s \in S} |g(s)| < \infty.$$

For any set $A \subset S$ we have

$$\tilde{m}(f\varphi_A) < \varepsilon/2 + \tilde{m}(g\varphi_A) < \varepsilon/2 + a\tilde{m}(A) \leq +\infty.$$

Let $\delta = \varepsilon/(2a)$. If $\tilde{m}(A) < \delta$, then $\tilde{m}(f\varphi_A) < \varepsilon$ and this proves the proposition. ■

D. The integral

In the special case $D = F$, we can define the integral $\int f \, dm$ for functions $f \in \mathcal{F}_F(\tilde{m}_{F,G})$. To somewhat simplify the notation, we shall write $\mathcal{F}_{F,G}(m)$ or $\mathcal{F}_{F,G}(\tilde{m})$ instead of $\mathcal{F}_F(\tilde{m}_{F,G})$.

30. The construction of the integral $\int f \, dm$ is as follows:

Let $f \in \mathcal{F}_{F,G}(m)$. Then $f \in L^1_F(|m_z|)$ for every $z \in Z$, hence, the integral $\int f \, dm_z$ is defined, in the sense of §2 and is a real number.

The mapping $z \mapsto \int f \, dm_z$ is a continuous linear functional on Z:

$$\left| \int f \, dm_z \right| \leq \int |f| \, d|m_z| \leq |z| \tilde{m}(f),$$

hence this mapping belongs to Z^*. We denote it by $\int f \, dm$ and we call it the integral of f with respect to m. We have therefore:

$$\int f \, dm \in Z^*,$$

$$\langle \int f \, dm, z \rangle = \int f \, dm_z, \text{ for } z \in Z$$

and
$$\left|\int f dm\right| \leq \tilde{m}(f).$$

If $f \in \mathcal{F}_{F,G}(m)$ and $A \in \Sigma$, then $\varphi_A f \in \mathcal{F}_{F,G}(m)$. We denote, as usual,
$$\int_A f dm = \int \varphi_A f dm.$$

If m is σ-additive, then the semivariation $\tilde{m}_{\mathbb{R},E}$ is finite and taking $Z = E^*$, the above construction yields, in particular, the integral $\int \varphi dm$ for real-valued functions $\varphi \in \mathcal{F}_{\mathbb{R},E}(m) = \mathcal{F}_{\mathbb{R}}(\tilde{m}_{\mathbb{R},E})$. We have
$$\int \varphi dm \in E^{**},$$
$$\left\langle \int \varphi dm, x^* \right\rangle = \int \varphi d(x^* m), \text{ for } x^* \in E^*$$

and
$$\left|\int \varphi dm\right| \leq \tilde{m}_{\mathbb{R},E}(\varphi).$$

From the inequality $|\int f dm| \leq \tilde{m}_{F,G}(f)$, it follows that the integral mapping $f \mapsto \int f dm$ of $\mathcal{F}_{F,G}(m)$ into Z^* is continuous. Then, any theorem stating the convergence $f_n \to f$ in $\mathcal{F}_{F,G}(m)$ can be completed by stating the convergence of the integrals, $\int f_n dm \to \int f dm$ in Z^*.

31. Remarks. a) The space $\mathcal{F}_{F,G}(m)$ of m-integrable functions does not depend on the norming space Z. But the integral $\int f dm \in Z^*$ depends on Z. Let us denote it for the moment, by $(Z) \int f dm$.

If $Y \subset Z$ is another norming space for G, then $(Y) \int f dm \in Y^*$.

The relationship between the two integrals is that $(Y) \int f dm$ is the restriction to Y of the integral $(Z) \int f dm$:
$$\left\langle (Y) \int f dm, y \right\rangle = \left\langle (Z) \int f dm, y \right\rangle, \text{ for } y \in Y.$$

In particular, if m is σ-additive, we can take G^* as a norming space for G and then $(G^*) \int f dm \in G^{**}$. Then for any other norming space Z, we have
$$\left\langle (Z) \int f dm, z \right\rangle = \left\langle (G^*) \int f dm, z \right\rangle, \text{ for } z \in Z.$$

From now on we shall omit (Z) in the notation of the integral.

b) If m is σ-additive and has finite variation $|m|$, we can consider the space $L_F^1(m)$ and the integral $\int f dm$ for $f \in L_F^1(m)$, defined in §2. In this case, m

has finite semivariation $\tilde{m}_{F,G} \leq |m|$ and for each $z \in G_1^*$ we have $|m_z| \leq |m|$. Then, for any m-measurable function $f : S \to F$ we have

$$\tilde{m}_{F,G}(f) \leq \sup_{|z|\leq 1} \int |f|\,d|m_z| \leq \int |f|\,d|m| \leq \|f\|_1,$$

therefore $L_F^1(m) \subset \mathcal{F}_{F,G}(m)$ and the embedding is continuous.

If $f \in L_F^1(m)$, there is a sequence (f_n) of Σ-step functions converging to f in $L_F^1(m)$. Then $\int f_n\,dm \to \int f\,dm$, where $\int f\,dm$ is the integral of §2.

But we have also $f_n \to f$ in $\mathcal{F}_{F,G}(m)$, hence $\int f_n\,dm \to \int f\,dm$, where $\int f\,dm$ is the integral defined in this paragraph.

It follows that for functions $f \in L_F^1(m)$, the integral $\int f\,dm$ is the same, whether we consider f in $L_F^1(m)$ or in $\mathcal{F}_{F,G}(m)$.

32. We are particularly interested in the case when $\int f\,dm$ belongs to G rather than Z^*. Evidently, this is true if $Z = G^*$ and G is reflexive. We shall see that this is also true if $c_0 \not\subset G$ and $S = \bigcup_n S_n$ with $S_n \in \mathcal{D}$. (Theorem 48).

In general, if $\mathcal{C}$ is a subset of $\mathcal{F}_{F,G}(m)$ such that $\int f\,dm \in G$ for every $f \in \mathcal{C}$, then, by the continuity of the integral, we have also $\int f\,dm \in G$ for f in the closure of $\mathcal{C}$ in $\mathcal{F}_{F,G}(m)$.

For example, if $\mathcal{C} = \mathcal{S}_F(\mathcal{D})$, then $\int f\,dm \in G$ for $f \in \mathcal{S}_F(\mathcal{D})$, hence $\int f\,dm \in G$ for any function f in the closure of $\mathcal{S}_F(\mathcal{D})$ in $\mathcal{F}_{F,G}(m)$.

Since the $\mathcal{D}$-step functions are not necessarily dense in the set $\mathcal{F}_{F,G}(\mathcal{B},m)$, the integral $\int f\,dm$ of functions $f \in \mathcal{F}_{F,G}(\mathcal{B},m)$ does not necessarily belong to G.

However, we shall see in Theorem 48 that if m_{F,G^*} is uniformly σ-additive and if $S \in \mathcal{D}$, then $\int f\,dm \in G$ for every $f \in \mathcal{B}_F(\mathcal{D})$.

In particular, if m is σ-additive, then $m_{\mathbb{R},E^*}$ is uniformly σ-additive, hence we have $\int \varphi\,dm \in E$ for $\varphi \in \mathcal{B}_{\mathbb{R}}(\mathcal{D})$ if $S \in \mathcal{D}$.

33. Remark. In the preceding section we called $\mathcal{F}_{F,G}(m)$ the space of F-valued, m-integrable functions. But it is reasonable to restrict ourselves to the subset of functions $f \in \mathcal{F}_{F,G}(m)$ for which the integral $\int f\,dm$ belongs to G and satisfy additional conditions imposed by the specific problem to which the integration theory is applied.

For example, in the case of the stochastic integral in Chapter 2, we impose the condition that the stochastic integral is cadlag (right continuous, with left limits).

E. Convergence theorems

The Egorov theorem is not valid, in general. However, using a control measure, it is valid whenever $m_{F,Z}$ is uniformly σ-additive.

34. Theorem. (Egorov) *Assume that the set of measures $m_{F,Z}$ is uniformly σ-additive and let $f_n, f : S \to D$, $n \in \mathbb{N}$, be m-measurable functions such that $f_n \to f$, m-a.e. Then*

a) *For every set $A \in \mathcal{D}$ and for every $\varepsilon > 0$, there is a set $B \in \mathcal{D}$ with $B \subset A$ such that $\tilde{m}(A - B) < \varepsilon$ and $f_n \to f$ uniformly on B.*
  If $\mathcal{D}$ is a σ-algebra, we can take $A = S$.
b) *For every set $A \in \mathcal{D}$ we have*
$$f_n \varphi_A \to f \varphi_A, \text{ in } \tilde{m}\text{-measure}.$$
If $\mathcal{D}$ is a σ-algebra then
$$f_n \to f, \text{ in } \tilde{m}\text{-measure}.$$

Proof. Let $A \in \mathcal{D}$ and consider the σ-algebra $\mathcal{D} \cap A$ of subsets of A. Since the set $m_{F,Z}$ of measures is uniformly σ-additive on $\mathcal{D} \cap A$, by Theorem 4.25, there is a positive, finite, σ-additive measure λ on $\mathcal{D} \cap A$ such that $\tilde{m}_{F,G} \ll \lambda$.

Let $\varepsilon > 0$. There is a $\delta > 0$ such that for every set $C \in \mathcal{D} \cap A$ with $\lambda(C) < \delta$ we have $\tilde{m}(C) < \varepsilon$.

We can apply now the Egorov theorem for λ: for this $\delta > 0$, there is a set $B \in \mathcal{D} \cap A$ such that $\lambda(A - B) < \delta$ and $f_n \to f$ uniformly on B; at the same time, $f_n \varphi_A \to f \varphi_A$ in λ-measure. We deduce that $\tilde{m}(A - B) < \varepsilon$ and that $f_n \varphi_A \to f \varphi_A$ in $\tilde{m}$-measure. ∎

Uniform convergence implies convergence in $\mathcal{F}_\mathcal{D}(\tilde{m})$.

35. Theorem. *Let (f_n) be a sequence from $\mathcal{F}_\mathcal{D}(\tilde{m})$ converging uniformly to a function $f : S \to D$. Assume there is a set $A \in \Sigma$ with $\tilde{m}_{F,G}(A) < \infty$, such that all functions f_n vanish outside A. Then $f \in \mathcal{F}_{F,G}(\tilde{m})$ and $\tilde{m}(f_n - f) \to 0$.*
If $D = F$, then $\int f_n dm \to \int f dm$ in Z^*.

Proof. Let $\varepsilon > 0$ and let N be such that for any $n \geq N$ and any $s \in S$ we have $|f_n(s) - f(s)| < \varepsilon$. Then, for $n \geq N$ we have $\tilde{m}(f_n - f) \leq \varepsilon \tilde{m}(A) < \infty$. It follows that for $n \geq N$ we have $f_n - f \in \mathcal{F}_\mathcal{D}(\tilde{m})$, hence $f \in \mathcal{F}_\mathcal{D}(\tilde{m})$. We deduce also that $\tilde{m}(f_n - f) \to 0$. ∎

We state now the analog of Vitali's theorem.

36. Theorem. (Vitali) *Let (f_n) be a sequence from $\mathcal{F}_\mathcal{D}(\tilde{m})$ and $f : S \to D$ an m-measurable function. Assume that the following conditions are satisfied:*
a) $f_n \to f$ *in $\tilde{m}$-measure,*
or
a') $f_n \to f$, m-*a.e.*, $S \in \mathcal{D}$ *and $m_{F,Z}$ is uniformly σ-additive.*
b) $\lim_{\tilde{m}(A) \to 0} \tilde{m}(f_n \varphi_A) = 0$, *uniformly for $n \in \mathbb{N}$.*
c) *For every $\varepsilon > 0$, there is a set $A_\varepsilon \in \Sigma$ with $\tilde{m}_{F,G}(A_\varepsilon) < \infty$, such that*
$$\tilde{m}_{F,G}(f_n \varphi_{S - A_\varepsilon}) < \varepsilon, \text{ for all } n \in \mathbb{N}.$$

(condition c) is automically satisfied if $\tilde{m}_{F,G}(S) < \infty$; in particular, if condition a') is satisfied).
Then $f \in \mathcal{F}_\mathcal{D}(\tilde{m}_{F,G})$ and $\tilde{m}_{F,G}(f_n - f) \to 0$.

If $D = F$, then
$$\int f_n \, dm \to \int f \, dm, \text{ in } Z^*.$$

Conversely, if $f \in \mathcal{F}_D(\mathcal{B}, \tilde{m}_{F,G})$ and $\tilde{m}_{F,G}(f_n - f) \to 0$, then conditions a) and b) are satisfied.

Proof. By Egorov's theorem, condition a') implies condition a). Assume conditions a), b) and c) satisfied and prove that (f_n) is a Cauchy sequence in $\mathcal{F}_D(\tilde{m}_{F,G})$. Let $\varepsilon > 0$ and let $A_\varepsilon \in \Sigma$ be a set satisfying condition c). By conditon b) there is a $\delta > 0$ such that for every set $A \in \Sigma$ with $\tilde{m}_{F,G}(A) < \delta$ we have $\tilde{m}_{F,G}(f_n \varphi_A) < \varepsilon$ for all $n \in \mathbb{N}$. By condition a), there is an N_ε such that, if we set
$$B_{nm} = \{s \in A_\varepsilon : |f_n(s) - f_m(s)| > \varepsilon / \tilde{m}_{F,G}(A_\varepsilon)\},$$
then for any $n, m \geq N_\varepsilon$ we have $B_{nm} \in \Sigma$ and
$$\tilde{m}_{F,G}(B_{nm}) < \delta.$$

Then, for $n, m \geq N_\varepsilon$ we have
$$\tilde{m}(f_n - f_m) \leq \tilde{m}((f_n - f_m)\varphi_{B_{nm}})$$
$$+ \tilde{m}((f_n - f_m)\varphi_{A_\varepsilon - B_{nm}}) + \tilde{m}((f_n - f_m)\varphi_{S - A_\varepsilon})$$
$$\leq \tilde{m}(f_n \varphi_{B_{nm}}) + \tilde{m}(f_m \varphi_{B_{nm}}) + \tilde{m}((f_n - f_m)\varphi_{A_\varepsilon - B_{nm}}) + \tilde{m}(f_n \varphi_{S - A_\varepsilon})$$
$$+ \tilde{m}(f_m \varphi_{S - A_\varepsilon}) \leq 5\varepsilon,$$

hence (f_n) is a Cauchy sequence in $\mathcal{F}_D(\tilde{m}_{F,G})$. Since this space is complete, there is a Σ-measurable function $g \in \mathcal{F}_D(\tilde{m}_{F,G})$ such that $\tilde{m}_{F,G}(f_n - g) \to 0$. Then, by Proposition 19 we have $f_n \to g$ in $\tilde{m}_{F,G}$-measure. Since, by hypothesis a) we have also $f_n \to f$ in $\tilde{m}_{F,G}$-measure, it follows that $f = g$, m-a.e., hence $f \in \mathcal{F}_D(\tilde{m}_{F,G})$ and $\tilde{m}(f_n - f) \to 0$.

If $D = F$, it follows that $\int f \, dm \to \int f \, dm$ in Z^*.

Conversely, assume that $f \in \mathcal{F}_D(\tilde{m}_{F,G})$ and $\tilde{m}_{F,G}(f_n - f) \to 0$. By Proposition 19 we have $f_n \to f$ in $\tilde{m}_{F,G}$-measure, so, condition a) is satisfied. To prove b), assume $f \in \mathcal{F}_D(\mathcal{B}, \tilde{m}_{F,G})$. Let $\varepsilon > 0$ and let N be such that for every $n \geq N$ we have $\tilde{m}(f_n - f) < \varepsilon/2$. Then, for every $A \in \Sigma$ and $n \geq N$ we have $\tilde{m}(f_n \varphi_A - f \varphi_A) < \varepsilon/2$; therefore
$$\tilde{m}(f_n \varphi_A) \leq \tilde{m}(f \varphi_A) + \varepsilon/2.$$

By Proposition 29 there is a $\delta_0 > 0$ such that $A \in \Sigma$ and $\tilde{m}(A) < \delta_0$ implies $\tilde{m}(f \varphi_A) < \varepsilon/2$. Then for $n \geq N$ and $\tilde{m}(A) < \delta_0$ we have $\tilde{m}(f_n \varphi_A) < \varepsilon$.

For $n \leq N$, using again Proposition 29, we can find a common $\delta_1 > 0$ such that $A \in \Sigma$ and $\tilde{m}(A) < \delta_1$ implies $\tilde{m}(f_n \varphi_A) < \varepsilon$, for all $n \leq N$.

If we take $\delta = \inf(\delta_0, \delta_1)$, then for any $A \in \Sigma$ with $\tilde{m}(A) < \delta$ we have $\tilde{m}(f_n \varphi_A) < \varepsilon$ for all $n \in \mathbb{N}$, and this proves assertion b). ∎

From Vitali's Theorem we deduce now Lebesgue's Theorem, but only for functions from $\mathcal{F}_D(\mathcal{B}, m)$ and only if m has bounded semivariation.

37. Theorem. (Lebesgue) *Assume $S \in \mathcal{D}$. Let (f_n) be a sequence from $\mathcal{F}_D(\mathcal{B}, m)$, $f \in \mathcal{F}_D(m)$ a function, and $g \in \mathcal{F}(\mathcal{B}, m)$ a positive function.*
Assume the following conditions are satisfied:
a) $f_n \to f$ in $\tilde{m}$-measure
or
a') $f_n \to f$, m-a.e. and $m_{F,Z}$ is uniformly σ-additive.
b) $|f_n| \le g$, m-a.e., for each $n \in \mathbb{N}$.
Then $f \in \mathcal{F}_D(\mathcal{B}, m)$ and $\tilde{m}(f_n - f) \to 0$.
If $D = F$, then $\int f_n dm \to \int f dm$ in Z^*.

Proof. Conditions a) and a') are the same in Lebesgue's Theorem and Vitali's Theorem. We shall prove now that condition b) in Lebesgue's Theorem implies condition b) in Vitali's Theorem.

Since $g \in \mathcal{F}_{\mathbb{R}}(\mathcal{B}, m)$, by Proposition 29 we have $\tilde{m}(g\varphi_A) \to 0$ as $\tilde{m}(A) \to 0$. From the inequality $|f_n| \le g$, m-a.e., for all $n \in \mathbb{N}$, we deduce that $\tilde{m}(f_n \varphi_A) \to 0$ as $\tilde{m}(A) \to 0$, uniformly for $n \in \mathbb{N}$ and this is condition b) in Vitali's Theorem.

Condition c) in Vitali's Theorem is automatically satisfied, since $\tilde{m}_{F,G}(S) < \infty$. The conclusion of Vitali's Theorem is then valid for Lebesgue's Theorem. ∎

F. Properties of the space $\mathcal{F}_D(\mathcal{B}, \tilde{m}_{F,G})$

We already proved a property of the space $\mathcal{F}_D(\mathcal{B}, \tilde{m}_{F,G})$ in Proposition 29, which was used in the proof of Vitali's Convergence Theorem 36 and of Lebesgue's Convergence Theorem 37.

As we mentioned before, the space $\mathcal{F}_D(\mathcal{B}, \tilde{m}_{F,G})$ has some properties similar to the space of Bochner-integrable functions.

The following theorem is a variation of Proposition 29.

38. Proposition. *Let $f \in \mathcal{F}_D(\mathcal{B}, \tilde{m}_{F,G})$ and assume m_{F,G^*} is uniformly σ-additive. Then*

$$\tilde{m}_{F,G}(f\varphi_{A_n}) \to 0 \text{ as } A_n \downarrow \phi \text{ with } A_n \in \mathcal{D}.$$

Proof. Let $A_n \downarrow \phi$ with $A_n \in \mathcal{D}$. Since m_{F,G^*} is uniformly σ-additive, by Proposition 4.23 we have $\tilde{m}_{F,G}(A_n) \to 0$. By Proposition 29 we have then $\tilde{m}_{F,G}(f\varphi_{A_n}) \to 0$. ∎

39. Corollary. *Assume m is σ-additive. If $f \in \mathcal{F}_D(\mathcal{B}, \tilde{m}_{\mathbb{R},E})$, then*

$$\tilde{m}_{\mathbb{R},E}(f\varphi_{A_n}) \to 0 \text{ as } A_n \downarrow \phi \text{ with } A_n \in \mathcal{D}.$$

Proof. By Theorem 4.27, $m_{\mathbb{R},E^*}$ is uniformly σ-additive and we can apply Proposition 38. ∎

§5. INTEGRATION WITH RESPECT TO A MEASURE WITH FINITE SEMIVARIATION

The following proposition is a partial converse of Propositions 29 and 38.

40. Proposition. *Let $f \in \mathcal{F}_D(\tilde{m})$. If $\tilde{m}(f\varphi_{A_n}) \to 0$ as $A_n \downarrow \phi$ with $A_n \in \Sigma$, then $f \in \mathcal{F}_D(\mathcal{B}, \tilde{m})$.*

Proof. Since f is m-measurable, it is equal m-a.e. to a Σ-measurable function $g : S \to D$. Denote $A = \{g \neq 0\}$ and $A_n = \{|g| \leq n\}$ for each n. The sets A and A_n belong to Σ and we have $g\varphi_{A_n} \in \mathcal{F}_D(\mathcal{B}, \tilde{m})$ for each n. The sets $B_n = S - A_n = \{|g| > n\}$ belong to Σ and $B_n \downarrow \phi$, hence, by hypothesis we have
$$\tilde{m}(g - g\varphi_{A_n}) = \tilde{m}(g\varphi_{B_n}) \to 0,$$
consequently $g \in \mathcal{F}_D(\mathcal{B}, m)$. Then we have also $f \in \mathcal{F}_D(\mathcal{B}, m)$. ∎

41. Proposition. *If $f : S \to D$ is m-measurable and $|f| \leq g \in \mathcal{F}_{\mathbb{R}}(\mathcal{B}, \tilde{m})$, then $f \in \mathcal{F}_D(\mathcal{B}, \tilde{m})$.*

Proof. We can assume f and g are Σ-measurable. Let $A_n = \{g \geq n\} \in \Sigma$. By Property 16 of the seminorm $\tilde{m}$ we have
$$\tilde{m}(A_n) \leq \frac{1}{n}\tilde{m}(g),$$
therefore $\tilde{m}(A_n) \to 0$. Then by proposition 29 we have $\tilde{m}(g\varphi_{A_n}) \to 0$. If we set $B_n = S - A_n$, then $g\varphi_{B_n}$ is bounded by n, hence $g\varphi_{B_n} \in \mathcal{F}_D(\mathcal{B}, \tilde{m})$. Then
$$\tilde{m}(f - f\varphi_{B_n}) = \tilde{m}(f\varphi_{A_n}) \leq \tilde{m}(g\varphi_{A_n}) \to 0,$$
therefore $f \in \mathcal{F}_D(\mathcal{B}, \tilde{m})$. ∎

42. Corollary. *A function $f : S \to D$ belongs to $\mathcal{F}_D(\mathcal{B}, \tilde{m})$ iff f is m-measurable and $|f| \in \mathcal{F}_{\mathbb{R}}(\mathcal{B}, \tilde{m})$.*

The following proposition gives sufficient conditions for the Σ-step functions to be dense in $\mathcal{F}_D(\mathcal{B}, m)$.

43. Proposition. a) *If $\tilde{m}_{F,G}(S) < \infty$, then the set $\mathcal{S}_{\mathbb{R}}(\Sigma)$ of real-valued, Σ-step functions is dense in $\mathcal{F}_{\mathbb{R}}(\mathcal{B}, \tilde{m}_{F,G})$.*
b) *If $S \in \mathcal{D}$ and $m_{F,Z}$ is uniformly σ-additive, then the set $\mathcal{S}_D(\Sigma)$ of D-valued, Σ-step functions is dense in $\mathcal{F}_D(\mathcal{B}, \tilde{m}_{F,G})$.*

Proof. To prove a) let $f \in \mathcal{F}_{\mathbb{R}}(\mathcal{B}, \tilde{m}_{F,G})$ be a bounded function. We can assume f is Σ-measurable. Then there is a sequence (f_n) of real-valued, Σ-step functions such that $f_n \to f$ uniformly and $|f_n| \leq |f|$. By Theorem 35 we have $\tilde{m}_{F,G}(f_n - f) \to 0$; therefore the set $\mathcal{S}_{\mathbb{R}}(\Sigma)$ is dense in the set of bounded functions of $\mathcal{F}_{\mathbb{R}}(\mathcal{B}, \tilde{m}_{F,G})$, hence $\mathcal{S}_{\mathbb{R}}(\Sigma)$ is dense in $\mathcal{F}_{\mathbb{R}}(\mathcal{B}, \tilde{m}_{F,G})$.

To prove b), assume $m_{F,Z}$ is uniformly σ-additive and let $f \in \mathcal{F}_D(\mathcal{B}, \tilde{m}_{F,G})$ be a Σ-measurable function. Let (f_n) be a sequence of D-valued, Σ-step functions such that $f_n \to f$ and $|f_n| \leq |f|$. We can apply Lebesgue's Theorem

37 and deduce that $\tilde{m}_{F,G}(f_n - f) \to 0$, hence $\mathcal{S}_D(\Sigma)$ is dense in $\mathcal{F}_D(\mathcal{B}, \tilde{m}_{F,G})$. ∎

The following proposition gives sufficient conditions for the step functions over a ring $\mathcal{R}$ to be dense in $\mathcal{F}_D(\mathcal{B}, \tilde{m}_{F,G})$.

44. Proposition. *Assume that $S \in \mathcal{D}$ and m_{F,G^*} is uniformly σ-additive. If $\mathcal{R}$ is a ring generating the σ-algebra $\mathcal{D}$, then the set $\mathcal{S}_D(\mathcal{R})$ of D-valued, $\mathcal{R}$-step functions is dense in $\mathcal{F}_D(\mathcal{B}, \tilde{m}_{F,G})$.*

Proof. Let $\mathcal{R}$ be a ring generating the σ-algebra $\Sigma = \mathcal{D}$. Since m_{F,G^*} is uniformly σ-additive, by Theorem 4.25 there is a positive, *finite*, σ-additive measure λ on Σ such that $\tilde{m}_{F,G} \ll \lambda$ and $\lambda \leq \tilde{m}_{F,G}$ on Σ.

Let f be a D-valued, *bounded*, Σ-measurable function. Then $f \in L_D^1(\lambda)$. Since the set $\mathcal{S}_D(\mathcal{R})$ is dense in $L_D^1(\lambda)$, there is a sequence (f_n) from $\mathcal{S}_D(\mathcal{R})$ such that $f_n \to f$ in $L_D^1(\lambda)$ and pointwise, λ-a.e. If $a > \sup_{s \in S} |f(s)|$, we can replace each function f_n by the function g_n defined by:

$$g_n(s) = f_n(s) \text{ if } |f_n(s)| \leq a \text{ and } f_n(s) = 0 \text{ if } |f_n(s)| > a.$$

Then $g_n \in \mathcal{S}_D(\mathcal{R})$ and we still have $g_n \to f$, λ-a.e.; therefore $g_n \to f$, m-a.e. We can apply now Lebesgue's Theorem 37 and deduce that $\tilde{m}(g_n - f) \to 0$; therefore $\mathcal{S}_D(\mathcal{R})$ is dense in the set $\mathcal{F}_D(\mathcal{B}, \tilde{m}_{F,G})$. ∎

G. Relationship between the spaces $\mathcal{F}_D(m)$

In this section we state some relationship properties of different $\mathcal{F}_D(m)$-type spaces.

45. Proposition. *Assume the embedding $E \subset L(F, G)$ is an isometry and m is σ-additive. Then:*
a) *For any real-valued, Σ-measurable function $\varphi : S \to \mathbb{R}$ we have*

$$\tilde{m}_{\mathbb{R}, E}(\varphi) \leq \tilde{m}_{F,G}(\varphi) \leq \infty.$$

b) $\mathcal{F}_{\mathbb{R}}(m_{F,G}) \subset \mathcal{F}_{\mathbb{R}}(m_{\mathbb{R}, E})$.

Proof. Let $\varphi : S \to \mathbb{R}$ be a Σ-measurable function, $s : S \to \mathbb{R}$ be a $\mathcal{D}$-step function such that $|s| \leq |\varphi|$ and $x \in F$ with $|x| \leq 1$. Then $sx : S \to F$ is a $\mathcal{D}$-step function and we have $|sx| \leq |s||x| \leq |\varphi|$. Then

$$\left|(\int s dm)x\right| = \left|\int sx \, dm\right| \leq \tilde{m}_{F,G}(\varphi).$$

Taking the supremum for $x \in F_1$ we get

$$\left|\int s \, dm\right| \leq \tilde{m}_{F,G}(\varphi).$$

Taking the supremum for $|s| \leq |\varphi|$ we obtain

$$\tilde{m}_{\mathbb{R},E}(\varphi) \leq \tilde{m}_{F,G}(\varphi).$$

∎

The following proposition is similar to Corollary 1.37 and Theorem 2.26.

46. Proposition. *Let $x \in F$ and define the measure $mx : \mathcal{D} \to G$ by*

$$(mx)(A) = m(A)x, \text{ for } A \in \mathcal{D}.$$

Then
a) For every $z \in Z$, the measure $(mx)_z$ is σ-additive, mx has finite semivariation $(mx)^{\sim}_{\mathbb{R},G}$, and for every positive, Σ-measurable function $\varphi : S \to \mathbb{R}_+$ we have

$$(mx)^{\sim}_{\mathbb{R},G}(\varphi) \leq |x|\tilde{m}_{F,G}(\varphi)$$

and

$$\tilde{m}_{F,G}(\varphi x) = |x|\tilde{m}_{F,G}(\varphi).$$

If $Z = G^*$, then mx is σ-additive.
b) $\mathcal{F}_{\mathbb{R}}(\tilde{m}_{F,G}) \subset \mathcal{F}_{\mathbb{R}}((mx)^{\sim}_{\mathbb{R},G})$ and $x\mathcal{F}_{\mathbb{R}}(\tilde{m}_{F,G}) \subset \mathcal{F}_F(\tilde{m}_{F,G})$.
c) Assume $S = \bigcup_{n \in \mathbb{N}} S_n$ with $S_n \in \mathcal{D}$.
If $\varphi \in \mathcal{F}_{\mathbb{R}}(\tilde{m}_{F,G})$, then

$$\int \varphi x \, dm = \int \varphi \, d(mx).$$

d) Assume $S = \bigcup_{n \in \mathbb{N}} S_n$ with $S_n \in \mathcal{D}$ and m is σ-additive. If $\varphi \in \mathcal{F}_{\mathbb{R}}(\tilde{m}_{\mathbb{R},E})$ and $\int \varphi \, dm \in E$, then

$$(\int \varphi \, dm)x = \int \varphi x \, dm = \int \varphi \, d(mx)$$

and

$$\langle (\int \varphi \, dm)x, z \rangle = \langle \int \varphi x \, dm, z \rangle = \int \varphi \, d\langle mx, z \rangle, \text{ for } z \in Z.$$

Proof. To prove assertion a), let $z \in Z$ and $A \in \mathcal{D}$. Then

$$|(mx)_z(A)| = |\langle m(A)x, z \rangle| = |\langle x, m_z(A) \rangle|$$
$$\leq |x| \, |m_z(A)| \leq |x| \, |m_z|(A),$$

hence

$$|(mx)_z| \leq |x| \, |m_z|.$$

It follows that $(mx)_z$ is σ-additive and has finite variation $|(mx)_z|$. Then, for every Σ-measurable function $\varphi \geq 0$ we have

$$\int \varphi d|(mx)_z| \leq |x| \int \varphi d|m_z| \leq |x|\, \tilde{m}_{F,G}(\varphi).$$

Taking the supremum for $z \in Z_1$ we obtain

$$(mx)^{\sim}_{\mathbb{R},G}(\varphi) \leq |x|\tilde{m}_{F,G}(\varphi) \leq \infty.$$

In particular, taking $\varphi = \varphi_A$ with $A \in \mathcal{D}$, we have

$$(mx)^{\sim}_{\mathbb{R},G}(A) \leq |x|\tilde{m}_{F,G}(A) < \infty,$$

hence mx has finite semivariation $(mx)^{\sim}_{\mathbb{R},G}$. If $Z = G^*$, then by the Pettis Theorem 3.19, mx is σ-additive.

The second equality of assertion a) is immediate. From assertion a) we deduce assertion b).

To prove assertion c), assume first $\varphi = \varphi_A$ with $A \in \mathcal{D}$. Then

$$\left(\int \varphi dm\right)x = m(A)x = \int \varphi x \, dm$$

and

$$m(A)x = \int \varphi d(mx),$$

hence, the equalities in both assertions c) and d) hold if φ is a $\mathcal{D}$-step function.

Assume now $S = \bigcup_{n \in \mathbb{N}} S_n$ with $S_n \in \mathcal{D}$ and let $\varphi \in \mathcal{F}_{\mathbb{R}}(\tilde{m}_{F,G})$. There is a sequence (φ_n) of real-valued $\mathcal{D}$-step functions such that $\varphi_n \to \varphi$ and $|\varphi_n| \leq |\varphi|$, m-a.e. By b) we have $\varphi \in L^1_{\mathbb{R}}((mx)_z)$ for $z \in Z_1$; we can apply Lebesgue's Theorem and deduce that

$$\left\langle \int \varphi_n d(mx), z \right\rangle \to \left\langle \int \varphi d(mx), z \right\rangle.$$

Similarly, since $\varphi x \in \mathcal{F}_F(\tilde{m}_{F,G})$, we have $\varphi x \in L^1_F(m_z)$ for $z \in Z$ and

$$\left\langle \int \varphi_n x\, dm, z \right\rangle \to \left\langle \int \varphi x\, dm, z \right\rangle.$$

For each n we have

$$\int \varphi_n x\, dm = \int \varphi_n d(mx).$$

It follows that

$$\int \varphi x\, dm = \int \varphi d(mx)$$

and assertion c) is proved.

§5. INTEGRATION WITH RESPECT TO A MEASURE WITH FINITE SEMIVARIATION 95

To prove assertion d), assume $S = \bigcup_{n \in N} S_n$ with $S_n \in \mathcal{D}$ and m is σ-additive. Then m_{x^*} is σ-additive for every $x^* \in E^*$.

Let $\varphi \in \mathcal{F}_{\mathbb{R}}(\tilde{m}_{\mathbb{R},E})$ with $\int \varphi dm \in E$. Then using the above sequence (φ_n), for each $x^* \in E^*$ we apply Lebesgue's Theorem in the space $L^1(x^*m)$ and deduce that

$$\int \varphi_n d(x^*m) \to \int \varphi d(x^*m),$$

that is,

$$\langle \int \varphi_n dm, x^* \rangle \to \langle \int \varphi dm, x^* \rangle.$$

We choose now $x^* \in E^*$ defined by

$$x^*(e) = \langle x(e), z \rangle, \text{ for } e \in E \text{ and } z \in G^*.$$

Since $\int \varphi dm \in E$, the above convergence can be written

$$\langle (\int \varphi_n dm)x, z \rangle \to \langle (\int \varphi dm)x, z \rangle.$$

By the above we have

$$\langle \int \varphi_n x dm, z \rangle \to \langle \int \varphi x dm, z \rangle.$$

For each n we have

$$\langle (\int \varphi_n dm)x, z \rangle = \langle \int \varphi_n x dm, z \rangle.$$

It follows that

$$(\int \varphi dm)x = \int \varphi x dm,$$

and the first equalities in assertion d) are proved.

Replacing m with mx in assertion c) and considering $G \subset L(G^*, \mathbb{R})$, we obtain for every $z \in G^*$,

$$\langle \int \varphi d(mx), z \rangle = \int \varphi d\langle mx, z \rangle,$$

and the second equalities in assertion d) follow. ∎

H. The indefinite integral of measures with finite semivariation

The definition of the measure gm is the same as for measures with finite variation (Definition 2.27).

47. Definition. Let $g \in \mathcal{F}_{F,G}(m)$. We denote by $gm : \Sigma \to Z^*$ the additive measure defined by

$$(gm)(A) = \int_A gdm, \text{ for } A \in \Sigma$$

and we call it the indefinite integral of g with respect to m, or the measure with density g and base m, or the product of g and m.

Remark. The measure gm is not necessarily σ-additive. But for each $z \in Z$ and $A \in \Sigma$ we have

$$(gm)_z(A) = \langle z, (gm)(A) \rangle = \langle z, \int_A gdm \rangle = \int_A gdm_z.$$

Since m_z is σ-additive and $g \in L^1_F(m_z)$, the last integral is σ-additive as a function of $A \in \Sigma$; consequently the measure $(gm)_z : \Sigma \to \mathbb{R}$ is σ-additive. It follows also that

$$(gm)_z = gm_z.$$

The following proposition shows that if $\int_A gdm \in G$ for every $A \in \Sigma$, then gm is σ-additive. It also gives sufficient conditions for the integral to belong to G, rather than to Z^* or G^{**}.

48. Proposition. a) Assume $\mathcal{D} = \Sigma$ and that the set of measure m_{F,G^*} is uniformly σ-additive. Then

$$\int gdm \in G, \text{ for every } g \in \mathcal{F}_{F,G}(\mathcal{B}, m).$$

a') Assume $\mathcal{D} = \Sigma$ and m is σ-additive. Then

$$\int gdm \in E, \text{ for every } g \in \mathcal{F}_{\mathbb{R},E}(\mathcal{B}, m).$$

b) Assume $c_o \not\subset G$ and $S = \bigcup S_n$ with $S_n \in \mathcal{D}$. Then

$$\int gdm \in G, \text{ for every } g \in \mathcal{F}_{F,G}(m).$$

b') Assume $c_0 \not\subset E$ and $S = \bigcup S_n$ with $S_n \in \mathcal{D}$. Then

$$\int gdm \in E, \text{ for every } g \in \mathcal{F}_{\mathbb{R},E}(m).$$

c) If $g \in \mathcal{F}_{F,G}(m)$ and $\int_A gdm \in G$ for every $A \in \Sigma$, then the measure gm is σ-additive.

c') If $g \in \mathcal{F}_{\mathbb{R},E}(m)$ and $\int_A gdm \in E$ for every $A \in \Sigma$, then the measure gm is σ-additive.

§5. INTEGRATION WITH RESPECT TO A MEASURE WITH FINITE SEMIVARIATION 97

Proof. Assume the hypothesis of assertion a). By Proposition 43 b), the set $\mathcal{S}_F(\Sigma)$ is dense in $\mathcal{F}_{F,G}(\mathcal{B}, m)$. Since $\int g\, dm \in G$ for $g \in \mathcal{S}_F(\Sigma)$ and since the integral is continuous, it follows that $\int g\, dm \in G$ for every $g \in \mathcal{F}_{F,G}(\mathcal{B}, m)$.

Assertion a') follows from a) since the set of measures $m_{\mathbb{R}, E^*}$ is uniformly σ-additive.

To prove b), assume $c_0 \not\subset G$ and $S = \bigcup S_n$ with $S_n \in \mathcal{D}$. Assume first g is a σ-step function from $\mathcal{F}_{F,G}(m)$, of the form

$$g = \sum_{1 \leq n < \infty} \varphi_{A_n} x_n$$

with $A_n \in \mathcal{D}$ mutually disjoint and $x_n \in F$.

Let $z \in G^*$. Since $g \in L^1_F(|m_z|)$, we have

$$\sum_n |\langle m(A_n) x_n, z \rangle| \leq \sum_n |m_z|(A_n)|x_n|$$
$$= \int |g|\, d|m_z| < \infty,$$

hence the series $\sum m(A_n)x_n$ is weakly unconditionally convergent. Since $c_0 \not\subset G$, by the Bessaga–Pelczynski theorem, the series $\sum m(A_n)x_n$ converges to an element $y \in G$. Thus

$$\langle \int g\, dm, z \rangle = \int g\, dm_z = \sum_n m_z(A_n) x_n$$
$$= \langle \sum_n m(A_n) x_n, z \rangle = \langle y, z \rangle,$$

for every $z \in G^*$; consequently $\int g\, dm = y \in G$.

Let now $g \in \mathcal{F}_{F,G}(m)$ with $B = \{g \neq 0\} \in \mathcal{D}$. We can assume g is Σ-measurable. Then there is a sequence (g_n) of F-valued, Σ-measurable, σ-step functions such that $g_n \to g$ uniformly and $|g_n| \leq |g|$. From this last inequality and from the assumption that $B = \{g \neq 0\} \in \mathcal{D}$, it follows that each g_n vanishes outside a set $B \in \mathcal{D}$ with $\tilde{m}_{F,G}(B) < \infty$. By Theorem 35, we have $g_n \to g$ in $\mathcal{F}_{F,G}(m)$; therefore $\int g_n\, dm \to \int g\, dm$, consequently $\int g\, dm \in G$.

If $g \in \mathcal{F}_{F,G}(m)$ is arbitrary, let (S_n) be a sequence of disjoint sets from $\mathcal{D}$ with union S. Then $f = \sum_{1 \leq n < \infty} f \varphi_{S_n}$. Let $z \in G^*$. Then $f \in L^1_F(|m_z|)$ and

$$\sum_n |\langle \int f \varphi_{S_n}\, dm, z \rangle| = \sum_n |\int f \varphi_{S_n}\, dm_z|$$
$$\leq \sum_n \int |f| \varphi_{S_n}\, d|m_z| = \int |f|\, d|m_z| < \infty,$$

that is, the series $\sum \int f\varphi_{S_n} dm$ is weakly unconditionally convergent. By the Bessaga–Pelczynski theorem, the series is convergent and has a sum $y \in G$. Then

$$\langle \int f dm, z \rangle = \int f dm_z = \sum_n \int f\varphi_{S_n} dm_z$$
$$= \sum_n \langle \int f\varphi_{S_n} dm, z \rangle = \langle \sum_n \int f\varphi_{S_n} dm, z \rangle = \langle y, z \rangle,$$

consequently, $\int f dm = y \in G$. This proves assertion b). Assertion b') is a particular case of b), with $F = \mathbb{R}$.

To prove assertion c), let $g \in \mathcal{F}_{F,G}(m)$ such that $\int_A g dm \in G$ for every $A \in \Sigma$. For each $z \in G^*$, the measure m_z is σ-additive and has finite variation $|m_z|$, hence gm_z is σ-additive and

$$\langle \int_A g dm, z \rangle = \int_A g dm_z, \text{ for } A \in \Sigma.$$

It follows that gm is weakly σ-additive. By the Pettis Theorem 3.19, gm is strongly σ-additive. Assertion c') is a particular case of assertion c). ∎

We prove first the associativity formula in case the density is real valued.

49. Theorem. *Assume m is σ-additive and $S = \bigcup_n S_n$ with $S_n \in \mathcal{D}$. Let $\varphi \in \mathcal{F}_\mathbb{R}(\tilde{m}_{F,G}) \cap \mathcal{F}_\mathbb{R}(\tilde{m}_{\mathbb{R},E})$ and assume $\int_A \varphi dm \in E$ for every $A \in \Sigma$. Then:*
a) *The measure φm is σ-additive and has finite semivariation $(\varphi m)\tilde{}_{F,G}$ on Σ;*
b) *If $f \geq 0$ is Σ-measurable, then*

$$(\varphi m)\tilde{}_{\mathbb{R},E}(f) = \tilde{m}_{\mathbb{R},E}(\varphi f)$$

and

$$(\varphi m)\tilde{}_{F,G}(f) = \tilde{m}_{F,G}(\varphi f);$$

c) *We have $f \in \mathcal{F}_{F,G}((\varphi m)^\sim)$ iff $f\varphi \in \mathcal{F}_{F,G}(\tilde{m})$ and in this case we have*

$$\int f d(\varphi m) = \int f\varphi dm$$

and the associativity formula

$$f(\varphi m) = (f\varphi)m.$$

d) *Suppose m_{F,G^*} is uniformly σ-additive on $\mathcal{D}$ and $\varphi \in \mathcal{F}_\mathbb{R}(\mathcal{B}, \tilde{m}_{F,G})$. Then $(\varphi m)_{F,G^*}$ is uniformly σ-additive on $\mathcal{D}$.*

Conversely, if $(\varphi m)_{F,G^*}$ is uniformly σ-additive on Σ then $\varphi \in \mathcal{F}_\mathbb{R}(\mathcal{B}, \tilde{m}_{F,G})$.

§5. INTEGRATION WITH RESPECT TO A MEASURE WITH FINITE SEMIVARIATION

Proof. Denote $n = \varphi m$. The measure n is σ-additive by Proposition 48, using the hypothesis $\int_A \varphi dm \in E$ for $A \in \Sigma$. The fact that $\tilde{n}_{F,G}$ is finite on Σ will follows from assertion b).

To prove b), let $z \in G^*$, $y \in F$, and $A \in \Sigma$. Then using Proposition 46 d) for the measures m and m_z we get

$$\langle y, n_z(A) \rangle = \langle n(A)y, z \rangle = \langle (\int 1_A \varphi dm) y, z \rangle$$

$$= \langle \int 1_A \varphi y dm, z \rangle = \int 1_A \varphi y dm_z = \langle y, \int_A \varphi dm_z \rangle,$$

therefore

$$n_z(A) = \int_A \varphi dm_z, \text{ for } A \in \Sigma,$$

that is, $n_z = \varphi m_z$. By Theorem 2.29 we have

$$|n_z|(A) = \int_A |\varphi| d|m_z|, \text{ for } A \in \Sigma.$$

If $f \geq 0$ is Σ-measurable, then by Theorem 2.30 we have

$$\int f d|n_z| = \int |f\varphi| d|m_z|.$$

Taking the supremum for $z \in G_1^*$ we obtain the second equality in assertion b):

$$\tilde{n}_{F,G}(f) = \tilde{m}_{F,G}(f\varphi).$$

The first equality in assertion b) follows by taking $F = \mathbb{R}$ and $G = E$.

To prove c), let $f : S \to F$ be a Σ-measurable function. By Theorem 2.31 c) and from the equality $n_z = \varphi m_z$ proved above, it follows that f is $|n_z|$-integrable iff $f\varphi$ is $|m_z|$-integrable and in this case we have

$$\int f dn_z = \int f\varphi dm_z$$

that is,

$$\langle \int f dn, z \rangle = \langle \int f\varphi dm, z \rangle.$$

It follows that $f \in \mathcal{F}_{F,G}(\tilde{n})$ iff $f\varphi \in \mathcal{F}_{F,G}(\tilde{m})$ and in this case

$$\int f dn = \int f\varphi dm.$$

Replacing f with $f\varphi_A$ with $A \in \Sigma$, we get the equality $f(\varphi m) = (f\varphi)m$.

To prove assertion d), assume m_{F,G^*} is uniformly σ-additive and that $\varphi \in \mathcal{F}_{\mathbb{R}}(\mathcal{B}, \tilde{m}_{F,G})$. If we take $f = \varphi_A$ with $A \in \Sigma$ in the second equality of assertion b) we get

$$(\varphi m)\widetilde{_{F,G}}(A) = \tilde{m}_{F,G}(\varphi_A \varphi).$$

Let (A_n) be a decreasing sequence from $\mathcal{D}$ with $A_n \downarrow \phi$. Since m_{F,G^*} is uniformly σ-additive on $\mathcal{D}$, by Theorem 4.23, we have $\tilde{m}_{F,G}(A_n) \to 0$; therefore, by Proposition 29, $\tilde{m}_{F,G}(\varphi_{A_n}\varphi) \to 0$, consequently $(\varphi m)\tilde{}_{F,G}(A_n) \to 0$. Then, by Theorem 4.23, the set $(\varphi m)_{F,G^*}$ is uniformly σ-additive on $\mathcal{D}$.

Conversely, assume $(\varphi m)_{F,G^*}$ is uniformly σ-additive on Σ. Then, for every decreasing sequence (A_n) from Σ with $A_n \downarrow \phi$ we have $(\varphi m)\tilde{}_{F,G}(A_n) \to 0$ (Theorem 4.23). By the above equality, we have $\tilde{m}_{F,G}(\varphi_{A_n}\varphi) \to 0$. Then, by Proposition 40, we have $\varphi \in \mathcal{F}_{\mathbb{R}}(\mathcal{B}, \tilde{m}_{F,G})$. ∎

Next we state a theorem concerning the associativity formula in case the density is vector-valued.

50. Theorem. *Assume m is σ-additive and $S = \bigcup_n S_n$, with $S_n \in \mathcal{D}$. Let $f \in \mathcal{F}_{F,G}(\tilde{m})$ and assume $\int_A f dm \in G$ for every $A \in \Sigma$. Then*
a) *The measure fm is σ-additive and has finite semivariation $(fm)\tilde{}_{\mathbb{R},G}$ on Σ;*
b) *If $\varphi \geq 0$ is Σ-measurable, then*

$$(fm)\tilde{}_{\mathbb{R},G}(\varphi) \leq \tilde{m}_{F,G}(\varphi f).$$

If $f \in \mathcal{F}_F(\tilde{m}_{\mathbb{R},E}) \cap \mathcal{F}_F(\tilde{m}_{F,G})$, then

$$(|f|m)\tilde{}_{F,G}(\varphi) = \tilde{m}_{F,G}(\varphi f)$$

c) *If φ is real-valued and Σ-measurable and if $\varphi f \in \mathcal{F}_{F,G}(m)$, then $\varphi \in \mathcal{F}_{\mathbb{R},G}(fm)$ and we have*

$$\int \varphi d(fm) = \int \varphi fm$$

and

$$\varphi(fm) = (\varphi f)m.$$

Proof. The proof is similar to that of Theorem 49, interchanging f and φ. In the proof of assertion b) we take $y \in \mathbb{R}$ and from the equality

$$n_z(A) = \int_A f dm_z, \text{ for } A \in \Sigma$$

we can only deduce the inequality

$$|n_z|(A) \leq \int_A |f|d|m_z|, \text{ for } A \in \Sigma,$$

therefore

$$\int \varphi d|n_z| \leq \int_A |\varphi f|d|m_z|$$

for $\varphi \geq 0$, Σ-measurable. It follows then that

$$\tilde{m}_{\mathbb{R},G}(\varphi) \leq \tilde{m}_{F,G}(f\varphi).$$

The equality
$$(|f|m)\widetilde{_{F,G}}(\varphi) = \tilde{m}_{F,G}(f\varphi)$$
follows from Theorem 49 b). ∎

§6. STRONG ADDITIVITY

Strong additivity will be used in the next paragraph to prove extension theorems.

1. Definition. *A set function $m : \mathcal{R} \to E$ defined on a ring $\mathcal{R}$ is said to be a strongly additive measure, if it is additive and if for any sequence (A_n) of disjoint sets from $\mathcal{R}$, the series $\sum m(A_n)$ is convergent in E (or, equivalently, $\lim_n \sum_{i>n} m(A_i) = 0$).*

m is said to be locally strongly additive on $\mathcal{R}$, if for every $A \in \mathcal{R}$, the restriction of m to the ring $\mathcal{R} \cap A$ is strongly additive, i.e., for every sequence (A_n) of disjoint sets from $\mathcal{R}$ with union contained in a set of $\mathcal{R}$, the series $\sum m(A_n)$ is convergent.

A set K of additive measures $m : \mathcal{R} \to E$ on a ring $\mathcal{R}$ is said to be uniformly (resp. locally uniformly) strongly additive, if for any sequence (A_n) of disjoint sets from $\mathcal{R}$ (resp. with union contained in a set of $\mathcal{R}$), the series $\sum m(A_n)$ is convergent in E, uniformly for $m \in K$, or, equivalently,

$$\lim_n \sum_{i>n} m(A_i) = 0, \text{ uniformly for } m \in K.$$

Strongly additive measures have many properties of σ-additive measures.

We list some of the properties of strongly additive measures that will be used in the sequel. For some of the proofs, the reader is referred to [D–U].

2. *An additive measure $m : \mathcal{R} \to E$ on a ring $\mathcal{R}$ is strongly additive, iff for every sequence (A_n) of disjoint sets from $\mathcal{R}$ we have $m(A_n) \to 0$.*

3. *An additive measure $m : \mathcal{R} \to E$ on a ring $\mathcal{R}$ is strongly additive, iff for any increasing (resp. decreasing) sequence (A_n) from $\mathcal{R}$, the limit $\lim_n m(A_n)$ exists in E.*

4. *Any bounded (resp. finite), positive, additive measure on a ring is strongly additive (resp. locally strongly additive).*

5. *Any additive measure with bounded (resp. finite) variation on a ring is strongly additive (resp. locally strongly additive).*

6. *Any real-valued, bounded, additive measure on a ring is strongly additive.*

In fact, such a measure has bounded variation (Proposition 2.16).

7. *A σ-additive measure on a σ-ring (resp. δ-ring) is strongly additive (resp. locally strongly additive).*

A σ-additive measure on a ring or on a δ-ring is not necessarily strongly additive, even if it is bounded.

8. Theorem. *Assume $c_0 \not\subset E$. If $m : \mathcal{R} \to E$ is a bounded (resp. locally bounded), additive measure on a ring $\mathcal{R}$, then m is strongly additive (resp. locally strongly additive).*

Proof. Assume m is bounded on $\mathcal{R}$. Let (A_n) be a sequence of disjoint sets from $\mathcal{R}$ and show that the series $\sum m(A_n)$ is convergent in E.

For each $x^* \in E^*$, the scalar measure x^*m is bounded on $\mathcal{R}$, hence it has bounded variation $|x^*m|$, (Proposition 2.16). Thus

$$\sum_{1 \leq i \leq n} |x^*m(A_i)| \leq |x^*m|(\bigcup_{1 \leq i \leq n} A_i)$$
$$\leq 2\sup\{|x^*m|(B) : B \in \mathcal{R}\} < \infty;$$

hence the series $\sum_{1 \leq i < \infty} x^*m(A_i)$ is unconditionally convergent. Since $c_0 \not\subset E$, by the Bessaga–Pelczynski theorem [B–P], the series $\sum_{1 \leq i < \infty} m(A_i)$ is convergent, thus m is strongly additive. ∎

The converse of Theorem 8 is valid for any Banach space E:

9. Proposition. *Any strongly additive measure on an algebra is bounded. Any locally strongly additive measure on a ring is locally bounded.*

The proof is the same as that of Theorem 2.14, for σ-additive measures on a σ-ring or an a δ-ring.

10. Proposition. *If $m : \mathcal{R} \to E$ is a strongly additive measure on a ring $\mathcal{R}$, then there exists a sequence (A_n) from $\mathcal{R}$ such that $m(B) = 0$ for any set $B \in \mathcal{R}$ disjoint from the union $\bigcup_n A_n$.*

If $m : \mathcal{S} \to E$ is a strongly additive measure on a σ-ring $\mathcal{S}$, then there is a set $S_0 \in \mathcal{S}$ such that m vanishes outside S_0.

The proof is the same as that of Theorem 3.1 for σ-additive measures on a σ-ring.

The main part of the following theorem is that strong additivity of m implies uniform strong additivity of the set $m_{\mathbb{R},E^*} = \{|x^*m| : x^* \in E_1^*\}$ of positive measures.

11. Theorem. *Let K be a set of additive measures $m : \mathcal{R} \to E$ on a ring $\mathcal{R}$ and $Z \subset E^*$ a norming space for E.*

The first six assertions below are equivalent and each of them implies assertion (vii).

If $\mathcal{R}$ is an algebra or a σ-ring, then all seven assertions below are equivalent.
(i) *K is uniformly strongly additive.*
(ii) *The set $\{x^*m : m \in K, x^* \in Z_1\}$ is uniformly strongly additive.*
(iii) *For any sequence (A_n) of disjoint sets from $\mathcal{R}$ we have $m(A_n) \to 0$, uniformly for $m \in K$.*
(iv) *For any sequence (A_n) of disjoint sets from $\mathcal{R}$ we have $\tilde{m}_{\mathbb{R},E}(A_n) \to 0$, uniformly for $m \in K$.*
(v) *The set of variations $\{|x^*m| : m \in K, x^* \in Z_1\}$ is uniformly strongly additive.*

(vi) *For every increasing sequence (A_n) of sets from $\mathcal{R}$, the limit $\lim_n m(A_n)$ exists, uniformly for $m \in K$.*
(vii) *For every decreasing sequence (A_n) of sets from $\mathcal{R}$, the limit $\lim_n m(A_n)$ exists, uniformly for $m \in K$.*

Proof. The equivalence of (i) and (ii) is immediate. The implication (i) $\Longrightarrow$ (iii) is evident. To prove the implication (iii) $\Longrightarrow$ (iv), assume (iv) is not true. Then there is a sequence (A_n) of disjoint sets from $\mathcal{R}$ and an $\varepsilon > 0$ such that $\tilde{m}_{\mathbb{R},E}(A_n) > \varepsilon$ for all $m \in K$ and for all n. Let $a > 2$. By Proposition 4.14, for each n there is a set $B_n \in \mathcal{R}$ with $B_n \subset A_n$ and

$$\tilde{m}_{\mathbb{R},E}(A_n) < a|m(B_n)|, \text{ for all } m \in K.$$

The sets B_n are mutually disjoint and $|m(B_n)| > \varepsilon/a$ for every $m \in K$ and for every n, which contradicts (iii). Therefore (iii) $\Longrightarrow$ (iv).

To prove the implication (iv) $\Longrightarrow$ (v), assume (v) is not satisfied. Then there is a sequence (A_n) of disjoint sets from $\mathcal{R}$ and an $\varepsilon > 0$ such that for every $n \in \mathbb{N}$ we have

$$\sup_{m \in K} \sup_{|x^*| \leq 1} \sum_{i \geq n} |x^*m|(A_i)| > \varepsilon.$$

Then, for each $n \in \mathbb{N}$, there is an $m(n) \in \mathbb{N}$ such that

$$\sup_{m \in K} \sup_{|x^*| \leq 1} \sum_{n \leq i < m(n)} |x^*m|(A_i) > \varepsilon,$$

hence

$$\sup_{m \in K} \sup_{|x^*| \leq 1} |x^*m|(\bigcup_{n \leq i < m(n)} A_i) > \varepsilon.$$

If we set $B_n = \bigcup_{n \leq i < m(n)}(A_i)$, then $B_n \in \mathcal{R}$ and

$$\sup_{m \in K} \tilde{m}_{\mathbb{R},E}(B_n) = \sup_{m \in K} \sup_{|x^*| \leq 1} |x_n^*m|(B_n) > \varepsilon.$$

If we take $n_1 = 1$, $n_2 = m(n_1)$, $n_3 = m(n_2), \ldots$, then the sequence $(B_{n_k})_k$ consists of mutually disjoint sets from $\mathcal{R}$ and

$$\sup_{m \in K} \tilde{m}_{\mathbb{R},E}(B_{n_k}) > \varepsilon,$$

which contradicts (iv). Therefore (iv) $\Longrightarrow$ (v). The implication (v) $\Longrightarrow$ (i) is evident.

Assume now (i) and prove (vi). Let (A_n) be an increasing sequence of sets from $\mathcal{R}$. Denote $B_1 = A_1$ and $B_n = A_n - A_{n-1}$ for $n \geq 2$. Then B_n are mutually disjoint sets from $\mathcal{R}$ and $A_n = \bigcup_{1 \leq i \leq n} B_i$; therefore

$$m(A_n) = \sum_{1 \leq i \leq n} m(B_i).$$

Since K is uniformly strongly additive, the series
$$\sum_{1\leq i\leq\infty} m(B_i) = \lim_n \sum_{1\leq i\leq n} m(B_i)$$
is convergent, uniformly for $m \in K$. It follows that $\lim_n m(A_n)$ exists, uniformly for $m \in K$, which is assertion (vi).

Assume now assertion (vi) and prove (i). Let (A_n) be a sequence of disjoint sets from $\mathcal{R}$. Denote $B_n = \bigcup_{i\leq n} A_i$. Then (B_n) is an increasing sequence of sets from $\mathcal{R}$ and we have $\sum_{1\leq i\leq n} m(A_i) = m(B_n)$. By assumption (vi), $\lim_n m(B_n)$ exists, uniformly for $m \in K$. Then $\lim_n \sum_{i\leq n} m(A_i) = \sum_{1\leq i<\infty} m(A_i)$ exists, uniformly for $m \in K$ and this is assertion (i). We proved that assertions (i)–(vi) are equivalent.

Assume now (vi) and prove (vii). Let (A_n) be a decreasing sequence from $\mathcal{R}$. Denote $B_n = A_1 - A_n$. Then (B_n) is an decreasing sequence from $\mathcal{R}$ and we have $m(B_n) = m(A_1) - m(A_n)$. By assumption (vi), $\lim_n m(B_n)$ exists, uniformly for $m \in K$. Then $\lim_n m(A_n)$ exists, uniformly for $m \in K$, which is assertion (vii).

Finally, assume that $\mathcal{R}$ is an algebra or a σ-ring and that assertion (vii) is true and prove assertion (vi). Let (A_n) be a increasing sequence from $\mathcal{R}$. There is a set $A \in \mathcal{R}$ such that $A_n \subset A$ for every n. In fact, if $\mathcal{R}$ is an algebra, we take $A = S$; if $\mathcal{R}$ is a σ-ring, we take $A = \bigcup_n A_n$. Denote $B_n = A - A_n$. Then (B_n) is decreasing sequence from $\mathcal{R}$ and we have $m(B_n) = m(A) - m(A_n)$. Since, by assumption (vii), $\lim_n m(B_n)$ exists, uniformly for $m \in K$, it follows that $\lim_n m(A_n)$ exists, uniformly for $m \in K$, which is assertion (vi). ∎

For a family consisting of one measure we have the following corollary:

12. Corollary. *Let $m : \mathcal{R} \to E$ be an additive measure on a ring $\mathcal{R}$ and $Z \subset E^*$ a norming space for E.*

The first six assertions below are equivalent and each of them implies assertion (vii).

If $\mathcal{R}$ is an algebra or a σ-ring, then all seven assertions below are equivalent.
(i) *m is strongly additive.*
(ii) *The set of measures $\{x^*m : x^* \in Z_1\}$ is uniformly strongly additive.*
(iii) *$m(A_n) \to 0$ for any sequence (A_n) of disjoint sets from $\mathcal{R}$.*
(iv) *$\tilde{m}_{\mathbb{R},E}(A_n) \to 0$ for any sequence (A_n) of disjoint sets from $\mathcal{R}$.*
(v) *The set of variations $m_{\mathbb{R},Z} = \{|x^*m| : x^* \in Z_1\}$ is uniformly strongly additive.*
(vi) *For every increasing sequence (A_n) of sets from $\mathcal{R}$, the limit $\lim_n m(A_n)$ exists.*
(vii) *For every decreasing sequence (A_n) of sets from $\mathcal{R}$, the limit $\lim_n m(A_n)$ exists.*

The following theorem will be used in the proof of the extension Theorem 7.7.

13. Theorem. *Let K be a set of σ-additive measures $m: \mathcal{S} \to E$ on a σ-ring $\mathcal{S}$ and let $\mathcal{R}$ be a ring generating $\mathcal{S}$.*

If the restrictions to $\mathcal{R}$ of the measures of K are uniformly strongly additive on $\mathcal{R}$, then K is uniformly σ-additive on $\mathcal{S}$.

Proof. Assume K is uniformly strongly additive on $\mathcal{R}$. By Theorem 11, the set of positive measures $K^* = \{|x^*m| : m \in K, x^* \in Z_1\}$ is uniformly strongly additive on $\mathcal{R}$.

To prove that K is uniformly σ-additive on $\mathcal{S}$, it is enough to prove that K^* is uniformly σ-additive on $\mathcal{S}$. Assume not. Then there is an $\varepsilon > 0$, a sequence (A_n) from $\mathcal{S}$ with $A_n \downarrow \phi$, and a sequence (μ_n) from K^* such that $\mu_n(A_n) > 2\varepsilon$ for each n.

Consider first the measure μ_1 and the set A_1. Since $\mu_1(A_1)$ is equal to the outer measure $\mu_1^*(A_1)$, by the Caratheodory extension process, there is a sequence (B_i) of disjoint sets from $\mathcal{R}$ such that

$$A_1 \subset \bigcup_i B_i \text{ and } \sum_i \mu_1(B_i) < \mu_1(A_1) + \varepsilon/2.$$

Since the sequence (μ_n) is uniformly strongly additive on $\mathcal{R}$, we have

$$\lim_{k \to \infty} \sum_{i > k} \mu_n(B_i) = 0, \text{ uniformly for } n.$$

There is a k_0 such that

$$\sum_{i > k_0} \mu_n(B_i) \leq \varepsilon/2, \text{ for all } n.$$

Set $C_1 = \bigcup_{i \leq k_0} B_i \in \mathcal{R}$. For each $m \geq 1$ we have $A_m \subset A_1 \subset \bigcup_i B_i$, hence

$$A_m = (\bigcup_{i \leq k_0} B_i \cap A_m) \cup (\bigcup_{i > k_0} B_i \cap A_m)$$
$$\subset (C_1 \cap A_m) \cup (\bigcup_{i > k_0} B_i),$$

hence, for every n we have

$$\mu_n(A_m) \leq \mu_n(C_1 \cap A_m) + \sum_{i > k_0} \mu_n(B_i)$$
$$\leq \mu_n(C_1 \cap A_m) + \varepsilon/2.$$

We have then

$$\mu_n(C_1 \cap A_m) \geq \mu_n(A_m) - \varepsilon/2, \text{ for all } n \text{ and } m$$

and

$$\mu_1(A_1) + \varepsilon/2 > \mu_1(C_1).$$

Consider now the measures μ_1, μ_2 and the set A_2. We can find a common sequence (B_i') of disjoint sets from $\mathcal{R}$ such that $E_2 \subset \bigcup_i B_i'$,

$$\sum_i \mu_1(B_i') < \mu_1(E_2) + \varepsilon/4$$

and

$$\sum_i \mu_2(B_i') < \mu_2(E_2) + \varepsilon/4.$$

In fact, we can find two sequences (B_j^1) and (B_k^2) of disjoint sets from $\mathcal{R}$, covering E_2 and such that

$$\sum_j \mu_1(B_j^1) < \mu_1(E_2) + \varepsilon/4$$

and

$$\sum_k \mu_2(B_k^2) < \mu_2(E_2) + \varepsilon/4.$$

The countable family $(B_j^1 \cap B_k^2)_{j,k}$ consists of disjoint sets from $\mathcal{R}$, covers E_2, and we have

$$\sum_{j,k} \mu_1(B_j^1 \cap B_k^2) = \sum_j \sum_k \mu_1(B_j^1 \cap B_k^2)$$
$$= \sum_j \mu_1(B_j^1 \cap (\bigcup_k B_k^2)) \leq \sum_j \mu_1(B_j^1) < \mu_1(E) + \varepsilon/4$$

and, similarly,

$$\sum_{j,k} \mu_2(B_j^1 \cap B_k^2) < \mu_2(E) + \varepsilon/4.$$

We can take (B_i') the family $(B_j^1 \cap B_k^2)_{j,k}$ arranged as a sequence.

Proceeding as in the first part of the proof, since (μ_n) are uniformly strongly additive on $\mathcal{R}$ there is a k_1 such that

$$\sum_{i > k_1} \mu_n(B_i') \leq \varepsilon/4, \text{ for all } n.$$

Set $C_2 = \bigcup_{i \geq k_1} B_i' \in \mathcal{R}$. Then for each $m \geq 2$ we have $A_m \cap C_1 \subset \bigcup_i B_i'$, hence

$$A_m \cap C_1 = (\bigcup_{i \leq k_1} B_i' \cap A_m \cap C_1) \cup (\bigcup_{i > k_1} (B_i' \cap A_m \cap C_1))$$
$$\subset (C_1 \cap A_m \cap C_2) \cup (\bigcup_{i > k_1} B_i'),$$

therefore, for each n we have

$$\mu_n(A_m \cap C_1) \leq \mu_n(A_m \cap C_1 \cap C_2) + \varepsilon/4.$$

We have then, for all $n \geq 1$ and $m \geq 2$,

$$\mu_n(A_m \cap C_1 \cap C_2) \geq \mu_n(A_m \cap C_1) - \varepsilon/4$$
$$\geq \mu_n(A_m) - \varepsilon/2 - \varepsilon/4.$$

We have also $\mu_1(A_2) + \varepsilon/4 > \mu_1(C_2)$ and $\mu_2(A_2) + \varepsilon/4 > \mu_2(C_2)$.

Continuing this way, by induction, we can find a sequence (C_n) from $\mathcal{R}$ such that for all $n \geq 1$ and $m \geq k$ we have

$$\mu_n(A_m \cap C_1 \cap \cdots \cap C_k) \geq \mu_n(A_m) - \sum_{i \leq k} \varepsilon/2^i$$

and

$$\mu_j(A_k) + \varepsilon/2^k > \mu_j(C_k), \text{ for } 1 \leq j \leq k.$$

Since $A_k \downarrow \phi$, from the last inequality we deduce that $\lim_k \mu_j(C_k) = 0$; therefore

$$\lim_k \mu_j(C_1 \cap C_2 \cap \cdots \cap C_k) = 0, \text{ for every } j.$$

Since the sequence (μ_n) is uniformly strongly additive on $\mathcal{R}$ and since the sequence $(C_1 \cap C_2 \cap \cdots \cap C_k)_k$ is decreasing, we have, by Theorem 11,

$$\lim_k \mu_j(C_1 \cap C_2 \cap \cdots \cap C_k) = 0, \text{ uniformly in } j.$$

From the above construction we have

$$0 = \limsup_k {}_j \mu_j(C_1 \cap C_2 \cap \cdots \cap C_k)$$
$$\geq \lim_k \mu_k(C_1 \cap C_2 \cap \cdots \cap C_k)$$
$$\geq \limsup_{k \to \infty} \mu_k(A_k) - \varepsilon$$
$$\geq 2\varepsilon - \varepsilon = \varepsilon > 0$$

and we have reached a contradiction.

It follows that the set K^* of positive measures is uniformly σ-additive on $\mathcal{S}$, hence K is uniformly σ-additive on $\mathcal{S}$. ∎

§7. EXTENSION OF MEASURES

1. We start with the extension of a positive, σ-additive measure $\mu : \mathcal{R} \to [0, +\infty]$ on a ring $\mathcal{R}$. By the Caratheodory procedure, μ has a σ-additive extension $\mu' : \mathcal{S} \to [0, +\infty]$ on the σ-ring $\mathcal{S} = \sigma r(\mathcal{R})$ generated by $\mathcal{R}$; namely, μ' is the restriction to $\mathcal{S}$ of the outer measure μ^*. The extension is unique if μ is σ-finite on $\mathcal{R}$, i.e., if every set A of $\mathcal{R}$ is the union of a sequence (A_n) from $\mathcal{R}$ with $\mu(A_n) < \infty$. If $\mathcal{S}$ is not a σ-algebra, we can perform a second σ-additive extension $\mu'' : \Sigma \to [0, +\infty]$ on the σ-algebra $\Sigma = \sigma a(\mathcal{R})$ generated by $\mathcal{R}$; namely, we set

$$\mu''(B) = \sup\{\mu'(A) : A \in \mathcal{S} \cap B\}.$$

If μ is finite on $\mathcal{R}$, then μ'' has the finite measure property (FMP) (see 1.34). μ'' is the only σ-additive extension of μ to Σ with the FMP.

If $\mathcal{S} = \bigcup_{1 \leq n < \infty} \mathcal{S}_n$ with $\mathcal{S}_n \in \mathcal{R}$, then $\Sigma = \mathcal{S}$ and the second extension is no longer necessary.

We shall continue to denote μ' and μ'' by μ.

If μ is a positive, σ-additive measure on $\mathcal{R}$, it has the following important property:

For every set $A \in \Sigma$ with $\mu(A) < \infty$ and for every $\varepsilon > 0$, there is a set $B \in \mathcal{R}$ with $\mu(A \Delta B) < \varepsilon$.

If μ is *finite* on $\mathcal{R}$, then it is *finite* also on the δ-ring $\mathcal{D} = \delta r(\mathcal{R})$ generated by $\mathcal{R}$.

If μ is *bounded* on $\mathcal{R}$, then it is *bounded* on the σ-algebra $\Sigma = \sigma a(\mathcal{R})$.

If $m : \mathcal{R} \to E$ is a vector-valued, σ-additive measure, it does not necessarily have a σ-additive extension on the δ-ring $\mathcal{D} = \delta r(\mathcal{R})$.

In many cases, a measure is defined initially on a semiring and is first extended to the ring generated by the semiring. This is the case, for example, of a function $g : \mathbb{R} \to E$ and the finitely additive measure m_g defined for each interval of the form $[a, b)$ by

$$m_g([a, b)) = g(b) - g(a).$$

The class of the intervals $[a, b)$ is a semiring $\mathcal{P}$ and we extend m_g to an additive measure on the ring $\mathcal{R} = r(\mathcal{P})$ generated by $\mathcal{P}$. This extension is possible due to the following proposition:

2. Proposition. *Let $m : \mathcal{P} \to E$ (or $\overline{\mathbb{R}}_+$) be a finitely additive measure defined on a semiring $\mathcal{P}$. Then m can be extended uniquely to an additive measure $m' : \mathcal{R} \to E$ (or $\overline{\mathbb{R}}_+$) on the ring $\mathcal{R} = r(\mathcal{P})$ generated by $\mathcal{P}$.*

If m is σ-additive, then m' also is σ-additive.

Proof. Let $A \in \mathcal{R}$ and let $(A_i)_{1 \leq i \leq n}$ and $(B_j)_{1 \leq j \leq m}$ be two families of mutually disjoint sets from $\mathcal{P}$, each of them with union A.

For each i we have
$$A_i = \bigcup_{1 \le j \le m} (A_i \cap B_j)$$
and $(A_i \cap B_j)_{1 \le j \le m}$ is a finite family of mutually disjoint sets from $\mathcal{P}$. Since m is finitely additive we have
$$m(A_i) = \sum_{1 \le j \le m} m(A_i \cap B_j),$$
hence
$$\sum_{1 \le i \le n} m(A_i) = \sum_i \sum_j m(A_i \cap B_j).$$
In the same way we deduce that
$$\sum_{1 \le j \le m} m(B_j) = \sum_j \sum_i m(A_i \cap B_j).$$
It follows that
$$\sum_i m(A_i) = \sum_j m(B_j).$$
We define then,
$$m'(A) = \sum_i m(A_i)$$
and the definition of $m'(A)$ is independent of the particular family (A_i) of disjoint sets from $\mathcal{P}$ with union A.

Evidently, if $A \in \mathcal{P}$, then $m'(A) = m(A)$. From the definition of m', it follows that m' is additive on $\mathcal{R}$.

To prove the uniqueness of the extension m', assume $m'' : \mathcal{R} \to E$ (or $\overline{\mathbb{R}}_+$) is another additive measure extending m. For every set $A \in \mathcal{R}$ and every finite family $(A_i)_{1 \le i \le n}$ of disjoint sets from $\mathcal{P}$ with union A we have
$$m'(A) = \sum_i m'(A_i) = \sum_i m''(A_i) = m''(A),$$
hence $m'' = m'$.

Assume now m is σ-additive on $\mathcal{P}$ and prove that m' is σ-additive on $\mathcal{R}$. Let (A_n) be a sequence of mutually disjoint sets from $\mathcal{R}$ with union $A \in \mathcal{R}$. Each set A_n is of the form
$$A_n = \bigcup_j A_{nj}$$
where $(A_{nj})_j$ is a finite family of disjoint sets from $\mathcal{P}$, therefore
$$m'(A_n) = \sum_j m(A_{nj}).$$

If $A \in \mathcal{P}$, then
$$A = \bigcup_n A_n = \bigcup_{n,j} A_{nj}$$
and the sets $(A_{nj})_{nj}$ are mutually disjoint. Since m is σ-additive on $\mathcal{P}$ we have
$$m(A) = \sum_{nj} m(A_{nj}) = \sum_n \sum_j m(A_{nj}) = \sum_n m'(A_n).$$

If A does not belong to $\mathcal{P}$, then
$$A = \bigcup_k B_k$$
where $(B_k)_k$ is a finite family of disjoint sets from $\mathcal{P}$; therefore
$$m'(A) = \sum_k m(B_k).$$

For each k we have
$$B_k = \bigcup_n (A_n \cap B_k)$$
and $(A_n \cap B_k)_n$ is a sequence of disjoint sets from $\mathcal{R}$. By the above we have then
$$m(B_k) = \sum_n m'(A_n \cap B_k).$$

Similarly, for each n we have
$$A_n = \bigcup_k (A_n \cap B_k)$$
and $(A_n \cap B_k)_k$ is a finite family of disjoint sets from $\mathcal{R}$. As m' is finitely additive, we have
$$m'(A_n) = \sum_k m'(A_n \cap B_k).$$

We deduce that
$$m'(A) = \sum_k m(B_k) = \sum_k \sum_n m'(A_n \cap B_k)$$
$$= \sum_n \sum_k m'(A_n \cap B_k) = \sum_n m'(A_n),$$
therefore m' is σ-additive. ■

We present now some theorems about the extension of a σ-additive measure from a ring $\mathcal{R}$ to the δ-ring $\mathcal{D} = \delta r(\mathcal{R})$ or to the σ-algebra $\Sigma = \sigma a(\mathcal{R})$ generated by $\mathcal{R}$.

First we consider the extension of measures that are absolutely continuous with respect to a positive measure.

3. Theorem. *Let $m : \mathcal{R} \to E$ be a σ-additive measure on a ring $\mathcal{R}$ and assume there is a positive, finite, σ-additive measure $\mu : \mathcal{R} \to \mathbb{R}_+$ such that $m \ll \mu$. Then:*
a) m and μ can be extended uniquely to σ-additive measures m' and μ' respectively, on the δ-ring $\mathcal{D} = \delta r(\mathcal{R})$ generated by $\mathcal{R}$ and we still have $m' \ll \mu'$.
b) If μ is bounded on $\mathcal{R}$, then m and μ can be extended to σ-additive measures m' and μ' respectively on the σ-algebra $\Sigma = \sigma a(\mathcal{R})$ generated by $\mathcal{R}$ and we still have $m' \ll \mu'$.

Proof. The positive measure μ can be extended to a positive, finite, σ-additive measure μ' on the δ-ring $\mathcal{D}$. We define on $\mathcal{D}$ the semimetric $\rho(A, B) = \mu'(A \Delta B)$. Since for every $A \in \mathcal{D}$ and $\varepsilon > 0$, there is $B \in \mathcal{R}$ with $\mu'(A \Delta B) < \varepsilon$, it follows that $\mathcal{R}$ is dense in $\mathcal{D}$ for the semimetric ρ.

For an additive measure $n : \mathcal{R} \to E$ we have $n \ll \mu$ iff n is uniformly continuous on $\mathcal{R}$ for the semimetric ρ.

Since, by hypothesis, we have $m \ll \mu$, we deduce that m is uniformly continuous on $\mathcal{R}$; therefore m can be extended uniquely to a uniformly continuous mapping $m' : \mathcal{D} \to E$.

The functions $f(A, B) = A \cup B$, $g(A, B) = A \cap B$ and $h(A, B) = A - B$ defined on $\mathcal{D} \times \mathcal{D}$ with values in $\mathcal{D}$ are uniformly continuous for the semimetric ρ on $\mathcal{D}$ and the semimetric $\rho + \rho$ on $\mathcal{D} \times \mathcal{D}$. We deduce then that m' is additive on $\mathcal{D}$; and since m' is uniformly continuous, it folows that $m' \ll \mu'$; consequently m' is σ-additive.

If μ is bounded on $\mathcal{R}$, then it can be extended to a positive, bounded, σ-additive measure on on the σ-algebra Σ. We define, as above, the semimetric ρ on Σ and $\mathcal{R}$ is still dense in Σ for ρ. We proceed then as above to extend m to a σ-additive measure $m' : \Sigma \to E$ such that $m' \ll \mu'$. ∎

If m has finite variation $|m|$, we can take $\mu = |m|$ and obtain the following theorem.

4. Theorem. *Let $m : \mathcal{R} \to E$ be a σ-additive measure with finite variation $|m|$. Then*
a) m can be extended uniquely to a σ-additive measure $m' : \mathcal{D} \to E$ with finite variation $|m'|$ on the δ-ring $\mathcal{D} = \delta r(\mathcal{R})$ generated by $\mathcal{R}$ and we have

$$|m'|(A) = |m|(A), \text{ for } A \in \mathcal{R}.$$

b) If m has bounded variation $|m|$ on $\mathcal{R}$, then m can be extended to a σ-additive measure $m' : \Sigma \to E$ with finite variation $|m'|$ on the σ-algebra $\Sigma = \sigma a(\mathcal{R})$ generated by $\mathcal{R}$, and we have

$$|m'|(A) = |m|(A), \text{ for } A \in \mathcal{R}.$$

Proof. Denote by μ the extension of the positive, σ-additive measure $|m|$, from $\mathcal{R}$ to $\mathcal{D}$. We have

$$|m(A)| \leq |m|(A), \text{ for } A \in \mathcal{R},$$

therefore $m \ll \mu$ on $\mathcal{R}$. By Theorem 3, m can be extended to a σ-additive measure $m' : \mathcal{D} \to E$ such that $m' \ll \mu$. We want to prove that

$$|m'(A)| \leq \mu(A), \text{ for } A \in \mathcal{D}.$$

In fact, let $A \in \mathcal{D}$. Since $\mathcal{R}$ is dense in $\mathcal{D}$ for the semimetric $\rho(A, B) = \mu(A \Delta B)$, there is a sequence (A_n) from $\mathcal{R}$ such that $\rho(A_n, A) \to 0$. Since m' and μ are continuous on $\mathcal{D}$ we have $m'(A_n) \to m'(A)$ and $\mu(A_n) \to \mu(A)$, that is, $m(A_n) \to m'(A)$ and $|m|(A_n) \to \mu(A)$. Since for each A_n we have $|m(A_n)| \leq |m|(A_n)$, we deduce that $|m'(A)| \leq \mu(A)$.

From this inequality we deduce that m' has finite variation $|m'|$ on $\mathcal{D}$ and we have $|m'|(A) \leq \mu(A)$, for $A \in \mathcal{D}$.

In particular,

$$|m'|(A) \leq |m|(A), \text{ for } A \in \mathcal{R}.$$

On the other hand,

$$|m(A)| = |m'(A)| \leq |m'|(A), \text{ for } A \in \mathcal{R},$$

therefore

$$|m|(A) \leq |m'|(A), \text{ for } A \in \mathcal{R},$$

consequently $|m'| = |m|$ on $\mathcal{R}$.

If m has bounded variation $|m|$ on $\mathcal{R}$, then $|m|$ can be extended to a positive, bounded, σ-additive measure μ on the σ-algebra $\Sigma = \sigma a(\mathcal{R})$ and the above reasoning remains valid if we replace $\mathcal{D}$ by Σ. ∎

5. Corollary. *Let $\mu : \mathcal{R} \to \mathbb{R}$ be a real valued, σ-additive measure on a ring $\mathcal{R}$.*
a) *If μ is locally bounded on $\mathcal{R}$, then μ can be extended to a σ-additive measure $\mu' : \mathcal{D} \to \mathbb{R}$ with finite variation $|\mu'|$ on the δ-ring $\mathcal{D} = \delta r(\mathcal{R})$ and we have $|\mu'| = |\mu|$ on $\mathcal{R}$.*
b) *If μ is bounded on $\mathcal{R}$, then μ can be extended to a σ-additive measure $\mu' : \Sigma \to \mathbb{R}$ with finite variation $|\mu'|$ on the σ-algebra $\Sigma = \sigma a(\mathcal{R})$ and we have*

$$|\mu'| = |\mu| \text{ on } \mathcal{R}.$$

In fact, a locally bounded (resp. bounded), real-valued, σ-additive measure on a ring has finite (resp. bounded) variation (Proposition 2.16)

6. Corollary. *Let $m : \mathcal{R} \to E \subset L(F, G)$ be a σ-additive measure with finite (resp. bounded) semivariation $\tilde{m}_{F,G}$ and assume m can be extended to a σ-additive measure $m' : \mathcal{D} \to E$ on the δ-ring $\mathcal{D}$ generated by $\mathcal{R}$ (resp. to a σ-additive measure $m' : \Sigma \to E$ on the σ-algebra Σ generated by $\mathcal{R}$).*

Then m' has finite (resp. bounded) semivariation $\tilde{m}'_{F,G}$ and

$$\tilde{m}'_{F,G} = \tilde{m}_{F,G}, \ on \ \mathcal{R}.$$

Proof. Let $z \in G^*$. Then m'_z is an F^*-valued, σ-additive measure on $\mathcal{D}$ (resp. Σ) and is the unique extension of m_z. Since $|m_z| \leq \tilde{m}_{F,G}$, we deduce that m_z has finite (resp. bounded) variation $|m_z|$. Then, by Theorem 4, m'_z has finite (resp. bounded) variation $|m'_z|$ on $\mathcal{D}$ (resp. Σ) and we have

$$|m'_z| = |m_z|, \ on \ \mathcal{R}.$$

By Proposition 4.15, the measure m' has finite (resp. bounded) semivariation $\tilde{m}'_{F,G}$ on $\mathcal{D}$ (resp. Σ). Then, by Proposition 4.13 we have

$$\tilde{m}'_{F,G} = \tilde{m}_{F,G}, \ on \ \mathcal{R}.$$

∎

In the absence of a positive measure, we have the following important extension theorem.

7. Theorem. *Let $m : \mathcal{R} \to E$ be an additive measure on a ring $\mathcal{R}$ and $Z \subset E^*$ a norming space for E.*
a) *Assume m is locally strongly additive on $\mathcal{R}$ and the real measure x^*m is σ-additive for every $x^* \in Z$.*

Then m is σ-additive on $\mathcal{R}$ and can be extended to a σ-additive measure $m' : \mathcal{D} \to E$ on the δ-ring $\mathcal{D} = \delta r(\mathcal{R})$.
b) *Assume m is strongly additive and bounded on $\mathcal{R}$ and x^*m is σ-additive for every $x^* \in Z$.*

Then m can be extended to a σ-additive measure $m' : \Sigma \to E$ on the σ-algebra $\Sigma = \sigma a(\mathcal{R})$.

Proof. We prove first that m is σ-additive on $\mathcal{R}$. Let (A_n) be a decreasing sequence from $\mathcal{R}$ with $A_n \downarrow \phi$. Since m is locally strongly additive, $\lim_n m(A_n) = x$ exists in E. For each $x^* \in Z$, the measure x^*m is σ-additive, hence $\lim_n x^*m(A_n) = 0$. At the same time, $\lim_n x^*m(A_n) = x^*x$. It follows that $x^*x = 0$ for every $x^* \in Z$. Since Z is norming for E, we have $x = 0$; therefore $m(A_n) \to 0$; consequently m is σ-additive on $\mathcal{R}$.

Now we prove assertion b). Since m is strongly additive, there is a sequence (S_n) from $\mathcal{R}$ such that m vanishes outside the union $\bigcup_n S_n$ (Proposition 6.10), so we can consider that $S = \bigcup_n S_n$ and $\Sigma = \sigma r(\mathcal{R})$, the σ-ring generated by $\mathcal{R}$.

Since m is bounded, for each $x^* \in E_1^*$, the real measure x^*m is σ-additive and bounded on $\mathcal{R}$. By Proposition 2.16 b), x^*m has bounded variation $|x^*m|$ and vanishes outside the union $\bigcup S_n$.

Since m is strongly additive, by Proposition 6.12, the set $m_{\mathbb{R},E^*} = \{|x^*m| : x^* \in E_1^*\}$ is uniformly strongly additive.

Since x^*m is σ-additive on $\mathcal{R}$ and has bounded variation $|x^*m|$, by Theorem 4, x^*m can be extended to a σ-additive measure $m_{x^*} : \Sigma \to \mathbb{R}$ with finite variation $|m_{x^*}|$, which is equal to $|x^*m|$ on $\mathcal{R}$. In addition, $|m_{x^*}|(B) = 0$ for every set $B \in \Sigma$ disjoint from the union $\bigcup_n S_n$.

For each set $A \in \Sigma$ we define $m'(A) : E^* \to \mathbb{R}$ by

$$\langle x^*, m'(A) \rangle = m_{x^*}(A), \text{ for } x^* \in E^*$$

Then $m'(A)$ is a continuous linear functional on E^*: for $|x^*| \leq 1$ we have

$$|\langle x^*, m'(A) \rangle| = |m_{x^*}(A)| \leq |m_{x^*}|(S)$$
$$\leq 2\sup\{|x^*m(B)| : B \in \mathcal{R}\} \leq 2c|x^*|,$$

where $c = \sup\{|m(B)| : B \in \mathcal{R}\} < \infty$ (Proposition 2.16). Then $m'(A) \in E^{**}$, m' is additive and $m'(A) = m(A)$ for $A \in \mathcal{R}$.

To prove that m' is σ-additive on Σ, it is enough to prove that the family $\{|m_{x^*}| : x^* \in E_1^*\}$ is uniformly σ-additive on Σ. This follows from Theorem 6.11, taking $K = \{x^*m' : x^* \in E_1^*\} = \{m_{x^*} : x^* \in E_1^*\}$. In fact, K consists of σ-additive measures on $\Sigma = \sigma r(\mathcal{R})$ and the measures $x^*m' = m_{x^*} = x^*m$ with $|x^*| \leq 1$ are uniformly strongly additive on $\mathcal{R}$.

To prove that m' takes on values in E rather than E^{**}, consider the class $\mathcal{C}$ of sets $A \in \Sigma$ with $m'(A) \in E$. Then $\mathcal{R} \subset \mathcal{C}$ and $\mathcal{C}$ is a σ-ring; therefore $\mathcal{C}$ contains the σ-ring $\mathcal{S} = \sigma r(\mathcal{R})$ generated by $\mathcal{R}$. But for every set $B \in \Sigma$ disjoint from $\bigcup S_n$ we have

$$m_{x^*}(B) = 0 \text{ for all } x^* \in E_1^*,$$

therefore $m(B) = 0$. It follows that $m(A) \in E$ for every $A \in \Sigma$. Assertion b) is proved.

To prove assertion a) assume m is locally strongly additive. Then m is locally bounded on $\mathcal{R}$ (Proposition 6.9).

For each set $A \in \mathcal{R}$, the restriction m_A of m to the algebra $\mathcal{R} \cap A$ of subsets of A is bounded, strongly additive and x^*m_A is σ-additive for $x^* \in Z$.

By assertion b) proved above, there is a unique σ-additive extension m'_A of m_A to the σ-ring $\sigma r(\mathcal{R} \cap A) = \sigma r(\mathcal{R}) \cap A$.

If $A, B \in \mathcal{R}$ and $A \subset B$, then, by the uniqueness of the extension, we have $m'_A = m'_B$ on $\sigma r(\mathcal{R}) \cap A$.

Let now $B \in \mathcal{D} = \delta r(\mathcal{R})$. There is a set $A \in \mathcal{R}$ such that $B \subset A$. Then $B \in \sigma r(\mathcal{R}) \cap A$. We define $m'(B) = m'_A(B)$ and the definition of $m'(B)$ is independent on the set $A \in \mathcal{R}$ with $B \subset A$. The set function $m' : \mathcal{D} \to E$ is evidently additive. Let $B_n \downarrow \phi$ in $\mathcal{D}$ and let $A \in \mathcal{R}$ with $B_1 \subset A$. Then $m'(B_n) = m'_A(B_n) \to 0$; therefore m' is σ-additive on $\mathcal{D}$ and the theorem is proved. ∎

Combining Theorem 6.8 and Theorem 7 we obtain the following theorem:

8. Theorem. *Assume $c_0 \not\subset E$ and let $Z \subset E^*$ be a norming space for E.*

Let $m : \mathcal{R} \to E$ be an additive measure on a ring $\mathcal{R}$ such that x^*m is σ-additive for every $x^* \in Z$.
a) If m is locally bounded on $\mathcal{R}$, then m is σ-additive on $\mathcal{R}$ and can be extended to a σ-additive measure $m' : \mathcal{D} \to E$ on the δ-ring $\mathcal{D} = \delta r(\mathcal{R})$.
b) If m is bounded on $\mathcal{R}$, then m can be extended to a σ-additive measure $m' : \Sigma \to E$ on the σ-algebra $\Sigma = \sigma a(\mathcal{R})$.

The following particular case of the preceding theorem is used in the construction of the stochastic integral.

9. Theorem. *Assume $c_0 \not\subset E$ and let $Z \subset E^*$ be a norming space for E. Let $(\Omega, \mathcal{F}, P)$ be a probability space, $1 \leq p < \infty$ and let $m : \mathcal{R} \to L_E^p(P)$ be a locally bounded, additive measure on a ring $\mathcal{R}$. For each $x^* \in Z$ define the measure $x^*m : \mathcal{R} \to L^p(\mu)$ by $(x^*m)(A) = \langle m(A), x^* \rangle$, for $A \in \mathcal{R}$.*

Assume that x^*m is σ-additive on $\mathcal{R}$ for each $x^* \in Z$. Then
a) *m is σ-additive on $\mathcal{R}$ and can be extended to a σ-additive measure $m' : \mathcal{D} \to L_E^p(P)$ on the δ-ring $\mathcal{D} = \delta r(\mathcal{R})$.*
b) *If, in addition, m is bounded on $\mathcal{R}$, then m can be extended to a σ-additive measure $m' : \Sigma \to L_E^p(P)$ on the σ-algebra $\Sigma = \sigma a(\mathcal{R})$.*

Proof. Since $c_0 \not\subset E$, by Theorem 1.40, $L_E^p(P)$ does not contain a copy of c_0. Let M be the space of Z-valued, $\mathcal{R}$-step functions. Then

$$M \subset L_{E^*}^q(P) \subset (L_E^p(P))^*, \text{ with } \frac{1}{p} + \frac{1}{q} = 1.$$

Since the embedding $E \subset L(Z, \mathbb{R})$ is an isometry, by Theorem 1.38, M is norming for $L_E^p(P)$. Let $f \in M$. Consider the real-valued measure fm defined on $\mathcal{R}$ by

$$(fm)(A) = \int \langle m(A), f \rangle dP, \text{ for } A \in \mathcal{R}.$$

If $A_n \in \mathcal{R}$ and $A_n \downarrow \phi$, then for each $x^* \in Z$ we have $\langle m(A_n), x^* \rangle \to 0$ in $L^p(P)$, as $n \to \infty$. Then $(fm)(A_n) \to 0$; that is, fm is σ-additive on $\mathcal{R}$. We can apply now Theorem 8, replacing E and Z with $L_E^p(P)$ and M respectively. ∎

§8. APPLICATIONS

We give here some immediate applications of the integration theory presented in §5 to the Riesz representation theorem, to the integral representation of continuous linear operations on L^p-spaces, and to Random Gaussian measures. Other applications will be given in §20 and §32 to the Stieltjes integral with respect to functions with finite *semivariation* (rather than finite *variation*). The rest of the book is devoted to the stochastic integral, as the main application of the integration theory of §5.

A. The Riesz representation theorem

1. Let K be a compact Haussdorf space and $C_F(K)$ the space of the continuous functions $f : K \to F$, endowed with the sup norm. Let $\mathcal{B}(K)$ be the σ-algebra of the Borel subsets of K.

Any function $f \in C_F(K)$ can be approximated uniformly by a sequence (f_n) of F-valued, Borel step functions.

Let $m : \mathcal{B}(K) \to E \subset L(F, G^{**})$ be an additive measure with finite semivariation relative to (F, G^{**}), such that for each $z \in G^*$, the measure $m_z : \mathcal{B}(K) \to F^*$ defined by

$$\langle x, m_z(A) \rangle = \langle m(A)x, z \rangle, \text{ for } A \in \mathcal{B}(K) \text{ and } x \in F,$$

is σ-additive.

We can apply to the measure m the theory presented in §5 and define the space $\mathcal{F}_{F,G^{**}}(m)$ and the integral $\int f dm \in G^{**}$ for functions $f \in \mathcal{F}_{F,G^{**}}(m)$. The space $\mathcal{F}_{F,G^{**}}(m)$ contains the space $\mathcal{S}_F(\mathcal{B}(K))$ of F-valued, Borel step functions. By Theorem 5.35, $\mathcal{F}_{F,G^{**}}(m)$ contains the uniform limits of sequences of Borel step functions; in particular, $\mathcal{F}_{F,G^{**}}(m)$ contains the space $C_F(K)$. Moreover, if $f \in C_F(K)$ and (f_n) is a sequence from $\mathcal{S}_F(\mathcal{B}(K))$, converging uniformly to f, then, by Theorem 5.35, we have $\int f_n dm \to \int f dm$.

It follows that for functions $f \in C_F(K)$, the integral $\int f dm$ of §5 coincides with the "immediate integral" defined in 1.27.

The Riesz Representation Theorem 1.29 can, therefore, be stated in terms of the integral of §5 rather than in terms of the immediate integral.

The advantage is that the integral of §5 is defined on a much larger class of functions than the immediate integral and has better properties.

B. Integral representation of continuous linear operations on L^p-spaces

2. Let (X, Σ, μ) be a measure space and consider the space $L^p_F(\mu)$ for $1 \leq p < \infty$. In [D.1], Theorem 1, p 259, we gave an integral representation of continuous linear operators $U : L^p_E(\mu) \to G$, in the form

$$U(f) = \int f dm, \text{ for } f \in L^p_F,$$

where $m : \Sigma \to L(F, G)$ is a σ-additive measure with finite semivariation on the δ-ring Σ_f of sets $A \in \Sigma$ with $\mu(A) < \infty$. The integral $\int f dm$ was defined in a special way ([D.1], Definition 1, p. 254).

We can now give an integral representation of U in terms of the integral of §5.

3. Theorem. *Let (S, Σ, μ) be a measure space, Σ_f the δ-ring of the sets $A \in \Sigma$ with $\mu(A) < \infty$, F and G two Banach spaces, $1 \leq p < \infty$, and $B : L_F^p(\mu) \to G$ a continuous linear mapping.*

Then there is a σ-additive measure $m : \Sigma_f \to L(F, G)$ with finite semivariation $\tilde{m}_{F,G}$ such that:

a) $\tilde{m}_{F,G}(A) \leq \|B\| \mu(A)^{\frac{1}{p}}$, *for $A \in \Sigma_f$;*
b) $\tilde{m}_{F,G}(f) \leq \|B\| \|f\|_p \leq \infty$, *for Σ-measurable functions $f : S \to F$;*
c) $L_F^p(\mu) \subset \mathcal{F}_{F,G}(m)$, *continuously;*
and
d) $B(f) = \int f dm$, *for $f \in L_F^p(\mu)$.*

If B is an isometry, then:

a') $\tilde{m}_{F,G}(A) = \mu(A)^{\frac{1}{p}}$, *for $A \in \Sigma_f$;*
b') $\tilde{m}_{F,G}(f) = \|f\|_p \leq \infty$, *for Σ-measurable functions $f : S \to F$;*
and
c') $L_F^p(\mu) = \mathcal{F}_{F,G}(m)$, *isometrically.*

Proof. Without loss of generality, we can assume $\|B\| \leq 1$. For $A \in \Sigma_f$ and $x \in F$ we have $\varphi_A x \in L_F^p(\mu)$. Set

$$m_x(A) = B(\varphi_A x) \in G.$$

If $A \in \Sigma_f$, the mapping $m(A) : F \to G$ defined by

$$m(A)x = m_x(A), \text{ for } x \in F,$$

is linear and continuous:

$$|m(A)x| = |m_x(A)| = |B(\varphi_A x)| \leq \|\varphi_A x\|_p = |x| \mu(A)^{1/p},$$

therefore $m(A) \in L(F, G)$ and

$$|m(A)| \leq \mu(A)^{1/p}.$$

If B is an isometry, we have

$$|m(A)x| = |x| \mu(A)^{1/p}$$

and

$$|m(A)| = \mu(A)^{1/p}, \text{ for } A \in \Sigma_f.$$

The mapping $m : \Sigma_f \to L(F, G)$ is evidently additive. It is also σ-additive: if $A_n \downarrow \phi$ in Σ_f, then $\mu(A_n) \to 0$, hence $m(A_n) \to 0$.

If $s = \sum_i \varphi_{A_i} x_i$ is a Σ_f-step function with $A_i \in \Sigma_f$ and $x_i \in F$, then

$$\int s\, dm = \sum_i m(A_i) x_i = \sum_i B(1_{A_i} x_i)$$
$$= B(s) \in G,$$

hence

$$\left| \int s\, dm \right| = |B(s)| \leq \|s\|_p,$$

with equality of B is an isometry. For any Σ-measurable function $f : S \to F$ we have

$$\tilde{m}_{F,G}(f) = \sup \left| \int s\, dm \right| \leq \sup \|s\|_p = \|f\|_p \leq \infty,$$

the supremum being taken for all Σ_f-step functions $s : S \to F$ with $|s| \leq |f|$. If B is an isometry, then

$$\tilde{m}_{F,G}(f) = \|f\|_p \leq \infty.$$

This proves assertions b) and b').

Assertion c) follows from b) and assertion c') follows from b').

For $f = 1_A x$ with $A \in \Sigma_f$ and $x \in F$ with $|x| = 1$, we get

$$\tilde{m}_{F,G}(A) \leq \mu(A)^{1/p} < \infty$$

with equality if B is an isometry. This shows that m has finite semivariation $\tilde{m}_{F,G}$ and proves assertions a) and a').

It remains to prove assertion d):

$$B(f) = \int f\, dm, \text{ for } f \in L_F^p(\mu).$$

This equality was already established for Σ_f-step functions $f \in L_F^p(\mu)$.

Let $f \in L_F^p(\mu)$; we can assume that f is Σ-measurable. There is a sequence (s_n) of F-valued, Σ_f-step functions such that $s_n \to f$ pointwise and in $L_F^p(\mu)$. Then $B(s_n) \to B(f)$ in G and

$$\tilde{m}_{F,G}(s_n - f) \leq \|s_n - f\|_p \to 0;$$

therefore $s_n \to f$ in $\mathcal{F}_{F,G}(m)$. By the continuity of the integral we have $\int s_n dm \to \int f dm$. Since for each n we have

$$B(s_n) = \int s_n dm,$$

we deduce that

$$B(f) = \int f\, dm.$$

∎

Remark. By the above proof, if B is an isometry, the set $\mathcal{S}_F(\Sigma_f)$ of Σ_f-step functions is dense in $\mathcal{F}_{F,G}(m)$ and the integral $\int f\, dm$ belongs to G rather than G^{**}.

C. Random Gaussian measures

This section is motivated by the problem of representing a Gaussian measure as a genuine integral.

Let $(\Omega, \mathcal{F}, P)$ be a probability space and (S, Σ, μ) a measure space. Let Σ_f be the δ-ring of sets $A \in \Sigma$ with $\mu(A) < \infty$.

4. A real-valued random variable X is called a *centered Gaussian random variable* if its law on $\mathbb{R}$ is the probability with density of the form

$$a(2\pi)^{-\frac{1}{2}} \exp(a^2 x^2 / 2).$$

This means that if $F : \mathbb{R} \to \mathbb{R}$ is the distribution function of X defined by $F(x) = P(\{X \leq x\})$ for $x \in \mathbb{R}$, then

$$F(x) = \int_{-\infty}^{x} a(2\pi)^{-\frac{1}{2}} \exp(a^2 t^2 / 2) dt, \text{ for } x \in \mathbb{R}.$$

A subspace of $L^2(P)$ consisting of centered Gaussian random variables is called a *Gaussian space*.

5. Definition. *A Gaussian measure on the measure space (S, Σ, μ) is an isomorphism $B : L^2(\mu) \to L^2(P)$ having as range a Gaussian space.*

If μ is the Lebesgue measure on $\mathbb{R}^n$, a Gaussian measure on $(\mathbb{R}^n, \mathcal{B}(\mathbb{R}^n), \mu)$ is called a Brownian measure.

A Gaussian measure B is a linear mapping preserving the inner product. If B is a Gaussian measure on (S, Σ, μ) and if for every $A \in \Sigma_f$ we denote

$$m(A) = B(\varphi_A),$$

we obtain a σ-aditive measure $m : \Sigma_f \to L^2(P)$. This is the reason why B is called a measure ([N], p. 64, 65).

It was an unsolved problem whether we can respresent B as an integral with respect to m:

$$B(f) = \int f dm, \text{ for } f \in L^2(\mu).$$

Using the integration theory with respect to measures with finite semivariation, the answer to this problem is affirmative.

6. Theorem. *Let $B : L^2(\mu) \to L^2(P)$ be a Gaussian measure on the measure space (S, Σ, μ). Then there is a σ-additive measure $m : \Sigma_f \to L^2(P)$ with finite semivariation $\tilde{m}_{\mathbb{R}, L^2(P)}$ such that*

$$\mathcal{F}_{\mathbb{R}, L^2(P)}(m) = L^2(\mu), \text{ isometrically}$$

and

$$B(f) = \int f dm, \text{ for } f \in L^2(\mu).$$

Theorem 6 is obtained from Theorem 3 by taking $p = 2$, $F = \mathbb{R}$ and $G = L^2(P)$. The fact that the range of B is a Gaussian space is not used in the proof of this theorem.

Chapter 2
The Stochastic Integral

In this chapter we apply the integration theory developed in §5 to construct the stochastic integral of one-parameter processes. Given a cadlag, adapted process X with $X_t \in L_E^p$ for $t \geq 0$, we associate to X an additive measure $I_X : \mathcal{R} \to L_E^p$, defined on the ring $\mathcal{R}$ generated by the predictable rectangles. The process X is said to be p-summable if I_X can be extended to a σ-additive measure on the σ-algebra $\mathcal{P}$ of predictable sets, with finite semivariation relative to (F, L_E^p). Then we can use the construction presented in §5 to define an integral $\int H dI_X$, for F-valued, predictable processes H. The stochastic integral of H with respect to X is a process denoted by $H \cdot X$ and defined by

$$(H \cdot X)_t(\omega) = \left(\int_{[0,t]} H dI_X \right)(\omega), \text{ a.s.,}$$

provided that the integral belongs to L_G^p and it has a cadlag modification.

In Chapter 3 we shall prove that the square integrable martingales with values in a Hilbert space are 2-summable. In Chapters 4 and 5 it is shown that the processes with integrable variation or with integrable semivariation are also summable.

§9. SUMMABLE PROCESSES

A. Notations

The reader is supposed to be familiar with the general theory of stochastic processes presented, for example, in [D–M], or any elementary book on stochastic processes. In the sequel we shall use the following notations:

1. E, F, G are Banach spaces with $E \subset L(F, G)$.

2. $(\Omega, \mathcal{F}, P)$ is a *probability space*. The P-negligible (resp. P-integrable) sets or functions will be called, simply, negligible (resp. integrable). Instead of P-a.e. we shall write a.s. (almost surely). A set $M \subset \mathbb{R}_+ \times \Omega$ is called *evanescent* if it is contained in a set of the form $\mathbb{R}_+ \times A$ with $A \subset \Omega$, negligible.

$(\mathcal{F}_t)_{t \in \mathbb{R}_+}$ is a filtration, i.e., each $\mathcal{F}_t$ is a σ-algebra contained in $\mathcal{F}$ and $\mathcal{F}_s \subset \mathcal{F}_t$ if $s \leq t$. We assume the filtration satisfies the *usual conditions*, i.e., $\mathcal{F}_t = \bigcap_{s>t} \mathcal{F}_s$ for every $t \geq 0$ and $\mathcal{F}_0$ contains all negligible sets.

A *stopping time* (or an *optional stopping time*) is a function $T : \Omega \to \overline{\mathbb{R}}_+$ such that $\{T \leq t\} \in \mathcal{F}_t$ for every $t \geq 0$. If $S \leq T$ are two stopping times we define the *stochastic interval*

$$(S, T] = \{(t, \omega) \in \mathbb{R}_+ \times \Omega : S(\omega) < t \leq T(\omega)\}.$$

The other stochastic intervals $[S, T), (S, T), [S, T]$ are defined similarly. The graph of the stopping time T is denoted by $[T]$ and is defined by $[T] = [T, T]$.

3. $1 \leq p < \infty$. For any Banach space D we denote $L_D^p = L_D^p(P)$. We have $L_E^p \subset L(F, L_G^p)$.

4. $\mathcal{R}$ is the ring of subsets of $\mathbb{R}_+ \times \Omega$ generated by the semiring of predictable rectangles of the form $\{0\} \times A$ with $A \in \mathcal{F}_0$ and $(s, t] \times A$ with $A \in \mathcal{F}_s$. The σ-algebra generated by $\mathcal{R}$ is called the *predictable* σ-algebra and is denoted by $\mathcal{P}$. The predictable σ-algebra is also generated by the adapted, left continuous, real-valued processes.

A $\mathcal{P}$-measurable process is called a *predictable* process.

A stopping time T is said to be *predictable* if the stochastic interval $[T, \infty)$ is predictable.

The *optional* σ-algebra $\mathcal{O}$ is the σ-algebra generated by the adapted, right continuous processes. It is also generated by the stochastic intervals $(S, T]$ with $S \leq T$ simple stopping times. An $\mathcal{O}$-measurable process is called an *optional process*.

Other σ-algebras of interest are $\mathcal{F}_T$ and $\mathcal{F}_{T-}$, where T is a stopping time. $\mathcal{F}_T$ is the σ-algebra of all the sets $A \in \mathcal{F}$ with $A \cap \{T \leq t\} \in \mathcal{F}_t$ for every $t \geq 0$. The σ-algebra $\mathcal{F}_{T-}$ is generated by the sets of the form $A \cap \{T < t\}$ with $t \geq 0$ and $A \in \mathcal{F}_t$

5. $X : \mathbb{R}_+ \times \Omega \to E$ is a *cadlag* (i.e., right continuous, with left limits), *adapted* process (i.e., X_t is $\mathcal{F}_t$-measurable for every $t \geq 0$), with $X_t \in L_E^p$ for every $t \geq 0$. Such a process has separable range.

We consider X automatically extended on $\mathbb{R} \times \Omega$, with $X_t = 0$ for $t < 0$. Then $X_{0-} = 0$. We extend also the filtration with $\mathcal{F}_t = \mathcal{F}_0$ for $t < 0$.

Let $H : \mathbb{R}_+ \times \Omega \longrightarrow D$ be a process and T a stopping time. We denote by H_T the function defined by $H_T(\omega) = H_{T(\omega)}(\omega)$ for $\omega \in \Omega$ with $T(\omega) < \infty$ and by $H_T I_{\{T < \infty\}}$ the function equal to H_T on $\{T < \infty\}$ and equal to 0 on $\{T = \infty\}$.

The stopped process H^T, obtained by stopping the process H at time T, is defined by $H_t^T = H_{t \wedge T}$. The process H^{T-}, obtained by stopping H before T, is defined by $H_t^{T-} = H_t$ if $t < T$ and $H_t^{T-} = H_{T-}$ for $t \geq T$.

A process Y is said to be a *modification* of a process Z if for every $t \geq 0$ we have $Y_t = Z_t$, a.s., the negligible set depending on t.

A process Y is called *evanescent* if the set $\{Y \neq 0\}$ is evanescent.

Two processes Y and Z, with values in the same Banach space, are said to be *indistinguishable* if $Y - Z$ is evanescent.

B. The measure I_X

6. We define the stochastic measure $I_X \colon \mathcal{R} \to L_E^p$, first for the predictable rectangles of $\mathcal{R}$ by

$$I_X(\{0\} \times A) = 1_A X_0, \text{ for } A \in \mathcal{F}_0$$

and

$$I_X((s, t] \times A) = 1_A (X_t - X_s), \text{ for } A \in \mathcal{F}_s.$$

Since I_X is finitely additive on the semiring of the above predictable rectangles, it can be extended uniquely to a finitely additive measure on $\mathcal{R}$ (Proposition 7.2).

We have then

$$I_X([0, t] \times A) = 1_A X_t, \text{ for } t \geq 0 \text{ and } A \in \mathcal{F}_0.$$

In particular,

$$I_X([0, t] \times \Omega) = X_t, \text{ for } t \geq 0.$$

If the process X is understood, we can write I instead of I_X.

7. Since $L_E^p \subset L(F, L_G^p)$, we can consider the semivariation of I_X relative to the pair (F, L_G^p). To simplify the notation, we shall write $\tilde{I}_{F,G}$ or $(\tilde{I}_X)_{F,G}$ instead of $(\tilde{I}_X)_{F, L_G^p}$ and we shall call it the semivariation of I relative to (F, G):

$$\tilde{I}_{F,G}(A) = \sup \| \sum I_X(A_i) x_i \|_p, \text{ for } A \in \mathcal{R},$$

where the supremum is taken for all finite families $(A_i)_{i \in I}$ of mutually disjoint sets from $\mathcal{R}$ contained in A and all families $(x_i)_{i \in I}$ of elements from F_1.

C. Summable processes

8. Definition. *Let $X : \mathbb{R}_+ \times \Omega \to E \subset L(F,G)$ be a cadlag, adapted process, with $X_t \in L_E^p$ for $t \geq 0$.*

We say that X is p-summable relative to the pair (F,G) if I_X has a σ-additive extension $I_X : \mathcal{P} \to L_E^p$ with finite semivariation relative to (F,G).

If $p = 1$, we say, simply, that X is summable relative to (F,G).

The stochastic integral $H \cdot X$ will be defined with respect to p-summable processes X.

We shall prove in the next chapters that the following classes of processes are p-summable:

a) Processes X with integrable variation (Theorem 19.13)
b) Processes X with integrable semivariation, provided that $c_0 \not\subset E$ and $c_0 \not\subset G$ (Theorem 21.12)
c) Square integrable martingales X, in case E and G are Hilbert spaces (Theorem 17.7).

9. Remarks. a) *X is p-summable relative to $(\mathbb{R}, E)$ iff I_X has a σ-additive extension to $\mathcal{P}$.*

In fact, by Theorem 4.20, the extension I_X to $\mathcal{P}$ has bounded semivariation $\tilde{I}_{\mathbb{R},E}$ on $\mathcal{P}$.

b) *If $1 \leq p' < p < \infty$ and if X is p-summable relative to (F,G), then X is p'-summable relative to (F,G).*

In fact, $L_E^p \subset L_E^{p'}$ and $I_X : \mathcal{P} \to L_E^p$ is σ-additive, therefore $I_X : \mathcal{P} \to L_E^{p'}$ is also σ-additive. For finite families $(A_i)_{i \in I}$ of disjoint sets from $\mathcal{R}$ and families $(x_i)_{i \in I}$ of elements from F_1 we have

$$\|\sum_i I(A_i)x_i\|_{p'} \leq \|\sum_i I(A_i)x_i\|_p,$$

hence $\tilde{I}_{F, L_G^{p'}} \leq \tilde{I}_{F, L_G^p}$.

In particular, p-summability implies 1–summability.

c) *If X is p-summable relative to (F,G), then X is p-summable relative to $(\mathbb{R}, E)$.*

In fact, if X is p-summable, then $I_X : \mathcal{P} \to L_E^p$ is σ-additive and we use Remark a).

d) *X is p-summable relative to (F,G) iff I_X has a σ-additive extension to $\mathcal{P}$ and I_X has bounded semivariation on $\mathcal{R}$ (rather than on $\mathcal{P}$), relative to (F,G).*

Use Proposition 4.19

It follows that the problem of summability reduces to a great extent to that of the σ-additive extension of I_X from $\mathcal{R}$ to $\mathcal{P}$.

D. Computation of I_X for predictable rectangles

In the sequel we shall assume that X is p-summable relative to (F, G), unless specified otherwise.

The following theorem states that a p-summable process is cadlag in L_E^p and gives the computation of the measure I_X for other types of predictable rectangles from $\mathcal{P}$ (Compare with 6 above).

10. Proposition. a) *For every $t \in \mathbb{R}_+$ we have $X_{t-} \in L_E^p$,*

$$\lim_{s \downarrow t} X_s = X_t, \text{ in } L_E^p$$

and

$$\lim_{s \uparrow t} X_s = X_{t-}, \text{ in } L_E^p.$$

b) *We have*

$$I_X((s,t) \times A) = 1_A(X_{t-} - X_s), \text{ for } 0 \leq s < t \text{ and } A \in \mathcal{F}_s;$$
$$I_X([s,t] \times A) = 1_A(X_t - X_{s-}), \text{ for } 0 \leq s < t \text{ and } A \in \mathcal{F}_{s-};$$
$$I_X(\{s\} \times A\} = 1_A(X_s - X_{s-}) = 1_A \Delta X_s, \text{ for } s \geq 0 \text{ and } A \in \mathcal{F}_{s-};$$
$$I_X([s,t) \times A) = 1_A(X_{t-} - X_{s-}) \text{ for } 0 \leq s < t \text{ and } A \in \mathcal{F}_{s-}.$$

c) *For every $t \in \mathbb{R}_+$ and $A \in \mathcal{F}_0$ we have*

$$I_X([0,t) \times A) = 1_A X_{t-};$$

in particular,

$$I_X([0,t) \times \Omega) = X_{t-}.$$

Proof. If $t_n \downarrow t$, then $[0, t_n] \times \Omega \downarrow [0, t] \times \Omega$; since I_X is σ-additive we have

$$X_{t_n} = I_X([0, t_n] \times \Omega) \to I_X([0, t] \times \Omega) = X_t, \text{ in } L_E^p$$

and this proves the first equality in assertion a).

If $t = 0$, then $X_{0-} = 0$ and for $s < 0$ we have $X_s = 0$, hence $\lim_{s \uparrow 0} X_s = X_{0-}$.

Assume $t > 0$ and let $0 < t_n \uparrow t$. Then $[0, t_n] \times \Omega \uparrow [0, t) \times \Omega$; hence

$$X_{t_n} = I_X([0, t_n] \times \Omega) \to I_X([0, t) \times \Omega), \text{ in } L_E^p.$$

On the other hand, $X_{t_n} \to X_{t-}$ pointwise. It follows that $X_{t-} \in L_E^p$ and $X_{t_n} \to X_{t-}$, in L_E^p and the second equality in assertion a) follows.

To prove the first equality in assertion b), let $s < t_n \uparrow t$ and $A \in \mathcal{F}_s$. Then $(s, t_n] \times A \uparrow (s, t) \times A$, hence

$$1_A(X_{t_n} - X_s) = I_X((s, t_n] \times A) \to I_X((s, t) \times A), \text{ in } L_E^p.$$

At the same time, $1_A(X_{t_n} - X_s) \to 1_A(X_{t-} - X_s)$ pointwise; therefore the two limits are equal:

$$I_X((s,t) \times A) = 1_A(X_{t-} - X_s).$$

To prove the second equality in assertion b), assume first that $s = 0$ and $A \in \mathcal{F}_{0-} = \mathcal{F}_0$. Then

$$I_X([0,t] \times A) = 1_A X_t = 1_A(X_t - X_{0-}).$$

Assume now that $0 < s < t$ and $A \in \mathcal{F}_r$ with $r < s$. Let $r < s_n \uparrow s$. Then $(s_n, t] \times A \downarrow [s,t] \times A$ and these sets are predictable. It follows that

$$1_A(X_t - X_{s_n}) = I_X((s, t_n] \times A) \to I_X([s,t] \times A), \text{ in } L^p_E.$$

We have also $1_A(X_t - X_{s_n}) \to 1_A(X_t - X_{s-})$, pointwise; therefore the two limits are equal:

$$I_X([s,t] \times A) = 1_A(X_t - X_{s-}).$$

This equality remains valid for $A \in \bigcup_{r<s} \mathcal{F}_r$ and $\bigcup_{r<s} \mathcal{F}_r$ generates $\mathcal{F}_{s-}$.

The two mappings $A \mapsto I_X([s,t] \times A)$ and $A \mapsto 1_A(X_t - X_{s-})$ from $\mathcal{F}_{s-}$ into L^p_E are σ-additive and coincide on a class generating $\mathcal{F}_{s-}$. It follows that the equality is valid for every $A \in \mathcal{F}_{s-}$ and the second equality in assertion b) is proved.

In particular, for $s = t$ and $A \in \mathcal{F}_{s-}$ we have

$$I_X(\{s\} \times A) = 1_A(X_s - X_{s-}) = 1_A \Delta X_s.$$

Then, for $A \in \mathcal{F}_{s-}$ we have

$$I_X([s,t) \times A) = I_X(\{s\} \times A) + I_X((s,t) \times A) = 1_A(X_{t-} - X_{s-})$$

and assertion b) is proved.

Assertion c) follows from b). ■

A p-summable process has limit at ∞ in L^p_E which coincides with the pointwise limits $X_{\infty-}$, if this limit exists. This allows us to compute I_X for unbounded predictable rectangles.

11. Proposition. *There is a random variable $X_\infty \in L^p_E$ such that*

$$\lim_{t \to \infty} X_t = X_\infty, \text{ in } L^p_E.$$

If X has pointwise left limit $X_{\infty-}$ on a set $A \in \mathcal{F}$, then

$$1_A X_\infty = 1_A X_{\infty-}, \text{ a.s.}$$

We have

$$I_X([0, \infty) \times A) = 1_A X_\infty, \text{ for } A \in \mathcal{F}_0$$

and
$$I_X((t,\infty) \times A) = 1_A(X_\infty - X_t), \text{ for } A \in \mathcal{F}_t.$$

In particular
$$I_X([0,\infty) \times \Omega) = X_\infty.$$

Proof. Let $t_n \uparrow \infty$ and $A \in \mathcal{F}_0$. Then
$$I_X([0,\infty) \times A) = \lim I_X([0,t_n] \times A)$$
$$= \lim 1_A X_{t_n} = 1_A \lim X_{t_n}, \text{ in } L_E^p.$$

Set $X_\infty = I_X([0,\infty) \times \Omega) \in L_E^p$. Then, taking $A = \Omega$ above we get
$$X_\infty = I_X([0,\infty) \times \Omega) = \lim X_{t_n}, \text{ in } L_E^p,$$

hence
$$X_\infty = \lim_{t \to \infty} X_t, \text{ in } L_E^p.$$

If $X_{\infty-}$ exists pointwise, then the two limits X_∞ and $X_{\infty-}$ are equal a.s.

From the above we deduce that
$$I_X([0,\infty) \times A) = 1_A \lim X_{t_n} = 1_A X_\infty.$$

Finally, if $t \geq 0$ and $A \in \mathcal{F}_t$, then
$$I_X((t,\infty) \times A) = \lim I_X((t,t_n] \times A)$$
$$= \lim 1_A(X_{t_n} - X_t) = 1_A(X_\infty - X_t), \text{ in } L_E^p.$$

∎

E. Computation of I_X for stochastic intervals

Assume X is p-summable relative to (F, G).

12. The σ-algebra $\mathcal{P}$ of predictable subsets of $\mathbb{R}_+ \times \Omega$ contains the stochastic intervals of the form
$$(S,T] = \{(t,\omega) \in \mathbb{R}_+ \times \Omega : S(\omega) < t \leq T(\omega)\},$$

where $S \leq T$ are stopping times (with finite or infinite values). Other stochastic intervals are defined similarly.

The ring $\mathcal{R}$ is generated by the stochastic intervals $(S,T]$ with $S \leq T$ simple stopping times.

If I_X is extended to a σ-additive measure $I_X : \mathcal{P} \to L_E^p$, it is convenient to extend it further to sets of the form $\{\infty\} \times A$ with $A \in \mathcal{F}_\infty := \bigvee_{t \geq 0} \mathcal{F}_t$, by setting
$$I_X(\{\infty\} \times A) = 0.$$

Then $\mathcal{P}[0,\infty] := \mathcal{P} \cup (\{\infty\} \times \mathcal{F}_\infty)$ is the σ-algebra of predictable subsets of $\overline{\mathbb{R}}_+ \times \Omega$. The extension I_X is still σ-additive on $\mathcal{P}[0,\infty]$. Then $I_X((S,T])$ has the same value whether $(S,T]$ is regarded as a subset of $\mathbb{R}_+ \times \Omega$ as defined above, or as a subset of $\overline{\mathbb{R}}_+ \times \Omega$, defined by

$$(S,T] = \{(t,\omega) \in \overline{\mathbb{R}}_+ \times \Omega : S(\omega) < t \leq T(\omega)\}.$$

Similar considerations hold for other types of predictable stochastic intervals: $[S,T]$ if S is predictable, (S,T) if T is predictable, $[S,T)$ if both S and T are predictable. The graph $[T]$ of T is also predictable if T is predictable.

13. Consider the process X extended at ∞ by X_∞, according to Proposition 11 and let T be a stopping time. Then $X_T(\omega)$ is defined for every $\omega \in \Omega$, even if $T(\omega) = \infty$.

If $A \in \mathcal{F}_T$, the stopping time T_A is defined by $T_A(\omega) = T(\omega)$ if $\omega \in A$ and $T_A(\omega) = \infty$ if $\omega \notin A$.

Then the predictable rectangles can be written as stochastic intervals:

$$\{0\} \times A = [0_A], \text{ if } A \in \mathcal{F}_0,$$
$$(s,t] \times A = (s_A, t_A], \text{ if } A \in \mathcal{F}_s$$
$$[s,t] \times A = [s_A, t_A], \text{ if } A \in \mathcal{F}_{s-}$$
$$\{s\} \times A = [s_A], \text{ if } A \in \mathcal{F}_{s-}$$

and

$$[s,t] \times A = [s_A, t_A], \text{ if } A \in \mathcal{F}_{s-}.$$

We can write also

$$1_A(X_t - X_s) = X_{t_A} - X_{s_A}, \text{ if } s < t \text{ and } A \in \mathcal{F}_s$$

and so on. With these notations, the measure I_X can be defined as

$$I_X([0_A]) = X_{0_A}, \text{ for } A \in \mathcal{F}_0$$

and

$$I_X((s_A, t_A]) = X_{t_A} - X_{s_A}, \text{ for } s < t \text{ and } A \in \mathcal{F}_s$$

and we can rewrite all the properties in Proposition 10 in terms of stopping times.

14. Proposition. a) *For any stopping time T we have $X_T \in L_E^p$,*

$$I_X([0,T]) = X_T$$

and

$$I_X((T,\infty)) = X_\infty - X_T.$$

b) If (T_n) is a decreasing sequence of stopping times with $T_n \downarrow T$, then

$$X_{T_n} \to X_T, \text{ in } L_E^p.$$

c) If $S \leq T$ are stopping times, then

$$I_X((S,T]) = X_T - X_S.$$

Assume that the pointwise limit $X_{\infty-}$ exists a.s. Then
a') For any predictable stopping time T we have $X_{T-} \in L_E^p$,

$$I_X([0,T)) = X_{T-}$$

and

$$I_X([T,\infty)) = X_\infty - X_{T-}.$$

b') If (T_n) is an increasing sequence of stopping times with $T_n \uparrow T$, then

$$X_{T_n} \to X_{T-}, \text{ in } L_E^p.$$

c') If $S \leq T$ are stopping times, then

$$I_X([S,T]) = X_T - X_{S-}, \text{ if } S \text{ is predictable};$$

in particular

$$I_X([S]) = X_S - X_{S-} = \Delta X_S, \text{ if } S \text{ is predictable}.$$
$$I_X((S,T)) = X_{T-} - X_S, \text{ if } T \text{ is predictable},$$

and

$$I_X([S,T)) = X_{T-} - X_{S-}, \text{ if both } S \text{ and } T \text{ are predictable}.$$

Proof. To prove assertion a), assume first that T is a simple stopping time, of the form

$$T = \sum_{1 \leq i \leq n} 1_{A_i} t_i$$

with $0 < t_i \leq \infty$, $t_i \neq t_j$ for $i \neq j$, $A_i \in \mathcal{F}_{t_i}$ for each i, the sets A_i are mutually disjoint and $\bigcup_{1 \leq i \leq n} A_i = \Omega$. Then

$$(T,\infty) = \bigcup_{1 \leq i \leq n} (t_i, \infty) \times A_i, \text{ disjoint union,}$$

hence

$$I_X((T,\infty)) = \sum_i I_X((t_i,\infty) \times A_i)$$
$$= \sum_i 1_{A_i}(X_\infty - X_{t_i}) = X_\infty - \sum_i 1_{A_i} X_{t_i} = X_\infty - X_T.$$

If T is an arbitrary stopping time, there is a decreasing sequence $T_n \downarrow T$ of simple stopping times T_n. Then

$$I_X((T,\infty)) = \lim I_X((T_n,\infty)) = \lim(X_\infty - X_{T_n}), \text{ in } L_E^p.$$

Since X is right continuous, we have

$$\lim(X_\infty - X_{T_n}) = X_\infty - X_T, \text{ pointwise.}$$

It follows that $X_\infty - X_T \in L_E^p$ and

$$\lim(X_\infty - X_{T_n}) = X_\infty - X_T, \text{ in } L_E^p.$$

We deduce then that $X_T \in L_E^p$,

$$\lim X_{T_n} = X_T, \text{ in } L_E^p$$

and

$$I_X((T,\infty)) = X_\infty - X_T.$$

Then

$$I_X([0,T]) = I_X([0,\infty) \times \Omega) - I_X((T,\infty))$$
$$= X_\infty - (X_\infty - X_T) = X_T$$

and assertion a) is proved. To prove b), let (T_n) be a decreasing sequence of stopping times with $T_n \downarrow T$. Then

$$X_T = I_X([0,T]) = \lim I_X([0,T_n]) = \lim X_{T_n}, \text{ in } L_E^p.$$

If now $S \leq T$ are stopping times, then

$$I_X((S,T]) = I_X([0,T]) - I_X([0,S]) = X_T - X_S$$

and assertion c) is proved.

To prove the remaining assertions, assume that $X_{\infty-}$ exists pointwise. Let T be a predictable stopping time and (T_n) an increasing sequence of stopping times with $T_n \uparrow T$. Then

$$I_X([0,T)) = \lim I_X([0,T_n]) = \lim X_{T_n}, \text{ in } L_E^p.$$

Since $\lim X_{T_n} = X_{T-}$ pointwise, we deduce that $X_{T-} \in L_E^p$ and $X_{T_n} \to X_{T-}$ in L_E^p and assertions a') and b') are proved.

If $S \leq T$ are stopping times and S is predictable, then

$$I_X([S,T]) = I_X([0,T]) - I_X([0,S)) = X_T - X_{S-}$$

and the rest of the equalities in assertion c') are proved similarly. ∎

§10. THE STOCHASTIC INTEGRAL

A. The space $\mathcal{F}_D(\tilde{I}_{F,L_G^p})$

1. Let $X : \mathbb{R}_+ \times \Omega \to E \subset L(F,G)$ be a cadlag, adapted process and assume X is p-summable relative to (F,G).

Consider the σ-additive measure $I_X : \mathcal{P} \to L_E^p \subset L(F, L_G^p)$ with bounded semivariation $\tilde{I}_{F,G}$ relative to (F, L_G^p). We consider I_X extended to $\mathcal{P}[0, \infty]$ with $I_X(\{\infty\} \times A) = 0$ for $A \in \mathcal{F}_\infty = \bigvee_{t \geq 0} \mathcal{F}_t$. As usual, we identify functions with their equivalence classes in L_E^p or L_G^p.

We can apply the integration theory developed in §5, replacing S, Σ, m with $\mathbb{R}_+ \times \Omega, \mathcal{P}, I_X$ respectively and E, F, G with L_E^p, F, L_G^p respectively.

The space $L_{G^*}^q$ with $1/p + 1/q = 1$ is a subspace of $(L_G^p)^*$ and is a norming space for L_G^p. The space of simple functions of $L_{G^*}^q$ is also a norming space for L_G^p, hence any space $Z \subset L_{G^*}^q$ containing the simple functions is a norming space for L_G^p.

Let $Z \subset L_{G^*}^q$ be a norming space for L_G^p. We denote $m = I_X$ and for any $z \in Z$ we consider the measure

$$m_z = (I_X)_z : \mathcal{P}[0, \infty] \to F^*$$

defined for $A \in \mathcal{P}[0,\infty]$ and $y \in F$ by

$$\langle y, m_z(A) \rangle = \langle m(A)y, z \rangle = \int \langle I_X(A)(\omega)y, z(\omega) \rangle dP(\omega),$$

where the bracket in the integral represents the duality between G and G^*. Then we have

$$(\tilde{I}_X)_{F, L_G^p} = \tilde{m}_{F, L_G^p} = \sup\{|m_z| : z \in Z, \|z\|_q \leq 1\}.$$

We note that $\{\infty\} \times \Omega$ is $|m_z|$-negligible for every $z \in L_{G^*}^q$.

If p is fixed, to simplify the notation, we can write $I = I_X$ and $\tilde{I}_{F,G} = \tilde{I}_{F,L_G^p}$.

We shall also write $I_{F,Z} = (I_X)_{F,Z}$ for the set of positive measures $|(I_X)_z| = |m_z|$ with $z \in Z$ and $\|z\|_q \leq 1$.

For any Banach space D we denote by $\mathcal{F}_D(\tilde{I}_{F,G}) = \mathcal{F}_D(\tilde{I}_{F,L_G^p})$ the space of predictable processes $H : \mathbb{R}_+ \times \Omega \to D$ such that

$$\tilde{I}_{F,G}(H) = \tilde{m}_{F, L_G^p}(H) = \sup\{\int |H| d|m_z| : \|z\|_q \leq 1\} < \infty.$$

For any extension of H to $\overline{\mathbb{R}}_+ \times \Omega$, the value of $\tilde{I}_{F,G}(H)$ is the same.

The set $\mathcal{F}_D(\tilde{I}_{F,G})$ is a vector space with seminorm $\tilde{I}_{F,G}$ and $\mathcal{F}_D(\tilde{I}_{F,G})$ is complete for this seminorm (Corollary 5.25). We have $\mathcal{F}_D(\tilde{I}_{F,G}) \subset \bigcap_{z \in Z} L_D^1(|I_z|)$ (See 5.21). For any set $\mathcal{C} \subset \mathcal{F}_D(\tilde{I}_{F,G})$ we denote by $\mathcal{F}_D(\mathcal{C}, \tilde{I}_{F,G})$ the closure of $\mathcal{C}$ in $\mathcal{F}_D(\tilde{I}_{F,G})$.

B. The integral $\int H dI_X$

2. If $D = F$ we can define the integral $\int H dI_X \in Z^*$ for any $H \in \mathcal{F}_F(\tilde{I}_{F,G})$. To simplify the notation we shall write

$$\mathcal{F}_{F,G}(X), \mathcal{F}_{F,L_G^p}(X), \mathcal{F}_{F,G}(I_X) \text{ or } \mathcal{F}_{F,L_G^p}(I_X)$$

instead of

$$\mathcal{F}_F((\tilde{I}_X)_{F,G}) \text{ or } \mathcal{F}_F((\tilde{I}_X)_{F,L_G^p}).$$

Let $Z \subset (L_G^p)^*$ be a norming space for L_G^p. Let $H \in \mathcal{F}_{F,G}(X)$; then $H \in L_F^1(|(I_X)_z|)$ for every $z \in Z$, hence the integral $\int H d(I_X)_z$ is defined and is a scalar. The mapping $z \mapsto \int H d(I_X)_z$ is a linear continuous functional on Z, denoted $\int H dI_X$. We have therefore, $\int H dI_X \in Z^*$,

$$\left\langle \int H dI_X, z \right\rangle = \int H d(I_X)_z, \text{ for } z \in Z$$

and

$$\left| \int H dI_X \right| \leq \tilde{I}_{F,G}(H).$$

If we take $Z = L_{G^*}^q$, then $\int H dI_X \in (L_{G^*}^q)^*$; if we take $Z = (L_G^p)^*$ then $\int H dI_X \in (L_G^p)^{**}$. The relationship between the integrals $\int H dI_X$ corresponding to different choices of Z is expressed by the following property:

Let $Z \subset Z' \subset (L_G^p)^*$ be two norming spaces for L_G^p. Then the integral $\int H dI_X$ corresponding to Z is the restriction to Z of the integral $\int' H dI_X$ corresponding to Z'.

C. A convergence theorem

We shall state now a very useful version of the Lebesgue Theorem for pointwise convergence $H^n \to H$, in which the conclusion involves the convergence of the integrals, $\int H^n dI_X \to \int H dI_X$, weakly in L_G^p, but not necessarily the convergence $H^n \to H$ in $\mathcal{F}_{F,G}(X)$. Other theorems about the convergence $H^n \to H$ in $\mathcal{F}_{F,G}(X)$ will be stated later (Theorems 12.1, 12.4, 12.5, 12.6).

3. Theorem. *Let $(H^n)_{0 \leq n < \infty}$ be a sequence of elements from $\mathcal{F}_{F,G}(X)$ such that $|H^n| \leq |H^0|$ for each n and $H^n \to H$ pointwise. Assume that*
(i) $\int H^n dI_X \in L_G^p$ *for every $n \geq 1$*
and
(ii) *The sequence $(\int H^n dI_X)_n$ converges pointwise on Ω, weakly in G.*
  Then
a) $\int H dI_X \in L_G^p$
and
b) $\int H^n dI_X \to \int H dI_X$,
in the $\sigma(L_G^p, L_{G^*}^q)$ topology of L_G^p, as well as pointwise, weakly in G.

c) If $(\int H^n dI_X)_n$ converges pointwise on Ω, strongly in G, then

$$\int H^n dI_X \to \int H dI_X,$$

strongly in L_G^1.

Proof. Since $|H| \leq |H^0|$, we deduce that $H \in \mathcal{F}_{F,G}(X)$. Let $z \in L_{G^*}^q$. We can apply Lebesgue's Theorem to (H^n) in the space $L_F^1(|I_z|)$ and deduce that $H^n \to H$ in $L_F^1(|I_z|)$ and thus $\int H^n dI_z \to \int H dI_z$, that is,

$$E(\langle (\int H^n dI)(\cdot), z(\cdot) \rangle) \to \langle \int H dI, z \rangle.$$

If $h \in L^\infty(P)$, then $hz \in L_{G^*}^q$, hence, replacing z with hz we obtain

$$E(h(\cdot)\langle (\int H^n dI_X)(\cdot), z(\cdot) \rangle) \to \langle \int H dI_X, hz \rangle.$$

Thus the sequence $(\langle (\int H^n dI_X)(\cdot), z(\cdot) \rangle)$ is weakly Cauchy in $L^1(P)$, hence it is conditionally weakly compact (Definition 3.24). By Corollary 3.27 we have

$$\lim_{P(A) \to 0} \int_A |\langle (\int H^n dI_X)(\cdot), z(\cdot) \rangle| dP = 0, \text{ uniformly for } n \in \mathbb{N}.$$

If we set

$$\phi(\omega) := \lim_n (\int H^n dI_X)(\omega), \text{ weakly in } G,$$

then, by the Vitali's Convergence Theorem 1.47 we have $\langle \phi(\cdot), z(\cdot) \rangle \in L^1(P)$ and

$$\langle (\int H^n dI_X)(\cdot), z(\cdot) \rangle \to \langle \phi(\cdot), z(\cdot) \rangle$$

in $L^1(P)$, hence

$$E(\langle (\int H^n dI_X)(\cdot), z(\cdot) \rangle) \to E(\langle \phi(\cdot), z(\cdot) \rangle).$$

Since $\langle \phi(\cdot), z(\cdot) \rangle \in L^1(P)$ for every $z \in L_{G^*}^q$, we deduce that $\phi \in L_G^p$ (Theorem 1.39). We then deduce that

$$\langle \phi, z \rangle = E(\langle \phi(\cdot), z(\cdot) \rangle)$$
$$= \lim E(\langle \int H^n dI_X(\cdot), z(\cdot) \rangle) = \langle \int H dI_X, z \rangle,$$

hence $\int H dI_X = \phi \in L_G^p$ (which is assertion a) and

$$\lim_n (\int H^n dI_X) = \int H dI_X, \text{ pointwise on } \Omega, \text{ weakly in } G.$$

From the above, it follows that

$$\int H^n dI_X \to \int H dI_X, \text{ in the } \sigma(L_G^p, L_{G^*}^q) \text{ topology of } L_G^p$$

and this proves assertion b).

In particular, the above convergence takes place in the $\sigma(L_G^1, L_{G^*}^\infty)$ topology of L_G^1, hence the sequence $(\int H^n dI_X)$ is conditionally compact in L_G^1 for the $\sigma(L_G^1, L_{G^*}^\infty)$ topology. Then, by Theorem 3.27 we have

$$\lim_{P(A) \to 0} \int_A \left| \int H^n dI_X \right| dP = 0, \text{ uniformly in } n.$$

If $\phi(\omega) = \lim_n (\int H^n dI_X)(\omega)$, strongly in G, we can apply the Vitali's Convergence Theorem 1.47 for L_G^1 and deduce that $\int H^n dI_X \to \phi$ in L_G^1; therefore

$$\int H^n dI_X \to \int H dI_X, \text{ in } L_G^1$$

and this proves assertion c). ∎

D. The stochastic integral $H \cdot X$

If $H \in \mathcal{F}_{F,G}(X)$, then, for every $t \geq 0$ we have $1_{[0,t]} H \in \mathcal{F}_{F,G}(X)$; we denote

$$\int_{[0,t]} H dI_X = \int 1_{[0,t]} H dI_X.$$

Also we define

$$\int_{[0,\infty]} H dI_X := \int_{[0,\infty)} H dI_X = \int H dI_X.$$

Taking $Z = (L_G^p)^*$, for each $H \in \mathcal{F}_{F,G}(X)$ we obtain a family $(\int_{[0,t]} H dI_X)_{t \in \mathbb{R}_+}$ of elements of $(L_G^p)^{**}$.

We are interested in processes H for which $\int_{[0,t]} H dI_X \in L_G^p$ for each $t \geq 0$. In this case we denote by the same symbol the equivalence class $\int_{[0,t]} H dI_X$ in L_G^p, as well as any random variable belonging to this equivalence class. We obtain in this way a process $(\int_{[0,t]} H dI_X)_{t \geq 0}$ with values in G. The problem is whether or not this process is adapted and cadlag. The following theorem gives a partial answer to this problem.

4. Theorem. *Let $X : \mathbb{R} \to E \subset L(F,G)$ be a cadlag, adapted, p-summable process relative to (F,G) and $H \in \mathcal{F}_{F,G}(X)$ such that $\int_{[0,t]} H dI_X \in L_E^p$ for every $t \geq 0$.*

Then the process $(\int_{[0,t]} H dI_X)_{t \geq 0}$ is adapted.

§10. THE STOCHASTIC INTEGRAL   137

Proof. Let $t \geq 0$ and prove that $\int_{[0,t]} H dI_X$ is $\mathcal{F}_t$-measurable. We divide the proof into several steps.
a) Assume first that $H = 1_M x$ with $M \in \mathcal{R}$ and $x \in F$. If $M = \{0\} \times A$ with $A \in \mathcal{F}_0$, then

$$\int_{[0,t]} H dI_X = \int 1_{\{0\} \times A} dI_X = I_X(\{0\} \times A) = 1_A X_0$$

and $1_A X_0$ is $\mathcal{F}_0$-measurable.

Assume $M = (u,v] \times A$ with $A \in \mathcal{F}_u$; if $t \leq u$, then

$$\int_{[0,t]} H dI_X = 0;$$

if $u < t$, then $([0,t] \times \Omega) \cap ((u,v] \times A) = (u, t \wedge v] \times A$ and

$$\int_{[0,t]} H dI_X = 1_A(X_{t \wedge v} - X_u).$$

It follows that in all cases $\int_{[0,t]} H dI_X$ is $\mathcal{F}_t$-measurable. We deduce then that $\int_{[0,t]} H dI_X$ is $\mathcal{F}_t$-measurable if $H = 1_M x$ with $M \in \mathcal{R}$.

b) Let $x \in F$ and denote by $\mathcal{P}_0$ the class of the sets $M \in \mathcal{P}$ such that $\int_{[0,t]} 1_M x dI_X$ is $\mathcal{F}_t$-measurable. By step a), $\mathcal{P}_0$ contains $\mathcal{R}$. We prove now that $\mathcal{P}_0$ is a monotone class. Let (M_n) be a monotone sequence from $\mathcal{P}_0$ with limit M and prove that $M \in \mathcal{P}_0$.

Let $z \in (L_G^p)^*$. Then $1_{[0,t]} 1_{M_n} x \to 1_{[0,t]} 1_M x$ in $L_F^1((I_X)_z)$; therefore

$$\int_{[0,t]} 1_{M_n} x d(I_X)_z \to \int_{[0,t]} 1_M x d(I_X)_z,$$

that is,

$$\langle \int_{[0,t]} 1_{M_n} x dI_X, z \rangle \to \langle \int_{[0,t]} 1_M x dI_X, z \rangle$$

where the bracket represents the duality between L_G^p and its dual.

It follows that

$$\int_{[0,t]} 1_{M_n} x dI_X \to \int_{[0,t]} 1_M x dI_X$$

weakly in $L_G^p = L_G^p(\mathcal{F}, P)$. For each n, the random variable $\int_{[0,t]} 1_{M_n} x dI_X$ is $\mathcal{F}_t$-measurable, by hypothesis; therefore it belongs to the subspace $L_G^p(\mathcal{F}_t, P)$. It follows that the weak limit $\int_{[0,t]} 1_M x dI_X$ belongs to the weak closure in $L_G^p(\mathcal{F}, P)$ of the convex set $L_G^p(\mathcal{F}_t, P)$, which is equal to the strong closure of $L_G^p(\mathcal{F}_t, P)$ in $L_G^p(\mathcal{F}, P)$ and this strong closure is $L_G^p(\mathcal{F}_t, P)$ itself. It follows that $\mathcal{P}_0$ is a monotone class; consequently, $\mathcal{P}_0 = \mathcal{P}$.

From step b) we deduce that $\int_{[0,t]} H dI_X$ is $\mathcal{F}_t$-measurable for every $\mathcal{P}$-step process $H : \mathbb{R}_+ \times \Omega \to F$.

c). Let now H be as in the statement of the theorem. Since H is predictable, there is a sequence (H^n) of predictable step processes such that $H^n \to H$ pointwise and $|H^n| \leq |H|$ everywhere, for every n.

For each $z \in (L_G^p)^*$ we have $H \in L_F^1((I_X)_z)$ and we can apply Lebesgue's Theorem in this space, to deduce that $1_{[0,t]}H^n \to 1_{[0,t]}H$ in $L_F^1((I_X)_z)$, hence

$$\int_{[0,t]} H^n d(I_X)_z \to \int_{[0,t]} H d(I_X)_z,$$

that is,

$$\langle \int_{[0,t]} H^n dI_X, z \rangle \to \langle \int_{[0,t]} H dI_X, z \rangle,$$

consequently

$$\int_{[0,t]} H^n dI_X \to \int_{[0,t]} H dI_X$$

weakly in $L_F^p(\mathcal{F}, P)$. Since, by step b), we have $\int_{[0,t]} H^n dI_X \in L_F^p(\mathcal{F}_t, P)$ and since $L_F^p(\mathcal{F}_t, P)$ is convex and weakly closed in $L_F^p(\mathcal{F}, P)$, we deduce by the same argument as in step b) that $\int_{[0,t]} H dI_X \in L_F^p(\mathcal{F}_t, P)$ that is, $\int_{[0,t]} H dI_X$ is $\mathcal{F}_t$-measurable. ∎

If X and H are as in the statement of Theorem 4, it is not certain that there is a cadlag choice of representatives $(\int_{[0,t]} H dI_X)(\omega)$.

This remark leads to the following definition.

5. Definition. *We denote by $L_{F,G}^1(X)$ the set of processes $H \in \mathcal{F}_{F,G}(I_X)$ satisfying the following two conditions:*
a) $\int_{[0,t]} H dI_X \in L_G^p$ *for every* $t \in \mathbb{R}_+$;
b) *The process* $(\int_{[0,t]} H dI_X)_{t \geq 0}$ *has a cadlag modification.*

The processes $H \in L_{F,G}^1(X)$ are said to be integrable with respect to X.

If $H \in L_{F,G}^1(X)$, then any cadlag modification of the process $(\int_{[0,t]} H dI_X)_{t \geq 0}$ is called the stochastic integral of H with respect to X and is denoted by $H \cdot X$ or $\int H dX$:

$$(H \cdot X)_t(\omega) = (\int H dX)_t(\omega) = (\int_{[0,t]} H dI_X)(\omega), \text{ a.s.}$$

It follows that the stochastic integral is defined up to an evanescent process. For $t = \infty$ we have

$$(H \cdot X)_\infty = \int_{[0,\infty]} H dI_X = \int_{[0,\infty)} H dI_X = \int H dI_X.$$

For an $\mathcal{R}$-step process $H : \mathbb{R}_+ \times \to \Omega \to F$ we have

$$(H \cdot X)_t(\omega) = \int_{[0,t]} H_s(\omega) dX_s(\omega),$$

that is, the stochastic integral can be computed pathwise.

There are other instances when the stochastic integral can be computed pathwise, as a Stieltjes integral.

From Theorem 4 and Definition 5 we deduce the following result:

6. Theorem. *If $X : \mathbb{R}_+ \times \Omega \to E \subset L(F,G)$ is a p-summable process relative to (F,G) and if $H \in L^1_{F,G}(X)$, then the stochastic integral $H \cdot X$ is a cadlag, adapted process.*

The following theorem states that the stochastic integral $H \cdot X$ is cadlag not only pathwise, but also in L^1_G, even if $H \cdot X$ is not summable. If $H \cdot X$ is itself p-summable, then it is cadlag in L^p_G(Proposition 9.10). In the proof we use the convergence Theorem 3.

7. Theorem. *Let $H \in L^1_{F,G}(X)$.*
a) *For every $t \in [0, \infty)$ we have $(H \cdot X)_{t-} \in L^p_G$ and*

$$(H \cdot X)_{t-} = \int_{[0,t)} H dI_X, \ a.s.$$

If $(H \cdot X)_{\infty-}(\omega)$ exists for each $\omega \in \Omega$, then

$$(H \cdot X)_{\infty-} = (H \cdot X)_\infty = \int H dI_X, \ a.s.$$

b) *The mapping $t \mapsto (H \cdot X)_t$ is cadlag in L^1_G.*

Proof. Let $t_n \uparrow t$. If $(H \cdot X)_{\infty-}$ exists, we allow $t = \infty$. Then $1_{[0,t_n]}H \to 1_{[0,t)}H$ pointwise, $|1_{[0,t_n]}H| \leq |H|$ for every n, $\int 1_{[0,t_n]} H dI_X = (H \cdot X)_{t_n} \in L^p_G$ and $(H \cdot X)_{t_n} \to (H \cdot X)_{t-}$ pointwise on Ω, strongly in G, that is, $\int 1_{[0,t_n]} H dI_X \to (H \cdot X)_{t-}$ pointwise on Ω, strongly in G. By Theorem 3, $\int 1_{[0,t)} H dI_X \in L^p_G$ and

$$\int 1_{[0,t_n]} H dI_X \to \int 1_{[0,t)} H dI_X, \text{ in } L^1_G.$$

It follows that $(H \cdot X)_{t-} = \int_{[0,t)} H dI_X$ a.s., hence $(H \cdot X)_{t-} \in L^p_G$. It follows also that the mapping $t \mapsto (H \cdot X)_t$ form $\overline{\mathbb{R}}_+$ to L^1_G has left limit at every point $t < \infty$ and even at $t = \infty$, if $(H \cdot X)_{\infty-}(\omega)$ exists for every $\omega \in \Omega$. This proves assertion a) and half of assertion b). Taking $t_n \downarrow t$, a similar proof shows that $(H \cdot X)_{t_n} \to (H \cdot X)_t$ in L^1_G, hence $t \mapsto (H \cdot X)_t$ is cadlag in L^1_G. ∎

§11. THE STOCHASTIC INTEGRAL AND STOPPING TIMES

In this paragraph we prove some properties of the stochastic integral in connection with stopping times.

A. Stochastic integral of elementary processes

Let $X : \mathbb{R}_+ \times \Omega \to E \subset L(F,G)$ be a p-summable process relative to (F, G).

We prove first an auxiliary proposition which is an extension of Proposition 9.14. In the proof we shall use the convergence Theorem 10.3.

1. Proposition. a) *Let $S \leq T$ be stopping times and $h : \Omega \to F$ be an $\mathcal{F}_S$-measurable, bounded random variable. Then*

$$\int h 1_{(S,T]} dI_X = h(X_T - X_S).$$

b) *If T is predictable, then*

$$\int h 1_{(S,T)} dI_X = h(X_{T-} - X_S)$$

c) *If S is predictable and h is $\mathcal{F}_{S-}$-measurable, then*

$$\int h 1_{[S,T]} dI_X = h(X_T - X_{S-})$$

and

$$\int h 1_{[S]} dI_X = h(X_S - X_{S-}) = h \Delta X_S.$$

d) *If both S and T are predictable and h is $\mathcal{F}_{S-}$-measurable, then*

$$\int h 1_{[S,T)} dI_X = h(X_{T-} - X_{S-}).$$

Proof. To prove assertion a) assume first that $h = 1_A y$ with $A \in \mathcal{F}_S$ and $y \in F$. Then, by Proposition 9.14,

$$\int h 1_{(S,T]} dI_X = I_X((S_A, T_A]) y = (X_{T_A} - X_{S_A}) y = h(X_T - X_S).$$

Then the equality holds for any $\mathcal{F}_S$-step function h. For the general case, let (h_n) be a sequence of $\mathcal{F}_S$-step functions with $h_n \to h$ and $|h_n| \leq |h|$ for every n. Then $h_n 1_{(S,T]} \to h 1_{(S,T]}$ pointwise and

$$\int h_n 1_{(S,T]} dI_X = h_n(X_T - X_S) \to h(X_T - X_S)$$

in G, pointwise on Ω. We have also

$$\int h_n 1_{(S,T]} dI_X = h_n(X_T - X_S) \in L^p_G.$$

We can apply the convergence Theorem 10.3 and deduce that $\int h 1_{(S,T]} dI_X \in L^p_G$ and

$$\int h_n 1_{(S,T]} dI_X \to \int h 1_{(S,T]} dI_X,$$

pointwise and in L^1_G. It follows that

$$\int h 1_{(S,T]} dI_X = h(X_T - X_S)$$

and assertion a) is proved. Assertion b) and the first equality of assertion c) are proved similarly, using Proposition 9.14 and the convergence Theorem 10.3. The second equality of assertion c) is obtained from the first equality of assertion c), taking $T = S$. Finally assertion d) is obtained using the equality $[S,T] = [S] \cup (S,T)$ and assertions b) and c). ∎

We deduce first the following corollary:

2. Corollary. *Let H be an F-valued, elementary process of the form*

$$H = H_0 1_{\{0\}} + \sum_{1 \le i \le n} H_i 1_{(T_i, T_{i+1}]},$$

where $0 = T_0 \le T_1 \le \cdots \le T_{n+1}$ are stopping times and for each $i = 0, 1, 2, \ldots, n$, H_i is an F-valued, $\mathcal{F}_{T_i}$-measurable, bounded random variable.

Then $H \in L^1_{F,G}(X)$ and the stochastic integral $H \cdot X$ can be computed pathwise:

$$(H \cdot X) = H_0 X_0 + \sum_{1 \le i \le n} H_i(X^{T_{i+1}} - X^{T_i}).$$

Proof. By Proposition 9.14, for $1 \le i \le n$ we have

$$X_t^{T_i} = X_{T_i \wedge t} \in L^p_E, \text{ hence } H_i X_t^{T_i} \in L^p_G.$$

By Proposition 1, for $1 \le 0 \le n$ we have

$$\int_{(0,t]} H_i 1_{(T_i, T_{i+1}]} dI_X = H_i(X_{T_{i+1} \wedge t} - X_{T_i \wedge t}).$$

Since X is cadlag, each process X^{T_i} is cadlag, hence $H_i 1_{(T_i, T_{i+1}]} \in L^1_{F,G}(X)$ and

$$(H_i 1_{(T_i, t_{i+1}]} \cdot X)_t = H_i(X_t^{T_{i+1}} - X_t^{T_i}).$$

For $i = 0$, taking $S = 0$ in Proposition 1 we get

$$\int_{[0,t]} H_0 1_{\{0\}} dI_X = H_0 X_0,$$

hence $H_0 1_{\{0\}} \cdot X \in L^1_{F,G}(X)$ and

$$(H_0 1_{\{0\}} \cdot X)_t = H_0 X_0.$$

It follows that $H \in L^1_{F,G}(X)$ and

$$(H \cdot X)_t = H_0 X_0 + \sum_{1 \leq i \leq n} H_i (X_t^{T_{i+1}} - X_t^{T_i}).$$

■

3. Corollary. *Every simple, F-valued process of the form*

$$H = y_0 1_{A_0 \times \{0\}} + \sum_{1 \leq i \leq n} 1_{A_i} 1_{(t_i, t_{i+1}]} y_i$$

with $0 = t_0 < t_1 < \cdots < t_n < t_{n+1} \leq \infty$, $y_i \in F$ and $A_i \in \mathcal{F}_{t_i}$, belongs to $L^1_{F,G}(X)$ and

$$(H \cdot X)_t = y_0 1_{A_0} X_0 + \sum_{1 \leq i \leq n} 1_{A_i} y_i (X_{t_{i+1} \wedge t} - X_{t_i \wedge t}).$$

4. Remark. For a step process of the form

$$H = \sum_{1 \leq i \leq n} 1_{A_i} y_i,$$

with $A_i \in \mathcal{P}$ and $y_i \in F$, we have

$$\int_{[0,t]} H dI_X = \sum_{1 \leq i \leq n} I_X(A_i \cap ([0,t] \times \Omega)) y_i \in L^p_G;$$

but it does not follow that the stochastic integral $H \cdot X$ is defined, since the mapping $t \mapsto (\int_{[0,t]} H dI_X)(\omega)$ is not necessarily cadlag.

The following proposition is an extension of Proposition 1, the random variable h being replaced by hH, where H is a process. We will use the convergence Theorem 10.3.

5. Proposition. *Let $S \leq T$ be stopping times and assume that either*
(i) $h : \Omega \to \mathbb{R}$ *is bounded, $\mathcal{F}_S$-measurable and $H \in \mathcal{F}_{F,G}(X)$,*
or
(ii) $h : \Omega \to F$ *is bounded, $\mathcal{F}_S$-measurable and*

$$H \in \mathcal{F}_{\mathbb{R}}(I_{F,G}) \cap \mathcal{F}_{\mathbb{R}}(I_{\mathbb{R},E}).$$

Then:
a) If $\int 1_{(S,T]}HdI_X \in L_G^p$, in case (i) and $\int 1_{(S,T]}HdI_X \in L_E^p$ in case (ii), then
$$\int h1_{(S,T]}HdI_X = h\int 1_{(S,T]}HdI_X.$$

b) If S is predictable, h is $\mathcal{F}_{S-}$-measurable and if $\int 1_{[S,T]}HdI_X \in L_G^p$, in case (i) and $\int 1_{[S,T]}HdI_X \in L_E^p$, in case (ii), then
$$\int h1_{[S,T]}HdI_X = h\int 1_{[S,T]}HdI_X$$

and
$$\int h1_{[S]}HdI_X = h\int 1_{[S]}HdI_X.$$

Proof. Assume first hypothesis (i). Let H be of the form $H = 1_{(s,t] \times A}y$ with $A \in \mathcal{F}_s$ and $y \in F$. Since $h1_A y$ is $\mathcal{F}_{S \vee s}$-measurable, by Proposition 1 we have
$$\int h1_{(S,T]}HdI_X = \int h1_A y 1_{(S \vee s, T \wedge t]}dI_X$$
$$= h1_A y(X_{T \wedge t} - X_{S \vee s}) = h\int 1_{(S,T]}HdI_X \in L_G^p.$$

If $H = 1_{\{0\} \times A}y$ with $A \in \mathcal{F}_0$, then $1_{(S,T]}H = 0$ and $h1_{(S,T]}H = 0$ and the equality of assertion a) holds. It follows that for $B \in \mathcal{R}$ we have
$$\int h1_{(S,T]}1_B ydI_X = h\int 1_{(S,T]}1_B ydI_X \in L_G^p.$$

For any $z \in L_{G^*}^q$, $\frac{1}{p} + \frac{1}{q} = 1$, we have then
$$\int h1_{(S,T]}1_B yd(I_X)_z = \int 1_{(S,T]}1_B yd(I_X)_{hz}.$$

The class of sets $B \in \mathcal{P}$ for which the above equality holds for all $z \in L_{G^*}^q$ is a monotone class which contains $\mathcal{R}$, hence the equality holds for all $B \in \mathcal{P}$ and $z \in L_{G^*}^q$.

It follows that for any predictable, simple process H we have
$$\int h1_{(S,T]}Hd(I_X)_z = \int 1_{(S,T]}Hd(I_X)_{hz}.$$

If $H \in \mathcal{F}_{F,G}(X)$, Lebesgue's Theorem implies that the above equality holds for every $z \in L_{G^*}^q$.

Assume now that $\int_{(S,T]} HdI_X \in L_G^p$. Then $h\int_{(S,T]} HdI_X \in L_G^p$ and
$$\langle h\int 1_{(S,T]}HdI_X, z\rangle = \langle \int 1_{(S,T]}HdI_X, hz\rangle$$
$$= \int 1_{(S,T]}Hd(I_X)_{hz} = \int h1_{(S,T]}Hd(I_X)_z$$
$$= \langle \int h1_{(S,T]}HdI_X, z\rangle,$$

for all $z \in L_{G^*}^q$. It follows that

$$\int h 1_{(S,T]} H dI_X = h \int 1_{(S,T]} H dI_X,$$

and this proves assertion a) in case (i).

Assume now hypothesis (ii) and let $H : \mathbb{R}_+ \times \Omega \to \mathbb{R}$ be predictable with $\tilde{I}_{F,G}(H) < \infty$ and $\tilde{I}_{\mathbb{R},E}(H) < \infty$, that is,

$$H \in \mathcal{F}_{\mathbb{R}}(I_{F,G}) \cap \mathcal{F}_{\mathbb{R}}(I_{\mathbb{R},E}) \subset \mathcal{F}_{\mathbb{R},E}(X).$$

Assume that $\int 1_{(S,T]} H dI_X \in L_E^p$. Consider first the case $h = h'y$ with $y \in F$ and h' a real-valued, bounded, $\mathcal{F}_S$-measurable random variable. By Proposition 5.46 we have

$$\int 1_{(S,T]} H y dI_X = y \int 1_{(S,T]} H dI_X.$$

By the first part of the proof we have

$$\int h 1_{(S,T]} H dI_X = h' \int 1_{(S,T]} H y dI_X = h \int 1_{(S,T]} H dI_X \in L_G^p.$$

This equality then holds for any F-valued, $\mathcal{F}_S$-step function h.

Assume now $h : \Omega \to F$ is $\mathcal{F}_S$-measurable and bounded and let (h_n) be a sequence of F-valued, $\mathcal{F}_S$-step functions such that $h_n \to h$ and $|h_n| \leq |h|$. We have $\int h_n 1_{(S,T]} H dI_X \in L_G^p$ for each n and

$$\int h_n 1_{(S,T]} H dI_X = h_n \int 1_{(S,T]} H dI_X \to h \int 1_{(S,T]} H dI_X,$$

pointwise on Ω, strongly in G. By the convergence Theorem 10.3 we have $\int h 1_{(S,T]} H dI_X \in L_G^p$ and

$$\int h_n 1_{(S,T]} H dI_X \to \int h 1_{(S,T]} H dI_X,$$

pointside on Ω, strongly in L_G^1. It follows that the two limits are equal a.s.:

$$\int h 1_{(S,T]} H dI_X = h \int 1_{(S,T]} H dI_X.$$

This proves assertion a) in case (ii).

To prove assertion b) under hypothesis (i), assume that S is predictable and h is $\mathcal{F}_{S-}$-measurable. Let H be of the form $H = 1_{(s,t] \times A} y$ with $A \in \mathcal{F}_s$ and $y \in F$. Then

$$1_{[S,T]} H = y 1_{A \cap \{s < S\}} 1_{[S, T \wedge t]} + y 1_{A \cap \{S \leq s\}} 1_{(s, T \wedge t]},$$

§11. THE STOCHASTIC INTEGRAL AND STOPPING TIMES   145

hence

$$h\int 1_{[S,T]} H \, dI_X = h\int y1_{A\cap\{s<S\}} 1_{[S,T\wedge t]} \, dI_X + h\int y1_{A\cap\{S\leq s\}} 1_{(s,T\wedge t]} \, dI_X.$$

On the other hand

$$h1_{[S,T]} H = yh1_{A\cap\{s<S\}} 1_{[S,T\wedge t]} + yh1_{A\cap\{S\leq s\}} 1_{(s,T\wedge t]}.$$

Since $A \in \mathcal{F}_s$, we have $A \cap \{s < S\} \in \mathcal{F}_{S-}$, hence $y1_{A\cap\{s<S\}}$ and $yh1_{A\cap\{s<S\}}$ are $\mathcal{F}_{S-}$-measurable. We can apply Proposition 1 c) twice and deduce that

$$\int yh1_{A\cap\{s<S\}} 1_{[S,T\wedge t]} \, dI_X = yh1_{A\cap\{s<S\}} (X_{T\wedge t} - X_{S-})$$

$$= h\int y1_{A\cap\{s<S\}} 1_{[S,T\wedge t]} \, dI_X.$$

Since h is $\mathcal{F}_{S-}$-measurable, it is $\mathcal{F}_S$-measurable, hence $h1_{\{S\leq s\}}$ is $\mathcal{F}_s$-measurable; therefore $yh1_{A\cap\{S\leq s\}}$ is $\mathcal{F}_s$-measurable; also $\{S \leq s\} \in \mathcal{F}_s$, hence $y1_{A\cap\{S\leq s\}}$ is $\mathcal{F}_s$-measurable. We can apply Proposition 1 a) twice and deduce that

$$\int yh1_{A\cap\{S\leq s\}} 1_{(s,T\wedge t]} \, dI_X = yh1_{A\cap\{S\leq s\}} (X_{T\wedge t} - X_s)$$

$$= h\int y1_{A\cap\{S\leq s\}} 1_{(s,T\wedge t]} \, dI_X.$$

It follows that

$$\int h1_{(S,T]} H \, dI_X$$
$$= \int yh1_{A\cap\{s<S\}} 1_{[S,T\wedge t]} \, dI_X + \int yh1_{A\cap\{S\leq s\}} 1_{(s,T\wedge t]} \, dI_X$$
$$= h\int y1_{A\cap\{s<S\}} 1_{[S,T\wedge t]} \, dI_X + h\int y1_{A\cap\{S\leq s\}} 1_{(s,T\wedge t]} \, dI_X$$
$$= h\int 1_{[S,T]} H \, dI_X.$$

If $H = 1_{\{0\}\times A} y$ with $A \in \mathcal{F}_0$ and $y \in F$, then $y1_{\{0\}\times A}$ and $hy1_{\{0\}\times A}$ are $\mathcal{F}_{S-}$-measurable. We can apply twice Proposition 1 c):

$$\int h1_{[S,T]} H \, dI_X = \int hy1_{\{0\}\times A} 1_{[S,T]} \, dI_X$$
$$= hy1_{\{0\}\times A} (X_T - X_{S-})$$
$$= h\int y1_{\{0\}\times A} 1_{[S,T]} \, dI_X = h\int 1_{[S,T]} H \, dI_X.$$

It follows that for $B \in \mathcal{R}$ we have

$$\int h 1_{[S,T]} 1_B y dI_X = h \int 1_{[S,T]} 1_B y dI_X.$$

The proof continues then exactly as that of assertion a).

If we assume hypothesis (ii), the proof of assertion b) is exactly as that of assertion a), taking h' and h_n to be $\mathcal{F}_{S_-}$-measurable. ∎

B. Stopping the stochastic integral

Let $X : \mathbb{R}_+ \times \Omega \to E \subset L(F,G)$ be a p-summable process relative to (F,G).

6. Theorem. *Let* $H \in L^1_{F,G}(X)$ *and let* T *be a stopping time. Then* $1_{[0,T]}H \in L^1_{F,G}(X)$ *and*

$$(1_{[0,T]}H) \cdot X = (H \cdot X)^T.$$

If T is predictable, then $1_{[0,T)}H \in L^1_{F,G}(X)$ *and*

$$(1_{[0,T)}H) \cdot X = (H \cdot X)^{T-}.$$

Proof. Suppose first that T is a simple stopping time of the form

$$T = \sum_{1 \le i \le n} 1_{A_i} t_i$$

with $0 \le t_1 \le t_2 \le \ldots t_n \le +\infty$, $A_i \in \mathcal{F}_{t_i}$ mutually disjoint and with union Ω. For each $\omega \in \Omega$ there is a unique i such that $\omega \in A_i$ and then $T(\omega) = t_i$. Then

$$(H \cdot X)_T(\omega) = (H \cdot X)_{t_i}(\omega) = (\int_{[0,t_i]} H dI_X)(\omega)$$

hence

$$(H \cdot X)_T = \sum_{1 \le i \le n} 1_{A_i} \int_{[0,t_i]} H dI_X$$

$$= \int_{[0,\infty]} H dI_X - \sum_{1 \le i \le n} 1_{A_i} \int_{(t_i,\infty]} H dI_X$$

$$= \int_{[0,\infty]} H dI_X - \sum_{1 \le i \le n} \int_{(t_i,\infty]} 1_{A_i} H dI_X$$

$$= \int_{[0,\infty]} H dI_X - \int 1_{(T,\infty)} H dI_X = \int 1_{[0,T]} H dI_X.$$

We used Proposition 5 a) with $h = 1_{A_i}$, since $A_i \in \mathcal{F}_{t_i}$.

Let now T be an arbitrary stopping time and (T_n) a decreasing sequence of simple stopping times with $T_n \downarrow T$. Then $1_{[0,T_n]}H \to 1_{[0,T]}H$ pointwise,

$$\int 1_{[0,T_n]}H\, dI_X = (H \cdot X)_{T_n} \in L_G^p$$

and

$$\int 1_{[0,T_n]}H\, dI_X = (H \cdot X)_{T_n} \to (H \cdot X)_T,$$

since $H \cdot X$ is cadlag. We can apply the convergence Theorem 10.3 and deduce that $\int 1_{[0,T]}H\, dI_X \in L_G^p$ and

$$\int 1_{[0,T_n]}H\, dI_X \to \int 1_{[0,T]}H\, dI_X$$

pointwise and in L_G^1. It follows that the two limits are equal,

$$\int 1_{[0,T]}H\, dI_X = (H \cdot X)_T.$$

Replacing T with $T \wedge t$ we obtain

$$\int_{[0,t]} 1_{[0,T]}H\, dI_X = (H \cdot X)_{T \wedge t}.$$

It follows that $t \mapsto \int_{[0,t]} 1_{[0,T]}H\, dI_X$ is cadlag, hence $1_{[0,T]}H \in L^1_{F,G}(X)$ and

$$(1_{[0,T]}H \cdot X)_t = (H \cdot X)_{T \wedge t} = (H \cdot X)_t^T.$$

If T is predictable, let (T_n) be an increasing sequence of stopping times with $T_n \uparrow T$. Then $1_{[0,T_n]}H \to 1_{[0,T)}H$ and the above proof can be repeated, using the convergence Theorem 10.3, to deduce that $1_{[0,T)}H \in L^1_{F,G}(X)$ and

$$(1_{[0,T)}H \cdot X)_t = (H \cdot X)_t^{T-}.$$

∎

Taking $H = 1$, we obtain the following corollary:

7. Corollary. *For every stopping time T we have*

$$1_{[0,T]} \cdot X = X^T.$$

If T is predictable then
$$1_{[0,T)} \cdot X = X^{T-}.$$

Proof. We notice that the constant process $H = 1$ belongs to $L^1_{F,G}(X)$, hence, by Theorem 6, $1_{[0,T]} \in L^1_{F,G}(X)$; and in case T is predictable, then $1_{[0,T)} \in L^1_{F,G}(X)$. The corollary then follows from Theorem 6. ∎

We can extend now Proposition 5 for integrable processes H.

8. Theorem. *Let $S \leq T$ be stopping times and assume that either*
(i) $h : \Omega \to \mathbb{R}$ is bounded, $\mathcal{F}_S$-measurable and $H \in L^1_{F,G}(X)$,
or
(ii) $h : \Omega \to F$ is bounded, $\mathcal{F}_S$-measurable and $H \in L^1_{\mathbb{R},E}(X)$.
Then
a) $1_{(S,T]}H$ and $h1_{(S,T]}H$ are integrable with respect to X and

$$(h1_{(S,T]}H) \cdot X = h[(1_{(S,T]}H) \cdot X].$$

b) If S is predictable and h is $\mathcal{F}_{S_-}$-measurable, then $1_{[S,T]}H$ and $h1_{[S,T]}H$ are integrable with respect to X and

$$(h1_{[S,T]}H) \cdot X = h[(1_{[S,T]}H) \cdot X).$$

Proof. From Theorem 6 we deduce that the process

$$1_{(S,T]}H = 1_{[0,T]}H - 1_{[0,S]}H$$

belongs to $L^1_{F,G}(X)$ in case (i) and to $L^1_{\mathbb{R},E}(X)$ in case (ii); therefore, $\int_{[0,t]} 1_{(S,T]} H dI_X \in L^p_G$ in case (i) and $\int_{[0,t]} 1_{(S,T]} H dI_X \in L^p_E$ in case (ii). Then by Proposition 5 a) we have

$$\int_{[0,t]} h1_{(S,T]} H dI_X = h \int_{[0,t]} 1_{(S,T]} H dI_X.$$

Since the right-hand side term is cadlag, it follows that the left-hand side term is cadlag; therefore $h1_{(S,T]}H$ is integrable with respect to X. The above equality is written then

$$((h1_{(S,T]}H) \cdot X)_t = h((1_{(S,T]}H) \cdot X)_t$$

and assertion a) is proved.
Assertion b) is proved similarly, using Theorem 6 and Proposition 5 b). ∎

C. Summabilty of stopped processes

Let $X : \mathbb{R}_+ \times \Omega \to E \subset L(F,G)$ be a p-summable process relative to (F,G). The next theorem gives a more complete description of the properties of X^T. The proofs follow from the previous results and definitions.

9. Theorem. *Let T be a stopping time.*
a) X^T is p-summable relative to (F,G) and we have

$$I_{X^T}(A) = I_X([0,T] \cap A), \text{ for } A \in \mathcal{P}[0,\infty].$$

a') If T is predictable then X^{T-} is p-summable relative to (F,G) and we have

$$I_{X^{T-}}(A) = I_X([0,T) \cap A), \text{ for } A \in \mathcal{P}[0,\infty].$$

b) *For every predictable, F-valued process H we have*
$$svar_{F,L_G^p} I_{X^T}(H) = svar_{F,L_G^p} I_X(1_{[0,T]}H).$$

b') *If T is predictable then*
$$svar_{F,L_G^p} I_{X^{T-}}(H) = svar_{F,L_G^p} I_X(1_{[0,T)}H).$$

c) *We have $H \in \mathcal{F}_{F,G}(X^T)$ iff $1_{[0,T]}H \in \mathcal{F}_{F,G}(X)$ and in this case we have*
$$\int H dI_{X^T} = \int 1_{[0,T]} H dI_X.$$

c') *If T is predictable, then $H \in \mathcal{F}_{F,G}(X^{T-})$ iff $1_{[0,T)}H \in \mathcal{F}_{F,G}(X)$ and in this case we have*
$$\int H dI_{X^{T-}} = \int 1_{[0,T)} H dI_X.$$

d) *We have $H \in L^1_{F,G}(X^T)$ iff $1_{[0,T]}H \in L^1_{F,G}(X)$ and in this case we have*
$$H \cdot X^T = (1_{[0,T]}H) \cdot X.$$

If $H \in L^1_{F,G}(X)$, then $H \in L^1_{F,G}(X^T)$, $1_{[0,T]}H \in L^1_{F,G}(X)$ and
$$(H \cdot X)^T = H \cdot X^T = (1_{[0,T]}H) \cdot X.$$

d') *Assume T is predictable. We have $H \in L^1_{F,G}(X^{T-})$ iff $1_{[0,T)}H \in L^1_{F,G}(X)$ and in this case we have*
$$H \cdot X^{T-} = (1_{[0,T)}H) \cdot X$$

If $H \in L^1_{F,G}(X)$, then $H \in L^1_{F,G}(X^{T-})$, $1_{[0,T)}H \in L^1_{F,G}(X)$ and
$$(H \cdot X)^{T-} = H \cdot X^{T-} = (1_{[0,T)}H) \cdot X.$$

e) *If the set of measures $(I_X)_{F,L_{G^*}^q}$ is uniformly σ-additive, then so is $(I_{X^T})_{F,L_{G^*}^q}$.*

e') *If T is predictable and if $(I_X)_{F,L_{G^*}^q}$ is uniformly σ-additive, then so is $(I_{X^{T-}})_{F,L_{G^*}^q}$.*

Proof. a) For $A \in \mathcal{F}_0$ we have
$$I_{X^T}(\{0\} \times A) = 1_A X_0 = I_X(\{0\} \times A) = I_X([0,T] \cap (\{0\} \times A)).$$

For $s < t$ and $A \in \mathcal{F}_s$ we have, by Proposition 1,
$$I_{X^T}((s,t] \times A) = 1_A(X_t^T - X_s^T) = 1_A(X_{T \wedge t} - X_{T \wedge s})$$
$$= 1_A(I_X((T \wedge s, T \wedge t])) = 1_A \int 1_{(s,t]} 1_{[0,T]} dI_X$$
$$= \int 1_A 1_{(s,t]} 1_{[0,T]} dI_X = I_X([0,T] \cap ((s,t] \times A)).$$

We used above, Proposition 5 a) with $h = 1_A$, $(S, T] = (s, t]$ and $H = 1_{[0,T]}$. It follows that

$$I_{X^T}(B) = I_X([0, T] \cap B), \text{ for } B \in \mathcal{R}.$$

Since X is p-summable relative to (F, G), I_X can be extended to a σ-additive measure $I_X : \mathcal{P} \to L_G^p$. The equality

$$I_{X^T}(A) = I_X([0, T] \cap A), \text{ for } A \in \mathcal{P},$$

defines the unique σ-additive extension of I_{X^T} to $\mathcal{P}$. We have further

$$\operatorname{svar}_{F, L_G^p} I_{X^T}(A) = \operatorname{svar}_{F, L_G^p} I_X([0, T] \cap A), \text{ for } A \in \mathcal{P}.$$

In fact, let $A \in \mathcal{P}$, let (B_i) be a finite family of disjoint sets from $\mathcal{P}$ contained in A and (x_i) a family of elements from F_1. Then

$$|\sum_i I_{X^T}(B_i) x_i|_{L_G^p} = |\sum_i I_X([0, T] \cap B_i) x_i|_{L_G^p}$$
$$\leq \operatorname{svar}_{F, L_G^p} I_X([0, T] \cap A),$$

hence

$$\operatorname{svar}_{F, L_G^p} I_{X^T}(A) \leq \operatorname{svar}_{F, L_G^p} I_X([0, T] \cap A).$$

For the converse inequality, let (A_i) be a finite family of disjoint sets from $\mathcal{P}$ contained in $[0, T] \cap A$ and (x_i) be a family from F_1. Then $A_i = [0, T] \cap A_i \subset A$, hence

$$|\sum_i I_X(A_i) x_i|_{L_G^p} = |\sum_i I_X([0, T] \cap A_i) x_i|_{L_G^p}$$
$$= |\sum_i I_{X^T}(A_i) x_i|_{L_G^p} \leq \operatorname{svar}_{F, L_G^p} I_{X^T}(A),$$

therefore

$$\operatorname{svar}_{F, L_G^p} I_X([0, T] \cap A) \leq \operatorname{svar}_{F, L_G^p} I_{X^T}(A),$$

and the equality follows. Since I_X has finite semivariation relative to (F, L_G^p), it follows that I_{X^T} has finite semivariation relative to (F, L_G^p), hence X^T is p-summable relative to (F, L_G^p) and assertion a) is proved.

Assertion a') is proved similarly.

To prove assertion b), let $z \in L_{G*}^q$, $\frac{1}{p} + \frac{1}{q} = 1$, with $\|z\|_q \leq 1$. Then, from a) we deduce

$$(I_{X^T})_z(A) = (I_X)_z([0, T] \cap A), \text{ for } A \in \mathcal{P}.$$

Then we have the same equality for the variations:

$$|(I_{X^T})_z|(A) = |(I_X)_z|([0, T] \cap A), \text{ for } A \in \mathcal{P}.$$

It follows that for any predictable F-valued process H we have

$$\int |H|d|(I_{X^T})_z| = \int 1_{[0,T]}|H|d|(I_X)_z|.$$

Taking the supremum for $\|z\|_q \leq 1$ we obtain the equality in assertion b). Assertion b') is proved similarly.

It follows that $H \in \mathcal{F}_{F,G}(X^T)$ iff $1_{[0,T]}H \in \mathcal{F}_{F,G}(X)$.

Let $H \in \mathcal{F}_{F,G}(X^T) = \bigcap_{\|z\|_q \leq 1} L^1_F((I_{X^T})_z)$. From the equality

$$(I_{X^T})_z(A) = (I_X)_z([0,T] \cap A), \text{ for } A \in \mathcal{P}$$

we deduce that

$$\int H d(I_{X^T})_z = \int 1_{[0,T]} H d(I_X)_z,$$

therefore

$$\int H dI_{X^T} = \int 1_{[0,T]} H dI_X$$

and this is the equality in assertion c). Assertion c') is proved similarly.

To prove assertion d), we replace H with $1_{[0,t]}H$ in assertion c) and deduce that $1_{[0,t]}H \in \mathcal{F}_{F,G}(X^T)$ iff $1_{[0,t]}1_{[0,T]}H \in \mathcal{F}_{F,G}(X)$ and in this case we have

$$\int_{[0,t]} H dI_{X^T} = \int_{[0,t]} 1_{[0,T]} H dI_X.$$

It follows that $H \in L^1_{F,G}(X^T)$ iff $1_{[0,T]}H \in L^1_{F,G}(X)$ and in this case we have

$$(H \cdot X^T)_t = ((1_{[0,T]}H) \cdot X)_t.$$

If now $H \in L^1_{F,G}(X)$, then, from Theorem 6 we deduce that $1_{[0,T]}H \in L^1_{F,G}(X)$ and

$$(1_{[0,T]}H) \cdot X = (H \cdot X)^T.$$

Then assertion d) follows. Assertion d') is proved similarly.

Finally, from the equality

$$|(I_{X^T})_z|(A) = |(I_X)_z|([0,T] \cap A), \text{ for } A \in \mathcal{P}$$

it follows that if $(I_X)_{F,L^q_{G^*}}$ is uniformly σ-additive, then so is $(I_{X^T})_{F,L^q_{G^*}}$, which proves assertion e) and assertion e') is proved similarly. ∎

D. The jumps of the stochastic integral

Assume X is p-summable relative to (F, G).

10. Theorem. *For any process $H \in L^1_{F,G}(X)$ we have*

$$\Delta(H \cdot X) = H\Delta X.$$

Proof. Assume first H is bounded. By Proposition 9.10 we have

$$\Delta X_t = X_t - X_{t-} \in L^p_E$$

By Theorem 10.7 we have $(H \cdot X)_{t-} \in L^p_G$ and

$$(H \cdot X)_{t-} = \int_{[0,t)} H dI_X.$$

Then

$$\Delta(H \cdot X)_t = (H \cdot X)_t - (H \cdot X)_{t-} = \int_{[0,t]} H dI_X - \int_{[0,t)} H dI_X$$
$$= \int_{\{t\}} H dI_X = \int_{\{t\}} H_t dI_X = H_t \int_{\{t\}} dI_X = H_t \Delta X_t,$$

where we used Proposition 5 with $h = H_t$, since H_t is $\mathcal{F}_{t-}$-measurable.

Assume now $H \in L^1_{F,G}(X)$. For each n, the stopping time $T_n = \inf\{t : |H_t| \geq n\}$ is predictable and $1_{[0,T_n)}|H| \leq n$. By the first part of the proof we have

$$\Delta(1_{[0,T_n)} H \cdot X) = 1_{[0,T_n)} H \Delta X.$$

On the other hand

$$\Delta(1_{[0,T_n)} H \cdot X)_t = \int 1_{\{t\}} 1_{[0,T_n)} H dI_X$$
$$= \int 1_{\{t\}} 1_{\{t<T_n\}} H dI_X = 1_{\{t<T_n\}} \int 1_{\{t\}} H dI_X$$
$$= 1_{\{t<T_n\}} \Delta(H \cdot X)_t,$$

where we used again Proposition 5, since $1_{\{t<T_n\}}$ is $\mathcal{F}_{t-}$-measurable. Thus

$$1_{\{t<T_n\}} \Delta(H \cdot X)_t = 1_{\{[0,T_n)\}} H_t \Delta X_t$$

and the desired equality follows by letting $n \to \infty$. ∎

§12. CONVERGENCE THEOREMS

Assume X is p-summable relative to (F, G). We have already proved a useful Lebesgue-type convergence theorem (Theorem 10.3) for processes in $\mathcal{F}_{F,G}(X)$, concerning the convergence of the integrals.

In this paragraph we shall consider Lebesgue- and Vitali-type theorems for convergence in the topology of $L^1_{F,G}(X)$, as well as pointwise uniform convergence of the integrals on compact time intervals for a suitable subsequence.

A. The completeness of the space $L^1_{F,G}(X)$

The key result needed for the uniform convergence property is the following theorem, which will imply that the space $L^1_{F,G}(X)$ is complete.

1. Theorem. *Let (H^n) be a sequence from $L^1_{F,G}(X)$ and assume that $H^n \to H$ in $\mathcal{F}_{F,G}(X)$. Then:*
a) $H \in L^1_{F,G}(X)$.
b) $(H^n \cdot X)_t \to (H \cdot X)_t$, *in* L^p_G, *for* $t \in [0, \infty]$.
c) *There is a subsequence (r_n) such that*

$$(H^{r_n} \cdot X)_t \to (H \cdot X)_t, \ a.s., \ as \ n \to \infty,$$

uniformly on every bounded time interval.

Proof. Since $H^n \to H$ in $\mathcal{F}_{F,G}(X)$, for every $t \geq 0$ we have $1_{[0,t]} H^n \to 1_{[0,t]} H$ in $\mathcal{F}_{F,G}(X)$. Since the integral is continuous, we deduce that

$$(H^n \cdot X)_t = \int_{[0,t]} H^n dI_X \to \int_{[0,t]} H dI_X, \ in \ (L^p_{G^*})^*.$$

Since $H^n \in L^1_{F,G}(X)$ we have $\int_{[0,t]} H^n dI_X \in L^p_G$ and

$$(H^n \cdot X)_t \to \int_{[0,t]} H dI_X, \ in \ L^p_G.$$

To prove assertion a), we have to prove that the process $(\int_{[0,t]} H dI_X)_{t \geq 0}$ is cadlag; then it will follow that

$$(H \cdot X)_t = \int_{[0,t]} H dI_X, \ a.s.,$$

hence

$$(H^n \cdot X)_t \to (H \cdot X)_t, \ in \ L^p_G,$$

which is assertion b).

In order to prove that $(\int_{[0,t]} H dI_X)_{t \geq 0}$ is cadlag, we shall construct a subsequence (r_n) such that

$$(H^{r_n} \cdot X)_t \to \int_{[0,t]} H dI_X, \ a.s., \ as \ n \to \infty,$$

uniformly on any bounded interval; since $(H^{r_n} \cdot X)$ are cadlag, it will follow that the limit is also cadlag, hence $H \in L^1_{F,G}(X)$ and then

$$(H^{r_n} \cdot X)_t \to (H \cdot X)_t, \text{ a.s., as } n \to \infty,$$

uniformly on every bounded interval, which is assertion c).

Since (H^n) is a Cauchy sequence in $\mathcal{F}_{F,G}(X)$, there is a subseqeunce (H^{r_n}) of (H^n) such that

$$\tilde{I}_{F,G}(H^{r_n} - H^{r_{n+1}}) \leq 4^{-n}, \text{ for each } n.$$

Let $t_0 > 0$ be arbitrary. For each n denote $K^n = H^{r_n}$ and $Z^n = K^n \cdot X$ and define the stopping time

$$u_n = \inf\{t : |Z^n_t - Z^{n+1}_t| > 2^{-n}\} \wedge t_0.$$

Let $G_n = \{u_n < t_0\}$. By Theorem 11.9 d), for each stopping time v we have

$$Z^n_v = (K^n \cdot X)_v = (K^n \cdot X)^v_\infty = ((1_{[0,v]} K^n) \cdot X)_\infty = \int_{[0,v]} K^n dI_X,$$

hence

$$E(|Z^n_v - Z^{n+1}_v|) = E(|\int_{[0,v]} (K^n - K^{n+1}) dI_X|)$$
$$= (\|\int_{[0,v]} (K^n - K^{n+1}) dI_X\|_{L^1_G}$$
$$\leq \|\int_{[0,v]} (K^n - K^{n+1}) dI_X\|_{L^p_G} \leq \tilde{I}_{F,G}(K^n - K^{n+1}) \leq 4^{-n}.$$

In particular, for $v = u_n$ we have

$$E(|Z^n_{u_n} - Z^{n+1}_{u_n}|) \leq 4^{-n}.$$

We have also
$$P(G_n) \leq 2^n E(|Z^n_{u_n} - Z^{n+1}_{u_n}|) \leq 2^{-n}.$$

In fact, if $\omega \in G_n$, we have $u_n(\omega) < t_0$ and there is a sequence $t_i \downarrow u_n(\omega)$ with $t_i < t_0$ such that $|Z^n_{t_i}(\omega) - Z^{n+1}_{t_i}(\omega)| > 2^{-n}$ for each i; by right continuity of Z^n and Z^{n+1} we deduce that

$$|Z^n_{u_n}(\omega) - Z^{n+1}_{u_n}(\omega)| \geq 2^{-n};$$

therefore
$$E(|Z^n_{u_n} - Z^{n+1}_{u_n}|) \geq 2^{-n} P(G_n)$$

and the desired inequality follows.

Let $G_0 = \limsup_n G_n$. Then $P(G_0) = 0$. For $\omega \notin G_0$, there is an N such that, if $n \geq N$, we have $\omega \notin G_n$, hence $u_n(\omega) = t_0$. Thus

$$\sup_{t<t_0} |Z_t^n(\omega) - Z_t^{n+1}(\omega)| \leq 2^{-n}.$$

It follows that for $\omega \notin G_0$, the sequence $(Z_t^n(\omega))$ is Cauchy in G, uniformly for $t < t_0$.

We have also

$$\|Z_t^n - Z_t^{n+1}\|_{L_G^p} \leq 2^{-n},$$

hence (Z_t^n) is a Cauchy sequence in L_G^p. The process Z defined for $t < t_0$ by

$$Z_t(\omega) = \lim_n Z_t^n(\omega) \text{ if } \omega \notin G_0 \text{ and } Z_t(\omega) = 0 \text{ if } \omega \in G_0,$$

is cadlag.

At the beginning of the proof we noticed that

$$(H^n \cdot X)_t \to \int_{[0,t]} H dI_X, \text{ in } L_G^p, \text{ for } t \geq 0.$$

Then, for $t < t_0$ we have

$$Z_t^n = (H^{r_n} \cdot X)_t \to \int_{[0,t]} H dI_X, \text{ in } L_G^p.$$

It follows that the two limits of (Z_t^n) are equal a.s.:

$$\int_{[0,t]} H dI_X = Z_t, \text{ a.s. for } t < t_0.$$

Since Z is cadlag at every $t < t_0$ it follows that the process $(\int_{[0,t]} H dI_X)_{t \geq 0}$ is cadlag at every $t < t_0$. Since t_0 was arbitrary, this process is cadlag at every $t \geq 0$. It follows that $H \in L_{F,G}^1(X)$,

$$(H \cdot X)_t = \int_{[0,t]} H dI_X,$$

and

$$(H \cdot X)_t = Z_t, \text{ a.s. for } t < t_0$$

and assertions b) and c) follow. ■

2. Corollary. $L_{F,G}^1(X)$ *is complete.*

3. Corollary. *If $I_{F,L_{G^*}^q}$ is uniformly σ-additive, then $L_{F,G}^1(X)$ contains all the F-valued, bounded, predictable processes. (In particular, this is the case if $F = \mathbb{R}$).*

Proof. In fact, by Proposition 5.44, the space of F-valued, $\mathcal{R}$-step functions is dense in the subspace of bounded processes of $\mathcal{F}_{F,G}(X)$. Since the F-valued, $\mathcal{R}$-step functions belong to $L^1_{F,G}(X)$, the F-valued, bounded, predictable processes belong to $L^1_{F,G}(X)$. ∎

Remark. We shall see that we have $L^1_{F,G}(X) = \mathcal{F}_{F,G}(X)$ if X has integrable variation (Theorem 19.14) or if X is a p-summable martingale and if L^p_G is reflexive (Corollary 16.3 and Theorem 17.13).

B. The Uniform Convergence Theorem

Uniform convergence of processes yields convergence in $L^1_{F,G}(X)$, as the next theorem shows.

4. Theorem. *Let (H^n) be a sequence from $\mathcal{F}_{F,G}(X)$ which converges pointwise uniformly to a process H. Then:*
a) $H \in \mathcal{F}_{F,G}(X)$ *and* $H^n \to H$ *in* $\mathcal{F}_{F,G}(X)$.
  Assume, in addition, that $H^n \in L^1_{F,G}(X)$ for each n. Then
b) $H \in L^1_{F,G}(X)$ *and* $H^n \to H$ *in* $L^1_{F,G}(X)$;
c) *For every $t \in [0, \infty]$ we have $(H^n \cdot X)_t \to (H \cdot X)_t$, in L^p_G.*
d) *There is a subsequence (r_n) such that $(H^{r_n} \cdot X)_t \to (H \cdot X)_t$, a.s. as $n \to \infty$, uniformly on any bounded interval.*

Proof. Assertion a) is immediate. Assertions b), c) and d) follow from Theorem 1. ∎

C. The Vitali and the Lebesgue Convergence Theorems

The following two theorems follow from Theorems 1 and 4 and from the general Vitali and Lebesgue Convergence Theorems (Theorems 5.36 and 5.37).

5. Theorem. (Vitali). *Let (H^n) be a sequence from $\mathcal{F}_{F,G}(X)$ and let H be an F-valued, predictable process. Assume that*
(i) $\tilde{I}_{F,G}(H^n 1_A) \to 0$ *as* $\tilde{I}_{F,G}(A) \to 0$, *uniformly in n*
and that any one of the conditions (ii) *or* (iii) *below is true:*
(ii) $H^n \to H$ *in* $\tilde{I}_{F,G}$-*measure;*
(iii) $H^n \to H$ *pointwise and $I_{F,L^q_{G^*}}$ is uniformly σ-additive (this is the case if H^n are real-valued, i.e., $F = \mathbb{R}$).*
  Then:
a) $H \in \mathcal{F}_{F,G}(X)$ *and* $H^n \to H$ *in* $\mathcal{F}_{F,G}(X)$.
  Conversely, if $H^n, H \in \mathcal{F}_{F,G}(\mathcal{B}, X)$ and $H^n \to H$ in $\mathcal{F}_{F,G}(X)$, then conditions (i) *and* (ii) *are satisfied.*
  Under the hypotheses (i) *and* (ii) *or* (iii), *assume, in addition, that $H^n \in L^1_{F,G}(X)$ for each n. Then*
b) $H \in L^1_{F,G}(X)$ *and* $H^n \to H$ *in* $L^1_{F,G}(X)$;
c) *For every $t \in [0, \infty]$ we have $(H^n \cdot X)_t \to (H \cdot X)_t$, in L^p_G;*

d) There is a subsequence (r_n) such that $(H^{r_n} \cdot X)_t \to (H \cdot X)_t$, a.s., as $n \to \infty$, uniformly on any bounded interval.

6. Theorem. (Lebesgue). Let (H^n) be a sequence from $\mathcal{F}_{F,G}(X)$ and let H be an F-valued predictable process. Assume that
(i) There is a process $\phi \in \mathcal{F}_\mathbb{R}(\mathcal{B}, I_{F,G})$ such that

$$|H^n| \leq \phi \text{ for each } n;$$

and that any one of the conditions (ii) or (iii) below is true:
(ii) $H^n \to H$ in $\tilde{I}_{F,G}$-measure;
(iii) $H^n \to H$ pointwise and $I_{F,L^q_{G^*}}$ is uniformly σ-additive (this is the case if H^n are real valued, i.e., $F = \mathbb{R}$).
   Then:
a) $H \in \mathcal{F}_{F,G}(\mathcal{B}, X)$ and $H^n \to H$ in $\mathcal{F}_{F,G}(X)$.
   Assume, in addition that $H^n \in L^1_{F,G}(X)$ for each n. Then
b) $H \in L^1_{F,G}(X)$ and $H^n \to H$ in $L^1_{F,G}(X)$;
c) For every $t \in [0, \infty]$ we have $(H^n \cdot X)_t \to (H \cdot X)_t$, in L^p_G;
d) There is a subsequence (r_n) such that $(H^{r_n} \cdot X)_t \to (H \cdot X)_t$, a.s., as $n \to \infty$, uniformly on any bounded interval.

D. The stochastic integral of σ-elementary and of caglad processes as a pathwise Stieltjes integral

The stochastic integral can be computed pathwise for elementary processes of the form

$$H = H_0 1_{\{0\}} + \sum_{1 \leq i \leq n} H_i 1_{(T_i, T_{i+1}]},$$

where $0 = T_0 \leq T_1 \leq T_2 \leq \cdots \leq T_{n+1}$ are stopping times and for $0 \leq i \leq n$, H_i is an F-valued, bounded, $\mathcal{F}_{T_i}$-measurable random variable (Corollary 11.2). A σ-elementary process is of the form

$$H = H_0 1_{\{0\}} + \sum_{1 \leq n < \infty} H_n 1_{(T_n, T_{n+1}]},$$

where $0 = T_0 \leq T_1 \leq T_2 \leq \ldots$ is a sequence of stopping times and for each $n \geq 0$, H_n is an F-valued, bounded, $\mathcal{F}_{T_n}$-measurable random variable.

A σ-elementary process does not necessarily belong to $\mathcal{F}_{F,G}(X)$; but if it belongs to $\mathcal{F}_{F,G}(X)$, then it belongs to $L^1_{F,G}(X)$ and its stochastic integral can be computed pathwise. This result will follow from the following general theorem.

7. Theorem. Let $H \in \mathcal{F}_{F,G}(X)$ and assume there is a sequence $T_n \uparrow \infty$ of stopping times such that $1_{[0,T_n]} H \in L^1_{F,G}(X)$, for every n. Then $H \in L^1_{F,G}(X)$ and

$$(1_{[0,T_n]} H) \cdot X \to H \cdot X, \text{ pointwise.}$$

Proof. We remark first that by Theorem 11.9 d) we have $1_{[0,T_n]}H \in L^1_{F,G}(X^{T_n})$ for each n. For each $t \geq 0$ we have

$$1_{[0,t]}1_{[0,T_n]}H \to 1_{[0,t]}H, \text{ pointwise,}$$
$$|1_{[0,t]}1_{[0,T_n]}H| \leq |H|, \text{ for each } n,$$
$$\int_{[0,t]} 1_{[0,T_n]}HdI_{X^{T_n}} \in L^p_G$$

and

$$\int_{[0,t]} 1_{[0,T_n]}HdI_X = ((1_{[0,T_n]}H) \cdot X)_t = ((1_{[0,T_n]}H) \cdot X^{T_n})_t.$$

We shall prove that this last sequence converges pointwise, as $n \to \infty$. In fact, for $m \leq n$ and $t < T_m$ we have

$$(1_{[0,T_m]}H \cdot X)^{T_n}_t = (1_{[0,T_n]}H \cdot X)^{T_m}_t = (1_{[0,T_m]}H \cdot X)_t;$$

For each $t \geq 0$ and $\omega \in \Omega$, we choose $m = m_\omega$ such that $t < T_m(\omega)$. Then, for $n \geq m$ we have

$$(1_{[0,T_m]}H \cdot X)_t(\omega) = (1_{[0,T_n]}H \cdot X)^{T_m}_t(\omega) = (1_{[0,T_n]}H \cdot X)^{T_n}_t(\omega),$$

hence

$$(1_{[0,T_m]}H \cdot X)_t(\omega) = \lim_n (1_{[0,T_n]}H \cdot X)^{T_n}_t(\omega) = \lim_n \left(\int_{[0,t]} 1_{[0,T_n]}HdI_X \right)(\omega).$$

This proves the pointwise convergence asserted above.

The hypotheses of the convergence Theorem 10.3 are satisfied for the sequence $H^n = 1_{[0,t]}1_{[0,T_n]}H$. It follows that $\int 1_{[0,t]}HdI_X \in L^p_G$ and

$$\int_{[0,t]} 1_{[0,T_n]}HdI_X \to \int_{[0,t]} HdI_X, \text{ pointwise.}$$

For each $\omega \in \Omega$ and $m = m_\omega$ as above, we have

$$(\int_{[0,t]} HdI_X)(\omega) = (1_{[0,T_m]}H \cdot X)_t(\omega),$$

hence, the process $(\int_{[0,t]} HdI_X)_{t \geq 0}$ is cadlag. It follows that $H \in L^1_{F,G}(X)$ and

$$(H \cdot X)_t = \int_{[0,t]} HdI_X = \lim_n (1_{[0,T_n]}H \cdot X)_t.$$

∎

8. Corollary. $L^1_{F,G}(X)$ *contains all the σ-elementary processes of $\mathcal{F}_{F,G}(X)$. If H is a σ-elementary process in standard form:*

$$H = H_0 1_{\{0\}} + \sum_{1 \leq n < \infty} H_n 1_{(T_n, T_{n+1}]};$$

then the stochastic integral $H \cdot X$ can be computed pathwise:

$$(H \cdot X)_t(\omega) = H_0(\omega) X_0(\omega) + \sum_{1 \leq n < \infty} H_n(\omega)(X_{T_{n+1} \wedge t}(\omega) - X_{T_n \wedge t}(\omega)).$$

Proof. We have $T_n \uparrow \infty$ and

$$1_{[0,T_n]} H = H_0 1_{\{0\}} + \sum_{1 \leq i < n} H_i 1_{(T_{i+1}, T_i]} \in L^1_{F,G}(X).$$

If $H \in \mathcal{F}_{F,G}(X)$, by Theorem 7 we have $H \in L^1_{F,G}(X)$ and

$$(H \cdot X)_t(\omega) = \lim_n (1_{[0,T_n]} H \cdot X)_t(\omega)$$

$$= \lim_n (H_0 X_0 + \sum_{1 \leq i < n} H_i(X_{T_{i+1} \wedge t} - X_{T_i \wedge t}))(\omega)$$

$$= H_0(\omega) X_0(\omega) + \sum_{1 \leq i < \infty} H_i(\omega)(X_{T_{i+1} \wedge t}(\omega) - X_{T_i \wedge t}(\omega)).$$

∎

Remark. There are σ-elementary processes which do not belong to $\mathcal{F}_{F,G}(X)$; such processes are not integrable with respect to X. However, we shall see (Theorem 15.12) that such processes are "locally integrable" with respect to any "locally summable process", even if the random variables H_n are not bounded.

The next theorem considers all caglad processes of $\mathcal{F}_{F,G}(X)$, not just the σ-elementary processes.

9. Theorem. $L^1_{F,G}(X)$ *contains all caglad processes of $\mathcal{F}_{F,G}(X)$.*

In particular, $L^1_{F,G}(X)$ contians all bounded, caglad, adapted, F-valued processes.

Proof. Assume first H is a bounded, caglad, adapted, F-valued process. Then the right limits process H_+ is cadlag, bounded and adapted. For each n define the sequence $T(n,k)$ of stopping times by

$$T(n,0) = 0$$

and for $k \geq 0$,

$$T(n, k+1) = \inf\{t > T(n,k) : |H_{t+} - H_{T(n,k)+}| > \frac{1}{n}\} \wedge (T(n,k) + \frac{1}{n}).$$

For each n define the σ-elementary process
$$H^n = \sum_{k \geq 0} H_{T(n,k)+1}1_{(T(n,k),\ T(n,k+1)]}.$$

We note that if $|H| \leq M$, then $|H^n| \leq M$ for each n, hence $H^n \in \mathcal{F}_{F,G}(X)$. By Corollary 8 we have $H^n \in L^1_{F,G}(X)$. Since H is caglad, from the definition of the above family of stopping times we deduce that $H^n \to H$ uniformly. By Theorem 4 we deduce that $H \in L^1_{F,G}(X)$.

Now assume $H \in \mathcal{F}_{F,G}(X)$ and that H is caglad; hence H is locally bounded. Let $S_n \uparrow \infty$ be a sequence of stopping times such that each process $1_{[0,S_n]}H$ is bounded. Since each such process is caglad, we have, by the above, $1_{[0,S_n]}H \in L^1_{F,G}(X)$. Then, by Theorem 7, we deduce that $H \in L^1_{F,G}(X)$. ∎

§13. SUMMABILITY OF THE STOCHASTIC INTEGRAL

Assume X is p-summable relative to (F,G). The following two theorems state that under certain conditions, the stochastic integral $H\cdot X$ is itself p-summable and the associativity formula $K\cdot(H\cdot X)=(KH)\cdot X$ holds.

This property follows from the associativity formula proved in Theorems 5.49 and 5.50 for the general integration theory with respect to vector measures with finite semivariation.

We consider first the case of real-valued processes H.

1. Theorem. Let $H\in\mathcal{F}_{\mathbb{R}}(\tilde{I}_{F,G})$. Assume that $H\in L^1_{\mathbb{R},E}(X)$ and $\int_A H\,dI_X\in L^p_E$ for $A\in\mathcal{P}$. Then:

a) $H\cdot X$ is p-summable relative to (F,G) and
$$dI_{H\cdot X}=d(HI_X)$$
where HI_X is the measure defined by
$$(HI_X)(A)=\int_A H\,dI_X,\text{ for }A\in\mathcal{P}.$$

b) For any predictable process $K\geq 0$ we have
$$(\tilde{I}_{H\cdot X})_{F,G}(K)=(\tilde{I}_X)_{F,G}(KH).$$

c) $K\in L^1_{F,G}(H\cdot X)$ iff $KH\in L^1_{F,G}(X)$, and in this case we have
$$K\cdot(H\cdot X)=(KH)\cdot X.$$

d) Assume $(I_X)_{F,L^q_{G^*}}$ is uniformly σ-additive. Then $(I_{H\cdot X})_{F,L^q_{G^*}}$ is uniformly σ-additve iff $H\in\mathcal{F}_{\mathbb{R}}(\mathcal{B},(I_X)_{F,G})$.

Proof. We note first that since $\int_A H\,dI_X\in L^p_E$ for every $A\in\mathcal{P}$, by Proposition 5.48, the measure HI_X is σ-additive on $\mathcal{P}$.

For any predictable rectangle $A\in\mathcal{R}$ we have

(i) $$I_{H\cdot X}(A)=\int_A H\,dI_X.$$

In fact, if $A=(s,t]\times B$ with $B\in\mathcal{F}_s$, then using Proposition 11.5 a) we have
$$I_{H\cdot X}((s,t]\times B)=1_B((H\cdot X)_t-(H\cdot X)_s)$$
$$=1_B\int_{(s,t]}H\,dI_X=\int_{(s,t]}1_B H\,dI_X=\int_{(s,t]\times B}H\,dI_X;$$

if $A=\{0\}\times B$ with $B\in\mathcal{F}_0$, then using Proposition 11.5 b) we have
$$I_{H\cdot X}(\{0\}\times B)=1_B(H\cdot X)_0=1_B\int_{\{0\}}H\,dI_X$$
$$=\int_{\{0\}}1_B H\,dI_X=\int_{\{0\}\times B}H\,dI_X.$$

It follows that the equality $I_{H \cdot X}(A) = \int_A H dI_X$ is valid for every $A \in \mathcal{R}$.

Since the measure $A \mapsto \int_A H dI_X$ is σ-additive for $A \in \mathcal{P}$, it follows that $I_{H \cdot X}$ can be extended to a σ-additive measure on $\mathcal{P}$, by the same equality,

(i') $$(I_{H \cdot X})(A) = \int_A H dI_X, \text{ for } A \in \mathcal{P},$$

that is, $d(I_{H \cdot X}) = d(H I_X)$.

To prove that $H \cdot X$ is p-summable relative to (F, G) we have to prove that $I_{H \cdot X}$ has finite semivariation relative to (F, G).

Let $z \in L^q_{G^*}, \frac{1}{p} + \frac{1}{q} = 1$, with $\|z\|_q \leq 1$ and $x \in F$. From the above equality (i') we deduce that for every $A \in \mathcal{P}$ we have

$$\langle x, (I_{H \cdot X})_z(A) \rangle = \langle I_{H \cdot X}(A) x, z \rangle = \langle (\int_A H dI_X) x, z \rangle$$
$$= \langle \int_A x H dI_X, z \rangle = \int_A x H d(I_X)_z = \langle x, \int_A H d(I_X)_z \rangle,$$

therefore,

(ii) $$(I_{H \cdot X})_z(A) = \int_A H d(I_X)_z.$$

Since $H \in \mathcal{F}_\mathbb{R}(\tilde{I}_{F,G})$, we have $H \in L^1_\mathbb{R}((I_X)_z)$. By Theorem 2.29, we have the following equality for the variations:

(iii) $$|(I_{H \cdot X})_z|(A) = \int_A |H| d|(I_X)_z|, \text{ for } A \in \mathcal{P}.$$

Taking the supremum for $\|z\|_q \leq 1$ we obtain

(iv) $$(\tilde{I}_{H \cdot X})_{F,G}(A) = (\tilde{I}_X)_{F,G}(1_A H) < \infty, \text{ for } A \in \mathcal{P}.$$

It follows that $H \cdot X$ is p-summable relative to (F, G) and this proves assertion a).

To prove assertion b), let $K \geq 0$ be a predictable process and let (K^n) be an increasing sequence of positive, simple, predictable processes such that $K^n \uparrow K$. From the above equality (iii) of the variations and using Theorem 2.31 b) we get, for each n,

$$\int K^n d|(I_{H \cdot X})_z| = \int K^n |H| d|(I_X)_z|.$$

Letting $n \to \infty$ we get

$$\int K d|(I_{H \cdot X})_z| = \int K |H| d|(I_X)_z|.$$

Taking the supremum for $\|z\|_q \leq 1$ we obtain the equality in assertion b).

To prove assertion c), let K be an F-valued, predictable process. From b) it follows that $K \in \mathcal{F}_{F,G}(H \cdot X)$ iff $KH \in \mathcal{F}_{F,G}(X)$.

Assume $K \in \mathcal{F}_{F,G}(H \cdot X)$. From the equality (ii):

$$(I_{H \cdot X})_z(A) = \int_A H d(I_X)_z, \text{ for } A \in \mathcal{P},$$

proved above, using Theorem 2.31 a) and b) we deduce that $1_{[0,t]} K \in L_F^1((I_{H \cdot X})_z)$ and $1_{[0,t]} KH \in L_F^1((I_X)_z)$ and we have

$$\int_{[0,t]} K d(I_{H \cdot X})_z = \int_{[0,t]} KH d(I_X)_z,$$

that is,

$$\langle \int_{[0,t]} K dI_{H \cdot X}, z \rangle = \langle \int_{[0,t]} KH dI_X, z \rangle.$$

Then

$$\int_{[0,t]} K dI_{H \cdot X} = \int_{[0,t]} KH dI_X \in (L_{G^*}^p)^*.$$

It follows that the left-hand side belongs to L_G^p and is cadlag iff the right-hand side has the same property. This means that $K \in L_{F,G}^1(H \cdot X)$ iff $KH \in L_{F,G}^1(X)$ and the equality of the integrals means

$$K \cdot (H \cdot X) = (KH) \cdot X,$$

which is assertion c).

Assertion d) follows from Theorem 5.49 d). ■

We consider now the case of vector-valued processes H.

2. Theorem. *Let* $H \in L_{F,G}^1(X)$ *be such that* $\int_A H dI_X \in L_G^p$ *for* $A \in \mathcal{P}$. *Then:*
a) $H \cdot X$ *is p-summable relative to* $(\mathbb{R}, G)$ *and*

$$dI_{H \cdot X} = d(HI_X).$$

b) *For any predictable process* $K \geq 0$ *we have*

$$(\tilde{I}_{H \cdot X})_{R,G}(K) \leq (\tilde{I}_X)_{F,G}(KH).$$

c) *If* K *is a real-valued predictable process and if* $KH \in L_{F,G}^1(X)$, *then* $K \in L_{\mathbb{R},G}^1(H \cdot X)$ *and we have*

$$K \cdot (H \cdot X) = (KH) \cdot X.$$

Proof. The proof is similar to that of Theorem 1, using the inequality

$$|(I_{H \cdot X})_z|(A) \leq \int_A |H| d|(I_X)_z|, \text{ for } A \in \mathcal{P},$$

and Theorem 5.50. ■

§14. SUMMABILITY CRITERION

A fundamental question concerning the stochastic integral is: When is X summable?

We shall present a rather unexpected result, that if $c_0 \not\subset E$, the mere boundedness of the measure I_X on the ring $\mathcal{R}$ implies that I_X is σ-additive and can be extended to a σ-additive measure on the σ-algebra $\mathcal{P}$. Then X is p-summable relative to $(\mathbb{R}, E)$; if I_X has bounded semivariation on $\mathcal{R}$ relative to (F, L_G^p), then X is p-summable relative to (F, G).

A. Quasimartingales and the Doléans measure

In order to prove the above-mentioned result, we have to prove that the Doléans measure of a quasimartingale of class (D) is σ-additive. For this purpose we need the following lemma.

1. Lemma. *Let $\mu : \mathcal{R} \to E$ be a finitely additive measure with bounded variation $|\mu|$. Then for any $s \geq 0$ the following two assertions are equivalent:*
a) $\lim_{t \downarrow s} \mu((s,t] \times F) = 0$, *for any* $F \in \mathcal{F}_s$;
b) $\lim_{t \downarrow s} |\mu|((s,t] \times \Omega) = 0$.

Proof. (cf. [Me]). Obviously, b) implies a). Assume now assertion a) is true. To prove b), let $t_n \downarrow s$ and $\varepsilon > 0$. Let (R_k) be a finite partition of $(s, t_1] \times \Omega$ consisiting of predictable rectangles, such that

$$|\mu|((s,t_1] \times \Omega) \leq \sum_k |\mu(R_k)| + \varepsilon.$$

For any $A \in \mathcal{R}$ contained in $(s, t_1] \times \Omega$ we have

$$\sum_k |\mu(A \cap R_k)| \geq \sum_k |\mu(R_k)| - \sum_k |\mu(R_k - A)|$$
$$\geq |\mu|((s,t_1] \times \Omega) - \varepsilon - |\mu|((s,t_1] \times (\Omega - A))$$
$$= |\mu|(A) - \varepsilon.$$

If we apply the above inequality to the set $A = (s, t_n] \times \Omega$ we obtain

$$|\mu|((s,t_n] \times \Omega) \leq \sum_k |\mu\left(((s,t_n] \times \Omega) \cap R_k\right)| + \varepsilon.$$

Each term of the right-hand side has limit 0. In fact, let $R_k = (a_k, b_k] \times A_k$ with $A_k \in \mathcal{F}_{a_k}$. If $a_k > s$, then there is an n_0 such that for $n \geq n_0$ we have $t_n < a_k$, consequently

$$(s, t_n] \times \Omega \cap R_k = \phi,$$

hence

$$\lim_n \mu((s,t_n] \times \Omega \cap R_k) = 0.$$

If $a_k = s$, then there is an n_0 such that for $n \geq n_0$ we have $t_n < b_k$; consequently
$$((s, t_n] \times \Omega) \cap R_k = (s, t_n] \times A_k$$
and $A_k \in \mathcal{F}_{a_k} = \mathcal{F}_s$. Then, by assertion a) we have
$$\lim \mu\left(((s, t_n] \times \Omega) \cap R_k\right) = \lim \mu((s, t_n] \times A_k) = 0.$$
From the above inequality we deduce that
$$\limsup_n |\mu|((s, t_n] \times \Omega) \leq \varepsilon.$$
Since $\varepsilon > 0$ is arbitrary, it follows that
$$\lim |\mu|((s, t_n] \times \Omega) = 0.$$
∎

We shall use the following notation:
If $Y : \mathbb{R}_+ \times \Omega \to E$ is a cadlag, adapted process with $Y_t \in L_E^1$ for every $t \geq 0$, we associate to Y the *Doléans measure* $\mu_Y : \mathcal{R} \to E$ defined first for predictable rectangles by
$$\mu_Y(\{0\} \times A) = E(1_A Y_0), \text{ for } A \in \mathcal{F}_0$$
and
$$\mu_Y((s, t] \times A) = E(1_A (Y_t - Y_s)), \text{ for } A \in \mathcal{F}_s,$$
and then extended by additivity to $\mathcal{R}$.

We deduce that
$$\mu_Y(B) = E(I_Y(B)), \text{ for } B \in \mathcal{R}.$$

We say that Y is a *quasimartingale* if the measure μ_Y has *bounded variation* on $\mathcal{R}$. For an extensive treatment of quasimartingales see ([B–D.4]).

2. Theorem. *Let Y be an E-valued, cadlag quasimartingale of class (D). Then the Doléans measure μ_Y is σ-additive on $\mathcal{R}$.*

Proof. (cf. [Me]). The proof will be divided into several steps. We shall write μ instead of μ_Y.
a) $\lim_{t \downarrow s} |\mu|((s, t] \times \Omega) = 0$.
In fact, since Y is a cadlag quasimartingale, it is right continuous in L_E^1. For $s \geq 0$ and $A \in \mathcal{F}_s$ we have $\lim_{t \downarrow s} 1_A (Y_t - Y_s) = 0$, in L_E^1, hence
$$\lim_{t \downarrow s} \int_A (Y_t - Y_s) dP = 0,$$
that is,
$$\lim_{t \downarrow s} \mu((s, t] \times A) = 0, \text{ for } A \in \mathcal{F}_s.$$

We apply then the preceding lemma to deduce assertion a) above.

b) For any decreasing sequence (H_n) from $\mathcal{F}$ with $H_n \downarrow \phi$ we have

$$\limsup_{n} {}_{\Delta_n} |\mu((S,T])| = 0,$$

where Δ_n is the family of all stochastic intervals $(S,T] \subset \mathbb{R}_+ \times H_n$ with S, T, simple stopping times.

In fact, for H_n, S, T as above we have

$$|\mu((S,T])| = |E(1_{H_n}(Y_T - Y_S))|$$
$$\leq E(1_{H_n}(|Y_T| + |Y_S|)).$$

Since Y is of class (D), the last term tends to 0, as $n \to \infty$, uniformly with respect to T and S. Then assertion b) follows.

c) For any $a > 0$ and for any decreasing sequence (H_n) form $\mathcal{F}$ with $H_n \downarrow \phi$ we have

$$\limsup_{n} \{|\mu|(A) : A \in \mathcal{R} \cap ([0,a] \times H_n)\} = 0.$$

In fact, let $a > 0$ and $H_n \in \mathcal{F}$ with $H_n \downarrow \phi$ and let $\varepsilon > 0$. There is a finite partition (R_k) of $[0,a] \times \Omega$ consisting of sets from $\mathcal{R}$ such that

$$|\mu|([0,a] \times \Omega) \leq \sum_k |\mu(R_k)| + \varepsilon.$$

Then, for any set $A \in \mathcal{R}$ with $A \subset [0,a] \times \Omega$, we get, as in the proof of Lemma 1,

$$|\mu|(A) \leq \sum_k |\mu(R_k \cap A)| + \varepsilon.$$

For each n and each set $A \subset [0,a] \times H_n$ from $\mathcal{R}$, there are simple stopping times $S \leq T$ such that $A \subset (S,T] \subset [0,a] \times H_n$; therefore

$$|\mu|(A) \leq |\mu|((S,T]).$$

Then

$$\sup\{|\mu|(A) : A \in \mathcal{R} \cap ([0,a] \times H_n)\}$$
$$\leq \sup\{|\mu|((S,T]) : (S,T] \subset [0,a] \times H_n\}$$
$$\leq \sum_k |\mu(R_k \cap (S,T])| + \varepsilon.$$

By assertion b), each term of the sum tends to 0 uniformly with respect to $(S,T]$. This proves c).

To state the next step, we define the class $\mathcal{C}[0,a]$ of finite unions of rectangles of the form $[s,t] \times H$ with $s \leq t \leq a$ and $H \in \mathcal{F}_s$. The class $\mathcal{C}[0,a]$ is closed under finite intersections.

d) For any decreasing sequence $(\overline{D}_n)$ from $\mathcal{C}[0,a]$ with $\overline{D}_n \downarrow \phi$ and for any sequence (D_n) from $\mathcal{R}$ with $D_n \subset \overline{D}_n$, we have

$$\lim_n |\mu|(D_n) = 0.$$

In fact, let D_n and $\overline{D}_n$ be as above. Let

$$H_n = pr_\Omega \overline{D}_n = \{\omega : \overline{D}_n \cap (\mathbb{R}_+ \times \{\omega\}) \neq \phi\}.$$

Then $H_n \in \mathcal{F}$. Let $\omega \in \Omega$. The ω-section $\overline{D}_n(\omega)$ is a finite union of compact intervals. Since $\overline{D}_n(\omega) \downarrow \phi$, there is a k such that $\overline{D}_k(\omega) = \phi$, hence $\omega \notin H_k$. Thus $H_n \downarrow \phi$. From the inclusion $D_n \subset \overline{D}_n \subset [0,a] \times H_n$ and from assertion c) we deduce that

$$\limsup_n |\mu|(D_n) \leq \limsup_n \{|\mu|(A) : A \in \mathcal{R} \cap ([0,a] \times H_n)\} = 0.$$

It follows that $\lim_n |\mu|(D_n) = 0$.

e) $|\mu|$ is σ-additive on $\mathcal{R}$.

Assume the contrary: there is a decreasing sequence (A_n) from $\mathcal{R}$ with $A_n \downarrow \phi$ and a number $a > 0$ such that

$$\lim_n |\mu|(A_n) = 4a.$$

For each n, A_n is the union of a finite family $(R(n,k))_{1 \leq k \leq b(n)}$ of rectangles of the form $R(n,k) = (s(n,k), t(n,k)] \times F(n,k)$ with $F(n,k) \in \mathcal{F}_{s(n,k)}$. By step a), for each pair (n,k) we can find an $s'(n,k)$ with $s(n,k) < s'(n,k) < t(n,k)$, such that

$$|\mu|((s(n,k), s'(n,k)] \times \Omega) \leq a 2^{-n} / b(n).$$

Let $\alpha > 0$ be such that $A_1 \subset [0, \alpha] \times \Omega$. Set

$$S_n = \bigcup_{1 \leq k \leq b(n)} ((s(n,k), s'(n,k)] \times \Omega).$$

Then

$$|\mu|(S_n) \leq \sum_k a 2^{-n}/b(n) = a 2^{-n},$$

therefore

$$|\mu|(\bigcup_{1 \leq i \leq n} S_i) \leq a, \text{ for each } n.$$

For every n let

$$C_n = \bigcup_{1 \leq k \leq b(n)} (s'(n,k), t(n,k)] \times F(n,k) \in \mathcal{R},$$

$$\overline{C}_n = \bigcup_{1 \leq k \leq b(n)} [s'(n,k), t(n,k)] \times F(n,k) \in \mathcal{C}[0, a],$$

$$D_n = \bigcap_{1 \leq i \leq n} C_i \in \mathcal{R}$$

$$\overline{D}_n = \bigcap_{1 \leq i \leq n} \overline{C}_i \in \mathcal{C}[0, a].$$

Then $A_n \subset S_n \cup C_n$. Since (A_n) is decreasing, we have

$$A_n \subset \left(\bigcup_{1 \leq i \leq n} S_i\right) \cup \left(\bigcap_{1 \leq i \leq n} C_i\right) = \left(\bigcup_{1 \leq i \leq n} S_i\right) \cup D_n.$$

Therefore

$$|\mu|(A_n) \leq |\mu|(D_n) + a.$$

By step d), we have $\lim_n |\mu|(D_n) = 0$. Passing to the limit in the above equality we obtain $4a \leq a$, a contradiction. Thus $|\mu|$ is σ-additive on $\mathcal{R}$. ∎

B. The summability criterion

If $g \in L^q_{E^*}$, $\frac{1}{p} + \frac{1}{p} = 1$, we denote by $G = (G_t)_{t \geq 0}$ the E^*-valued, uniformly integrable martingale defined by the conditional expectations $G_t = E(g|\mathcal{F}_t)$, for $t \geq 0$. If X is an E-valued process, we denote by XG the real-valued process $(\langle X_t, G_t \rangle)_{t \geq 0}$, where the bracket $\langle , \rangle$ denotes the duality between G and G^*.

For $f \in L^p_G$ and $g \in L^q_{G^*}$ we set

$$\langle f, g \rangle = E(\langle f(\cdot), g(\cdot) \rangle) = \int \langle f(\omega), g(\omega) \rangle dP(\omega).$$

3. Theorem. *Let $X : \mathbb{R}_+ \times \Omega \to E$ be an adapted, cadlag process such that $X_t \in L^p_E$ for every $t \geq 0$. If $c_0 \not\subset E$, assertions a)–d) below are equivalent.*

If E is any Banach space, assertions b), c) and d) are equivalent and a) implies b).
a) $I_X : \mathcal{R} \to L^p_E$ can be extended to a σ-additive measure on $\mathcal{P}$;
b) I_X is bounded on $\mathcal{R}$;
   Let $Z \subset L^q_{E^*}$ be any closed subspace norming for L^p_E.
c) *For every $g \in Z$, the real-valued measure $\langle I_X, g \rangle$ is bounded on $\mathcal{R}$;*
d) *For every $g \in Z$, the real-valued measure $\langle I_X, g \rangle$ is σ-additive and bounded on $\mathcal{R}$.*

Proof. The proof will be done as follows: a $\Longrightarrow$ b $\Longleftrightarrow$ c $\Longleftrightarrow$ d $\Longrightarrow$ a. The implication d $\Longrightarrow$ a is the only one that requires $c_0 \not\subset E$. All other implications are valid for any Banach space E.

The implication a $\Longrightarrow$ b is evident, since any σ-additive measure on a σ-algebra is bounded (Theorem 2.14). The implication b $\Longrightarrow$ c is obvious. To prove c $\Longrightarrow$ b, we note that for each $A \in \mathcal{R}$, the linear functional $g \mapsto \langle I_X(A), g \rangle$ on Z is continuous. Since Z is norming for L_E^p, we can embedd L_E^p into Z^* isometrically. If we assume c), then

$$\sup\{|<I_X(A), g>| : A \in \mathcal{R}\} < \infty, \text{ for each } g \in Z.$$

By the Banach–Steinhauss theorem, we deduce that

$$\sup\{|I_X(A)|_p : A \in \mathcal{R}\} < \infty,$$

which is assertion b).

The implication d $\Longrightarrow$ c is evident. Assume c) and prove d). Let $g \in L_{E^*}^q$. Consider the real-valued measure $\langle I_X, g \rangle$ on $\mathcal{R}$ defined by

$$\langle I_X, g \rangle(A) = \langle I_X(A), g \rangle = \int \langle I_X(A), g \rangle dP, \text{ for } A \in \mathcal{R}.$$

By assumption c), $\langle I_X, g \rangle$ is bounded on $\mathcal{R}$. Consider the real-valued process $XG = (\langle X_t, G_t \rangle)_{t \geq 0}$, where $G_t = E(g|\mathcal{F}_t)$. As before, μ_{XG} is the Doléans measure associated with XG. We have $\mu_{XG} = \langle I_X, g \rangle$ on $\mathcal{R}$.

In fact, for $B \in \mathcal{F}_0$ we have

$$\langle I_X, g \rangle(\{0\} \times B) = \int 1_B \langle X_0, g \rangle dP = \int 1_B X_0 G_0 dP$$
$$= \mu_{XG}(\{0\} \times B).$$

If $s < t$ and $B \in \mathcal{F}_s$, then

$$\langle I_X, g \rangle((s, t] \times B) = \int 1_B \langle X_t - X_s, g \rangle dP$$
$$= \int_B \langle X_t, G_t \rangle dP - \int_B \langle X_s, G_s \rangle dP = \mu_{XG}((s, t] \times B).$$

It follows that the real-valued measure μ_{XG} is bounded on $\mathcal{R}$. By Theorem 2.16, μ_{XG} has bounded variation on $\mathcal{R}$, hence XG is a quasimartingale.

For each n, define the stopping time $T_n = \inf\{t : |X_t| > n\}$. Then $T_n \uparrow \infty$ and $|X_t| \leq n$ on $[0, T_n)$. At this stage we do not know whether $X_{T_n} \in L_E^p$, but since XG is a quasimartingale, we know that $(XG)_{T_n} \in L^1$ and

$$|(XG)_t^{T_n}| \leq n|G_t|1_{\{t < T_n\}} + |(XG)_{T_n}|1_{\{t \geq T_n\}}.$$

Since G is a uniformly integrable martingale, it follows that $(XG)^{T_n}$ is a quasimartingale of class (D). This, in turn, implies that the corresponding Doléans measure $\mu_{(XG)^{T_n}}$ is σ-additive (Theorem 2) and has bounded variation on $\mathcal{R}$, hence it can be extended to a σ-additive measure with bounded variations on $\mathcal{P}$ (Theorem 7.4 b)).

For each predictable rectangle $(s,t] \times A$ with $A \in \mathcal{F}_s$ we have

$$\mu_{(XG)^{T_n}}((s,t] \times A) = \mu_{XG}(((s,t] \times A) \cap [0, T_n]).$$

We have also, for $A \in \mathcal{F}_0$,

$$\mu_{(XG)^{T_n}}(\{0\} \times A) = \mu_{XG}((\{0\} \times A) \cap [0, T_n]).$$

It follows that

$$\mu_{(XG)^{T_n}}(B) = \mu_{XG}(B \cap [0, T_n]), \text{ for } B \in \mathcal{P}.$$

Thus, μ_{XG} is σ-additive on the σ-ring $\mathcal{P} \cap [0, T_n]$; consequently, μ_{XG} is σ-additive on the ring

$$\mathcal{B} = \bigcup_{1 \leq n < \infty} \mathcal{P} \cap [0, T_n].$$

On the other hand, μ_{XG} is bounded on $\mathcal{R}$, hence it has bounded variation on $\mathcal{R}$. It follows that μ_{XG} is σ-additive and has bounded variation on the ring $\mathcal{B} \cap \mathcal{R}$, which generates $\mathcal{P}$; hence μ_{XG} can be extended to a σ-additive measure with bounded variation on $\mathcal{P}$ (Theorem 7.4 b).

Since $\langle I_X, g \rangle = \mu_{XG}$, it follows that $\langle I_X, g \rangle$ is bounded and σ-additive on $\mathcal{R}$, thus d) holds.

To prove d $\Longrightarrow$ a, we assume $c_0 \not\subset E$. If $\langle I_X, g \rangle$ is bounded and σ-additive on $\mathcal{R}$ for each $g \in Z$, then, by Theorem 7.8 b), I_X can be extended to a σ-additive measure on $\mathcal{P}$, which is assertion a). ∎

Remark. It can be proved that assertions b), c) and d) are equivalent with each one of the following assertions:
e) For each $g \in Z$, XG is a quasimartingale;
f) For each $g \in Z$, XG is a quasimartingale and $(XG)^* := \sup_t |(XG)_t|$ is integrable;
g) For each $g \in Z$, XG is a quasimartingale of class (D).

In the proof of the above theorem we already proved that d) implies e) and g).

§15. LOCAL SUMMABILITY AND LOCAL INTEGRABILITY

A. Definitions

Assume $X : \mathbb{R}_+ \times \Omega \to E \subset L(F,G)$ is a cadlag, adapted process with $X_t \in L_E^p$ for $t \geq 0$.

We shall define and study the properties of the stochastic integral $H \cdot X$ in case X is locally p-summable relative to (F,G) and H is locally integrable with respect to X.

1. Definition. *We say X is locally p-summable relative to (F,G) if there is an increasing sequence (T_n) of stopping times with $T_n \uparrow \infty$, such that for each n, the stopped process X^{T_n} is p-summable relative to (F,G).*

The sequence (T_n) is called a determining sequence for the local summability of X.

Examples of locally summable processes are processes with locally integrable variation or semivariation and the local martingales in Hilbert spaces. (§§17, 19, 21).

2. Definition. *Assume X is locally p-summable relative to (F,G) and let D be a Banach space.*

We denote by $\mathcal{F}_D(I_{F,G})_{loc}$ the space of all D-valued, predictable processes H for which there is an increasing sequence (T_n) of stopping times with $T_n \uparrow \infty$ such that for each n, X^{T_n} is p-summable relative to (F,G) and $1_{[0,T_n]}H \in \mathcal{F}_D((I_{X^{T_n}})_{F,G})$, that is,

$$(\tilde{I}_{X^{T_n}})_{F,G}(1_{[0,T_n]}H) < \infty.$$

A predictable process $H : \mathbb{R}_+ \times \Omega \to F$ is said to be locally integrable with respect to X, if there is an increasing sequence (T_n) of stopping times with $T_n \uparrow \infty$, such that, for each n, X^{T_n} is p-summable relative to (F,G) and $1_{[0,T_n]}H$ is integrable with respect to X^{T_n}.

We say that (T_n) is a determining sequence for the local integrability of H with respect to X.

The set of all F-valued, predictable processes H which are locally integrable with respect to X is denoted by $L^1_{F,G}(X)_{loc}$.

3. Definition. *Assume X is locally p-summable relative to (F,G).*

We say the set of measure $(I_X)_{F,L^q_{G^*}}$ is locally uniformly σ-additive, if there is an increasing sequence (T_n) of stopping times with $T_n \uparrow \infty$ such that, for each n, the set of measures $(I_{X^{T_n}})_{F,L^q_{G^*}}$ is uniformly σ-additive.

4. Definition. *If H^n and H are D-valued processes we say that $H^n \to H$ locally uniformly, if there is an increasing sequence (T_k) of stopping times with $T_k \uparrow \infty$, such that for each k, $H^n \to H$ uniformly on $[0, T_k]$.*

B. Basic properties

5. *If X is p-summable relative to (F, G), then X is locally p-summable relative to (F, G).*

In fact, for every sequence $T_n \uparrow \infty$ of stopping times and for every n, X^{T_n} is summable (Theorem 11.9 a)).

6. *If X is locally p-summable relative to (F, G), then X is locally p-summable relative to $(\mathbb{R}, E)$.*

Use Remark 9.9 c).

7. *If (T_n) is a sequence of stopping times, determining for the local p-summability of X relative to (F, G) and if $S_n \uparrow \infty$ is another sequence of stopping times, then $(T_n \wedge S_n)$ is determining for the local p-summability of X.*

In fact, for each n, X^{T_n} is p-summable relative to (F, G), hence $X^{T_n \wedge S_n} = (X^{T_n})^{S_n}$ is p-summable relative to (F, G) (Theorem 11.9 a)).

8. *If (T_n) is a determining sequence for the local integrability of H with respect to X and if $S_n \uparrow \infty$ is a sequence of stopping times, then $(T_n \wedge S_n)$ is determining for the local integrability of H with respect to X.*

In fact, $1_{[0,T_n]} H \in L^1_{F,G}(X^{T_n})$ for each n. By Theorem 11.9 d) we have $1_{[0,S_n]} 1_{[0,T_n]} H \in L^1_{F,G}(X^{T_n})$ and $1_{[0,S_n]} 1_{[0,T_n]} H \in L^1_{F,G}((X^{T_n})^{S_n})$, that is,

$$1_{[0, T_n \wedge S_n]} H \in L^1_{F,G}(X^{T_n \wedge S_n}).$$

9. Proposition. *Let H be an F-valued, predictable process, locally integrable with respect to X and let (T_n) be a sequence determining for the local integrability of H with respect to X.*

Then the limit

$$\lim_n (1_{[0,T_n]} H) \cdot X^{T_n}$$

exists pointwise outside an evanescent set, is cadlag, adapted and independent of the sequence (T_n).

Proof. For every n we have

$$((1_{[0,T_{n+1}]} H) \cdot X^{T_{n+1}})^{T_n} = (1_{[0,T_n]} H) \cdot X^{T_n}$$

outside an evanescent set A_n (Theorem 11.9 d)), hence the limit exists. If we denote

$$Y = \lim_n (1_{[0,T_n]} H) \cdot X^{T_n},$$

outside the evanescent set $A = \bigcup_n A_n$ and $Y = 0$ on A, then for each n we have

$$Y^{T_n} = (1_{[0,T_n]} H) \cdot X^{T_n},$$

§15. LOCAL SUMMABILITY AND LOCAL INTEGRABILITY   173

hence Y is cadlag. Since each process $(1_{[0,T_n]}H) \cdot X^{T_n}$ is adapted, so is the limit Y. Finally, if (S_n) is another sequence determining for the local integrability of H with respect to X, and if we denote

$$Y' = \lim_n (1_{[0,S_n]}H) \cdot X^{S_n},$$

then

$$Y' = \lim_n Y'^{T_n} = \lim_n (1_{[0, S_n \wedge T_n]}H) \cdot X^{S_n \wedge T_n}$$

and

$$Y = \lim_n Y^{S_n} = \lim_n (1_{[0, T_n \wedge S_n]}H) \cdot X^{T_n \wedge S},$$

hence $Y = Y'$, outside an evanescent set. ∎

The preceding proposition leads to the following definition:

10. Definition. *If X is locally p-summable relative to (F, G) and if H is an F-valued, predictable process, locally integrable with respect to X, then the stochastic integral of H with respect to X is the process denoted by $H \cdot X$ or $\int H dX$ and defined up to an evanescent set by the equality*

$$H \cdot X = \int H dX = \lim_n (1_{[0,T_n]}H) \cdot X^{T_n},$$

for any sequence (T_n) of stopping times which is determining for the local integrability of H with respect to X.

It follows that for each n we have

$$(H \cdot X)^{T_n} = (1_{[0,T_n]}H) \cdot X^{T_n}.$$

The following theorem states that integrability and local integrability are equivalent for processes of $\mathcal{F}_{F,G}(X)$, in case X is p-summable.

11. Theorem. *Let X be a p-summable process relative to (F, G) and $H \in \mathcal{F}_{F,G}(X)$. Then H is integrable with respect to X iff H is locally integrable with respect to X.*

In this case, the stochastic integral $H \cdot X$ is the same, whether H is considered integrable or locally integrable with respect to X.

Proof. Assume first H is integrable with respect to X and let $T_n \uparrow \infty$ be a sequence of stopping times.

By Theorem 11.9 a), X^{T_n} is p-summable relative to (F, G) and by Theorem 11.9 d), we have $1_{[0,T_n]}H \in L^1_{F,G}(X)$ and $1_{[0,T_n]}H \in L^1_{F,G}(X^{T_n})$; therefore H is locally integrable with respect to X. Then

$$\lim_n (1_{[0,T_n]}H \cdot X^{T_n}) = \lim_n (H \cdot X)^{T_n} = H \cdot X,$$

hence the two stochastic integrals coincide. Conversely assume H is locally integrable with respect to X and let (T_n) be a determining sequence of stopping times. Then $1_{[0,T_n]}H \in L^1_{F,G}(X^{T_n})$. By Theorem 11.9 d) we have also $1_{[0,T_n]}H \in L^1_{F,G}(X)$. By Theorem 12.7 we deduce that $H \in L^1_{F,G}(X)$. ∎

A σ-elementary process is not necessarily integrable with respect to a summable process; but it is locally integrable with respect to any locally summable process, as the following theorem shows.

12. Theorem. *Assume X is locally p-summable relative to (F,G) and let H be a σ-elementary process of the form*

$$H = H_0 1_{\{0\}} + \sum_{1 \leq i < \infty} H_i 1_{(T_i, T_{i+1}]},$$

where $0 = T_0 \leq T_1 \leq T_2 \leq \ldots$ is a sequence of stopping times with $T_i \uparrow \infty$ and for $0 \leq i < \infty$, H_i is $\mathcal{F}_{T_i}$-measurable but not necessarily bounded.

Then H is locally integrable with respect to X and the stochastic integral can be computed pathwise:

$$(H \cdot X)_t = H_0 X_0 + \sum_{1 \leq i < \infty} H_i (X_{T_{i+1} \wedge t} - X_{T_i \wedge t}).$$

Proof. We note first that for each t and ω, the above series reduces to a finite sum. For each n consider the stopping time $S_n = \inf\{t : |H_t| > n\}$. Since H is caglad, we have $S_n \uparrow \infty$ and $1_{[0,S_n]}|H| \leq n$. Then for each i we have $1_{[0,S_n]}|H_i| \leq n$. We observe that $1_{[0,S_n \wedge T_n]}H$ is an elementary process.

Let $U_n \uparrow \infty$ be a sequence of stopping times, determining for the local p-summability of X and set $R_n = U_n \wedge S_n \wedge T_n$. Then X^{U_n} is p-summable. By Theorem 11.9 a), $X^{R_n} = (X^{U_n})^{S_n \wedge T_n}$ is p-summable. The process $1_{[0,R_n]}H$ is an elementary process, hence it is integrable with respect to X^{R_n}. It follows that H is locally integrable with respect to X.

To compute the stochastic integral

$$(H \cdot X)_t = \lim_n (1_{[0,R_n]}H \cdot X^{R_n})_t,$$

we observe first that

$$1_{[0,R_n]}H = 1_{\{0\}}H_0 + \sum_{1 \leq i < \infty} H_i 1_{(T_i \wedge R_n, T_{i+1} \wedge R_n]}$$

$$= 1_{\{0\}}H_0 + \sum_{1 \leq i < n} H_i 1_{(T_i \wedge R_n, T_{i+1} \wedge R_n]},$$

hence

$$(1_{[0,R_n]}H) \cdot X^{R_n} = H_0 X_0 + \sum_{1 \leq i < n} H_i (X^{R_n}_{T_{i+1} \wedge R_n} - X^{R_n}_{T_i})$$

$$= H_0 X_0 + \sum_{1 \leq i < n} H_i (X_{T_{i+1} \wedge R_n} - X_{T_i \wedge R_n}).$$

Let $t \in \mathbb{R}_+$ and $\omega \in \Omega$ and take n such that $t < R_n(\omega)$. Then

$$(1_{[0,R_n]} H \cdot X^{R_n})_t = H_0 X_0 + \sum_{1 \leq i < n} H_i(X_{T_{i+1} \wedge t} - X_{T_i \wedge t}),$$

hence

$$(H \cdot X)_t = \lim_n ((1_{[0,R_n]} H) \cdot X^{R_n})_t$$
$$= H_0 X_0 + \sum_{1 \leq i < \infty} H_i(X_{T_{i+1} \wedge t} - X_{T_i \wedge t}).$$

■

C. Convergence theorems

13. Theorem. *Assume X is locally p-summable relative to (F,G) and let $H^n, H \in L^1_{F,G}(X)_{loc}$.*

Let $T_k \uparrow \infty$ be stopping times such that:
(i) *for each k, X^{T_k} is p-summable relative to (F,G);*
(ii) *for each k, the processes $1_{[0,T_k]} H^n$ and $1_{[0,T_k]} H$ belong to $L^1_{F,G}(X^{T_k})$;*
(iii) *for each k we have*

$$1_{[0,T_k]} H^n \to 1_{[0,T_k]} H, \text{ in } L^1_{F,G}(X^{T_k}), \text{ as } n \to \infty.$$

Then
a) *For each $t \geq 0$, $(H^n \cdot X)_t \to (H \cdot X)_t$ in probability;*
b) *There is a subsequence (r_n) such that*

$$(H^{r_n} \cdot X)_t \to (H \cdot X)_t, \text{ a.s. as } n \to \infty,$$

uniformly on every bounded interval.

Proof. To prove a), let $t \geq 0$ and $\varepsilon > 0$. Since $\{T_k \leq t\} \downarrow \phi$, we have $P(\{T_k \leq t\}) \downarrow 0$. Fix k_0 such that $P(\{T_{k_0} \leq t\}) < \varepsilon$. If $\eta > 0$, we have

$$P(\{|(H^n \cdot X)_t - (H \cdot X)_t| > \eta\})$$
$$\leq \varepsilon + P(\{t < T_{k_0}\} \cap \{|(H^n \cdot X)_t - (H \cdot X)_t| > \eta\}).$$

From hypothesis (iii) we deduce that

$$1_{[0,t]} 1_{[0,T_{k_0}]} H^n \to 1_{[0,t]} 1_{[0,T_{k_0}]} H, \text{ in } L^1_{F,G}(X^{T_{k_0}}),$$

which implies that

$$(H^n \cdot X)_t^{T_{k_0}} \to (H \cdot X)_t^{T_{k_0}},$$

in L^p_G, hence in probability.

There exists an N such that for $n \geq N$ we have
$$P(\{|(H^n \cdot X)_t^{T_{k_0}} - (H \cdot X)_t^{T_{k_0}}| > \eta\}) > \varepsilon,$$
thus
$$P(\{|(H^n \cdot X)_t - (H \cdot X)_t| > \eta\}) < 2\varepsilon, \text{ for } n \geq N,$$
and this proves assertion a).

Using hypothesis (iii), from Theorem 12.1 we deduce the existence of a sequence $(r(n,k))_n$ such that $\lim_n r(n,k) = \infty$ and
$$(1_{[0,T_k]}H^{r(n,k)} \cdot X^{T_k})_t \to (1_{[0,T_k]}H \cdot X^{T_k})_t = (H \cdot X)_t^{T_k},$$
a.s., as $n \to \infty$, uniformly on every bounded interval. Let A_k be the negligible exceptional set and A the union of the sets A_k. By induction with respect to k, we can construct the sequence $(r(n, k+1))_n$ to be a subsequence of $(r(n,k))_n$ for each k. Denote $r_n = r(n,n)$. Then $(r_n)_n$ is a subsequence of $(r(n,k))_n$ for each k; therefore, for each k we have
$$(H^{r_n} \cdot X)_t^{T_k} = (1_{[0,T_k]}H^{r_n} \cdot X^{T_k})_t \to (H \cdot X)_t^{T_k},$$
uniformly on every bounded interval.

Let $[0,T]$ be a bounded interval and $\omega \notin A$. Since $T_k \uparrow \infty$, there is a K such that for every $k \geq K$ we have $T_k(\omega) > T$, hence, for $t \leq T$ we have
$$(H^{r_n} \cdot X)_t^{T_k}(\omega) = (H^{r_n} \cdot X)_t(\omega)$$
and
$$(H \cdot X)_t^{T_k}(\omega) = (H \cdot X)_t(\omega).$$
It follows then that
$$(H^{r_n} \cdot X)_t(\omega) \to (H \cdot X)_t(\omega), \text{ uniformly for } t \in [0,T],$$
and this proves assertion b). ∎

14. Theorem. *Assume X is locally p-summable relative to (F, G) and let (H^n) be a sequence from $L^1_{F,G}(X)_{loc}$ converging locally uniformly on $\mathbb{R}_+ \times \Omega$ to a process H. Then:*
a) *H is locally integrable with respect to X;*
b) *For each $t \geq 0$, $(H^n \cdot X)_t \to (H \cdot X)_t$ in probability;*
c) *There is a subsequence (r_n) such that $(H^{r_n} \cdot X)_t \to (H \cdot X)_t$, as $n \to \infty$, uniformly on every bounded interval.*

Proof. Assume first that $H^n \to H$ uniformly. We choose N such that $|H^n - H^N| \leq 1$ for $n \geq N$. Let (T_k) be a determining sequence for the local integrability of $H^1, H^2, \cdots, H^N$ with respect to X. It follows that $1_{[0,T_k]}H^n \in L^1_{F,G}(X^{T_k})$ for every k and every $n \leq N$. This property is also valid for $n \geq N$. In fact, since $1_{[0,T_k]}H^N \in L^1_{F,G}(X^{T_k})$ for each k, we deduce that $1_{[0,T_k]}H^n \in \mathcal{F}_{F,G}(X^{T_k})$ for $n \geq N$. Since H^n is locally integrable with respect to X, using Theorem 11.9 d) we deduce that $1_{[0,T_k]}H^n$ is locally integrable with respect to X^{T_k}, hence by Theorem 11, $1_{[0,T_k]}H^n$ is integrable with respect to X^{T_k}.

Since $1_{[0,T_k]}H^n \to 1_{[0,T_k]}H$ uniformly, as $n \to \infty$, by the convergence Theorem 12.4, we deduce that for each k we have

$$1_{[0,T_k]}H \in L^1_{F,G}(X^{T_k}) \text{ and } 1_{[0,T_k]}H^n \to 1_{[0,T_k]}H,$$

in $L^1_{F,G}(X^{T_k})$, as $n \to \infty$.

Assume now $H^n \to H$ locally uniformly and let (S_k) be a sequence of stopping times with $S_k \uparrow \infty$, such that for each k, $H^n \to H$ uniformly on $[0, S_k]$.

For each k replace H^n and H with $1_{[0,S_k]}H^n$ and $1_{[0,S_k]}H$ respectively; by the above, there is a sequence $(T_{kl})_l$ of stopping times with $T_{kl} \uparrow \infty$ as $l \to \infty$, such that for each l, $X^{T_{kl}}$ is p-summable, $1_{[0,T_{kl}]}1_{[0,S_k]}H^n$ and $1_{[0,T_{kl}]}1_{[0,S_k]}H$ belong to $L^1_{F,G}(X^{T_k})$ and we have

$$1_{[0,T_{kl}]}1_{[0,S_k]}H^n \to 1_{[0,T_{kl}]}1_{[0,S_k]}H, \text{ in } L^1_{F,G}(X^{T_k}).$$

Consider the diagonal sequence (T_k) with $T_k = T_{kk}$. Then $S_k \wedge T_k \uparrow \infty$ and the processes $1_{[0,S_k \wedge T_k]}H^n$ and $1_{[0,S_k \wedge T_k]}H$ belong to $L^1_{F,G}(X^{T_k})$, therefore to $L^1_{F,G}(X^{S_k \wedge T_k})$, by Theorem 11.9 d). We have also

$$1_{[0,S_k \wedge T_k]}H^n \to 1_{[0,S_k \wedge T_k]}H,$$

in $L^1_{F,G}(X^{S_k \wedge T_k})$, by Theorem 11.9 b). Then we can apply Theorem 13 with T_k replaced by $S_k \wedge T_k$. Assertions b) and c) of the present theorem follow from assertions a) and b) of Theorem 13. ∎

Another application of Theorem 13 is the Lebesgue Convergence Theorem for locally integrable processes. A Vitali-type convergence theorem can also be proved along the same lines.

15. Theorem. (Lebesgue) *Assume X is locally p-summable relative to (F, G). Let (H^n) be a sequence from $L^1_{F,G}(X)_{loc}$, $H \in \mathbb{R}_+ \times \Omega \to F$ a predictable process and $\phi \in \mathcal{F}_\mathbb{R}(\mathcal{B}, (I_X)_{F,G})_{loc}$.*

Assume that:
(i) $|H^n| \leq \phi$, *for each n;*
and that any one of the conditions (ii) *or* (iii) *below is true:*
(ii) $H^n \to H$, *locally uniformly;*
(iii) $H^n \to H$ *pointwise and the family of measures $(I_X)_{F,L^q_{G^*}}$ is locally, uniformly σ-additive.*

Then
a) $H \in L^1_{F,G}(X)_{loc}$;
b) *For every $t \geq 0$ we have $(H^n \cdot X)_t \to (H \cdot X)_t$, in probability.*

c) There is a subsequence (r_n) such that

$$(H^{r_n} \cdot X)_t \to (H \cdot X)_t, \text{ a.s., as } n \to \infty,$$

uniformly on every bounded interval.

Proof. We choose a sequence of stopping times $T_k \uparrow \infty$ which is determining for the local p-summability of X and at the same time, for each k we have $H^n \to H$ uniformly on $[0, T_k]$, in case (ii), or $(I_{X^{T_k}})_{F, L^q_{G^*}}$ is uniformly σ-additive, in case (iii).

We can take the sequence (T_k) such that we have also

$$1_{[0, T_k]} \phi \in \mathcal{F}_\mathbb{R}(\mathcal{B}, (I_{X^{T_k}})_{F, G}).$$

Since each H^n is locally integrable with respect to X, it is locally integrable with respect to X^{T_k}, for each k; and by Theorem 11, each H^n is integrable with respect to each X^{T_k}. Then, by Theorem 11.9 d), $1_{[0, T_k]} H^n$ is integrable with respect to X^{T_k} for each n and k.

We can then apply Theorem 12.4 in case (ii) and the Lebesgue Theorem 12.6 in case (iii) and deduce that for each k we have $1_{[0, T_k]} H \in L^1_{F, G}(X^{T_k})$ and

$$1_{[0, T_k]} H^n \to 1_{[0, T_k]} H, \text{ in } L^1_{F, G}(X^{T_k}), \text{ as } n \to \infty.$$

It follows that $H \in L^1_{F, G}(X)_{\text{loc}}$. The hypotheses of Theorem 13 are satisfied and the conclusion of Theorem 13 coincides with assertions b) and c) of the present theorem. ∎

As an application of Theorem 14 we shall deduce the local integrability of any caglad, adapted process, with respect to any locally p-summable process.

16. Theorem. *Assume X is locally p-summable relative to (F, G) and let $H : \mathbb{R}_+ \times \Omega$ be a caglad, adapted process. Then:*
a) $H \in L^1_{F, G}(X)_{loc}$;
  There is a sequence (H^n) of F-valued, σ-elementary processes such that
b) $H^n \to H$ *uniformly*;
c) $(H^n \cdot X)_t \to (H \cdot X)_t$ *in probability, for each $t \geq 0$*;
d) *there is a sequence (r_n) such that*

$$(H^{r_n} \cdot X)_t \to (H \cdot X), \text{ a.s, as } n \to \infty,$$

uniformly on every bounded interval.

Proof. The process $K = H_+$ is cadlag, adapted and $H = K_-$. Let $b_n \downarrow 0$ and define the stopping times

$$v(n, 0) = 0$$

and

$$v(n, k+1) = \inf\{t > v(n, k) : |K_t - K_{v(n,k)}| > b_n\} \wedge (b_n + v(n, k)).$$

These stopping times have the following properties:
(i) for each n, we have $v(n,k) \uparrow \infty$ as $k \to \infty$;
(ii) $\lim_n \sup_k (v(n, k+1) - v(n,k)) = 0$;
(iii) $|K_t - K_{v(n,k)}| \leq a_n$, for $t \in [v(n,k), v(n, k+1))$.

For each n, define the σ-elementary process

$$H^n = \sum_{k \geq 0} H_{v(n,k)} 1_{(v(n,k), v(n,k+1)]}.$$

It folllows that $H^n \to K_- = H$, uniformly. The conclusion follows from Theorems 12 and 14. ∎

D. Additional properties

We shall state some properties that are extensions of corresponding theorems for integrable processes.

17. Theorem. *Assume X is locally p-summable relative to (F,G) and let $S \leq T$ be stopping times. Assume that either:*
(i) $h : \Omega \to \mathbb{R}$ *is $\mathcal{F}_S$-measurable and $H \in L^1_{F,G}(X)_{loc}$,*
or
(ii) $h : \Omega \to F$ *if $\mathcal{F}_S$-measurable and*

$$H \in L^1_{\mathbb{R}, E}(X)_{loc} \cap \mathcal{F}_{\mathbb{R}}(\tilde{I}_{F,G})_{loc}.$$

Then
a) $(h1_{(S,T]}H) \cdot X = h[1_{(S,T]}H \cdot X]$.
   If, in addition, S is predictable and h is $\mathcal{F}_{S-}$-measurable, then
b) $(h1_{[S,T]}H) \cdot X = h[1_{[S,T]}H \cdot X]$.

Proof. We use the corresponding Theorem 11.5. ∎

18. Theorem. *Assume X is locally p-summable relative to (F,G) and let T be a stopping time. Then*
a) X^T *is locally p-summable relative to (F,G) and we have*

$$X^T = 1_{[0,T]} \cdot X.$$

a') *If T is predictable, then X^{T-} is locally p-summable relative to (F,G) and we have*

$$X^{T-} = 1_{[0,T)} \cdot X.$$

b) $H \in L^1_{F,G}(X^T)_{loc}$ *iff $1_{[0,T]}H \in L^1_{F,G}(X)_{loc}$ and in this case we have*

$$H \cdot X^T = (1_{[0,T]}H) \cdot X.$$

b') *Assume T is predictable. Then $H \in L^1_{F,G}(X^{T-})_{loc}$ iff $1_{[0,T)}H \in L^1_{F,G}(X)_{loc}$ and in this case we have*

$$H \cdot X^{T-} = (1_{[0,T)}H) \cdot X.$$

c) If $H \in L^1_{F,G}(X)_{loc}$, then $H \in L^1_{F,G}(X^T)_{loc}$ and $1_{[0,T]}H \in L^1_{F,G}(X)_{loc}$ and we have
$$(H \cdot X)^T = H \cdot X^T = (1_{[0,T]}H) \cdot X.$$

c') If T is predictable and $H \in L^1_{F,G}(X)_{loc}$, then $H \in L^1_{F,G}(X^{T-})_{loc}$ and we have
$$(H \cdot X)^{T-} = H \cdot X^{T-} = (1_{[0,T)}H) \cdot X.$$

For the proof we use Theorem 11.9.

The following associativity formula for real-valued H follows from Theorem 13.1.

19. Theorem. *Assume X is locally p-summable relative to (F,G) (hence relative to $(\mathbb{R}, E)$) and let $H \in L^1_{\mathbb{R},E}(X)_{loc} \cap \mathcal{F}_{\mathbb{R}}(\tilde{I}_{F,G})_{loc}$.*

Assume there is a sequence (T_n) of stopping times, determining for the local integrability of H with respect to X, such that

$$\int_A 1_{[0,T_n]} H dI_{X^{T_n}} \in L^p_E, \text{ for every } A \in \mathcal{P}.$$

Then
a) $H \cdot X \in L^1_{F,G}(X)_{loc}$;
b) $K \in L^1_{F,G}(H \cdot X)_{loc}$ iff $KH \in L^1_{F,G}(X)_{loc}$ and in this case we have

$$K \cdot (H \cdot X) = (KH) \cdot X.$$

The associativity formula for vector-valued H follows from Theorem 13.2.

20. Theorem. *Assume X is locally summable relative to (F,G) and let $H \in L^1_{F,G}(X)_{loc}$.*

Assume there is a sequence (T_n) of stopping times, determining for the local integrability of H with respect to X, such that

$$\int_A 1_{[0,T_n]} H dI_{X^{T_n}} \in L^p_G, \text{ for every } A \in \mathcal{P}.$$

Then
a) $H \cdot X \in L^1_{\mathbb{R},G}(X)_{loc}$;
b) *If K is a real-valued, predictable process and $KH \in L^1_{F,G}(X)_{loc}$, then $K \in L^1_{\mathbb{R},G}(H \cdot X)_{loc}$ and we have*

$$K \cdot (H \cdot X) = (KH) \cdot X.$$

The following theorem of jumps of the stochastic integral follows from the corresponding Theorem 11.10.

21. Theorem. *Assume X is locally summable relative to (F,G) and let $H \in L^1_{F,G}(X)_{loc}$. Then*
$$\Delta(H \cdot X) = H \Delta X.$$

Chapter 3
Martingales

In this chapter we study the properties of the stochastic integral with respect to a martingale.

A martingale is not necessarily summable; but if it is, then the stochastic integral is again a martingale (Theorem 16.2). In §17 we study the square integrable martingales with values in a Hilbert space and prove that they are 2-summable (Theorem 17.7).

§16. STOCHASTIC INTEGRAL OF MARTINGALES

The main result of this paragraph is that the stochastic integral with respect to a martingale (resp. a local martingale) is again a martingale (resp. a local martingale).

We shall prove first that a vector-valued martingale has a cadlag modification. This result is known for real-valued martingales ([D–M], VI. 4).

1. Theorem. *Every E-valued martingale X has a cadlag modification.*

Proof. If $X = \sum_{1 \leq i \leq n} \phi^i x_i$ where $x_i \in F$ and ϕ^i are real-valued martingales, then, obviously, X has a cadlag modification.

Assume X is a uniformly integrable martingale, $X_t = E(X_\infty | \mathcal{F}_t)$ for $t \geq 0$, where $X_\infty \in L^1_E$. Let (X^n_∞) be a sequence of simple functions from L^1_E, converging to X_∞ in L^1_E. By the first part of the proof, for each n, the martingale $E(X^n_\infty | \mathcal{F}_t)$ has a cadlag modification X^n and $X^n_t \to X_t$ in L^1_E, for each $t \geq 0$. Then

$$\sup_{t \geq 0} E(|X_t^n - X_t^m|) \leq E(|X_\infty^n - X_\infty^m|) \to 0$$

as $m, n \to \infty$. From Doob's inequality we deduce that for every $\lambda > 0$ we have

$$\lambda P\{\sup_{t \geq 0} |X_t^n - X_t^m| \geq \lambda\} \leq 3 \sup E(|X_t^n - X_t^m|) \to 0, \text{ as } n, m \to \infty.$$

Then there is a subsequence $(n_k)_{k \in \mathbb{N}}$ such that $(X_t^{n_k})_k$ is Cauchy a.s. uniformly in $t \geq 0$. Let $Y_t(\omega) = \lim_{k \to \infty} X_t^{n_k}(\omega)$ if the limit exists and $Y(\omega) = 0$ otherwise. Since each X^{n_k} is cadlag, we deduce that Y is cadlag.

On the other hand, for each $t \geq 0$, we have $\lim_{k \to \infty} X_t^{n_k} = X_t$, in L_E^1. It follows that for each $t \geq 0$ we have $X_t = Y_t$, a.s., that is, Y is a cadlag modification of X.

If X is an arbitrary martingale, then for each $n \in \mathbb{N}$, we apply the preceding result to the martingale X^n obtained by stopping X at n and we obtain a cadlag modification Y^n of X^n.

For $n < m$ we have $X_t^n = X_t^m$ for $t < n$, hence $Y_t^n = Y_t^m$, a.s. for $t < n$. It follows that the limit $\lim_n Y_t^n$ exists a.s. for each $t \geq 0$. If we set $Y_t(\omega) = \lim_n Y_t^n(\omega)$ if $Y_t^n(\omega) = Y_t^m(\omega)$ for all $n < m$ and $Y_t(\omega) = 0$ otherwise, then Y is cadlag and $X_t = \lim_n X_t^n = \lim Y_t^n = Y_t$, a.s., for each $t \geq 0$, hence Y is a cadlag modification of X. ∎

The stochastic integral with respect to a martingale is again a martingale:

2. Theorem. *Assume X is p-summable relative to (F, G) and let $H \in \mathcal{F}_{F, L_G^p}(X)$.*

If X is a martingale and if $\int_{[0,t]} H dI_X \in L_G^p$ for every $t \geq 0$, then $H \in L_{F, L_G^p}^1(X)$ and the stochastic integral $H \cdot X$ is a uniformly integrable martingale, bounded in L_G^p.

In particular, for $p = 2$, if X is a square integrable martingale, if $H \in \mathcal{F}_{F, L_G^2}(X)$ and $\int_{[0,t]} H dI_X \in L_G^2$ for every $t \geq 0$, then $H \in L_{F, L_G^2}^1(X)$ and $H \cdot X$ is a square integrable martingale.

Proof. Let $t \geq 0$ and $A \in \mathcal{F}_t$ and prove that

(*) $$E(1_A \int 1_{(t, \infty]} H dI_X) = 0.$$

If $H = 1_{\{0\} \times B} x$ with $B \in \mathcal{F}_0$ and $x \in F$, then (*) holds. Assume $H = 1_{(u,v] \times B} x$ with $B \in \mathcal{F}_u$ and $x \in F$. If $v \leq t$, then (*) holds. Assume $t < v$. Then

$$\int 1_{(t, \infty]} H dI_X = 1_B x (X_v - X_{t \vee u}),$$

thus
$$1_A \int 1_{(t,\infty]} H dI_X = 1_{A\cap B} x(X_v - X_{t\vee u}).$$

Since $A \cap B \in \mathcal{F}_{t\vee u}$, taking expectations on both sides we obtain (*). It follows that (*) holds for all F-valued, $\mathcal{R}$-step processes H.

Assume now that H is predictable, in $\mathcal{F}_{F,L_G^p}(X)$ and $\int_{[0,t]} H dI_X \in L_G^p$ for $t \geq 0$. Let $y^* \in G^*$ and denote $z = 1_A y^* \in L_{G^*}^q$, $\frac{1}{p} + \frac{1}{q} = 1$. Since the $\mathcal{R}$-simple processes are dense in $L_F^1((I_X)_z)$, there is a sequence (H^n) of $\mathcal{R}$-step processes converging to H in $L_F^1((I_X)_z)$. Then
$$\int 1_{(t,\infty]} H^n d(I_X)_z \to \int 1_{(t,\infty]} H d(I_X)_z,$$
that is,
$$\langle \int_{(t,\infty]} H^n dI_X, z \rangle \to \langle \int_{(t,\infty]} H dI_X, z \rangle.$$

This means
$$E(\langle 1_A \int_{(t,\infty]} H^n dI_X, y^* \rangle) \to E(\langle 1_A \int_{(t,\infty]} H dI_X, y^* \rangle),$$
that is,
$$\langle E(1_A \int_{(t,\infty]} H^n dI_X), y^* \rangle \to \langle E(1_A \int_{(t,\infty]} H dI_X), y^* \rangle.$$

Since H^n are $\mathcal{R}$-step processes, from the first part of the proof, all the terms of the converging sequence are 0, hence
$$\langle E(1_A \int_{(t,\infty]} H dI_X), y^* \rangle = 0, \text{ for every } y^* \in G^*,$$
consequently $E(1_A \int_{(t,\infty]} H dI_X) = 0$.

It follows that $(\int_{[0,t]} H dI_X)_{t\geq 0}$ is a uniformly integrable martingale. Since every martingale has a cadlag modification (Theorem 1), we deduce that $H \in L_{F,L_G^p}^1(X)$. ∎

3. Corollary. *If L_G^p is reflexive and if X is a martingale, p-summable relative to (F,G), then $L_{F,L_G^p}^1(X) = \mathcal{F}_{F,L_G^p}(X)$.*

We say that the process X is a *local martingale*, if there is an increasing sequence (T_n) of stopping times with $T_n \uparrow \infty$ such that for each n, the process $X_{t\wedge T_n} I_{\{T_n > 0\}}$ is a uniformly integrable martingale.

The property of being a local martingale is inherited by the stochastic integral, if X is locally summable.

4. Theorem. *Assume X is locally p-summable relative to (F,G) and let $H \in L_{F,G}^1(X)_{loc}$.*

a) *If X is a local martingale, then $H \cdot X$ is a local martingale.*
b) *If X is a martingale and if for each $t \geq 0$, the stopped process X^t is p-summable relative to (F,G) and $1_{[0,t]} H \in L^1_{F,G}(X^t)$, then $H \cdot X$ is a martingale.*

Remark. A martingale and even a square integrable martingale, is not necessarily summable. But if E and G are Hilbert spaces and if X is a square integrable martingale, then X is 2-summable (Theorem 17.7). By Theorem 2, for each $H \in L^1_{F,L^2_G}(X)$, the stochastic integral $H \cdot X$ is again a square integrable martingale. It will follow then that every local martingale with values in a Hilbert space is locally 1-summable (Corollary 22.20)).

§17. SQUARE INTEGRABLE MARTINGALES

In this paragraph, E and G are Hilbert spaces over the reals and F is a Banach space such that $E \subset L(F, G)$. The inner product in a Hilbert space is denoted by $\langle \cdot, \cdot \rangle$.

The main result of this paragraph is that any E-valued, square integrable martingale M is 2-summable relative to any embedding $E \subset L(F, G)$.

For this, we have to prove first that the stochastic measure I_M can be extended to a σ-additive measure $I_M : \mathcal{P} \to L_E^2$ and then that I_M has finite semivariation relative to (F, L_G^2).

A *square integrable martingale* is a martingale $M : \mathbb{R}_+ \times \Omega \to E$ such that $M_t \in L_E^2$ for every $t \geq 0$ and $\sup_t \|M_t\|_2 < \infty$. This is equivalent to the existence of a random variable $M_\infty \in L_E^2$ such that $M_t = E(M_\infty | \mathcal{F}_t)$, for every $t \geq 0$. We shall write $L^1_{F,L_G^2}(M)$, $(\tilde{I}_M)_{F, L_G^2}$ or $\tilde{I}_{F, L_G^2}$ in place of $L^1_{F,G}(M)$, $(\tilde{I}_M)_{F, G}$, or $\tilde{I}_{F, G}$, respectively, if we want to avoid any confusion.

A. Extension of the measure I_M

For a square integrable martingale, the measure I_M is σ-additive and can be extended to the σ-algebra $\mathcal{P}$ (Theorem 2). Also, the measure $\|I_M(A)\|_2^2$ is σ-additive on $\mathcal{P}$ (Theorem 4).

We prove first an orthogonality property of the measure I_M.

1. Lemma. *Let M and N be two E-valued, square integrable martingales. For any disjoint sets $A, B \in \mathcal{R}$ and for any elements $x, y \in F$ we have*

$$I_M(A) \perp I_N(B), \text{ in } L_E^2$$

and

$$I_M(A)x \perp I_N(B)y, \text{ in } L_G^2.$$

Proof. Assume first $A = \{0\} \times C$ with $C \in \mathcal{F}_0$ and $B = (s, t] \times D$ with $D \in \mathcal{F}_s$. Then

$$\begin{aligned} \langle I_M(A), I_N(B) \rangle_{L_E^2} &= E(\langle 1_C M_0, 1_B(N_t - N_s) \rangle_E) \\ &= E(E(\langle 1_C M_0, 1_B(N_t - N_s) \rangle_E | \mathcal{F}_s)) \\ &= E(1_C 1_B \langle M_0, E(N_t - N_s) | \mathcal{F}_s) \rangle) = 0. \end{aligned}$$

If $A = (s, t] \times C$ with $C \in \mathcal{F}_s$ and $B = (u, v] \times D$ with $D \in \mathcal{F}_u$, then

$$a := \langle I_M(A), I_N(B) \rangle_{L_E^2} = E(1_C 1_D \langle M_t - M_s, N_v - N_u \rangle_E).$$

If $C \cap D = \phi$ then $a = 0$; if $(s, t] \cap (u, v] = \phi$, then we can assume $s < t \leq u < v$ and then

$$a = E(1_C 1_D \langle M_t - M_s, E(N_v - N_u | \mathcal{F}_u) \rangle_E) = 0.$$

If $A, B \in \mathcal{R}$ are disjoint, then we can write A and B as unions,

$$A = \bigcup_{i \in I} A_i \text{ and } B = \bigcup_{j \in J} B_j$$

of disjoint predictable rectangles A_i and B_j. By the above we have $I_M(A_i) \perp I_N(B_j)$ for each i and j, hence

$$I_M(A) = \sum_{i \in I} I_M(A_i) \perp \sum_{j \in J} I_M(B_j) = I_M(B).$$

If $A, B \in \mathcal{R}$ are disjoint and $x, y \in F$, we can apply the first part of the proof to the G-valued, square integrable martingales Mx and Ny. ∎

Remark. The above lemma will be extended in Proposition 3, for any disjoint sets A, B from $\mathcal{P}$.

2. Theorem. *If M is a square integrable martingale, then I_M can be extended as a σ-additive measure $I_M : \mathcal{P} \to L_E^2$.*

Proof. We shall prove that I_M is bounded on $\mathcal{R}$. Let $A \in \mathcal{R}$. Then A is a disjoint union of predictable rectangles $\{0\} \times A_0$ with $A \in \mathcal{F}_0$ and $(s_i, t_i] \times A_i$, with $A_i \in \mathcal{F}_{s_i}$, $i = 1, 2, \ldots, n$. Let $T = \max\{t_i; 1 \leq i \leq n\}$ and $B = [0, T] \times \Omega$. Then A and $B - A$ are disjoint, hence, by Lemma 1, $I_M(A) \perp I_M(B - A)$; consequently,

$$\|I_M(B)\|_2^2 = \|I_M(B - A) + I_M(A)\|_2^2$$
$$= \|I_M(B - A)\|_2^2 + \|I_M(A)\|_2^2,$$

therefore

$$\|I_M(A)\|_2^2 \leq \|I_M(B)\|_2^2$$
$$= \|M_T\|_2^2 \leq \sup_{t \geq 0} \|M_t\|_2^2 < \infty.$$

This proves that I_M is bounded on $\mathcal{R}$.

Since E is a Hilbert space, it does not contain c_0. We can then apply the extension Theorem 14.3 (the summability criterion) and deduce that I_M can be extended as a σ-additive measure on $\mathcal{P}$. ∎

Remarks. a) Here is an alternative proof of Theorem 2: By Lemma 14 infra we have $\mu_{\langle M \rangle} = \|I_M(A)\|_2^2$ for $A \in \mathcal{R}$, where $\langle M \rangle$ is the sharp bracket of M and $\mu_{\langle M \rangle}$ is the Doleans measure of $\langle M \rangle$, satisfying

$$\mu_{\langle M \rangle}(A) = E\big(I_{\langle M \rangle}(A)\big) = E\Big(\int 1_A d\langle M \rangle\Big), \text{ for } A \in \mathcal{R}.$$

It follows that $I_M \ll \mu_{\langle M \rangle}$. Since $\langle M \rangle$ is cadlag and increasing, by Theorem 14.7 infra the measure $\mu_{\langle M \rangle}$ is σ-additive on $\mathcal{R}$. We can use then the extension

Theorem 7.3 b), rather than the extension Theorem 14.3 to extend I_M to a σ-additive measure on $\mathcal{P}$.

b) Since $I_M : \mathcal{P} \to L_E^2$ is σ-additive, it follows that I_M has finite semivariation relative to $(\mathbb{R}, L_E^2)$, hence M is 2-summable relative to the embedding $E \subset L(\mathbb{R}, E)$. We shall prove in Theorem 7 that I_M is 2-summable relative to any embedding $E \subset L(F, G)$.

We can now extend Lemma 1 for sets from $\mathcal{P}$.

3. Proposition. *Let M and N be two E-valued square integrable martingales.*

For any disjoint sets $A, B \in \mathcal{P}$ and for any elements $x, y \in F$ we have

$$I_M(A) \perp I_N(B), \text{ in } L_E^2$$

and

$$I_M(A)x \perp I_N(B)y, \text{ in } L_G^2.$$

Proof. By Theorem 2, we can consider I_M and I_N extended as σ-additive measures on $\mathcal{P}$. If $A \in \mathcal{R}$, let $\sum_A$ be the class of sets $B \in \mathcal{P}$ such that $I_M(A) \perp I_N(B - A)$. By Lemma 1, $\sum_A$ contains $\mathcal{R}$. The class $\sum_A$ is monotone: if (B_n) is a monotone sequence from $\sum_A$ with limit B, then $(B_n - A)$ is a monotone sequence with limit $B - A$. Since I_N is σ-additive, we have $I_N(B_n - A) \to I_N(B - A)$, in L_E^2; since $I_M(A) \perp I_N(B_n - A)$ for each n, we deduce that $I_M(A) \perp I_N(B - A)$, hence $B \in \sum_A$. It follows that $\sum_A = \mathcal{P}$. Therefore for any sets $A \in \mathcal{R}$ and $B \in \mathcal{P}$ with $A \cap B = \phi$ we have

$$I_M(A) \perp I_N(B - A), \text{ in } L_E^2.$$

Let now $B \in \mathcal{P}$ and denote by $\sum'_B$ the class of sets $A \in \mathcal{P}$ such that $I_M(A) \perp I_M(B - A)$. Then $\sum'_B$ contains $\mathcal{R}$ and one can prove as above that $\sum'_B$ is a monotone class. It follows that $\sum'_B = \mathcal{P}$. Therefore, if $A, B \in \mathcal{P}$ are disjoint, then $I_M(A) \perp I_N(B)$ in L_E^2. For the second assertion of the theorem, we apply the first assertion to the square integrable martingales Mx and Ny. ∎

4. Theorem. *If M is an E-valued, square integrable martingale, then the mapping $A \mapsto \|I_M(A)\|_{L_E^2}^2$ is a positive, σ-additive measure on $\mathcal{P}$.*

Proof. Denote $\mu(A) = \|I_M(A)\|_{L_E^2}^2$ for $A \in \mathcal{P}$. We prove first that μ is additive: if $A, B \in \mathcal{P}$ are disjoint, then, by Proposition 3, $I_M(A) \perp I_M(B)$ in L_E^2, hence

$$\|I_M(A \cup B)\|_{L_E^2}^2 = \|I_M(A) + I_B(B)\|_{L_E^2}^2$$
$$= \|I_M(A)\|_{L_E^2}^2 + \|I_M(B)\|_{L_E^2}^2.$$

We prove now that μ is σ-additive: if $A_n \downarrow \phi$ is $\mathcal{P}$, then $I_M(A_n) \to 0$ in L_E^2, hence $\|I_M(A_n)\|_{L_E^2} \to 0$, consequently $\mu(A_n) \to 0$. ∎

B. Summability of square integrable martingales

In this section we prove the main result of this chapter, the 2-summability of square integrable martingales (Theorem 7). We first prove some properties of the semivariation of I_M.

5. Theorem. *Let $M : \mathbb{R}_+ \times \Omega \to E$ be a square integrable martingale.*
a) *For any embedding $E \subset L(F,G)$, with G Hilbert space, I_M has finite semivariation relative to (F, L_G^2) and we have*

$$(\tilde{I}_M)_{F, L_G^2}(A) \leq \|I_M(A)\|_{L_E^2} \leq \sup_{t \geq 0} \|M_t\|_{L_E^2} < \infty, \text{ for } A \in \mathcal{P}.$$

b) *For the embedding $E = L(\mathbb{R}, E)$ we have equality:*

$$(\tilde{I}_M)_{\mathbb{R}, L_E^2}(A) = \|I_M(A)\|_{L_E^2}, \text{ for } A \in \mathcal{P}.$$

c) *If M is a real-valued, square integrable martingale and D is any Hilbert space, then for the embedding $\mathbb{R} \subset L(D, D)$ we have the equality*

$$(\tilde{I}_M)_{D, L_D^2}(A) = \|I_M(A)\|_{L_\mathbb{R}^2}, \text{ for } A \in \mathcal{P}.$$

Proof. Let $A \in \mathcal{P}$, (A_i) a finite family of disjoint sets from $\mathcal{P}$ with union equal to A and (x_i) a finite family of elements from F_1.

Using the orthogonality property in Proposition 3 we have

$$\|\sum_i I_M(A_i)x_i\|_{L_G^2}^2 = \sum_i \|I_M(A_i)x_i\|_{L_G^2}^2$$
$$\leq \sum_i \|I_M(A_i)\|_{L_E^2}^2 = \|\sum_i I_M(A_i)\|_{L_E^2}^2 = \|I_M(A)\|_{L_E^2}^2,$$

hence
$$(\tilde{I}_M)_{F, L_G^2}(A) \leq \|I_M(A)\|_{L_E^2}.$$

Since, by Theorem 4, $A \mapsto \|I_M(A)\|_{L_E^2}^2$ is a positive measure, we have

$$\|I_M(A)\|_{L_E^2} \leq \|I_M(\mathbb{R}_+ \times \Omega)\|_{L_E^2} = \|M_\infty\|_{L_E^2} = \sup_{t \geq 0} \|M_t\|_{L_E^2}$$

and this proves assertion a).

To prove assertion b), let $\alpha \in \mathbb{R}$ with $|\alpha| = 1$. Then

$$\|I_M(A)\|_{L_E^2} = \|I_M(A)\alpha\|_{L_E^2} \leq (\tilde{I}_M)_{\mathbb{R}, L_E^2}$$

and the equality in assertion b) follows.

To prove assertion c), let $E = \mathbb{R}$, let D be a Hilbert space and consider $\mathbb{R} \subset L(D, D)$. Let $x \in D$ with $|x| = 1$. Then

$$\|I_M(A)\|_{L_\mathbb{R}^2} = \|I_M(A)x\|_{L_D^2} \leq (\tilde{I}_M)_{D, L_D^2}$$

and the equality in assertion c) follows. ∎

The embedding $L_E^2 \subset L(F, L_G^2)$ is not necessarily an isometry, even if the embedding $E \subset L(F,G)$ is an isometry. In general, there is no relationship between the semivariations $(\tilde{I}_M)_{\mathbb{R},L_E^2}$ and $(\tilde{I}_M)_{F,L_G^2}$, if E and G are arbitrary Banach spaces, or if M is an arbitrary process. However, in the case of a square integrable martingale in Hilbert spaces, these two semivariations are comparable.

6. Corollary. *If $M : \mathbb{R}_+ \times \Omega \to E \subset L(F,G)$ is a square integrable martingale (and E, G are Hilbert spaces) then*

$$(\tilde{I}_M)_{F,L_G^2}(A) \leq (\tilde{I}_M)_{\mathbb{R},L_E^2}(A) = \|I_M(A)\|_{L_E^2}, \text{ for } A \in \mathcal{P}.$$

If M is a real-valued, square integrable martingale and D is a Hilbert space and if we consider the embedding $\mathbb{R} \subset L(D,D)$, then

$$(\tilde{I}_M)_{D,L_D^2}(A) = (\tilde{I}_M)_{\mathbb{R},L_\mathbb{R}^2}(A) = \|I_M(A)\|_{L_\mathbb{R}^2}, \text{ for } A \in \mathcal{P}.$$

Proof. The first inequality follows from Theorem 5 a) and b) and the second inequality follows from Theorem 5 b) and c). ∎

From Theorems 2 and 5 we deduce the main result of this paragraph:

7. Theorem. *A square integrable martingale $M : \mathbb{R}_+ \times \Omega \to E$ is 2-summable relative to any embedding $E \subset L(F,G)$ with G Hilbert space.*

A real-valued square integrable martingale M is 2-summable relative to the embedding $\mathbb{R} \subset L(D,D)$, for any Hilbert space D.

8. Corollary. *A square integrable martingale $M : \mathbb{R}_+ \times \Omega \to E \subset L(F,G)$ is 1-summable relative to (F,G), with G Hilbert space and we have*

$$(\tilde{I}_M)_{F,L_G^1} \leq (\tilde{I}_M)_{F,L_G^2}.$$

A real-valued, square integrable martingale M is 1-summable relative to the embedding $\mathbb{R} \subset L(D,D)$, for any Hilbert space D and we have

$$(\tilde{I}_M)_{\mathbb{R},L_\mathbb{R}^1} \leq (\tilde{I}_M)_{D,L_D^1} \leq (I_M)_{D,L_D^2}.$$

Proof. The inequality $(\tilde{I}_M)_{\mathbb{R},L_\mathbb{R}^1} \leq (\tilde{I}_M)_{D,L_D^1}$ follows from Proposition 4.12, since the embedding $L_\mathbb{R}^1 \subset L(D, L_D^1)$ is an isometry. The other inequalities follow from the inequality $\|f\|_1 \leq \|f\|_2$ for $f \in L^2(P)$. ∎

9. Corollary. *A locally square integrable local martingale $M : \mathbb{R}_+ \times \Omega \to E$ is locally 2-summable relative to (F,G).*

A real-valued locally square integrable local martingale is locally 2-summable relative to the embedding $R \subset L(D,D)$, for any Hilbert space D.

Remark. A local martingale M is not necessarily locally square integrable; but can be decomposed as a sum $M = U + V$ of a locally square integrable martingale U and a local martingale V with integrable variations (Theorem 22.19). It follows that a local martingale with values in a Hilbert space is locally 1-summable (Corollary 22.20).

C. Properties of the space $\mathcal{F}_{F,G}(M)$

We start with a property of uniform σ-additivity.

10. Theorem. *If M is an E-valued, square integrable martingale, then the set of measures $(I_M)_{F,L_G^2}$ is uniformly σ-additive.*

Proof. If $A_n \downarrow \phi$ in $\mathcal{P}$ then $I_M(A_n) \to 0$ in L_E^2, since I_M is σ-additive. Then $\|I_M(A)\|_{L_E^2} \to 0$ and from Proposition 5 it follows that $(\tilde{I}_M)_{F,L_G^2}(A_n) \to 0$. We can apply now Theorem 4.23 to deduce that $(I_M)_{F,L_G^2}$ is uniformly σ-additive. ∎

11. Corollary. *If M is an E-valued, square integrable martingale, then the set of measures $(I_M)_{F,L_{G^*}^\infty}$ is uniformly σ-additive.*

Proof. If $A_n \downarrow \phi$ in $\mathcal{P}$, then, by Corollary 8 we have $(\tilde{I}_M)_{F,L_G^1}(A_n) \to 0$. We use Theorem 4.23 to deduce that $(I_M)_{F,L_{G^*}^\infty}$ is uniformly σ-additive. ∎

The following theorem extends Corollary 6 for predictable processes H rather than predictable sets A.

12. Theorem. *Let M be an E-valued, square integrable martingale. Then*
a)
$$\mathcal{F}_\mathbb{R}((\tilde{I}_M)_{\mathbb{R},L_E^2}) \subset \mathcal{F}_\mathbb{R}((\tilde{I}_M)_{F,L_G^2})$$
and for every $H \in \mathcal{F}_\mathbb{R}((\tilde{I}_M)_{\mathbb{R},L_E^2})$ we have
$$(\tilde{I}_M)_{F,L_G^2}(H) \leq (\tilde{I}_M)_{\mathbb{R},L_E^2}(H) = \left\| \int H dI_M \right\|_{L_E^2}.$$

b) *If M is real-valued and we consider $\mathbb{R} \subset L(D,D)$ with D a Hilbert space, then*
$$\mathcal{F}_\mathbb{R}((\tilde{I}_M)_{\mathbb{R},L_\mathbb{R}^2}) = \mathcal{F}_\mathbb{R}((\tilde{I}_M)_{D,L_D^2})$$
and for every $H \in \mathcal{F}_\mathbb{R}((\tilde{I}_M)_{\mathbb{R},L_\mathbb{R}^2})$ we have the equality
$$(\tilde{I}_M)_{D,L_D^2}(H) = (\tilde{I}_M)_{\mathbb{R},L_\mathbb{R}^2}(H).$$

Proof. Let $H \in \mathcal{F}_\mathbb{R}((\tilde{I}_M)_{\mathbb{R},L_E^2})$ and let $K : \mathbb{R}_+ \times \Omega \to F$ be a $\mathcal{P}$-step process,
$$K = \sum_{i \in I} 1_{A_i} x_i,$$
with $A_i \in \mathcal{P}$ mutually disjoint and $x_i \in F$. Assume $|K| \leq |H|$. By Proposition 3, the family $(I_M(A_i))_{i \in I}$ is orthogonal in L_E^2 and the family $(I_M(A_i)x_i)_{i \in I}$

is orthogonal in L_G^2. Then

$$\|\int K dI_M\|_{L_G^2}^2 = \|\sum_i I_M(A_i)x_i\|_{L_G^2}^2$$
$$= \sum_i \|I_M(A_i)x_i\|_{L_G^2}^2 \leq \sum_i \|I_M(A_i)\|_{L_E^2}^2 |x_i|^2$$
$$= \sum_i \|I_M(A_i)|x_i|\ \|_{L_E^2}^2 = \|\sum_i I_M(A_i)|x_i|\ \|_{L_E^2}^2$$
$$= \|\int |K|dI_M\|_{L_E^2}^2,$$

therefore,

$$\|\int K dI_M\|_{L_G^2} \leq \|\int |K|dI_M\|_{L_E^2} \leq (\tilde{I}_M)_{\mathbb{R},L_E^2}(H).$$

Taking the supremum for $|K| \leq |H|$, we obtain (see remark after Definition 4.1)

$$(\tilde{I}_M)_{F,L_G^2}(H) \leq (\tilde{I}_M)_{\mathbb{R},L_E^2}(H) < \infty,$$

hence $H \in \mathcal{F}_{\mathbb{R}}((\tilde{I}_M)_{F,L_G^2}$ and this proves assertion a). To prove assertion b), assume M is real-valued and $\mathbb{R} \subset L(D, D)$ with D a Hilbert space. Let $H \in \mathcal{F}_{\mathbb{R}}((\tilde{I}_M)_{D,L_D^2})$ and let

$$K = \sum_{i \in I} 1_{A_i}\alpha_i$$

with $A_i \in \mathcal{P}$ mutually disjoint, $\alpha_i \in \mathbb{R}$ and $|K| \leq |H|$. Let $x \in D$ with $|x| = 1$. Then

$$\|\int K dI_M\|_{L_{\mathbb{R}}^2}^2 = \|\sum_i I_M(A_i)\alpha_i\|_{L_{\mathbb{R}}^2}^2$$
$$= \sum_i \|I_M(A_i)\alpha_i\|_{L_{\mathbb{R}}^2}^2 = \sum_i \|I_M(A_i)\alpha_i x\|_{L_D^2}^2$$
$$= \|\sum_i I_M(A_i)\alpha_i x\|_{L_D^2}^2 = \|\int K x dI_M\|_{L_D^2}^2,$$

therefore,

$$\|\int K dI_M\|_{L_{\mathbb{R}}^2} = \|\int K x dI_M\|_{L_D^2} \leq (\tilde{I}_M)_{D,L_D^2}(H).$$

Taking the supremum for $|K| \leq |H|$, we obtain

$$(\tilde{I}_M)_{\mathbb{R},L_{\mathbb{R}}^2}(H) \leq (\tilde{I}_M)_{D,L_D^2}(H) < \infty.$$

Using assertion a), we get the equality in assertion b). ∎

The following theorem shows that if $F = \mathbb{R}$ or if $E = \mathbb{R}$, then the $\mathcal{R}$-step functions are dense in $\mathcal{F}_{F,G}(M)$.

13. Theorem. a) *If $M : \mathbb{R}_+ \times \Omega \to E \subset L(\mathbb{R}, E)$ is a square integrable martingale, then*

$$L^1_{\mathbb{R}, L^2_E}(M) = \mathcal{F}_{\mathbb{R}, L^2_E}(M),$$

and the real-valued, $\mathcal{R}$-step processes are dense in $L^1_{\mathbb{R}, L^2_E}(M)$.
b) *If M is a real-valued square integrable martingale and $\mathbb{R} \subset L(D, D)$ with D Hilbert space, then*

$$L^1_{D, L^2_D}(M) = \mathcal{F}_{D, L^2_D}(M)$$

and the D-valued, $\mathcal{R}$-step processes are dense in $L^1_{D, L^2_D}(M)$.

Proof. a) Since L^2_E is reflexive and M is 2-summable, by Corollary 16.3 we have $L^1_{\mathbb{R}, L^2_E}(M) = \mathcal{F}_{\mathbb{R}, L^2_E}(I_M)$. Let $H \in L^1_{\mathbb{R}, L^2_E}(M)$ and prove that H can be approximated in $L^1_{\mathbb{R}, L^2_E}(M)$ by bounded processes. We have $|H| \in \mathcal{F}_{\mathbb{R}, L^2_E}(I_M)$, that is, $|H| \in L^1_{\mathbb{R}, L^2_E}(M)$.

By Theorem 16.2, $|H| \cdot M$ is an E-valued, square integrable martingale; therefore, by Theorem 5 b) we have

$$(\tilde{I}_{|H| \cdot M})_{\mathbb{R}, L^2_E}(A) = \|I_{|H| \cdot M}(A)\|_{L^2_E}, \text{ for } A \in \mathcal{P}.$$

By Theorem 13.1 b) we have

$$(\tilde{I}_{|H| \cdot M})_{\mathbb{R}, L^2_E}(A) = (\tilde{I}_M)_{\mathbb{R}, L^2_E}(1_A |H|), \text{ for } A \in \mathcal{P}.$$

It follows that for $A \in \mathcal{P}$ we have

$$(\tilde{I}_M)_{\mathbb{R}, L^2_E}(1_A H) = (\tilde{I}_M)_{\mathbb{R}, L^2_E}(1_A |H|)$$
$$= (\tilde{I}_{|H| \cdot M})_{\mathbb{R}, L^2_E}(A) = \|I_{|H| \cdot M}(A)\|_{L^2_E}.$$

Since $I_{|H| \cdot M}$ is σ-additive in L^2_E on $\mathcal{P}$, it follows that if $A_n \downarrow \phi$, then $I_{|H| \cdot M}(A_n) \to 0$ in L^2_E, therefore $(\tilde{I}_M)_{\mathbb{R}, L^2_E}(1_{A_n} H) \to 0$.

By Proposition 5.40, we have $H \in \mathcal{F}_{\mathbb{R}}(\mathcal{B}, (\tilde{I}_M)_{\mathbb{R}, L^2_E})$. Since $(I_M)_{\mathbb{R}, L^2_E}$ is uniformly σ-additive (Theorem 10) we can use Proposition 5.44 to deduce that the set $\mathcal{S}_{\mathbb{R}}(\mathcal{R})$ is dense in $\mathcal{F}_{\mathbb{R}}(\mathcal{B}, (\tilde{I}_M)_{\mathbb{R}, L^2_E}) = \mathcal{F}_{\mathbb{R}}((\tilde{I}_M)_{\mathbb{R}, L^2_E})$. This proves assertion a).
b) Assume M is real-valued and let $H \in \mathcal{F}_{D, L^2_D}(M)$ and prove that H can be approximated by bounded processes.
From Theorem 12 b) we deduce

$$(\tilde{I}_M)_{\mathbb{R}, L^2_{\mathbb{R}}}(|H|) = (\tilde{I}_M)_{D, L^2_D}(|H|) = (\tilde{I}_M)_{D, L^2_D}(H) < \infty,$$

hence, by Theorem 16.3,

$$|H| \in \mathcal{F}_{\mathbb{R}}((\tilde{I}_M)_{\mathbb{R}, L^2_{\mathbb{R}}}) = \mathcal{F}_{\mathbb{R}, L^2_{\mathbb{R}}}(M) = L^1_{\mathbb{R}, L^2_{\mathbb{R}}}(M).$$

§17. SQUARE INTEGRABLE MARTINGALES   193

By the above proof, with $E = \mathbb{R}$, we have $|H| \in \mathcal{F}_\mathbb{R}((\tilde{I}_M)_{\mathbb{R},L^2_\mathbb{R}})$. Since $(I_M)_{\mathbb{R},L^2_\mathbb{R}}$ is uniformly σ-additive, if $A_n \downarrow \phi$ we have $(\tilde{I}_M)_{\mathbb{R},L^2_\mathbb{R}}(A_n) \to 0$ (Theorem 4.23), hence $(\tilde{I}_M)_{\mathbb{R},L^2_\mathbb{R}}(H1_{A_n}) \to 0$ (Theorem 5.38). Then we have also $(\tilde{I}_M)_{D,L^2_D}(H1_{A_n}) \to 0$. Then, by Proposition 5.40 we have $H \in \mathcal{F}_\mathbb{R}(\mathcal{B}, (\tilde{I}_M)_{D,L^2_D})$.

Finally, since $(I_M)_{D,L^2_D}$ is uniformly σ-additive (Theorem 10), the bounded processes can be approximated in $\mathcal{F}_{D,L^2_D}(M)$ by $\mathcal{R}$-step processes (Proposition 5.44). ∎

D. Isometrical isomorphism of $L^1_{F,G}(M)$ and $L^2_F(\mu_{\langle M \rangle})$

Let $M : \mathbb{R}_+ \times \Omega \to E$ be a square integrable martingale. Then $|M|^2$ is a submartingale of class (D) and has a Doob–Meyer decomposition

$$|M|^2 = N + \langle M \rangle,$$

where N is a martingale of class (D) and $\langle M \rangle$ is a predictable, integrable, increasing process, called the sharp bracket of M (see, for example, [D–M], VII. 40, 41).

Consider the Doléans measures $\mu_{|M|^2}$ and $\mu_{\langle M \rangle}$ defined for $A \in \mathcal{R}$ by

$$\mu_{|M|^2}(A) = E(I_{|M|^2}(A))$$

and

$$\mu_{\langle M \rangle}(A) = E(I_{\langle M \rangle}(A)).$$

Since N is a martingale, its Doléans measure μ_N is identically 0, hence

$$\mu_{|M|^2} = \mu_{\langle M \rangle}, \text{ on } \mathcal{R}.$$

Since $M_\infty \in L^2_E$, we can consider the real-valued, σ-additive measure $\langle I_M, M_\infty \rangle$ on $\mathcal{P}$ (Theorem 2), where the bracket means the inner product in the Hilbert space L^2_E:

$$\langle I_M(A), M_\infty \rangle = E(\langle I_M(A), M_\infty \rangle_E), \text{ for } A \in \mathcal{P}.$$

This measure is positive, as the following lemma shows:

14. Lemma. *For every set $A \in \mathcal{R}$ we have*

$$\langle I_M(A), M_\infty \rangle = \mu_{\langle M \rangle}(A) = \|I_M(A)\|^2_{L^2_E}.$$

Proof. Let $A = \{0\} \times B$ with $B \in \mathcal{F}_0$. Then

$$\langle I_M(A), M_\infty \rangle = \langle 1_B M_0, M_\infty \rangle = E(\langle 1_B M_0, M_\infty \rangle_E)$$
$$= E(1_B |M_0|^2),$$
$$\mu_{\langle M \rangle}(A) = \mu_{|M|^2}(A) = E(I_{|M|^2}(A)) = E(1_B |M_0|^2),$$

and
$$\|I_M(A)\|_{L_E^2}^2 = E(|I_M(A)|^2) = E(1_B|M_0|^2),$$
therefore we have the desired equality.

Let now $A = (s,t] \times B$ with $B \in \mathcal{F}_s$. Then
$$\langle I_M(A), M_\infty \rangle = \langle 1_B(M_t - M_s), M_\infty \rangle = E(1_B \langle M_t - M_s, M_\infty \rangle_E)$$
$$= E(1_B \langle M_t, M_\infty \rangle - 1_B \langle M_s, M_\infty \rangle_E)$$
$$= E(1_B(|M_t|^2 - |M_s|^2)),$$
$$\mu_{\langle M \rangle}(A) = \mu_{|M|^2}(A) = E(I_{|M|^2}(A))$$
$$= E(1_B(|M_t|^2 - |M_s|^2))$$
and
$$\|I_M(A)\|_{L_E^2}^2 = E(|I_M(A)|^2) = E(1_B|M_t - M_s|^2)$$
$$= E(1_B(|M_t|^2 - |M_s|^2))$$
and the desired equality holds. It follows that the equality in the lemma is valid for every $A \in \mathcal{R}$. ∎

15. Corollary. *The positive measure $\mu_{\langle M \rangle}$ can be extended to a σ-additive, finite, positive measure on $\mathcal{P}$ and we have*
$$\langle I_M(A), M_\infty \rangle = \mu_{\langle M \rangle}(A) = \|I_M(A)\|_{L_E^2}^2, \text{ for } A \in \mathcal{P}.$$

16. Corollary. *The process $\langle M \rangle$ is 1-summable relative to $(\mathbb{R}, \mathbb{R})$ and we have*
$$\mu_{\langle M \rangle}(A) = E(I_{\langle M \rangle}(A)), \text{ for } A \in \mathcal{P}.$$

Proof. The measure $I_{\langle M \rangle} : \mathcal{R} \to L^1$ is bounded. In fact, for $A \in \mathcal{R}$ we have
$$\|I_{\langle M \rangle}(A)\|_{L^1} = E(|I_{\langle M \rangle}(A)|) = E(I_{\langle M \rangle}(A))$$
$$= \mu_{\langle M \rangle}(A) \leq \mu_{\langle M \rangle}(\mathbb{R}_+ \times \Omega) < \infty.$$

Since $c_0 \not\subset \mathbb{R}$, we can apply the extension Theorem 14.3 and deduce that $I_{\langle M \rangle}$ can be extended to a σ-additive measure $I_{\langle M \rangle} : \mathcal{P} \to L^1$, hence $\langle M \rangle$ is 1–summable relative to $(\mathbb{R}, \mathbb{R})$. The equality
$$\mu_{\langle M \rangle}(A) = E(I_{\langle M \rangle}(A)), \text{ for } A \in \mathcal{R}$$
can now be extended, by σ-additivity, for every $A \in \mathcal{P}$. ∎

Remark. Here again (see remark following Theorem 2), we can use the extension Theorem 7.3 b), rather than Theorem 14.3. In fact, $I_{\langle M \rangle} \ll \mu_{\langle M \rangle}$ and by Theorem 19.7 infra, $\mu_{\langle M \rangle}$ is σ-additive.

The inequality
$$\left\| \int H dI_X \right\|_{L_G^p} \leq (\tilde{I}_X)_{F, L_G^p}(H)$$

valid for an arbitrary process X, p-summable relative to (F,G) and for $H \in \mathcal{F}_{F,G}(X)$, is an isometry in case M is a square integrable martingale, between the spaces $L^2_F(\mu_{\langle M \rangle})$ and $L^1_{F,G}(M)$, if either E or F is equal to $\mathbb{R}$.

17. Theorem. *Let $M : \mathbb{R}_+ \times \Omega \to E \subset L(F,G)$ be a square integrable martingale. Then:*

a)
$$L^2_F(\mu_{\langle M \rangle}) \subset \mathcal{F}_{F,L^2_G}(M) = L^1_{F,G}(M)$$

and for $H \in L^2_F(\mu_{\langle M \rangle})$ we have

$$\Big\| \int H dI_M \Big\|_{L^2_G} \leq (\tilde{I}_M)_{F,L^2_G}(H) \leq \|H\|_{L^2_F(\mu_{\langle M \rangle})}.$$

b) *For the particular embedding $E = L(\mathbb{R}, E)$ we have*

$$L^2_{\mathbb{R}}(\mu_{\langle M \rangle}) = L^1_{\mathbb{R}, L^2_E}(M)$$

and for $H \in L^1_{\mathbb{R}, L^2_E}(M)$ we have

$$\Big\| \int H dI_M \Big\|_{L^2_E} = (\tilde{I}_M)_{\mathbb{R},L^2_E}(H) = \|H\|^2_{L^2(\mu_{\langle M \rangle})}.$$

c) *If M is real-valued and D is a Hilbert space, then*

$$L^2_D(\mu_{\langle M \rangle}) = L^1_{D,L^2_D}(M)$$

and for $H \in L^1_{D,L^2_D}(M)$ we have

$$\Big\| \int H dI_M \Big\|_{L^2_D} = (\tilde{I}_M)_{D,L^2_D}(H) = \|H\|_{L^2_D(\mu_{\langle M \rangle})}.$$

Proof. To prove a), let $H \in L^2_F(\mu_{\langle M \rangle})$ and let $K = \sum_{1 \leq i \leq n} 1_{A_i} x_i$ be a $\mathcal{P}$-step process with $A_i \in \mathcal{P}$ mutually disjoint and $x_i \in F$ such that $|K| \leq |H|$. By Proposition 3, the family $(I_M(A_i)x_i)_{1 \leq i \leq n}$ is orthogonal in L^2_G and we have

$$\Big\| \int K dI_M \Big\|^2_{L^2_G} = \Big\| \sum_i I_M(A_i)x_i \Big\|^2_{L^2_G} = \sum_i \|I_M(A_i)x_i\|^2_{L^2_G}$$

$$\leq \sum_i \|I_M(A_i)\|^2_{L^2_E} |x_i|^2 = \sum_i E(|I_M(A_i)|^2)|x_i|^2$$

$$= \sum_i E(I_{|M|^2}(A_i)|x_i|^2) = E\Big(\int |K|^2 dI_{|M|^2}\Big)$$

$$= \int |K|^2 d\mu_{|M|^2} = \int |K|^2 d\mu_{\langle M \rangle} = \|K\|^2_{L^2(\mu_{\langle M \rangle})} \leq \|H\|^2_{L^2(\mu_{\langle M \rangle})},$$

therefore
$$\Big\| \int K dI_M \Big\|_{L^2_G} \leq \|H\|_{L^2(\mu_{\langle M \rangle})}.$$

Taking the supremum for $|K| \leq |H|$ we obtain

$$(\tilde{I}_M)_{F,L_G^2}(H) \leq \|H\|_{L^2(\mu_{\langle M \rangle})},$$

hence $H \in \mathcal{F}_{F,L_G^2}(M)$. The inequality

$$\left\| \int H dI_M \right\|_{L_G^2} \leq (\tilde{I}_M)_{F,L_G^2}(H)$$

follows from the definition of $\int H dI_M$ (see 10.2). The equality $\mathcal{F}_{F,L_G^2}(M) = L_{F,L_G^2}^1(M)$ follows from Corollary 16.3. This proves assertion a).

To prove assertion b), let $K = \sum_{1 \leq i \leq n} 1_{A_i} \alpha_i$ be a real-valued, $\mathcal{P}$-step process, with $A_i \in \mathcal{P}$ mutually disjoint and $\alpha_i \in \mathbb{R}$. The computation in the previous case yields this time an equality:

$$\left\| \int K dI_M \right\|_{L_E^2} = \|K\|_{L^2(\mu_{\langle M \rangle})}.$$

The inequality

$$\left\| \int K dI_M \right\|_{L_E^2} \leq (\tilde{I}_M)_{\mathbb{R},L_E^2}(K)$$

follows from the definition of the integral $\int H dI_M$ (see 10.2). Using assertion a) we get

$$\left\| \int K dI_M \right\|_{L_E^2} = (\tilde{I}_M)_{\mathbb{R},L_E^2}(K) = \|K\|_{L^2(\mu_{\langle M \rangle})}.$$

This shows that the identity mapping on the set of $\mathcal{P}$-step processes is an isometry for the norms of the spaces $L^2(\mu_{\langle M \rangle})$ and $L_{\mathbb{R},E}^1(M)$. Since the $\mathcal{P}$-step processes are dense in $L^2(\mu_{\langle M \rangle})$ and in $L_{\mathbb{R},E}^1(M)$ (Theorem 13 a), it follows that $L^2(\mu_{\langle M \rangle}) = L_{\mathbb{R},E}^1(M)$ and

$$(\tilde{I}_M)_{\mathbb{R},L_E^2}(H) = \|H\|_{L^2(\mu_{\langle M \rangle})}, \text{ for } H \in L_{F,G}^2(M).$$

The isometry $K \mapsto \int K dI_M$ from the $\mathcal{P}$-step processes of $L_{F,G}^1(M)$ into L_E^2 is also extended, by the continuity of the integral, to an isometry of the whole space $L_{F,G}^1(M)$ into L_E^2:

$$\left\| \int H dI_M \right\|_{L_E^2} = (\tilde{I}_M)_{\mathbb{R},L_E^2}(H), \text{ for } H \in L_{F,G}^1(M).$$

This proves assertion b).

To prove assertion c), assume M is real-valued and let D be a Hilbert space. Let $K = \sum_{1 \leq i \leq n} 1_{A_i} x_i$ be a $\mathcal{P}$-step process, with $A_i \in \mathcal{P}$ mutually disjoint and $x_i \in D$. The computation in assertion a) yields this time too an equality,

$$\left\| \int K dI_M \right\|_{L_D^2} = \|K\|_{L_D^2(\mu_{\langle M \rangle})}$$

§17. SQUARE INTEGRABLE MARTINGALES   197

and also
$$\|\int K dI_M\|_{L_D^2} = (\tilde{I}_M)_{D,L_D^2}(K).$$

Using Theorem 13 b), we deduce as in the proof of assertion b), that $L_D^2(\mu_{\langle M \rangle}) = L_{D,L_D^2}^1(M)$ and

$$\|\int H dI_M\|_{L_D^2} = (\tilde{I}_M)_{D,L_D^2}(H) = \|H\|_{L_D^2(\mu_{\langle M \rangle})}, \text{ for } H \in L_{D,L_D^2}^1(M).$$

■

Remark. Here is the classical approach of the stochastic integral of a real-valued, square integrable martingale M:

We identify the space $\mathcal{M}^2$ of cadlag, square integrable martingales with the space $L^2(P)$, by identifying a martingale $M \in \mathcal{M}^2$ with $M_\infty \in L^2(P)$ and endow $\mathcal{M}^2$ with the norm of $L^2(P)$.

For a simple process
$$H = \alpha_0 1_{\{0\}} + \sum_{i=1}^n \alpha_i 1_{(s_i, t_i]}$$

we define the stochastic integral $(H \cdot M)_t$ by

$$(H \cdot M)_t = \alpha_0 M_0 + \sum_{i=1}^n \alpha_i (M_{t_i \wedge t} - M_{s_i \wedge t}).$$

Then $H \cdot M \in \mathcal{M}^2$. If we consider $H \in L^2(\mu_{\langle M \rangle}) = L^2(\mathcal{P}, \mu_{\langle M \rangle})$, then one can prove that the mapping $H \mapsto H \cdot M$ is an isometry:

$$\|H \cdot M\|_{\mathcal{M}^2} = \|H\|_{L^2(\mu_{\langle M \rangle})}.$$

Since the simple processes are dense in $L^2(\mu_{\langle M \rangle})$, one can extend the above isometry to an isometry of the whole space $L^2(\mu_{\langle M \rangle})$ into $\mathcal{M}^2$. The value of this extension for a process $H \in L^2(\langle \mu \rangle)$ is denoted $H \cdot M$ and is called the stochastic integral of H with respect to M. This approach can be extended for square integrable martingales with values in a Hilbert space.

The isometry between the space $\mathcal{M}^2$ and $L^2(P)$, which is the starting step for the classical approach of the stochastic integral, is also obtained in Theorem 17 b) and c), using the measure-theoretic approach presented in this book.

Chapter 4
Processes with Finite Variation

In this chapter we study the processes with finite variation and we prove the following main results:
a) Processes with integrable variation are summable (Theorem 19.13);
b) The stochastic integral can be computed pathwise as a Stieltjes integral (Theorem 19.16);
c) If X is a right continuous process with integrable variation, we can associated to it a stochastic measure μ_X satisfying

$$\int H d\mu_X = E(\int H_s dX_s),$$

for F-valued, μ_X-integrable processes H (Theorem 19.8), where $\int H_s dX_s$ is the Stieltjes integral.

The chapter consists of two paragraphs. In §18 we study the properties of functions with finite variation. We prove that such a function $g : \mathbb{R} \to E$ is right continuous iff its variation $|g|$ is right continuous (Theorem 18.11) and that the measure m_g defined by $m_g(a, b] = g(b) - g(a)$ is σ-additive and has finite variation iff g is right continuous and has finite variation (Proposition 18.17 and Theorem 18.18).

If g has finite variation and is right continuous, we can apply the third stage of integration presented in §2 and define the integral $\int f dm_g$. The Stieltjes integral is defined then by the equality $\int f dg = \int f dm_g$.

In §19 we study the processes with finite variation and prove the above-mentioned results.

§18. FUNCTIONS WITH FINITE VARIATION AND THEIR STIELTJES INTEGRAL

In this paragraph $\mathcal{R}$ is the ring generated by the semiring $\mathcal{P}$ of the intervals $(a, b]$ with $a, b \in \mathbb{R}$. The σ-ring generated by $\mathcal{R}$ is the Borel σ-algebra $\mathcal{B}(\mathbb{R})$; the δ-ring generated by $\mathcal{R}$ is the class of bounded Borel sets.

A. Functions with finite variation

Let $g : \mathbb{R} \to E$ be a function.

1. Definition. *For any interval $I \subset \mathbb{R}$ (bounded or not) the variation of g on I is a number, finite or $+\infty$, denoted by $var(g, I)$ or $v(g, I)$ and defined by the following equality:*

$$var(g, I) = \sup \sum_i |g(t_{i+1}) - g(t_i)|$$

where the supremum is taken for all finite divisions $d : t_0 < t_1 < \cdots < t_n$ consisting of points from I.

We say that g has *finite variation* if $v(g, I) < \infty$ for every bounded interval.
We say that g has *bounded variation* if $v(g, \mathbb{R}) < \infty$.
If $I = [a, b]$ with $a < b$, in the above definition it is enough to take only divisions of the form $a = t_0 < t_1 < \cdots < t_n = b$.
If $I = (a, b]$ with $a \geq -\infty$, we take divisions of the form $t_0 < t_1 < \cdots < t_n = b$ with $a < t_0$.
We state first some general properties:

2. *If $a < b$, then*
$$v(g, (a, b]) = \sup_{\alpha > a} v(g, [\alpha, b]);$$

In particular
$$v(g, (-\infty, b]) = \sup_{\alpha > -\infty} v(g, [\alpha, b]).$$

3. *If $a < b$ and if g is right continuous then*
$$v(g, [a, b]) = v(g, (a, b]).$$

4. *If $I_1 \subset I_2$ are two intervals, then*
$$var(g, I_1) \leq var(g, I_2).$$

The variation is additive for intervals with disjoint interiors:

5. *If $-\infty \leq \alpha < a < b$, then*
$$var(g, (\alpha, b]) = var(g, (\alpha, a]) + var(g, [a, b])$$

and if $-\infty < \alpha < a < b$, then
$$var(g, [\alpha, b]) = var(g, [\alpha, a]) + var(g, [a, b]).$$

Two more equalities are valid for the intervals (a, c) and $[a, c)$.

We state now some properties of functions with *finite* variation.

6. *If g has finite variation and if $-\infty < \alpha < a < b$, then*
$$\begin{aligned} v(g, [a, b]) &= v(g, [\alpha, b]) - v(g, [\alpha, a]) \\ &= v(g, (\alpha, b]) - v(g, (\alpha, a]). \end{aligned}$$

If $-\infty \leq \alpha < a < b$ and g is right continuous, then
$$v(g, (a, b]) = v(g, (\alpha, b]) - v(g, (\alpha, a]).$$

7. *If $a < b$, then*
$$|g(b) - g(a)| \leq var(g, [a, b]).$$

If g is right continuous, then
$$|g(b) - g(a)| \leq var(g(a, b]).$$

8. *If $g : \mathbb{R} \to \mathbb{R}$ is increasing, then g has finite variation and*
$$v(g, [a, b]) = g(b) - g(a), \text{ for } a < b.$$

If g is increasing and right continuous, then
$$v(g, (a, b]) = g(b) - g(a).$$

9. *If $g : \mathbb{R} \to E$ has finite variation, then g has lateral limits at every point of $\mathbb{R}$.*

If g has bounded variation, then g has limits at $-\infty$ and at $+\infty$.
In this case we define $g(\infty) = \lim_{t \to \infty} g(t)$.

B. The variation function $|g|$

Let $g : \mathbb{R} \to E$ be a function with finite variation and $\alpha \in \mathbb{R}$. We define the variation function $|g|_\alpha : \mathbb{R} \to \mathbb{R}_+$ by
$$\begin{aligned} |g|_\alpha(t) &= v(g, (\alpha, t]), \text{ if } t \geq \alpha, \\ |g|_\alpha(t) &= -v(g, (t, \alpha]), \text{ if } t < \alpha. \end{aligned}$$

We define the variation function $|g| = |g|_{-\infty} \colon \mathbb{R} \to \mathbb{R}_+$ by

$$|g|(t) = var(g, (-\infty, t]), \text{ for } t \in \mathbb{R}.$$

If $var(g, (-\infty, t]) < \infty$ for every $t \in \mathbb{R}$, we say that g has *finite variation function* $|g|$.

We translate some of the preceding properties in terms of $|g|_\alpha$ and $|g|$.

4'. $|g|_\alpha$ and $|g|$ are increasing.

5'. If $-\infty < \alpha < a < b$, then

$$|g|_\alpha(b) = |g|_\alpha(a) + var(g, [a, b])$$

and

$$|g|(b) = |g|(a) + var(g, [a, b]).$$

6'. If g has finite variation and if $\alpha < a < b$, then

$$|g|_\alpha(b) - |g|_\alpha(a) = var(g, [a, b]).$$

If, in addition, g is right continuous, then

$$|g|_\alpha(b) - |g|_\alpha(a) = var(g, (a, b]).$$

6''. If g has finite variation function $|g|$ and if $a < b$, then

$$|g|(b) - |g|(a) = var(g, [a, b]).$$

If, in addition, g is right continuous, then

$$|g|(b) - |g|(a) = var(g, (a, b]).$$

7'. If g has finite variation and if $a < b$, then

$$|g(b) - g(a)| \leq |g|_\alpha(b) - |g|_\alpha(a).$$

If g has finite variation function $|g|$ and if $a < b$, then

$$|g(b) - g(a)| \leq |g|(b) - |g|(a).$$

8'. If g is increasing and $\alpha < a < b$, then

$$|g|_\alpha(b) - |g|_\alpha(a) = g(b) - g(a).$$

If g is increasing and has finite variation function $|g|$ and if $a < b$, then

$$|g|(b) - |g|(a) = g(b) - g(a).$$

10. If g has finite variation and if $\alpha, \beta \in \mathbb{R}$, then $|g|_\beta - |g|_\alpha$ is constant. If g has finite variation function $|g|$ and if $\alpha \in \mathbb{R}$ then $|g| - |g|_\alpha$ is constant.

11. Theorem. *Assume g has finite variation $|g|_\alpha$ with $\alpha \geq -\infty$. Then g is right (resp. left) continuous iff the variation function $|g|_\alpha$ is right (resp. left) continuous.*

Proof. If $|g|_\alpha$ is right continuous, from Property 7' we deduce that g is right continuous.

Conversely assume g is right continuous at a point $a \in \mathbb{R}$ but $M = |g|_\alpha(a+) - |g|_\alpha(a) > 0$ and prove that this leads to a contradiction. Let $b > a$ and $\varepsilon > 0$. We observe first that for $a < x < b$ we have, by property 6',
$$var(g, [a,b]) \geq v(g, [a,x]) = |g|_\alpha(x) - |g|_\alpha(a).$$
Letting $x \downarrow a$ we get $v(g, [a,b]) \geq |g|_\alpha(a+) - |g|_\alpha(a) = M$.

We can find by induction a strictly decreasing sequence (x_n) in $\mathbb{R}$ with $x_1 = b$, with $x_n \to a$ and for each n we can find a division d_n of $[x_{n+1}, x_n]$ such that
$$V_{d_n} := \sum_{t_i \in d_n} |g(t_{i+1}) - g(t_i)| \geq M - \varepsilon/2^n.$$

In fact, let Δ_1 be a division of $[a,b]$ such that
$$V_{\Delta_1} > v(g, [a,b]) - \varepsilon/4 \geq M - \varepsilon/4.$$

Since g is right continuous at a, there is an $x_2 > a$ such that if $x \in [a, x_2]$, then $|g(x) - g(a)| < \varepsilon/4$. We can choose $x_2 \leq \inf\{t \in \Delta_1, t > a\}$ and add x_2 to Δ_1. Denote $d_1 = \Delta_1 - \{a\}$. Then d_1 is a division of $[x_2, x_1]$ and
$$V_{d_1} = V_{\Delta_1} - |g(x_2) - g(a)| \geq M - \varepsilon/4 - \varepsilon/4 = M - \varepsilon/2.$$

The second step of the induction process is proved similarly. For every n, using Properties 4 and 6, we have
$$v(g, [a,b]) \geq v(g, [x_{n+1}, x_1]) = \sum_{1 \leq i < n} v(g, [x_{i+1}, x_i])$$
$$\geq \sum_{1 \leq i < n} V_{d_i} \geq nM - \varepsilon,$$
hence $v(g, [a,b]) = \infty$, which contradicts the hypothesis. ∎

C. The measure associated to a function

To the function $g : \mathbb{R} \to E$ we associate a finitely additive measure $m_g : \mathcal{P} \to E$, by the equality
$$m_g(a, b] = g(b) - g(a), \text{ for } a \leq b \text{ in } \mathbb{R}.$$

Then we can extend m_g uniquely to an additive measure, still denoted by m_g on the ring $\mathcal{R}$ (Theorem 7.2).

12. *For two functions $g, g' : \mathbb{R} \to E$ we have $m_g = m_{g'}$ iff $g - g'$ is constant.*

13. *If $g : \mathbb{R} \to \mathbb{R}$ is a real-valued function then $m_g \geq 0$ iff g is increasing.*

14. *If $g : \mathbb{R} \to E$ is a function and $x^* \in E^*$, then*
$$m_{\langle g, x^* \rangle} = \langle m_g, x^* \rangle.$$

The following two propositions state the relationship between the variations of g and m_g.

15. Proposition. a) *m_g has finite (resp. bounded) variation iff g has finite (resp. bounded) variation.*
b) *If $I \subset \mathbb{R}$ is any interval, then*
$$var(g, I) \leq var(m_g, I) \leq var(g, \overline{I}).$$

c) *We have the equality*
$$var(m_g, I) = var(g, I),$$
if either $\inf I = -\infty$ or $\inf I \in I$.

Proof. Assertion a) follows from assertion b).

Let $t_0 < t_1 < \cdots < t_n$ be a family of points from I. Then
$$\sum_i |g(t_{i+1}) - g(t_i)| = \sum_i |m_g(t_i, t_{i+1}]| \leq var(m_g, I),$$
therefore
$$var(g, I) \leq var(m_g, I).$$

To prove the second inequality of assertion b) let $((a_k, b_k])_{k \in K}$ be a finite family of disjoint intervals contained in I. We arrange the points a_k, b_k in increasing order, $t_0 < t_1 < \cdots < t_n$. Each interval $(a_k, b_k]$ is of the form $(t_i, t_{i+1}]$ for some $i(k)$.

All the points $t_1, \ldots, t_n$ belong to I and $t_0 \geq \inf I$. If either $\inf I = -\infty$ or $\inf I$ is finite and $\inf I \in I$, then we have also $t_0 \in I$; therefore
$$\sum_k |m_g(a_k, b_k]| \leq \sum_i |m_g(t_i, t_{i+1}]|$$
$$= \sum_i |g(t_{i+1}) - g(t_i)| \leq var(g, I),$$

hence $var(m_g, I) \leq var(g, I)$ and this proves assertion c).

If $\inf I$ is finite and if $\inf I \notin I$, then we might have $t_0 = \inf I$. In any case, all the points $t_0, t_1, \ldots, t_n$ belong to $\overline{I}$, therefore
$$\sum_k |m_g(a_k, b_k)| \leq \sum_i |g(t_{i+1}) - g(t_i)| \leq var(g, \overline{I}),$$

hence
$$var(m_g, I) \leq var(g, \overline{I})$$
and this proves assertion b). ∎

16. Proposition. *If g is right continuous then for any interval $I \subset \mathbb{R}$ we have*
$$var(m_g, I) = var(g, I).$$

Proof. By proposition 15 c), we can assume $\inf I$ finite and $\inf I \notin I$. Using assertion b) of Proposition 15 we only have to prove the inequality
$$var(m_g, I) \leq var(g, I).$$

Let $((a_k, b_k])_{k \in K}$ be a finite family of disjoint intervals contained in I and arrange the points a_k, b_k in increasing order, $t_0 < t_1 < \cdots < t_n$.

Let $\varepsilon > 0$. Since g is right continuous, there is a point $t_0' \in I$ such that $t_0 < t_0' < t_1$ and such that
$$|g(t_0) - g(t_0')| < \varepsilon.$$

It follows that
$$|g(t_1) - g(t_0)| \leq |g(t_1) - g(t_0')| + |g(t_0') - g(t_0)|$$
$$< |g(t_1) - g(t_0')| + \varepsilon.$$

Then
$$\sum_k |m_g(a_k, b_k]| \leq \sum_{0 \leq i < n} |m_g(t_i, t_{i+1}]| = \sum_{0 \leq i < n} |g(t_{i+1}) - g(t_i)|$$
$$= |g(t_1) - g(t_0)| + \sum_{1 \leq i < n} |g(t_{i+1}) - g(t_i)|$$
$$< |g(t_1) - g(t_0')| + \sum_{1 \leq i < n} |g(t_{i+1}) - g(t_i)| + \varepsilon$$
$$\leq var(g, I) + \varepsilon,$$

therefore
$$var(m_g, I) \leq var(g, I).$$
∎

17. Proposition. *Let $\alpha \geq -\infty$. If g has finite variation $|g|_\alpha$ and is right continuous we have*
$$|m_g| = m_{|g|_\alpha}.$$

Proof. We remark first that if $\beta \in \mathbb{R}$, then by Property 10, $|g|_\alpha - |g|_\beta$ is constant; therefore, by Property 12, we have $m_{|g|_\alpha} = m_{|g|_\beta}$. By Proposition

15 a), m_g has finite variation $|m_g|$. Let $(a, b] \subset \mathbb{R}$ be an interval and $\varepsilon > 0$. Let $t_0 < t_1 < \cdots < t_n$ be points from $(a, b]$ such that

$$var(g, (a, b]) \leq \sum_i |g(t_{i+1}) - g(t_i)| + \varepsilon.$$

Since g is right continuous, by Property 3 we have $v(g, [a, b]) = v(g, (a, b])$. Then, by Property 6' we have

$$m_{|g|_\alpha}(a, b] = |g|_\alpha(b) - |g|_\alpha(a) = var(g, [a, b])$$
$$= var(g, (a, b]) \leq \sum_i |g(t_{i+1}) - g(t_i)| + \varepsilon$$
$$= \sum_i |m_g(t_i, t_{i+1}]| + \varepsilon \leq |m_g|(a, b] + \varepsilon.$$

Since $\varepsilon > 0$ is arbitrary, we deduce that

$$m_{|g|_\alpha}(a, b] \leq |m_g|(a, b],$$

hence

$$m_{|g|_\alpha}(A) \leq |m_g|(A), \text{ for } A \in \mathcal{R}.$$

Conversely, let $A \in \mathcal{R}$. Then A is of the form $A = \bigcup_{1 \leq i \leq n} (a_i, b_i]$ with $(a_i, b_i]$ mutually disjoint. Then, by Property 7', we have

$$|m_g(A)| = |\sum_i m_g(a_i, b_i]| \leq \sum_i |m_g(a_i, b_i]|$$
$$= \sum_i |g(b_i) - g(a_i)| \leq \sum_i [|g|_\alpha(b_i) - |g|_\alpha(a_i)]$$
$$= \sum_i m_{|g|_\alpha}(a_i, b_i] = m_{|g|_\alpha}(A),$$

consequently,

$$|m_g|(A) = m_{|g|_\alpha}(A), \text{ for } A \in \mathcal{R}.$$

■

The following two theorems ensure the σ-additivity of the measure m_g.

18. Theorem. *Assume $g : \mathbb{R} \to \mathbb{R}$ is increasing. Then the positive measure $m_g : \mathcal{R} \to \mathbb{R}_+$ is σ-additive if and only if g is right continuous.*

Proof. If m_g is σ-additive, then for every $t_0 \in \mathbb{R}$ and for every decreasing sequence $t_n \downarrow t_0$ we have $(t_0, t_n] \downarrow \phi$, hence $m_g(t_0, t_n] \to 0$, that is, $g(t_n) - g(t_0) \to 0$; therefore g is right continuous at t_0.

Conversely, assume g is right continuous and prove the measure $\mu = m_g$ is σ-additive on the semiring $\mathcal{P}$ of the intervals of the form $(a, b]$.

§18. FUNCTIONS WITH FINITE VARIATION AND THEIR STIELTJES INTEGRAL

Let $(a_n, b_n]$ be a sequence of disjoint intervals from $\mathcal{P}$ with union $(a, b] \in \mathcal{P}$ and prove that
$$\mu(a, b] = \sum_{1 \leq n < \infty} \mu(a_n, b_n].$$

We have
$$\sum_{1 \leq n < \infty} \mu(a_n, b_n] = \lim_n \sum_{1 \leq i \leq n} \mu(a_i, b_i]$$
$$= \lim_n \mu(\bigcup_{1 \leq i \leq n} (a_i, b_i]) \leq \mu(a, b].$$

To prove the converse inequality, let $\varepsilon > 0$. Since g is right continuous, there is an $\alpha > a$ with $g(\alpha) - g(a) < \varepsilon$. For each n there is a $\beta_n > b_n$ with $g(\beta_n) - g(b_n) < \varepsilon/2^n$. Then
$$[\alpha, b] \subset \bigcup_{1 \leq n < \infty} (a_n, \beta_n).$$

There is a finite cover of $[\alpha, b]$:
$$[\alpha, b] \subset \bigcup_{1 \leq i \leq N} (a_i, \beta_i).$$

Since $(\alpha, b] \subset [\alpha, b]$ and $(a_i, \beta_i) \subset (a_i, \beta_i]$, we deduce
$$(\alpha, b] \subset \bigcup_{1 \leq i \leq N} (a_i, \beta_i].$$

Then
$$\mu(\alpha, b] \leq \mu(\bigcup_{1 \leq i \leq N} (a_i, \beta_i]) \leq \sum_{1 \leq i \leq N} \mu(a_i, \beta_i] \leq \sum_{1 \leq i < \infty} \mu(a_i, \beta_i],$$

that is,
$$g(b) - g(\alpha) \leq \sum_{1 \leq i < \infty} (g(\beta_i) - g(a_i)).$$

It follows that
$$\mu(a, b] = g(b) - g(a) < g(b) - g(\alpha) + \varepsilon \leq \sum_{1 \leq i < \infty} (g(b_i) - g(a_i) + \varepsilon/2^i) + \varepsilon$$
$$\leq \sum_{1 \leq i < \infty} \mu(a_i, b_i] + 2\varepsilon;$$

therefore, ε being arbitrary,
$$\mu(a, b] \leq \sum_{1 \leq i < \infty} \mu(a_i, b_i].$$

We can use now the extension Theorem 7.4 to deduce the following theorem: Denote by $\mathcal{D}$ the δ-ring of bounded Borel subsets of $\mathbb{R}$.

19. Theorem. *Assume $g : \mathbb{R} \to E$ is right continuous and has finite (resp. bounded) variation.*

Then m_g can be extended uniquely to a σ-additive measure $m : \mathcal{D} \to E$ (resp. $m : \mathcal{B}(\mathbb{R}) \to E$) with finite variation $|m|$ and the variation $|m|$ is the unique extension of $|m_g| = m_{|g|_\alpha}$.

The measure m is still denoted by m_g and is called the *Stieltjes measure* induced by g.

The *jump* of a function $g : \mathbb{R} \to \mathbb{R}$ at a point $t \in \mathbb{R}$ is denoted by $\Delta g(t)$ and is defined by
$$\Delta g(t) = g(t+) - g(t-).$$
The following theorem states the relationship between the jumps of g and its variation $|g|$.

20. Theorem. *Let $g : \mathbb{R} \to E$ be a right continuous function with finite variation $|g|$. Then for every $t \in \mathbb{R}$ we have $\Delta |g|(t) = |\Delta g(t)|$.*

Proof. Let $\mathcal{D}$ be the δ-ring of bounded Borel sets and $m_g : \mathcal{D} \to E$ the σ-additive measure with finite variation $|m_g| = m_{|g|}$ satisfying $m_g(a, b] = g(b) - g(a)$ for $a < b$. Let $t \in \mathbb{R}$ and (s_n) an increasing sequence in $\mathbb{R}$ with $s_n < t$ and $s_n \uparrow t$. Then $(s_n, t] \downarrow \{t\}$, hence $m_g(s_n, t] \to m_g(\{t\})$ and $|m_g|(s_n, t] \to |m_g|(\{t\})$.

We have, evidently, $|m_g|(\{t\}) = |m_g(\{t\})|$. At the same time, $g(s_n) \to g(t-)$ and $|g|(s_n) \to |g|(t-)$. From
$$m_g(s_n, t] = g(t) - g(s_n) \text{ and } |m_g|(s_n, t) = m_{|g|}(s_n, t) = |g|(t) - |g|(s_n)$$
we deduce that
$$m_g(\{t\}) = g(t) - g(t-) = \Delta g(t)$$
and
$$|m_g|(\{t\}) = |g|(t) - |g|(t-) = \Delta |g|(t).$$
Finally, from $|m_g|(\{t\}) = |m_g(\{t\})|$ it follows that $\Delta|g|(t) = |\Delta g(t)|$. ∎

D. The Stieltjes integral

21. Assume $g : \mathbb{R} \to E$ is right continuous and has finite variation $|g|$ and assume $E \subset L(F, G)$. The space $L^1_F(m_g) := L^1_F(|m_g|)$ of Bochner $|m_g|$-integrable functions $f : \mathbb{R} \to F$ is also denoted $L^1_F(dg)$ or $L^1_F(g)$.

§18. FUNCTIONS WITH FINITE VARIATION AND THEIR STIELTJES INTEGRAL

If $f \in L^1_F(dg)$, we define the Stieltjes integral $\int f\,dg$ by the equality

$$\int f\,dg = \int f\,dm_g.$$

Then we have

$$\left|\int f\,dg\right| \le \int |f|\,d|g|.$$

For an $|m_g|$-measurable function f defined $|m_g|$-a.e. on $\mathbb{R}$ and with values in $[0, +\infty]$, we define the Stieltjes integral $\int f\,d|g|$ by the equality

$$\int f\,d|g| = \int f\,d|m_g| \le \infty.$$

22. Comments. a) There is a certain inconsistency between the definition of the variation $v(g, I)$ of the function g and that of the variation $v(m_g, I)$ of the corresponding measure m_g. Both definitions are classical and universally accepted.

The inconsistency consists in the fact that, while in the definition of $var(g, I)$, we take divisions $t_0 < t_1 < \cdots < t_n$ of points t_i *belonging* to I, in the definition of $var(m_g, I)$ we take families $(A_i)_{i \in I}$ of disjoint intervals A_i of the form $A_i = (a_i, b_i]$, *contained* in I, but the endpoints a_i do not necessarily belong to I; for example, if I is open to the left, one of the points a_i might not belong to I. This inconsistency leads to the inequalities between $var(g, I)$ and $var(m_g, I)$ in Proposition 15 b).

We could avoid this inconsistency if, for example, in the definition of $v(g, I)$, we take the supremum for all divisions $t_0 < t_1 < \ldots t_n$ with $(t_i, t_{i+1}] \subset I$ for $0 \le i < n$.

According to Proposition 16, if g is right continuous, this inconsistency disappears.

Since we are interested only in the case when the measure m_g is σ-additive and since according to Theorems 18 and 19, m_g is σ-additive iff g is right continuous, the inconsistency mentioned above is irrelevant.

b) We could consider the semiring $\mathcal{P}'$ of the intervals of the form $[a, b)$ and the ring $\mathcal{R}'$ generated by $\mathcal{P}'$ and define the measure $m'_g : \mathcal{R}' \to E$ by

$$m'_g[a, b) = g(b) - g(a).$$

Then m'_g is σ-additive and has finite variation on $\mathcal{R}'$ iff g is left continuous and has finite variation; in this case, m'_g can be extended to a σ-additive measure m' with finite variation on the δ-ring $\mathcal{D}$ (or on the σ-algebra $\mathcal{B}(\mathbb{R})$, if g has bounded variation).

c) If we start from a function $g : \mathbb{R} \to E$ with finite variation, but not necessarily left or right continuous, it is more appropriate to define the measure $m : \mathcal{R} \to E$ by

$$m(a, b] = g(b+) - g(a+)$$

and the measure $m' : \mathcal{R} \to E$ by

$$m'[a,b) = g(b-) - g(a-).$$

Both measures are σ-additive and can be extended to the same σ-additive measure on $\mathcal{D}$ (or on $\mathcal{B}(\mathbb{R})$, if g has bounded variation).

In fact, the function $g_+(t) = g(t+)$ is right continuous and the function $g_-(t) = g(t-)$ is left continuous and we have

$$m(a,b] = m_{g_+}(a,b]$$

and

$$m'[a,b) = m_{g_-}[a,b).$$

§19. PROCESSES WITH FINITE VARIATION

In this paragraph we combine the results on the Stieltjes integral obtained in §18 and the results on the stochastic integration of Chapter 2 and apply them to processes with integrable variation.

As we mentioned at the beginning of this chapter the main results we obtain here are:
a) Processes with integrable variation are summable (Theorem 19.13);
b) The stochastic integral can be computed pathwise as a Stieltjes integral (Theorem 19.16);
c) The equality

$$\mu_X(M) = E(\int 1_M(s,\omega)dX_s(\omega)), \text{ for } M \in \mathcal{B}(\mathbb{R}_+) \times \mathcal{F}$$

defines a σ-additive measure μ_X with finite variation $|\mu_X| = \mu_{|X|}$ (Theorem 19.8).

A. Definition and properties

In this paragraph we shall denote $\mathcal{M} = \mathcal{B}(\mathbb{R}_+) \times \mathcal{F}$. We say that a process X with values in a Banach space E is *measurable* if it is measurable with respect to $\mathcal{M}$.

A measurable process which is not adapted to the filtration $(\mathcal{F}_t)_{t \geq 0}$ is called a *raw process*.

If X is a measurable process, then for every $t \geq 0$, the path X_t is $\mathcal{F}$-measurable. Conversely, if X is *right continuous* and if X_t is $\mathcal{F}$-measurable for every $t \geq 0$, then X is measurable.

1. Definition. a) *We say that a process $X : \mathbb{R}_+ \times \Omega \to E$ has finite variation, if for each $\omega \in \Omega$, the path $t \mapsto X_t(\omega)$ has finite variation on each interval $[0,t]$.*
b) *If $1 \leq p < \infty$, we say X has p-integrable variation if the total variation $|X|_\infty = var(X, \mathbb{R}_+)$ on $\mathbb{R}_+$ belongs to L^p.*

In particular, an increasing process $X : \mathbb{R}_+ \times \Omega \to \mathbb{R}_+$ is p-integrable if $X_\infty = \sup_{t \geq 0} X_t$ is p-integrable.

If $p = 1$, we say that X has integrable variation instead of 1-integrable variation; and if X is increasing, we say that X is integrable instead of 1-integrable.

If X has p-integrable variation for some $p \geq 1$, then X has integrable variation.

Let $X : \mathbb{R}_+ \times \Omega \to E$ be a measurable process. We extend it to $\mathbb{R} \times \Omega$ with $X_t = 0$ for $t < 0$. To say that X has finite variation means that for every $\omega \in \Omega$, the path $X.(\omega)$ has finite variation on any interval $(-\infty, t]$.

We define the *variation process* $|X|$ by the equality
$$|X|_t(\omega) = var(X.(\omega), (-\infty, t]), \text{ for } t \in \mathbb{R} \text{ and } \omega \in \Omega.$$
It follows that:
$$|X|_t(\omega) = 0, \text{ for } t < 0;$$
$$|X|_0(\omega) = |X_0(\omega)|;$$
$$|X|_t(\omega) = |X_0(\omega)| + var(X.(\omega), [0, t]), \text{ for } t > 0.$$

From Theorem 18.11 we deduce that a process X with finite variation is right continuous iff its variation $|X|$ is right continuous.

2. Proposition. *Let $X : \mathbb{R}_+ \times \Omega \to E$ be a right continuous process with finite variation $|X|$.*

If X is measurable (resp. optional, predictable) then its variation $|X|$ has the same property.

Proof. Since X is right continuous, its variation can be computed by using divisions from a set dense in $\mathbb{R}_+$, for example, from the set
$$\{\frac{i}{2^n} : i = 0, 1, 2, \ldots, n = 1, 2, \ldots\}.$$

For each n fixed and $t \geq 0$, consider the division $\{\frac{i}{2^n} \wedge t : i = 0, 1, 2, \ldots\}$ of $[0, t]$ and the corresponding sum
$$A_t^n = \sum_{0 \leq i < \infty} |X_{\frac{i+1}{2^n} \wedge t} - X_{\frac{i}{2^n} \wedge t}|.$$

The series has only finitely many terms $\neq 0$. If X is measurable (resp. optional, predictable) then the process $(A_t^n)_{t \geq 0}$ has the same property; therefore the variation $|X|_t = |X_0| + \sup_n A_t^n$ has the same property. ∎

3. Theorem. *Let $X : \mathbb{R}_+ \times \Omega \to E \subset L(F, G)$ be a right continuous, measurable process with finite variation $|X|$ and let $H : \mathbb{R}_+ \times \Omega \to F$ be a measurable process. Then.*

a) *The function $\omega \mapsto \int |H_s(\omega)| d|X|_s(\omega) \leq \infty$ is $\mathcal{F}$-measurable.*

b) *If $\int |H_s(\omega)| d|X|_s(\omega) < \infty$, a.s. on Ω, then the function $\omega \mapsto \int H_s(\omega) dX_s(\omega)$, defined a.s., is $\mathcal{F}$-measurable.*

c) *If $E(\int |H_s(\omega)| d|X|_s(\omega)) < \infty$, then the function $\omega \mapsto \int H_s(\omega) dX_s(\omega)$, defined a.s., is integrable.*

d) *If $E(|X|_\infty(\omega)) < \infty$ and H is bounded, then $\int |H_s(\omega)| d|X|_s(\omega) < \infty$, a.s. and the function $\omega \mapsto \int H_s(\omega) dX_s(\omega)$, defined a.s., is integrable.*

Proof. α) We shall prove first assertions a) and b) for $H = 1_M$ with $M = \{0\} \times A$ or $M = (a, b] \times A$ with $A \in \mathcal{F}$ (considering $F = \mathbb{R}$ and $G = E$). We have
$$\int 1_{\{0\} \times A}(s, \omega) d|X|_s(\omega) = 1_A(\omega)|X|_0(\omega)$$

§19. PROCESSES WITH FINITE VARIATION

and
$$\int 1_{(a,b]\times A}(s,\omega)d|X|_s(\omega) = 1_A(\omega)(|X|_b(\omega) - |X|_a(\omega)).$$

By Proposition 2, $|X|$ is measurable; therefore assertion a) is true. Assertion b) follows from
$$\int 1_{\{0\}\times A}(s,\omega)dX_s(\omega) = 1_A(\omega)X_0(\omega)$$

and
$$\int 1_{[a,b]\times A}(s,\omega)dX_s(\omega) = 1_A(\omega)(X_b(\omega) - X_a(\omega)).$$

Assertions a) and b) remain valid for M in the ring r generated by the above sets $\{0\} \times A$ and $(a,b] \times A$ with $A \in \mathcal{F}$.

β) We prove now assertion a) for $H = 1_M$ with $M \in \mathcal{M}$. Denote by $\mathcal{M}_0$ the class of sets $M \in \mathcal{M}$ for which assertion a) is true and prove that $\mathcal{M}_0$ is a monotone class. Let (M_n) be a monotone sequence from $\mathcal{M}_0$ with limit M. To make a choice, assume (M_n) is increasing. For each $\omega \in \Omega$, the sequence $(M_n(\omega))$ of sections is increasing in $\mathcal{B}(\mathbb{R}_+)$ with limit $M(\omega)$. Since $X.(\omega)$ is right continuous and has finite variation $|X|.(\omega)$, the corresponding measures $m_{X(\omega)}$ and $|m_{X(\omega)}| = m_{|X|(\omega)}$ are σ-additive on the δ-ring $\mathcal{D}$ of bounded Borel sets of $\mathbb{R}_+$. Moreover, the positive measure $m_{|X|(\omega)}$ can be extended as a σ-additive measure on $\mathcal{B}(\mathbb{R}_+)$, possibly with infinite values. Then
$$m_{|X|(\omega)}(M_n(\omega)) \nearrow m_{|X|(\omega)}(M(\omega)).$$

For each n, the function
$$\omega \mapsto \int 1_{M_n(\omega)}d|X|_s(\omega) = m_{|X|(\omega)}(M_n(\omega))$$

is $\mathcal{F}$-measurable. It follows that the function
$$\omega \mapsto \int 1_M(s,\omega)d|X|_s(\omega) = m_{|X|(\omega)}(M(\omega))$$

is $\mathcal{F}$-measurable. If (M_n) is decreasing, the proof is similar.

It follows that $M \in \mathcal{M}_0$, hence $\mathcal{M}_0$ is a monotone class. Since by step α), $\mathcal{M}_0$ contains the ring r, we deduce that $\mathcal{M}_0 = \mathcal{M}$, hence assertion a) is true for $H = 1_M$ with $M \in \mathcal{M}$.

Then assertion a) remains valid for any real-valued $\mathcal{M}$-step process H.

γ) Assume now $H : \mathbb{R}_+ \times \Omega \to F$ is a measurable process and let (ϕ^n) be an increasing sequence of positive $\mathcal{M}$-step functions such that $\phi^n \to |H|$. Then $\int \phi^n_s(\omega)d|X|_s(\omega) \nearrow \int |H_s(\omega)|d|X|_s(\omega)$. Since for each n, the function $\omega \mapsto \int \phi^n_s(\omega)d|X|_s(\omega)$ is $\mathcal{F}$-measurable, it follows that the function $\omega \mapsto \int |H_s(\omega)|d|X|_s(\omega)$ is $\mathcal{F}$-measurable and assertion a) is completely proved.

α') Let $0 < a < \infty$ and denote $\mathcal{M}_a = \mathcal{B}([0,a]) \times \mathcal{F}$ and $r_a = r \cap ([0,a] \times \mathcal{F})$. Then $\mathcal{M}_a$ is σ-ring generated by the ring r_a.

We prove now assertion b) for $H = 1_M$ with $M \in \mathcal{M}_a$. We note that in this case we have $\int |H_s(\omega)| d|X|_s(\omega) < \infty$, a.s. Denote by $\mathcal{M}_1$ the class of sets $M \in \mathcal{M}_a$ for which assertion b) is true and prove that $\mathcal{M}_1$ is a monotone class. Let (M_n) be an increasing sequence from $\mathcal{M}_1$ with limit M. Since $M_n \in \mathcal{M}_a$ for each n, we have also $M \in \mathcal{M}_a$. For each $\omega \in \Omega$, we have $M_n(\omega) \uparrow M(\omega)$ in $\mathcal{D}$. Then

$$m_{X(\omega)}(M_n(\omega)) \to m_{X(\omega)}(M(\omega)).$$

Since for each n, the function

$$\omega \mapsto \int 1_{M_n(\omega)}(s,\omega) dX_s(\omega) = m_{X(\omega)}(M_n(\omega))$$

is $\mathcal{F}$-measurable, it follows that the function

$$\omega \mapsto \int 1_M(s,\omega) dX_s(\omega) = m_{X(\omega)}(M(\omega))$$

is $\mathcal{F}$-measurable. If (M_n) is decreasing, the proof is similar. We deduce that $M \in \mathcal{M}_1$, hence $\mathcal{M}_1$ is a monotone class, containing r_a; consequently, $\mathcal{M}_1 = \mathcal{M}_a$ and assertion b) is proved for $H = 1_M$ with $M \in \mathcal{M}_a$.

β') Let now $H = 1_M$ with $M \in \mathcal{M}$ and $\int |H_s(\omega)| d|X|_s(\omega) < \infty$, a.s., except a negligible set N. For each n denote

$$M_n = M \cap ([0,n] \times \Omega) \in \mathcal{M}_n = \mathcal{B}([0,n]) \times \mathcal{F}.$$

We have $H^n = 1_{M_n} \le H$. For each $\omega \notin N$, the function $H.(\omega)$ is $m_{X(\omega)}$-integrable; by Lebesgue's theorem in $L^1(m_{X(\omega)})$ we have $H^n \to H(\omega)$ in $L^1(m_{X(\omega)})$; therefore $\int H^n_s(\omega) dX_s(\omega) \to \int H_s(\omega) dX_s(\omega)$. Since by step α'), each function $\omega \mapsto \int H^n_s(\omega) dX_s(\omega)$ is $\mathcal{F}$-measurable, we deduce that the function $\omega \mapsto \int H_s(\omega) dX_s(\omega)$ is $\mathcal{F}$-measurable and this proves assertion b) for $H = 1_M$ with $M \in \mathcal{M}$. Assertion b) remains valid for an F-valued, $\mathcal{M}$-step process H with $\int |H_s(\omega)| d|X|_s(\omega) < \infty$, a.s. .

γ') Let now H be an F-valued, measurable process with

$$\int |H_s(\omega)| d|X|_s(\omega) < \infty, a.s.,$$

except a negligible set N and prove assertion b). Let (H^n) be a sequence of F-valued, $\mathcal{M}$-step functions such that $H^n \to H$ and $|H^n| \le |H|$. For $\omega \notin N$ we apply Lebesgue's Theorem in the space $L^1_F(m_{X(\omega)})$ and deduce that

$$\int H^n_s(\omega) dX_s(\omega) \to \int H_s(\omega) dX_s(\omega).$$

Since, by step β'), assertion b) is valid for each H^n, we deduce that assertion b) is valid for H.

δ) Assume now $E(\int |H_s|(\omega)d|X_s|(\omega)) < \infty$ and prove assertion c). It follows first that $\int |H_s(\omega)|d|X|_s(\omega) < \infty$, a.s., hence by assertion b), the function $\omega \mapsto \int H_s(\omega)dX_s(\omega)$ is defined a.s. and is $\mathcal{F}$-measurable. From

$$|\int H_s(\omega)dX_s(\omega)| \leq \int |H_s(\omega)|d|X|_s(\omega)$$

we deduce that $E(|\int H_s(\omega)dX_s(\omega)|) < \infty$ and assertion c) is proved.

ϵ) Assume now $E(|X|_\infty) < \infty$ and H bounded and prove assertion d). Let $M > 0$ be such that $|H_s(\omega)| \leq M$ for every $s \in \mathbb{R}_+$ and $\omega \in \Omega$. We have first, $|X|_\infty < \infty$, a.s. Then

$$\int |H_s(\omega)|d|X|_s(\omega)) \leq M|X|_\infty(\omega) < \infty, \text{ a.s.}$$

By assertion c), the function $\omega \mapsto \int H_s(\omega)dX_s(\omega)$ is integrable. ∎

We can now state a criterion for a measurable process with finite variation to be optional or predictable.

For each real-valued, bounded, measurable process ϕ we denote by $^o\phi$ and $^p\phi$, respectively, the optional projection and the predictable projection of ϕ ([D–M].VI.43):

$$E(\phi_T I_{\{T<\infty\}}|\mathcal{F}_T) = {^o\phi_T} I_{\{T<\infty\}}, \text{ a.s.},$$

for every stopping time T;

$$E(\phi_T I_{\{T<\infty\}}|\mathcal{F}_{T-}) = {^p\phi_T} I_{\{T<\infty\}}, \text{ a.s.},$$

for every predictable stopping time T.

4. Theorem. *Let $X : \mathbb{R}_+ \times \Omega \to E$ be a right continuous, adapted process with integrable variation. Then X is optional (resp. predictable) iff, for every real-valued, bounded, measurable process ϕ we have*

$$E(\int \phi_s dX_s) = E(\int {^o\phi_s} dX_s) \text{ (resp. } E(\int {^p\phi_s} dX_s)).$$

Proof. The integrals are defined, by Theorem 3 d). The theorem is true if X is increasing ([D–M], VI. 57–59).

Assume now that X is E-valued. Since X is measurable, it is separably valued; hence we can assume E is separable. Then X is optional (resp. predictable), iff $\langle X, x^* \rangle$ is optional (resp. predictable), for every $x^* \in E^*$ (Proposition 1.21). Let $x^* \in E^*$ be arbitrary. The real-valued process $\langle X, x^* \rangle$ is cadlag and has integrable variation $|\langle X, x^* \rangle| \leq |X||x^*|$. We can write it as a difference of two increasing, integrable processes and, using the known result mentioned above for increasing processes, we deduce that $\langle X, x^* \rangle$ is optional (resp. predictable) iff, for every real-valued, bounded, measurable process ϕ we have

$$E(\int \phi_s d\langle X_s, x^* \rangle) = E(\int {^o\phi_s} d\langle X_s, x^* \rangle)(\text{resp. } E(\int {^p\phi_s} d\langle X_s, x^* \rangle)).$$

Now, using Theorem 2.26 we have

$$E(\int \phi_s d\langle X_s, x^*\rangle) = E(\langle \int \phi_s dX_s, x^*\rangle) = \langle E(\int \phi_s dX_s), x^*\rangle$$

and similarly

$$E(\int {}^o\phi_s d\langle X_s, x^*\rangle) = \langle E(\int {}^o\phi_s dX_s), x^*\rangle.$$

It follows that

$$E(\int \phi_s d\langle X_s, x^*\rangle) = E(\int {}^o\phi_s d\langle X_s, x^*\rangle),$$

iff

$$\langle E(\int \phi_s dX_s), x^*\rangle = \langle E(\int {}^o\phi_s dX_s), x^*\rangle;$$

Since x^* is arbitrary in E^*, this is equivalent to

$$E(\int \phi_s dX_s) = E(\int {}^o\phi_s dX_s).$$

Similarly,

$$E(\int \phi_s d\langle X_s, x^*\rangle) = E(\int {}^p\phi_s d\langle X_s, x^*\rangle)$$

for every $x^* \in E$ iff

$$E(\int \phi_s dX_s) = E(\int {}^p\phi_s dX_s).$$

∎

B. Optional and predictable measures

Let $m : \mathcal{M} \to E$ be a σ-additive measure. By Theorem 4.20, m has bounded semivariation $\tilde{m}_{\mathbb{R},E}$; therefore the integral $\int \phi dm$ is defined for every real-valued, bounded, measurable process ϕ and the integral belongs to E (Proposition 5.48 a').

5. Definition. *Let $m : \mathcal{M} \to E$ be a σ-additive measure.*
a) *We say m is a stochastic measure if m vanishes on evanescent sets.*
b) *We say the measure m is optional (resp. predictable) if for every set $A \in \mathcal{M}$ we have*

$$m(A) = \int {}^o(1_A) dm$$

(resp.

$$m(A) = \int {}^p(1_A) dm).$$

In this definition one can replace the sets $A \in \mathcal{M}$ by real-valued, bounded, measurable processes ϕ:

6. Proposition. *Let $m : \mathcal{M} \to E$ be a σ-additive stochastic measure. Then m is optional (resp. predictable) iff for every real-valued, bounded, measurable process ϕ we have*

$$\int \phi \, dm = \int {}^o\phi \, dm \ (resp. \int {}^P\phi \, dm).$$

Proof. Assume m is optional. From Definition 5 we deduce that the equality in the statement is valid for every real-valued, $\mathcal{M}$-step process.

Let ϕ be a real-valued, bounded, measurable process and let (ϕ^n) be a sequence of real-valued, $\mathcal{M}$-step processes, such that $\phi^n \to \phi$ uniformly and $|\phi^n| \leq |\phi|$. Let $z \in E^*$. Then $\int \phi^n dm_z \to \int \phi \, dm_z$. At the same time, ${}^o\phi^n \to {}^o\phi$ a.s. and $|{}^o\phi^n| \leq \|\phi\|_{\sup}$. We deduce that $\int {}^o\phi^n dm_z \to \int {}^o\phi \, dm_z$. Since for each n we have

$$\int \phi^n dm_z = \int {}^o\phi^n dm_z,$$

we deduce that

$$\int \phi \, dm_z = \int {}^o\phi \, dm_z,$$

hence $\int \phi \, dm = \int {}^o\phi \, dm$. The converse implication is evident. The proof for the predictable case is the same. ∎

C. The measure μ_X

We shall establish a correspondence $X \mapsto \mu_X$ between right continuous, raw processes X with integrable variation $|X|$ and stochastic measures μ_X with finite variation $|\mu_X|$ and we shall prove that $|\mu_X| = \mu_{|X|}$.

We begin with *real-valued*, increasing processes X, for which the correspondence $X \mapsto \mu_X$ is 1–1 and which will be used in the proof of the general theorem.

7. Theorem. *There is a 1–1 correspondence $X \longleftrightarrow \mu_X$ between the set of right continuous, measurable, increasing, integrable processes X with $X_0 = 0$ and the set of positive, finite, σ-additive stochastic measures μ_X on $\mathcal{M}$, given by the following equality:*

$$\mu_X(M) = E\left(\int 1_M dX_s\right), \text{ for } M \in \mathcal{M}.$$

If ϕ is a real-valued, measurable process we have $\phi \in L^1(\mu_X)$ iff $E(\int |\phi_s| dX_s) < \infty$; in this case $E(\int \phi_s dX_s)$ is defined and we have

$$\int \phi \, d\mu_X = E\left(\int \phi_s dX_s\right).$$

Proof. Let X be a right continuous, increasing, measurable and integrable process. By Theorem 3 c), if ϕ is a real-valued, measurable process with $E(\int |\phi_s| dX_s) < \infty$, then the integral $E(\int \phi_s dX_s)$ is defined.

For $\phi = 1_M$ with $M \in \mathcal{M}$, we have $E(\int |\phi_s| dX_s) \leq E(X_\infty) < \infty$, hence the integral $E(\int 1_M dX_s)$ is defined. Set

$$\mu_X(M) = E(\int 1_M dX_s), \text{ for } M \in \mathcal{M}.$$

Using the monotone convergence theorem, we can prove that the measure $\mu_X : \mathcal{M} \to \mathbb{R}_+$ is σ-additive. It vanishes on evanescent sets: if $A \in \mathcal{F}$ is negligible, then

$$0 \leq \mu_X(\mathbb{R}_+ \times A) \leq E(1_A X_\infty) = 0;$$

therefore μ_X is a stochastic measure.

From the definition of μ_X it follows that if ϕ is a real-valued, $\mathcal{M}$-step process, then

$$\int \phi d\mu_X = E(\int \phi_s dX_s).$$

Let now ϕ be a real-valued, measurable process and let ϕ^+ and ϕ^- be the positive part and the negative part of ϕ respectively.

Let (ϕ^n) be an increasing sequence of positive, $\mathcal{M}$-step functions such that $\phi^n \uparrow \phi^+$. Then $\int \phi^n d\mu_X \to \int \phi^+ d\mu_X \leq +\infty$. On the other hand, for each $\omega \in \Omega$, the sequence $(\phi^n(\omega))$ of positive, $\mathcal{B}(\mathbb{R}_+)$-step functions is increasing and converges to $\phi^+(\omega)$; therefore

$$\int \phi_s^n(\omega) dX_s(\omega) \nearrow \int \phi_s^+(\omega) dX_s(\omega) \leq +\infty.$$

Then

$$E(\int \phi_s^n(\omega) dX_s(\omega)) \to E(\int \phi_s^+(\omega) dX_s(\omega)) \leq +\infty.$$

Since for each step function ϕ^n we have

$$\int \phi^n d\mu_X = E(\int \phi_s^n dX_s),$$

passing to the limit we deduce that

$$\int \phi^+ d\mu_X = E(\int \phi_s^+ dX_s) \leq +\infty.$$

It follows that $\phi^+ \in L^1(\mu_X)$ iff $E(\int \phi_s^+ dX_s) < \infty$. Similarly, $\phi^- \in L^1(\mu_X)$ iff $E(\int \phi_s^- dX_s) < \infty$. We deduce then that $\phi \in L^1(\mu_X)$ iff $E(\int |\phi_s| dX_s) < \infty$ and in this case

$$\int \phi d\mu_X = E(\int \phi_s dX_s).$$

We have $\mu_X = \mu_{X-X_0}$; therefore we can restrict ourselves to processes X with $X_0 = 0$. This proves the first implication of the theorem.

Let now μ be a positive, σ-additive stochastic measure and show that there is a right continuous, integrable, increasing process X with $X_0 = 0$ such that $\mu = \mu_X$. For each $t \in [0, \infty]$ and $A \in \mathcal{F}$ define

$$\mu^t(A) = \mu([0, t] \times A), \text{ if } t < \infty$$

and $\mu^\infty(A) = \mu(\mathbb{R}_+ \times A)$. Then μ^t is a positive, σ-additive measure on $\mathcal{F}$. Since μ is a stochastic measure, if $A \in \mathcal{F}$ and $P(A) = 0$, then $\mu^t(A) = 0$, hence $\mu^t \ll P$. By the Radon–Nikodym Theorem, there is an integrable, positive, $\mathcal{F}$-measurable function Y_t such that

$$\mu^t(A) = \int_A Y_t(\omega) dP(\omega), \text{ for } A \in \mathcal{F}.$$

If $s \leq t \leq \infty$, then $\mu^s \leq \mu^t$, hence $Y_s(\omega) \leq Y_t(\omega)$, a.s. Denote by $N(s,t) \in \mathcal{F}$ the negligible set outside of which we have $Y_s(\omega) \leq Y_t(\omega)$. Let N be the union of the countable family $N(s,t)$ with s, t rational, including $t = \infty$. Then N is negligible and for $\omega \notin N$ we have

$$Y_s(\omega) \leq Y_t(\omega), \text{ for all rational } s \leq t.$$

Set

$$X_\infty(\omega) = Y_\infty(\omega) \text{ if } \omega \notin N,$$
$$X_t(\omega) = \lim_{r \downarrow t} Y_r(\omega), \ r \text{ rational, if } t < \infty \text{ and } \omega \notin N$$

and

$$X_t(\omega) = 0, \text{ if } \omega \in N.$$

Then $X.(\omega)$ is right continuous and increasing for every $\omega \in \Omega$. For each $\omega \in \Omega$, X_t is $\mathcal{F}$-measurable. We shall prove now that $X_t = Y_t$, a.s., for each $t \geq 0$.

In fact, if $t_n \downarrow t$ and $A \in \mathcal{F}$, then $[0, t_n] \times A \downarrow [0, t] \times A$. Since μ is σ-additive we deduce that

$$\mu([0, t_n] \times A) \to \mu([0, t] \times A),$$

therefore,

$$\mu^{t_n}(A) \to \mu^t(A),$$

that is,

$$\int_A Y_{t_n} dP \to \int_A Y_t dP.$$

Then, taking t_n rational and using the Monotone Convergence Theorem we obtain

$$\int_A Y_t dP = \lim_n \int_A Y_{t_n} dP = \int_A X_t dP.$$

Since $A \in \mathcal{F}$ is arbitrary, we deduce that $X_t = Y_t$, a.s.

Since for each $t \geq 0$, X_t is $\mathcal{F}$-measurable and since X is also right continuous, we deduce that X is measurable and, for $A \in \mathcal{F}$, we have

$$\mu([0,t] \times A) = \mu^t(A) = \int_A X_t dP = E(\int 1_{[0,t] \times A} dX_s).$$

This equality remains valid if we replace X with $X - X_0$; therefore, we can take $X_0 = 0$. Since $X_\infty = Y_\infty$ is integrable, it follows that X is an integrable, increasing process. By the first part of the proof we have

$$\mu_X([0,t] \times A) = E(\int 1_{[0,t] \times A} dX_s) = \mu([0,t] \times A).$$

Since both measures μ and μ_X are σ-additive and coincide on the ring generated by the rectangular sets $[0,t] \times A$, we have $\mu = \mu_X$ on $\mathcal{M}$. ∎

We prove now the general theorem about the correspondence $X \mapsto \mu_X$.

8. Theorem. *Let $X : \mathbb{R}_+ \times \Omega \to E$ be a measurable, right continuous process with integrable variation (i.e., $|X|_\infty \in L^1$).*

There is a stochastic measure $\mu_X : \mathcal{M} \to E$ with finite variation $|\mu_X|$ satisfying the following conditions:

a)
$$\mu_X(M) = E(\int 1_M dX_s), \text{ for } M \in \mathcal{M}$$

and

$$|\mu_X|(M) = E(\int 1_M d|X|_s), \text{ for } M \in \mathcal{M},$$

that is,

$$|\mu_X| = \mu_{|X|}.$$

b) *If $E \subset L(F,G)$ and $H : \mathbb{R}_+ \times \Omega \to F$ is a process, then $H \in L^1_F(\mu_X)$ iff H is μ_X-measurable and $E(\int |H_s| d|X|_s) < \infty$.*

In this case $E(\int H_s dX_s)$ is defined and we have

$$\int H d\mu_X = E(\int H_s dX_s)$$

and

$$\int H d|\mu_X| = E(\int H_s d|X_s|).$$

c) *The measure μ_X is optional (resp. predictable) iff X is optional (resp. predictable).*

Proof. Using Theorem 3 d), we deduce that the integral $E(\int 1_M dX_s)$ is defined for every $M \in \mathcal{M}$. For each $M \in \mathcal{M}$ set

$$\mu_X(M) = E(\int 1_M dX_s).$$

Then $\mu_X : \mathcal{M} \to E$ is additive.

Since the variation $|X|$ is increasing, right continuous and integrable, by Theorem 7, the corresponding measure $\mu_{|X|}$ is σ-additive, positive, finite and vanishes on evanescent sets. From the definition of μ_X, for every $M \in \mathcal{M}$ we have

$$|\mu_X(M)| \leq E(\int 1_M d|X|_s) = \mu_{|X|}(M).$$

It follows that μ_X is σ-additive, has finite variation $|\mu_X| \leq \mu_{|X|}$ and vanishes on evanescent sets, hence μ_X is a stochastic measure.

We shall prove now that $|\mu_X| = \mu_{|X|}$.

Since X is measurable, it is separably valued. We can therefore assume that E is separable. Then there is a separable space $Z \subset E^*$, norming for E. Since $\mu_X \ll \mu_{|X|}$, we can apply the Radon–Nikodym Theorem 2.34, with $F = \mathbb{R}$ and deduce that there is a process $H : \mathbb{R}_+ \times \Omega \to Z^*$ such that:

α) For every $z \in Z$, the function $\langle H, z \rangle$ is $\mu_{|X|}$-integrable, $\mathcal{M}$-measurable and

$$\langle \mu_X(M), z \rangle = \int_M \langle H, z \rangle d\mu_{|X|}, \text{ for } M \in \mathcal{M};$$

β) $|H| \leq 1$, $\mu_{|X|}$-a.e;

γ) $|H|$ is $\mu_{|X|}$-integrable, $\mathcal{M}$-measurable and

$$|\mu_X|(M) = \int_M |H| d\mu_{|X|}, \text{ for } M \in \mathcal{M}.$$

The inequality β) follows from $|\mu_X| \leq \mu_{|X|}$. Taking $M = [0,t] \times A$ with $A \in \mathcal{F}$ we get

$$E(1_A \langle X_t, z \rangle) = E(1_A \int_{[0,t]} \langle H_s, z \rangle d|X|_s).$$

Then

$$\langle X_t, z \rangle = \int_{[0,t]} \langle H_s, z \rangle d|X|_s, \text{ a.s.}$$

Since both sides are right continuous, this equality is true outside a P-negligible set $N(z) \subset \Omega$, for all $t \geq 0$.

Let Z_0 be a countable dense subset of Z. The set $N = \bigcup \{N(z) : z \in Z_0\}$ is P-negligible. Then

$$\langle X_t(\omega), z \rangle = \int_{[0,t]} \langle H_s(\omega), z \rangle d|X|_s(\omega), \text{ for } t \geq 0, z \in Z_0 \text{ and } \omega \notin N.$$

The function $X.(\omega) : \mathbb{R}_+ \to E$ is right continuous and has bounded variation $|X(\omega)|. = |X|.(\omega) \leq |X|_\infty(\omega)$ outside a negligible set N_0 which we can choose such that $N_0 \supset N$. Let $\omega \notin N_0$ and let $m_{X(\omega)}$ be the Stieltjes measure associated to $X.(\omega)$. Then $m_{X(\omega)}$ has finite variation $|m_{X(\omega)}| = m_{|X|(\omega)}$ (Theorem 18.19). We can apply Theorem 2.34 again for $m = m_{X(\omega)}$ and

$\mu = m_{|X|(\omega)} = |m_{X(\omega)}|$ and deduce that there is a function $G_\omega : \mathbb{R}_+ \to Z^*$ such that

α') For every $z \in Z$, the function $\langle G_\omega, z \rangle$ is $m_{|X|(\omega)}$-integrable, $\mathcal{B}(\mathbb{R}_+)$-measurable and

$$\langle m_{X(\omega)}(M), z \rangle = \int_M \langle G_\omega, z \rangle dm_{|X|(\omega)}, \text{ for } M \in \mathcal{B}(\mathbb{R}_+);$$

β') $|G_\omega|$ is $\mathcal{B}(\mathbb{R}_+)$-measurable and

$$|G_\omega| = 1, \; m_{|X|(\omega)}\text{-a.e.}$$

Taking $M = [0, t]$ we deduce

$$\langle X_t(\omega), z \rangle = \int_{[0,t]} \langle G_\omega(s), z \rangle d|X|_s(\omega).$$

From the two representations of $\langle X(\omega), z \rangle$ we obtain, for all $t \geq 0$,

$$\int_{[0,t]} \langle H_s(\omega), z \rangle d|X|_s(\omega) = \int_{[0,t]} \langle G_\omega(s), z \rangle d|X|_s(\omega);$$

therefore there is a $\mu_{|X|(\omega)}$-negligible set $N'(\omega, z) \subset \mathbb{R}_+$ such that

$$\langle H_s(\omega), z \rangle = \langle G_\omega(s), z \rangle, \text{ for } s \notin N'(\omega, z).$$

The set $N'(\omega) = \bigcup \{N'(\omega, z) : z \in Z_0\}$ is $\mu_{|X|(\omega)}$-negligible and for $s \notin N'(\omega)$ we have

$$H_s(\omega) = G_\omega(s).$$

Let now $A = \{(t, \omega) : |H_t(\omega)| < 1\}$. Then $A \in \mathcal{M}$ and for each $\omega \in \Omega$ we have $A(\omega) = \{t : |H_t(\omega)| < 1\} \in \mathcal{B}(\mathbb{R}_+)$. Since for $\omega \notin N_0$ we have $|H_s(\omega)| = |G_\omega(s)| = 1$, $\mu_{|X|(\omega)}$-a.e., we deduce that $A(\omega)$ is $\mu_{|X|(\omega)}$-negligible; therefore

$$\mu_{|X|}(A) = E(\int 1_{A(\omega)}(s) d|X|_s(\omega)) = 0.$$

It follows that $|H_t(\omega)| = 1$ for $\omega \notin N_0$, hence, by β), we have $|\mu_X| = \mu_{|X|}$ and this proves completely assertion a).

To prove assertion b), let $H : \mathbb{R}_+ \times \Omega \to F$ be a μ_X-measurable process with $E(\int |H_s| d|X_s|) < \infty$ and prove that $H \in L^1_F(\mu_X)$. Modifying H on a μ_X-negligible set, we can assume that H is measurable. Let $N \subset \Omega$ be a negligible set such that for $\omega \notin N$ we have $\int |H_s| d|X_s| < \infty$. This means that for $\omega \notin N$, the section $H_s(\omega)$ is $m_{X(\omega)}$-integrable, where $m_{X(\omega)}$ is the measure associated to $X.(\omega)$ by Theorem 18.19. By Theorem 3 c), the function $\omega \mapsto \int H_s(\omega) dX_s(\omega)$, defined outside N, is integrable.

Let (H^n) be a sequence of $\mathcal{M}$-step processes such that $H^n \to H$ and $|H^n| \leq |H|$. Let $\omega \notin N$. Then $H^n(\omega) \to H(\omega)$ and $|H^n(\omega)| \leq |H(\omega)|$. Since $H(\omega)$ is $m_{X(\omega)}$-integrable, by Lebesgue's Theorem we deduce that

$H^n(\omega) \to H(\omega)$ in $L^1_F(m_{X(\omega)})$ and $|H^n(\omega)| \to |H(\omega)|$ in $L^1(m_{|X|(\omega)})$. It follows that

$$\int |H^n_s(\omega) - H^m_s(\omega)|d|X|_s(\omega) \to 0, \text{ as } n, m \to \infty$$

and

$$\int H^n_s(\omega)dX_s(\omega) \to \int H_s(\omega)dX_s(\omega), \text{ as } n \to \infty.$$

For each n we have

$$|\int H^n_s(\omega)dX_s(\omega)| \le \int |H^n_s(\omega)|d|X|_s(\omega) \le \int |H_s(\omega)|d|X|_s(\omega).$$

Since $E(\int |H_s(\omega)|d|X|_s(\omega)) < \infty$, by hypothesis, we can apply Lebesgue's Theorem and deduce that

$$\int H^n_s(\cdot)dX_s(\cdot) \to \int H_s(\cdot)dX_s(\cdot), \text{ in } L^1_G(P),$$

therefore

$$E(\int H^n_s dX_s) \to E(\int H_s dX_s), \text{ in } G.$$

On the other hand, from

$$\int |H^n_s(\omega) - H^m_s(\omega)|d|X|_s(\omega) \to 0, \text{ a.s., as } n, m \to \infty$$

and

$$\int |H^n_s(\omega) - H^m_s(\omega)|d|X|_s(\omega) \le 2\int |H_s(\omega)|d|X|_s(\omega) < \infty, \text{a.s.},$$

we deduce, by Lebesgue's Theorem, that

$$E(\int |H^n_s - H^m_s|d|X|_s) \to 0, \text{ as } n, m \to \infty.$$

For each n and m we have, by the definition of $\mu_{|X|}$,

$$\int |H^n - H^m|d\mu_{|X|} = E(\int |H^n_s - H^m_s|d|X|_s),$$

hence (H^n) is a Cauchy sequence in $L^1_F(\mu_{|X|})$. Since $H^n \to H$ pointwise, it follows that $H \in L^1_F(\mu_X)$ and $H^n \to H$ in $L^1_F(\mu_X)$; consequently

$$\int H^n d\mu_X \to \int H d\mu_X.$$

Since for each n we have

$$\int H^n d\mu_X = E(\int H^n_s dX_s)$$

and since, by the above,
$$E(\int H_s^n dX_s) \to E(\int H_s dX_s),$$
passing to limits in the preceding equality we get
$$\int H d\mu_X = E(\int H_s dX_s).$$
This proves the first implication and the first equality of assertion b).

Conversely, assume $H \in L_F^1(\mu_X)$. Then H is equal μ_X-a.e. to an $\mathcal{M}$-measurable function. Without loss of generality, we can assume that H itself is $\mathcal{M}$-measurable. We have $|H| \in L_F^1(\mu_{|X|})$. Let (ϕ^n) be an increasing sequence of positive, $\mathcal{M}$-step processes with $\phi^n \to |H|$ and $\phi^n \leq |H|$. Then $\phi^n \to |H|$ in $L_F^1(\mu_{|X|})$, hence
$$\int \phi^n d\mu_{|X|} \to \int |H| d\mu_{|X|}.$$
For each $\omega \in \Omega$, the sections $\phi^n(\omega)$ and $|H(\omega)|$ are Borel functions on $\mathbb{R}_+$ and $\phi^n(\omega) \uparrow |H(\omega)|$; therefore
$$\int \phi_s^n(\omega) d|X|_s(\omega) \nearrow \int |H_s(\omega)| d|X|_s(\omega) \leq +\infty,$$
hence
$$E(\int \phi_s^n d|X|_s) \to E(\int |H_s| d|X|_s) \leq +\infty.$$
Since for each n we have
$$\int \phi^n d\mu_{|X|} = E(\int \phi_s^n d|X|_s),$$
passing to limits we get
$$\int |H| d\mu_{|X|} = E(\int |H_s| d|X|_s),$$
hence $E(\int |H_s| d|X|_s) < \infty$ and the second implication in assertion b) is proved.

If we apply the first equality to $|X|$ we deduce the second equality in assertion b).

Assertion c) follows from Theorems 4 and 6: for any real-valued, bounded, measurable process ϕ we have
$$\int \phi d\mu_X = E(\int \phi_s dX_s) \text{ and } \int {}^o\phi d\mu_X = E(\int {}^o\phi_s dX_s).$$
We have
$$\int \phi d\mu_X = \int {}^o\phi d\mu_X$$

iff X is adapted and

$$E(\int \phi_s dX_s) = E(\int {}^o\phi_s dX_s),$$

that is, μ_X is optional iff X is optional. The proof for the predictable case is similar. ∎

Remarks. a) The measure μ_X is called the measure associated to the process X with integrable variation.
b) Taking $F = \mathbb{R}$ and considering the embedding $E \subset L(\mathbb{R}, E)$, from assertion b) of Theorem 8 we deduce that a real-valued process ϕ belongs to $L^1_{\mathbb{R}}(\mu_X)$ iff ϕ is μ_X-measurable and $E(\int |\phi_s| d|X|_s) < \infty$. In this case we have

$$\int \phi d\mu_X = E(\int \phi_s dX_s).$$

9. Corollary. *Let $A, B : \mathbb{R}_+ \times \Omega \to E$ be two adapted, right continuous processes with integrable variation.*
a) *If for every stopping time T we have*

$$E(A_\infty - A_{T-}|\mathcal{F}_T) = E(B_\infty - B_{T-}|\mathcal{F}_T), \quad a.s.$$

(and $A_{0-} = B_{0-} = 0$), then A and B are indistinguishable.
b) *If A and B are predictable, if $A_0 = B_0$ and*

$$E(A_\infty - A_t|\mathcal{F}_t) = E(B_\infty - B_t|\mathcal{F}_t), \quad a.s. \text{ for every } t \geq 0,$$

then A and B are indistinguishable.

Proof. The process $X = A - B$ is measurable, adapted, right continuous and has integrable variation $|X| \leq |A| + |B|$. Hypothesis a) can be written

$$E(X_\infty - X_{T-}|\mathcal{F}_T) = 0, \quad a.s.$$

for each stopping time T. Let μ_X be the measure associated to X by Theorem 8. Since A and B are optional, the measure μ_X is optional and for every stopping time T we have

$$\mu_X([T, \infty)) = E(\int 1_{[T,\infty)} dX) = E(X_\infty - X_{T-}) = 0.$$

Since the stochastic intervals $[T, \infty)$ generate the σ-algebra $\mathcal{O}$ of optional sets, it follows that $\mu_X(M) = 0$ for every $M \in \mathcal{O}$; therefore

$$\int H d\mu_X = 0$$

for every real-valued, bounded, optional process H.

If $M \in \mathcal{M}$ and if we take $H = {}^o(1_M)$, then

$$\mu_X(M) = \int {}^o(1_M)d\mu_X = 0,$$

since μ_X is optional.

It follows that $\mu_X = 0$ on $\mathcal{M}$. Taking $M = [0,t] \times C$ with $C \in \mathcal{F}$, we deduce that

$$E(1_C X_t) = E(\int 1_{[0,t] \times C} dX) = \mu_X([0,t] \times C) = 0.$$

Since $C \in \mathcal{F}$ is arbitrary, for each $t \geq 0$ we deduce that $X_t = 0$, a.s. Since X is right continuous, there is a negligible set $N \subset \Omega$ such that for $\omega \notin N$, we have $X_t(\omega) = 0$ for all $t \geq 0$; therefore the set $\{X \neq 0\}$ is evanescent; consequently A and B are indistinguishable.

The proof of assertion b) is similar. ∎

Corollary 9 can be stated without conditional expectations.

10. Corollary. *Let $A, B : \mathbb{R}_+ \times \Omega \to E$ be two adapted, right continuous processes with integrable variation.*
a′) *If for every stopping time T we have*

$$E(A_\infty - A_{T-}) = E(B_\infty - B_{T-}),$$

then A and B are indistinguishable.
b′) *If A and B are predictable and if for every stopping time T we have*

$$E(A_\infty - A_T) = E(B_\infty - B_T),$$

then A and B are indistinguishable.

Proof. Evidently, assertion a) of Corollary 9 implies a′). Conversely, assume a′) and let T be a stopping time. For every set $A \in \mathcal{F}_T$, apply a′) to the stopping time T_A to obtain a).

Assume now that A, B are predictable and satisfy b) and prove b′). Denote $X = A - B$ and let T be a simple stopping time, $T = \sum_{1 \leq i \leq n} 1_{A_i} t_i$ with $0 \leq t_1 \leq \cdots \leq t_n$ and $A_i \in \mathcal{F}_{t_i}$ disjoint, with union Ω. Then

$$E(X_\infty - X_T) = E(X_\infty - \sum_i 1_{A_i} X_{t_i})$$
$$= \sum_i E(1_{A_i}(X_\infty - X_{t_i})) = \sum_i E(1_{A_i} E(X_\infty - X_{t_i} | \mathcal{F}_{t_i})) = 0.$$

If T is an arbitrary stopping time, we take a decreasing sequence (T_n) of simple stopping times with $T_n \downarrow T$. Then, by right continuity of X we have $X_\infty - X_{T_n} \to X_\infty - X_T$ pointwise and $|X_\infty - X_{T_n}| \leq 2|X_\infty| \leq 2|X|_\infty \in L^1$.

By Lebesgue's Theorem we have $X_\infty - X_{T_n} \to X_\infty - X_T$ in L^1, hence $E(X_\infty - X_{T_n}) \to E(X_\infty - X_T)$; since $E(X_\infty - X_{T_n}) = 0$ for each n, it follows that $E(X_\infty - X_T) = 0$, which is assertion b').

Conversely, if we apply b') to the stopping time t_A with $A \in \mathcal{F}_t$, we obtain b). ∎

11. Corollary. *If $A_t = E(A_\infty | \mathcal{F}_t)$ is an E-valued, uniformly integrable, cadlag, predictable martingale with $A_0 = 0$ and with integrable variation, then $A = 0$ except on an evanescent set.*

Proof. We take $B_t = 0$ for all $t \geq 0$ and we have $A_0 = 0 = B_0$ and

$$E(A_\infty - A_t | \mathcal{F}_t) = 0 = E(B_\infty - B_t | \mathcal{F}_t), \text{ for } t \geq 0.$$

By Corollary 9, A and B are indistinguishable. ∎

Remark. Corollaries 9, 10, and 11 will be extended for processes with integrable semivariation (Corollaries 21.9, 21.10, and 21.11).

In the following theorem we start with a stochastic measure m with finite variation and associate to it a process X with integrable variation such that $|m| = \mu_{|X|}$.

12. Theorem. *Let $m : \mathcal{M} \to E \subset L(F, G)$ be a stochastic measure with finite variation $|m|$ and let $Z \subset G^*$ be a norming space for G. Assume F and Z are separable.*

Then, there is a right continuous process $X : \mathbb{R}_+ \times \Omega \to L(F, Z^*)$ with the following properties:
(i) *For every $x \in F$ and $z \in Z$, the real-valued process $\langle Xx, z \rangle$ is measurable and has integrable variation;*
(ii) *The process X has measurable, integrable variation $|X|$ (i.e., $|X|_\infty \in L^1$);*
a) *If $\phi : \mathbb{R}_+ \times \Omega \to \mathbb{R}$ is a measurable process and $\phi \in L^1_\mathbb{R}(m)$, then the integrals $E(\int \phi_s d\langle X_s x, z\rangle)$ and $E(\int |\phi_s| d|X|_s)$ are defined for every $x \in F$ and $z \in Z$ and we have*

$$\langle (\int \phi \, dm)x, z \rangle = E(\int \phi_s d\langle X_s x, z \rangle)$$

and

$$\int |\phi| d|m| = E(\int |\phi_s| d|X|_s),$$

that is,

$$|m| = \mu_{|X|}.$$

a') *Let $x \in X$. If Xx is measurable, then*

$$(\int \phi \, dm)x = E(\int \phi_s d(X_s x)), \text{ for } \phi \in L^1_\mathbb{R}(m).$$

a'') *If X is measurable, then*

$$\int \phi dm = E(\int \phi_s dX_s), \text{ for } \phi \in L^1_{\mathbb{R}}(m)$$

and

$$\int H dm = E(\int H_s dX_s), \text{ for } H \in L^1_F(m);$$

b) *m is optional (resp. predictable) iff $\langle Xx, z \rangle$ is optional (resp. predictable) for every $x \in F$ and $z \in Z$; If G is separable, then m is optional (resp. predictable) iff Xx is optional (resp predictable) for every $x \in F$.*

c) *We can choose X with values in $L(F, G)$ in each of the following two cases:*

c_1) *G is the dual of a separable Banach space D and we choose $Z = D$, hence $G = Z^*$.*

c_2) *G has the Radon–Nikodym Property (RNP). In this case, X can be chosen such that Xx is measurable for every $x \in X$ and*

$$(\int \phi dm)x = E(\int \phi_s d(X_s x)), \text{ for } \phi \in L^1_{\mathbb{R}}(m).$$

d) *If E has the RNP, then X can be chosen measurable, with values in E and we have*

$$\int \phi dm = E(\int \phi_s dX_s), \text{ for } \phi \in L^1_{\mathbb{R}}(m)$$

and

$$\int H dm = E(\int H_s dX_s), \text{ for } H \in L^1_F(m).$$

Proof. Let V be the integrable, increasing, right continuous process corresponding to the positive measure $|m|$ by Theorem 7:

$$|m|(M) = E(\int 1_M dV_s), \text{ for } M \in \mathcal{M}.$$

For every $t \geq 0$ set $m^t(A) = m([0, t] \times A)$, for $A \in \mathcal{F}$. Then $m^t : \mathcal{F} \to E \subset L(F, G)$ is σ-additive, with finite variation $|m^t|$, absolutely continuous with respect to P and satisfies $|m^t| \leq |m|^t$, where

$$|m|^t(A) = |m|([0, t] \times A), \text{ for } A \in \mathcal{F}.$$

By the Radon–Nikodym Theorem 2.34, there is a function $X^0_t : \Omega \to L(F, Z^*)$ satisfying the following conditions:

α) For every $x \in X$ and $z \in Z$, the function $\langle X^0_t x, z \rangle$ is $\mathcal{F}$-measurable, integrable and

$$\langle m^t(A)x, z \rangle = E(1_A \langle X^0_t x, z \rangle), \text{ for } A \in \mathcal{F};$$

β) The function $|X^0_t|$ is $\mathcal{F}$-measurable, integrable and

$$|m^t|(A) = E(1_A |X^0_t|), \text{ for } A \in \mathcal{F}.$$

From the inequality $|m^t| \leq |m|^t$ and from β) we deduce that

$$E(1_A|X_t^0|) \leq E(1_A V_t), \text{ for } A \in \mathcal{F},$$

therefore

$$|X_t^0| \leq V_t, \text{ a.s.}$$

The exceptional negligible set belongs to $\mathcal{F}$.

Let $s < t$ and consider the measure $m^t - m^s : \mathcal{F} \to L(F, G)$ with finite variation $|m^t - m^s| \leq |m^t| + |m^s| \ll P$. We apply the same Radon–Nikodym Theorem 2.34 to obtain a function $X_{st}^0 : \Omega \to L(F, Z^*)$ satisfying the following conditions:

α') For every $x \in F$ and $z \in Z$, the function $\langle X_{st}^0 x, z \rangle$ is $\mathcal{F}$-measurable, integrable and

$$\langle (m^t(A) - m^s(A))x, z \rangle = E(1_A \langle X_{st}^0 x, z \rangle), \text{ for } A \in \mathcal{F}.$$

β') The function $|X_{st}^0|$ is $\mathcal{F}$-measurable, integrable and

$$|m^t - m^s|(A) = E(1_A|X_{st}^0|), \text{ for } A \in \mathcal{F}.$$

From α) and α') we deduce that, for $x \in F$ and $z \in Z$, we have

$$E(1_A \langle (X_t^0 - X_s^0)x, z \rangle) = E(1_A \langle X_{st}^0 x, z \rangle), \text{ for } A \in \mathcal{F},$$

therefore

$$\langle (X_t^0 - X_s^0)x, z \rangle = \langle X_{st}^0 x, z \rangle, \text{ a.s.},$$

the negligible set depends on x and z and belongs to $\mathcal{F}$. Taking the supremum for z in a countable set dense in Z_1 and for x in a countable set dense in F_1 we obtain

$$|X_t^0 - X_s^0| = |X_{st}^0|, \text{ a.s.},$$

therefore, from β'),

$$|m^t - m^s|(A) = E(1_A|X_t^0 - X_s^0|), \text{ for } A \in \mathcal{F}.$$

But

$$|m^t(A) - m^s(A)| = |m((s,t] \times A)| \leq |m|((s,t] \times A), \text{ for } A \in \mathcal{F},$$

therefore

$$|m^t - m^s|(A) \leq |m|((s,t] \times A) = (|m|^t - |m|^s)(A), \text{ for } A \in \mathcal{F},$$

hence

$$E(1_A|X_t^0 - X_s^0|) \leq E(1_A(V_t - V_s)), \text{ for } A \in \mathcal{F}.$$

It follows that

$$|X_t^0 - X_s^0| \leq V_t - V_s, \text{ a.s.};$$

the negligible set belongs to $\mathcal{F}$.

We can modify each X_r^0 with r rational on a certain negligible set, in order to have
$$|X_r^0 - X_{r'}^0| \leq V_r - V_{r'}, \text{ everywhere on } \Omega,$$
for r and r' rational.

Since V is cadlag, the limit
$$X_t(\omega) = \lim_{r \downarrow t} X_r^0(\omega), \ r \text{ rational},$$
exists for every $\omega \in \Omega$ and X is a right continuous process satisfying
$$|X_t - X_s| \leq V_t - V_s, \text{ everywhere}.$$

Since V is increasing, from this inequality we deduce that X has finite variation $|X| \leq V$. From the inequality
$$|X_r^0 - X_t^0| \leq V_r - V_t, \text{ a.s., for } t < r,$$
letting $r \downarrow t$ with r rational, we deduce that
$$X_t = X_t^0, \text{ a.s.}$$
outside a negligible set from $\mathcal{F}$. Since the function $\langle X_t^0 x, z \rangle$ is $\mathcal{F}$-measurable, by α) we deduce that the function $\langle X_t x, z \rangle$ is $\mathcal{F}$-measurable. Since X is right continuous, $\langle X_t x, z \rangle$ is also right continuous.

From the inequality
$$|\langle (X_t - X_s)x, z \rangle| \leq |X_t - X_s| \, |x| \, |z| \leq (V_t - V_s)|x| \, |z|$$
we deduce that $\langle X x, z \rangle$ has bounded variation $|\langle X x, z \rangle| \leq V|x||z|$, hence $\langle X x, z \rangle$ has integrable variation and this proves assertion (i). Since X is right continuous, $|X|$ is also right continuous (Theorem 18.11). Then we can compute the variation $|X|_t$ by using partitions of $[0,t]$ consisting of rational numbers. We deduce that $|X|_t$ is $\mathcal{F}$-measurable for every $t \geq 0$, hence $|X|$ is measurable; and from the inequality $|X| \leq V$ proved above it follows that X has integrable variation $|X|$ and this proves assertion (ii).

Let $x \in X$ and $z \in Z$. From α) and the fact that $X_t = X_t^0$, a.s., we deduce that for $M = [0,t] \times A$ with $A \in \mathcal{F}$ we have
α'') $\qquad \langle m(M)x, z \rangle = E(\int 1_M d\langle X_s x, z \rangle)$
$\qquad\qquad\qquad \leq E(\int 1_M d|X|_s |x||z|),$
hence
$$|m(M)| \leq E\left(\int 1_M d|X|_s\right);$$
therefore
$$|m|(M) \leq E\left(\int 1_M d|X|_s\right).$$

On the other hand, from $|X|_t \leq V_t$ we deduce

$$E(\int 1_M d|X|_s) \leq E(\int 1_M dV_s) = |m|(M),$$

hence

β'') $\quad\quad\quad\quad\quad\quad |m|(M) = E(\int 1_M d|X|_s).$

The inequalities α'') and the equality β'') remain valid for M in the ring $\mathcal{R}$.

Using Theorem 7 we have

$$\mu_{|X|}(M) = E(\int 1_M d|X|_s), \text{ for } M \in \mathcal{M}.$$

From β'') it follows that

$$|m|(M) = \mu_{|X|}(M), \text{ for } M = [0,t] \times A \text{ with } A \in \mathcal{F}.$$

Since both measures are σ-additive on $\mathcal{M}$, it follows that $|m| = \mu_{|X|}$ on $\mathcal{M}$. Then for any process $\phi \in L^1_{\mathbb{R}}(m)$ we have

$$\int |\phi| d|m| = \int |\phi| d\mu_{|X|} = E(\int |\phi_s| d|X|_s).$$

This is the second equality in assertion a).

Since $\langle Xx, z \rangle$ is right continuous and has integrable variation $|\langle Xx, z\rangle|$, we can apply Theorem 8 and deduce the existence of a σ-additive measure $\mu_{\langle Xx,z\rangle}$ with bounded variation

$$|\mu_{\langle Xx,z\rangle}| = \mu_{|\langle Xx,z\rangle|}$$

such that

$$\mu_{\langle Xx,z\rangle}(M) = E(\int 1_M d\langle Xx,z\rangle), \text{ for } M \in \mathcal{M}.$$

From α'') we deduce that

$$\mu_{\langle Xx,z\rangle}(M) = \langle m(M)x, z\rangle, \text{ for } M = [0,t] \times A \text{ with } A \in \mathcal{F}.$$

Since both measures are σ-additive on $\mathcal{M}$, it follows that $\mu_{\langle Xx,z\rangle} = \langle m(\cdot)x, z\rangle$ on $\mathcal{M}$. If $\phi \in L^1_{\mathbb{R}}(m)$, then $\phi \in L^1_{\mathbb{R}}(\langle m(\cdot)x, z\rangle) = L^1_{\mathbb{R}}(\mu_{\langle Xx,z\rangle})$ and we have

$$\langle (\int \phi dm)x, z\rangle = \int \phi d\langle m(\cdot)x, z\rangle = \int \phi d\mu_{\langle Xx,z\rangle}$$
$$= E(\int \phi_s d\langle X_s x, z\rangle).$$

The first equality follows from Theorem 2.26. This proves assertion a).

To prove assertion a') let $x \in F$ and assume Xx is measurable. Since it is right continous, we can compute its variation using only rational points and deduce that its variation $|Xx|$ is measurable. From $|X_t x| \leq |X_t||x| \leq |X|_t|x|$ we deduce that $|Xx|_t \leq |X|_t|x|$, hence Xx has integrable variation. Then, for every $z \in Z$ and $\phi \in L^1_{\mathbb{R}}(m)$ we have, by a) and Theorem 2.26,

$$\langle (\int \phi dm)x, z \rangle = E(\int \phi_s d\langle X_s x, z \rangle)$$
$$= E(\langle \int \phi dX_s x, z \rangle) = \langle E(\int \phi dX_s x), z \rangle$$

and assertion a') follows.

Assertion a'') is proved similarly, using Theorem 2.26 and assertion a').

To prove assertion b), let ϕ be a real-valued, bounded, measurable process and $^o\phi$ its optional projection.

For every $x \in X$ and $z \in Z$, we have, by assertion a) and using Theorem 2.26,

$$\int \phi d\langle m(\cdot)x, z \rangle = \langle (\int \phi dm)x, z \rangle\rangle = E(\int \phi d\langle Xx, z\rangle)$$

and similarly

$$\int {}^o\phi d\langle m(\cdot)x, z \rangle = E(\int {}^o\phi d\langle Xx, z\rangle).$$

It follows that $\langle Xx, z \rangle$ is optional iff the measure $\langle m(\cdot)x, z \rangle$ is optional, i.e., iff

$$\int \phi d\langle m(\cdot)x, z\rangle = \int {}^o\phi d\langle m(\cdot)x, z\rangle,$$

that is, using again Theorem 2.26, iff

$$\langle (\int \phi dm)x, z\rangle = \langle (\int {}^o\phi dm)x, z\rangle,$$

which is equivalent to

$$\int \phi dm = \int {}^o\phi dm$$

and this is equivalent to m being optional.

If G is separable then $\langle Xx, z\rangle$ is optional for every $z \in Z$ iff Xx is optional. The proof for the predictable m is similar.

To prove c), assume first that G is the dual of a separable Banach space D and choose $Z = D$. Then $Z^* = D^* = G$, hence X takes on values in $L(F, G)$.

Assume now that G has the RNP. Let $x \in F$ and $t \geq 0$. The measure $\mu^t : \mathcal{F} \to G$ defined by $\mu^t(A) = m^t(A)x$, for $A \in \mathcal{F}$, is σ-additive and has finite variation $|\mu^t| \leq |m|^t|x|$; therefore $\mu^t \ll P$. Since G has the RNP, there is an $\mathcal{F}$-measurable, Bochner-integrable function $Y_t \in L^1_F$, depending on x such that $\mu^t(A) = E(1_A Y_t)$, for $A \in \mathcal{F}$. Since Y_t is measurable, its range is in a separable subspace $G_0 \subset G$ and we can choose a separable subspace $Z_0 \subset Z$, norming for G_0. For $z \in Z_0$ and for $A \in \mathcal{F}$ we have

$$E(1_A \langle Y_t, z\rangle) = \langle \mu^t(A), z\rangle = \langle m^t(A)x, z\rangle$$
$$= E(1_A \langle X_t x, z\rangle),$$

therefore $\langle X_t x, z \rangle = \langle Y_t, z \rangle$, a.s.

Taking z in a countable set dense in Z_0 we obtain $X_t x = Y_t \in G$, a.s. Moreover, since F is separable and X is right continuous, we can modify X on an evanescent set such that $X_t x \in G$ for every $x \in F$ and $t \geq 0$; hence X_t has values in $L(F, G)$.

Since $X_t x = Y_t$, a.s. and Y_t is $\mathcal{F}$-measurable, it follows that $X_t x$ is $\mathcal{F}$-measurable for each $t \geq 0$. Then, since Xx is right continuous, it is $\mathcal{M}$-measurable and we can apply assertion a') and c$_2$) follows.

Assume now that E has the RNP. If we consider the embedding $E = L(\mathbb{R}, E)$, by c$_2$) there is a right continuous process Y with values in $L(\mathbb{R}, E) = E$ which is measurable, has finite variation $(|Y|_t)_{t \geq 0}$ and satisfies the equalities in assertion a'').

Since Z is norming for G, we can embed isometrically G into Z^*. We can therefore consider Y with values in $E \subset L(F, G) \subset L(F, Z^*)$. Both processes X and Y are right continuous, with finite variation and for $x \in F$, $z \in Z$ and $\phi \in L^1_{\mathbb{R}}(m)$ satisfy

$$\langle (\int \phi dm)x, z \rangle = \langle E(\int \phi_s dY_s)x, z \rangle$$
$$= E(\int \phi_s d\langle Y_s x, z \rangle)$$

and

$$\langle (\int \phi dm)x, z \rangle = E(\int \phi_s d\langle X_s x, z \rangle).$$

Taking $\phi = 1_{[0,t] \times A}$ with $A \in \mathcal{F}$ we obtain

$$E(1_A \langle X_t x, z \rangle) = E(1_A \langle Y_t x, z \rangle), \text{ for } A \in \mathcal{F},$$

hence

$$\langle X_t x, z \rangle = \langle Y_t x, z \rangle, \text{ a.s.}$$

Since both X and Y are right continuous and since F and Z are separable, we deduce that X and Y are undistinguishable. If we modify X on an evanescent set, we obtain $X = Y$; hence X is measurable, with values in E and we can apply a'') to deduce d). ∎

D. Summability of processes with integrable variation

The following theorem states the summability of processes X with integrable variation and the relationship between the measure μ_X associated to X by Theorem 8 and the measure I_X.

13. Theorem. *Let $X : \mathbb{R}_+ \times \Omega \to E$ be a cadlag, adapted process with integrable variation $|X|$. Then*
a) *The measure $I_X : \mathcal{R} \to L^1_E$ can be extended to a σ-additive measure $I_X : \mathcal{P} \to L^1_E$ with bounded variations $|I_X|$.*

b) X is 1-summbale relative to any embedding $E \subset L(F,G)$.
c) For any set $M \in \mathcal{P}$ we have

$$(I_X(M))(\omega) = \int 1_M(s,\omega)dX_s(\omega), \text{ a.s.}$$

and

$$(|I_X|(M))(\omega) = \int 1_M(s,\omega)d|X|_s(\omega), \text{ a.s.}$$

d) For any set $M \in \mathcal{P}$ we have

$$\mu_X(M) = E(I_X(M))$$

and

$$|I_X|(M) = |\mu_X|(M) = \mu_{|X|}(M) = E(I_{|X|}(M)).$$

Proof. From the definition of I_X we deduce that for every $M \in \mathcal{R}$ we have

$$I_X(M)(\omega) = \int 1_M(s,\omega)dX_s(\omega), \text{ a.s.}$$

and

$$I_{|X|}(M)(\omega) = \int 1_M(s,\omega)d|X|_s(\omega), \text{ a.s.};$$

hence

$$\mu_X(M) = E(I_X(M))$$

and

$$\mu_{|X|}(M) = E(I_{|X|}(M)).$$

Then, for every $M \in \mathcal{R}$ we have

$$\|I_X(M)\|_{L_E^1} = E(|\int 1_M(s,\omega)dX_s(\omega)|)$$
$$\leq E(\int 1_M(s,\omega)d|X|_s(\omega)) = \mu_{|X|}(M).$$

It follows that I_X has bounded variation $|I_X|$ on $\mathcal{R}$ satisfying $|I_X| \leq \mu_{|X|} = |\mu_X|$.

To prove the converse inequality, for every $M \in \mathcal{R}$ we have

$$|\mu_X(M)| = |E(I_X(M))| \leq E(|I_X(M)|)$$
$$= \|I_X(M)\|_{L_E^1} \leq |I_X|(M);$$

therefore $|\mu_X| \leq |I_X|$ on $\mathcal{R}$. It follows that

$$|\mu_X| = |I_X|, \text{ on } \mathcal{R}.$$

Since $|\mu_X|$ is σ-additive on $\mathcal{M}$, it follows that $|I_X|$ is σ-additive and bounded on $\mathcal{R}$. By Theorem 7.4, I_X and $|I_X|$ can be extended to σ-additive

§19. PROCESSES WITH FINITE VARIATION   235

measures on $\mathcal{P}$, still denoted by I_X and $|I_X|$ respectively and $|I_X|$ is still the variation of I_X on $\mathcal{P}$; thus assertion a) is proved. Assertion b) follows from the inequality $(\tilde{I}_X)_{F,G} \leq |I_X|$. By a monotone class argument, the equalities of assertion c), stated above for $M \in \mathcal{R}$, remain valid for $M \in \mathcal{P}$ and assertion c) is proved. Moreover since both $|I_X|$ and $|\mu_X|$ are σ-additive on $\mathcal{P}$ and are equal on $\mathcal{R}$, it follows that $|I_X| = |\mu_X|$ on $\mathcal{P}$. Taking expectations in the equalities of assertion c) for each $M \in \mathcal{P}$ we get the equalities of assertion d):

$$E(I_X(M)) = E(\int 1_M(s,\omega) dX_s(\omega)) = \mu_X(M)$$

and

$$E(|I_X|(M)) = E(\int 1_M(s,\omega) d|X|_s(\omega)) = \mu_{|X|}(M)$$
$$= |\mu_X|(M) = |I_X|(M).$$

∎

Remark. Theorem 13 will be extended in Theorem 16 for 1_M replaced by processes H.

14. Theorem. *Let $X : \mathbb{R}_+ \times \Omega \to E$ be a cadlag, adapted process with integrable variation, that is, $|X|_\infty \in L^1$. Then:*
a) *The set of measures $(I_X)_{F,L_{G^*}^\infty}$ is uniformly σ-additive;*
b) *We have*

$$L_F^1(\mathcal{P}, \mu_X) = L_F^1(\mathcal{P}, I_X) \subset L_{F,L_G^1}^1(X) = \mathcal{F}_{F,L_G^1}(X)$$

and the $\mathcal{R}$-step processes are dense in $L_{F,L_G^1}^1(X)$.

Proof. Since $|I_X|$ is σ-additive, if $A_n \downarrow \phi$ in $\mathcal{P}$ we deduce that $(\tilde{I}_X)_{F,L_G^1}(A_n) \to 0$; therefore $(I_X)_{F,L_{G^*}^\infty}$ is uniformly σ-additive (Theorem 4.23). The first equality in assertion b) follows from the equality $|\mu_X| = |I_X|$. The inclusion follows from $(\tilde{I}_X)_{F,L_G^1} \leq |I_X|$.

Since $(I_X)_{F,L_{G^*}^\infty}$ is uniformly σ-additive, by Proposition 5.44, the set $\mathcal{S}_F(\mathcal{R})$ of F-valued, $\mathcal{R}$-step processes is dense in the space $\mathcal{F}_{F,G}(\mathcal{B}, X)$. On the other hand each $\mathcal{R}$-step process is integrable with respect to X; therefore $\mathcal{F}_{F,G}(\mathcal{B}, X) \subset L_{F,G}^1(X)$.

For any process $H \in \mathcal{F}_{F,G}(X)$ and any sequence $A_n \downarrow \phi$ in $\mathcal{P}$ we have

$$(\tilde{I}_X)_{F,L_G^1}(H 1_{A_n})$$
$$= \sup_{z \in G_1^*} \int_{A_n} |H| d|(I_X)_z| \leq \int_{A_n} |H| d|I_X| \to 0.$$

By Proposition 5.40 we have then $H \in \mathcal{F}_{F,G}(\mathcal{B}, X)$. It follows that

$$\mathcal{F}_{F,G}(X) = \mathcal{F}_{F,G}(\mathcal{B}, X) = L_{F,G}^1(X),$$

and that the $\mathcal{R}$-step processes are dense in $L^1_{F,G}(X)$. ∎

15. Remark. Assume X has integrable variation. We can define I_X for every (not necessarily predictable) rectangle $[0,t] \times A$ with $A \in \mathcal{F}$, by

$$I_X([0,t] \times A) = 1_A X_t$$

and we still have

$$\|I_X(B)\|_{L^1_G} \leq |\mu_X|(B),$$

for B in the algebra generated by these rectangles. Since this algebra generates the σ-algebra $\mathcal{M} = \mathcal{B}(\mathbb{R}_+) \times \mathcal{F}$, I_X can be extended as a σ-additive measure with finite variation on the whole σ-algebra $\mathcal{M}$, not only on $\mathcal{P}$ and we still have $|I_X| = |\mu_X|$ on $\mathcal{M}$. We can then apply the integration theory of §5 with $\Sigma = \mathcal{M}$ and obtain the space $\mathcal{F}_{F,L^1_G}(\mathcal{M}, X)$. Then we can define a "stochastic integral" $(H \cdot X)_t = \int_{[0,t]} H dI_X$, in case $\int_{[0,t]} H dI_X \in L^1_G$ for $t \geq 0$. This integral is still cadlag, but it is not necessarily adapted.

E. The stochastic integral as a Stieltjes integral

The following theorem gives a representation of the stochastic integral as a pathwise Stieltjes integral. It also extends Theorem 13 by replacing 1_M with processes H.

16. Theorem. *Let X be a cadlag, adapted process, p-summable relative to (F, G) and with integrable variation $|X|$.*
Let $H : \mathbb{R}_+ \times \Omega \to F$ be a predictable process.
a) *Assume $H \in \mathcal{F}_{F,G}(X)$ and $\int |H_s(\omega)| d|X|_s(\omega) < \infty$, a.s. Then:*
a_1) *The integrals $\int H dI_X$ and $\int H_s(\omega) dX_s(\omega)$ are defined a.s. and we have*

$$(\int H dI_X)(\omega) = \int H_s(\omega) dX_s(\omega), \text{ a.s.}$$

a_2) $H \in L^1_{F,G}(X)$ *and*

$$(H \cdot X)_t(\omega) = \int_{[0,t]} H_s(\omega) dX_s(\omega), \text{ a.s., for each } t \geq 0.$$

b) *Assume $\int |H| d|I_X| < \infty$ and $\int |H_s(\omega)| d|X|_s(\omega) < \infty$, a.s. Then*

$$(\int |H| d|I_X|)(\omega) = \int |H_s(\omega)| d|X|_s(\omega), \text{ a.s.}$$

c) *Assume $E(\int |H_s(\omega)| d|X|_s(\omega)) < \infty$. Then $H \in L^1_F(|\mu_X|) = L^1_F(|I_X|)$ and we have*

$$\int H d\mu_X = E(\int H dI_X) = E(\int H_s(\omega) dX_s(\omega))$$

and
$$\int |H|d|\mu_X|(\omega) = \int |H|d|I_X|(\omega) = \int |H_s(\omega)|d|X|_s(\omega), \text{ a.s.}$$

Proof. To prove assertion a), assume $H \in \mathcal{F}_{F,G}(X)$ and $\int |H_s(\omega)|d|X|_s(\omega) < \infty$, a.s. Using Theorem 13 c), assertion a_1) is true if $H = 1_M x$ with $M \in \mathcal{P}$ and $x \in F$. Then assertion a_1) remains valid if H is a simple, predictable process.

Assume now H is as stated in assertion a). Let (H^n) be a sequence of simple, predictable processes such that $H^n \to H$ pointwise and $|H^n| \leq |H|$ for each n. Let N be a negligible set such that for $\omega \notin N$ we have $\int |H_s(\omega)|d|X|_s(\omega) < \infty$ and

$$(\int H^n dI_X)(\omega) = \int H_s^n(\omega) dX_s(\omega), \text{ for each } n.$$

For $\omega \notin N$ we have $H.(\omega) \in L_F^1(m_{X(\omega)})$; since $H^n(\omega) \to H(\omega)$ and $|H^n(\omega)| \leq |H(\omega)|$, we can apply Lebesgue's theorem in the space $L_F^1(m_{X(\omega)})$ and deduce that

$$\int |H_s^n(\omega) - H_s(\omega)|d|X|_s(\omega) \to 0$$

and

$$\int H_s^n(\omega) dX_s(\omega) \to \int H_s(\omega) dX_s(\omega).$$

For each n we have $\int H^n dI_X \in L_G^1$ and

$$(\int H^n dI_X)(\omega) = \int H_s^n(\omega) dX_s(\omega) \to \int H_s(\omega) dX_s(\omega), \text{ for } \omega \notin N.$$

The hypothesis of Theorem 10.3 is satisfied. We deduce that $\int H dI_X \in L_G^1$ and

$$\int H^n dI_X \to \int H dI_X,$$

pointwise for $\omega \notin N$ and in L_G^1. It follows that the two limits are equal a.s.:

$$(\int H dI_X)(\omega) = \int H_s(\omega) dX_s(\omega), \text{ a.s.}$$

and assertion a_1) is proved. Replacing H with $1_{[0,t]} H$ for $t \geq 0$, we deduce that

$$(\int_{[0,t]} H dI_X)(\omega) = \int_{[0,t]} H_s(\omega) dX_s(\omega), \text{ a.s., for each } t \geq 0.$$

Since the Stieltjes integral is right continuous, for each ω, it follows that $H \in L_{F,G}^1(X)$ and

$$(H \cdot X)_t(\omega) = \int_{[0,t]} H_s(\omega) dX_s(\omega), \text{ a.s.}$$

and assertion a_2) is proved.

To prove assertion b), assume $\int |H|d|I_X| < \infty$ and $\int |H_s(\omega)|d|X|_s(\omega) < \infty$, a.s. Then $H \in L^1_F(|I_X|)$. Let (H^n) be a sequence of simple predictable processes such that $H^n \to H$ and $|H^n| \leq |H|$. From the proof of assertion a_1) we have

$$\int |H^n_s(\omega)|d|X|_s(\omega) \to \int |H_s(\omega)|d|X|_s(\omega).$$

We can apply the Lebesgue Theorem in the space $L^1_F(|I_X|)$ and deduce that

$$\int |H^n|d|I_X| \to \int |H|d|I_X|.$$

For each n we have by Theorem 13 c),

$$\left(\int |H^n|d|I_X|\right)(\omega) = \int |H^n_s(\omega)|d|X|_s(\omega), \text{ a.s., for each } n.$$

It follows that

$$\left(\int |H|d|I_X|\right)(\omega) = \int |H_s(\omega)|d|X|_s(\omega), \text{ a.s.}$$

and assertion b) is proved.

Finally, assume $E(\int |H_s(\omega)|d|X|_s(\omega)) < \infty$, that is, $\int |H|d|\mu_X| < \infty$ (Theorem 8 b)). By Theorem 13 d) we have $|I_X| = |\mu_X|$, hence

$$\int |H|d|I_X| = \int |H|d|\mu_X| < \infty,$$

therefore, by assertion b) we have

$$(\int |H|d|\mu_X|)(\omega) = (\int |H|d|I_X|)(\omega) = \int |H_s(\omega)|d|X|_s(\omega), \text{ a.s.},$$

which is the second equality in assertion c).

Taking a sequence (H^n) of simple predictable processes with $H^n \to H$ and $|H^n| \leq |H|$ and applying Lebesgue's Theorem in $L^1_F(\mu_X)$ and $L^1_F(I_X)$ we deduce that

$$\int H^n d\mu_X \to \int H d\mu_X$$

and

$$\int H^n dI_X \to \int H dI_X, \text{ in } L^1_G.$$

Then

$$E(\int H^n dI_X) \to E(\int H dI_X), \text{ in } G.$$

Since for each n we have, by Theorem 13 d),

$$\int H^n d\mu_X = E(\int H^n dI_X),$$

using assertion a_1) we get

$$\int H d\mu_X = E(\int H dI_X) = E(\int H_s(\omega) dX_s(\omega))$$

and this proves assertion c). ∎

Remark. In Theorem 21.12 b') assertion a) will be extended for $H \in \mathcal{F}_{F,G}(X)$ with $\tilde{m}_{X(\omega)}(H(\omega)) < \infty$, a.s. even for X with finite variation.

F. The pathwise stochastic integral

Let $X : \mathbb{R}_+ \times \Omega \to E$ be a cadlag process with *finite variation* $|X|$. If X does not have integrable variation $|X|_\infty$, then X is not necessarily summable and the stochastic integral cannot be defined.

However, a pathwise stochastic integral can still be defined in this case. The pathwise stochastic integral is equal to the usual stochastic integral, in case X is summable and has finite variation (Theorem 16 a_2).

17. Definition. *Let $H : \mathbb{R}_+ \times \Omega \to F$.*
a) *We say that H is pathwise integrable with respect to X if for every $\omega \in \Omega$, the function $s \mapsto H_s(\omega)$ is Stieltjes integrable with respect to the function $s \mapsto X_s(\omega)$, or equivalently, with respect to the variation $s \mapsto |X|_s(\omega)$.*
b) *We say that H is locally pathwise integrable with respect to X if for every $\omega \in \Omega$ and $t \geq 0$, the function $s \mapsto H_s(\omega)$ is Stieltjes integrable with respect to the function $s \mapsto X_s(\omega)$ (or with respect to the function $s \mapsto |X|_s(\omega)$) on the interval $[0,t]$.*

If H is pathwise integrable with respect to X, then, evidently, H is locally pathwise integrable with respect to X.

If H is measurable, to say that H is locally pathwise integrable with respect to X means that for each $\omega \in \Omega$ and $t \geq 0$ we have $\int_{[0,t]} |H_s(\omega)| d|X|_s(\omega) < \infty$.

18. Proposition. *If H is locally pathwise integrable with respect to X, then, for each $\omega \in \Omega$, the function $t \mapsto \int_{[0,t]} H_s(\omega) dX_s(\omega)$ is cadlag and has finite variation $\leq \int_{[0,t]} |H_s(\omega)| d|X|_s(\omega)$.*

The proof is immediate.

This means that the process $(\int_{[0,t]} H_s(\omega) dX_s(\omega))_{t \geq 0}$ is cadlag and has finite variation.

19. Proposition. *If X and H are adapted and if H is measurable and locally pathwise integrable with respect to X, then the process $(\int_{[0,t]} H_s(\omega) dX_s(\omega))_{t \geq 0}$ is adapted.*

In fact, if $t \geq 0$, then the processes $X' = 1_{[0,t]}X$ and $H' = 1_{[0,t]}H$ are $\mathcal{B}(\mathbb{R}_+) \times \mathcal{F}_t$-measurable and H' is pathwise integrable with respect to X'. By Theorem 3 b) with $\mathcal{F}$ replaced by $\mathcal{F}_t$, the function
$\omega \mapsto \int H'_s(\omega)dX'_s(\omega) = \int_{[0,t]} H_s(\omega)dX_s(\omega)$ is $\mathcal{F}_t$-measurable.

We extend the definition of the stochastic integral for (not necessarily summable) processes X with finite variation.

20. Definition. *Let $X : \mathbb{R}_+ \times \Omega \to E \subset L(F,G)$ be a cadlag adapted process with finite variation and $H : \mathbb{R}_+ \times \Omega \to F$ a measurable, adapted (not necessarily predictable) process, locally pathwise integrable with respect to X. The pathwise stochastic integral of H with respect to X is the stochastic process denoted by $H \cdot X$ and defined by*

$$(H \cdot X)_t(\omega) = \int_{[0,t]} H_s(\omega)dX_s(\omega), \text{ for } (t,\omega) \in \mathbb{R}_+ \times \Omega.$$

21. Proposition. *Let X be a cadlag, adapted, locally p-summable process relative to (F,G) and with finite variation $|X|$ and let H be an F-valued, predictable process. If $H \in L^1_{F,G}(X)_{loc}$ and $\int_{[0,t]} |H_s(\omega)|d|X|_s(\omega) < \infty$ for every $(t,\omega) \in \mathbb{R}_+ \times \Omega$, then the pathwise stochastic integral $H \cdot X$ of Definition 20 coincides with the stochastic integral of Definition 15.10.*

We use Theorem 16.

In view of Proposition 21, the pathwise stochastic integral will be called, simply, the stochastic integral, if no confusion is possible.

We list now some of the properties of the stochastic integral with respect to a process X with finite variation, which are similar to the properties of the stochastic integral with respect to a locally p-summable process X.

In what follows, X is a *cadlag, adapted* process with *finite variation* $|X|$ and H is an F-valued, *measurable, adapted* process, *locally integrable* with respect to X.

22. Proposition. *The stochastic integral $H \cdot X$ is a cadlag, adapted process with finite variation $|H \cdot X|$ satisfying*

$$|H \cdot X|_t(\omega) \leq (|H| \cdot |X|)_t(\omega) < \infty, \text{ for } (t,\omega) \in \mathbb{R}_+ \times \Omega,$$

where $|H| = (|H_t|)_{t \geq 0}$ and $|X| = (|X|_t)_{t \geq 0}$.

If one of the processes X or H is real-valued, then we have the equality

$$|H \cdot X| = |H| \cdot |X|.$$

Use Proposition 2.28 and Theorem 2.29.

23. Proposition. *If T is a stopping time, then X^T has finite variation and we have*
$$X^T = 1_{[0,T]}X \text{ and } X^{T-} = 1_{[0,T)} \cdot X.$$

We use the equality $X^T = X1_{[0,T]} + X_T 1_{(T,\infty)}$ and the equality of the variations $|X^T| = |X|1_{[0,T]}$ and similar equalities for X^{T-}.

24. Proposition. *Let T be a stopping time. Then H is locally integrable with respect to X^T (resp. X^{T-}) iff $1_{[0,T]}H$ (resp. $1_{[0,T)}H$) is locally integrable with respect to X. In this case we have*

$$H \cdot X^T = (1_{[0,T]}H) \cdot X = (H \cdot X)^T$$

(resp. $H \cdot X^{T-} = (1_{[0,T)}H) \cdot X = (H \cdot X)^{T-}$).

Proof. The integrability property follows from the equality

$$\int |H_s(\omega)|d|X^T|_s(\omega) = \int |H_s(\omega)|1_{[0,T]}(s,\omega)d|X|_s(\omega)$$

and a similar equality for X^{T-}.

Then for every $t \geq 0$ we have (for each $\omega \in \Omega$)

$$\int_{[0,t]} H_s dX_s^T = \int_{[0,t]} H_s 1_{[0,T]} dX_s = \int_{[0,t\wedge T]} H_s dX_s,$$

hence

$$(H \cdot X^T)_t = (H1_{[0,T]}X)_t = (H \cdot X)_t^T.$$

Similar proof for X^{T-}. ∎

25. Proposition. *If H is real-valued and K is an F-valued, measurable, adapted process, then K is locally integrable with respect to $H \cdot X$ iff KH is locally integrable with respect to X. In this case we have*

$$K \cdot (H \cdot X) = (KH) \cdot X.$$

We apply Theorem 2.31 c) for each $\omega \in \Omega$.

26. Proposition. *If H is F-valued and K is a real-valued, measurable, adapted process such that KH is locally integrable with respect to X, then K is locally integrable with respect to $H \cdot X$ and we have*

$$K \cdot (H \cdot X) = (KH) \cdot X.$$

We apply Theorem 2.31 b) for each $\omega \in \Omega$.

27. Proposition. *We have*

$$\Delta(H \cdot X) = H\Delta X,$$

where $\Delta X_t = X_t - X_{t-}$ is the jump of X at t.

In fact, for each $t \geq 0$ we have, for each $\omega \in \Omega$,

$$\Delta(H \cdot X)_t = (H \cdot X)_t - (H \cdot X)_{t-} = \int_{[0,t]} H_s dX_s - \int_{[0,t)} H_s dX_s$$

$$= \int_{\{t\}} H_s dX_s = H_t \Delta X_t.$$

G. Semilocally summable processes

Putting together the stochastic integral and the pathwise stochastic integral we get the stochastic integral with respect to semilocally summable processes.

28. Definition. *A cadlag, adapted process $Z : \mathbb{R}_+ \times \Omega \to E \subset L(F,G)$ is said to be semilocally p-summable relative to (F,G), if it is of the form $Z = X+Y$, where X is an E-valued process, locally p-summable relative to (F,G) and Y is an E-valued, cadlag, adapted process with finite variation.*

An F-valued, predictable process H is said to be locally integrable with respect to a semilocally p-summable process $Z = X + Y$, if both stochastic integrals $H \cdot X$ and $H \cdot Y$ are defined. In this case the stochastic integral $H \cdot Z$ is defined by the equality

$$H \cdot Z = H \cdot X + H \cdot Y.$$

The stochastic integral $H \cdot Z$ is independent of the decomposition $Z = X + Y$; that is, if $Z = X' + Y'$ is another decomposition and if $H \cdot X$, $H \cdot Y, H \cdot X', H \cdot Y'$ are all defined, then $H \cdot X + H \cdot Y = H \cdot X' + H \cdot Y'$.

The above definition is an extension of the classical concept of *semimartingale* in the scalar case. But even in the Hilbert space case, the concept of semi local summability is more general than that of semimartingale.

29. An example of F-valued processes which are locally integrable with respect to any semilocally p-summable process relative to (F,G) is that of the σ-elementary processes, of the form

$$H = H_0 1_{\{0\}} + \sum_{1 \leq n < \infty} H_n 1_{(T_n, T_{n+1}]},$$

where (T_n) is an increasing sequence of stopping times with $T_n \uparrow \infty$, H_0 is an $\mathcal{F}_0$-measurable random variable, and for each n, H_n is an $\mathcal{F}_{T_n}$-measurable random variable. We do not assume H_0 and H_n to be bounded. For such a σ-elementary process H, its stochastic integral with respect to a semilocally p-summable process Z relative to (F,G) can be computed pathwise, a.s.:

$$(H \cdot Z)_t = H_0 Z_0 + \sum_{1 \leq n < \infty} H_n(Z_{T_{n+1} \wedge t} - Z_{T_n \wedge t}).$$

Another example is that of caglad adapted processes.

30. Theorem. *Any caglad, adapted process $H : \mathbb{R}_+ \times \Omega \to F$ is locally integrable with respect to any semi locally p-summable process Z relative to (F,G).*

If fact, if $Z = X + Y$ is a decomposition of Z and if H is caglad and adapted, then H is locally integrable with respect to X (Theorem 15.16) and with respect to Y (since H is locally bounded and measurable).

31. Propositions 23–27 of the stochastic integral remain valid for the stochastic integral with respect to a semilocally p-summable process.

Chapter 5
Processes with Finite Semivariation

In this chapter we study first, in §20, the functions with finite semivariation, and define a new type of Stieltjes integral for such functions, as an application of the integration theory presented in §5.

Then, in §21, we combine the Stieltjes integral with the stochastic integration of Chapter 2, to study the processes with integrable semivariation. The main results are:

a) Processes with integrable semivariation are summable (provided $c_0 \not\subset E$ and $c_0 \not\subset G$) (Theorem 21.12);

b) The stochastic integral can be computed pathwise, as a Stieltjes integral (Corollary 21.13);

c) The equality

$$\mu_X(M) = E(\int 1_M(s,\omega) dX_s(\omega)), \text{ for } M \in \mathcal{M},$$

defines a σ-additive measure μ_X with finite semivariation (Theorem 21.8).

Finally, in §22 we study the dual projections of processes with integrable semivariation; in particular the projections of processes with integrable variation. The main results are the existence of dual projections of processes with integrable variation (Theorem 22.8), and the decomposition of local martingales (Theorem 22.19).

§20. FUNCTIONS WITH FINITE SEMIVARIATION AND THEIR STIELTJES INTEGRAL

In this paragraph we study the functions with finite semivariation and define a new type of Stieltjes integral with respect to such functions. The results about functions with finite semivariation will be used in §21 in the study of processes with finite semivariation.

In this paragraph, we denote by $\mathcal{P}$ the semiring of the intervals $(a, b]$ with $a, b \in \mathbb{R}$ and by the $\mathcal{R}$ the ring generated by $\mathcal{P}$.

A. Functions with finite semivariation

Let $g : \mathbb{R} \to E \subset L(F, G)$ be a function.

1. Definition. *For any interval I (bounded or not) the semivariation of g on I, relative to the embedding $E \subset L(F, G)$, or relative to the pair (F, G), is a number denoted by $svar_{F,G}(g, I)$ or $sv_{F,G}(g, I)$ and defined by the following equality:*

$$svar_{F,G}(g, I) = \sup | \sum [g(t_{i+1}) - g(t_i)] x_i |,$$

the supremum being taken for all finite divisions $d : t_0 < t_1 < \cdots < t_n$ consisting of points from I and all finite families $(x_0, x_1, \ldots, x_{n-1})$ of elements of F with $|x_i| \leq 1$.

If the pair (F, G) is understood we shall write $sv(g, I)$ instead of $sv_{F,G}(g, I)$.

We say g has *finite* (resp. *bounded*) *semivariation* relative to (F, G) if $sv_{F,G}(g, I) < \infty$ for every bounded interval I (resp. if $sv_{F,G}(\mathbb{R}) < \infty$).

We state some of the properties of the semivariation of g, which are similar to those of the semivaration of a measure.

2. *If $I \subset J$ are two intervals, then*

$$svar(g, I) \leq svar(g, J).$$

3. *For every interval I we have*

$$svar(g, I) \leq var(g, I).$$

In some cases the variation and the semivariation are equal. For measures, this property was proved in Proposition 4.10.

4. Proposition. *Assume $E \subset L(F, \mathbb{R})$ isometrically. Then for any interval $I \subset \mathbb{R}$ we have*

$$svar(g, I) = var(g, I).$$

Proof. Let $I \subset \mathbb{R}$ be an interval, $t_0 < t_1, \cdots < t_n$ a family of points from I and $\varepsilon > 0$. For each $i < n$ there is an $x_i \in F_1$ such that $\langle x_i, g(t_{i+1}) - g(t_i) \rangle \geq 0$ and

$$|g(t_{i+1}) - g(t_i)| \leq \langle x_i, g(t_{i+1}) - g(t_i) \rangle + \frac{\varepsilon}{n}.$$

Then
$$\sum_{1 \le i \le n} |g(t_{i+1}) - g(t_i)| \le \sum_{1 \le i \le n} \langle x_i, g(t_{i+1}) - g(t_i) \rangle + \varepsilon$$
$$= |\sum_{1 \le i \le n} [g(t_{i+1}) - g(t_i)] x_i | + \varepsilon \le svar(g, I) + \varepsilon.$$

We deduce then that
$$var(g, I) \le svar(g, I).$$

Using inequality 3 we get the desired equality. ∎

5. Corollary. *If $g : \mathbb{R} \to \mathbb{R}$ is a real-valued function and if we consider $\mathbb{R} = L(\mathbb{R}, \mathbb{R})$, then the variation and the semivariation of g are equal.*

6. Proposition. *If $g : \mathbb{R} \to E \subset L(F, G)$ is a function, then for any interval $I \subset \mathbb{R}$ we have*
$$svar_{F,G}(g, I) \le svar_{E^*, \mathbb{R}}(g, I) = var(g, I).$$

If the embedding $E \subset L(F, G)$ is an isometry, then for any interval $I \subset \mathbb{R}$ we have
$$svar_{\mathbb{R}, E}(g, I) \le svar_{F, G}(g, I).$$

The proof is similar to that of Proposition 4.12, concerning the semivariation of measures.

B. Semivariation and norming spaces

As in the case of the semivariation of measures, there is an alternative definition of the semivariation of functions, in terms of norming spaces.

Let $g : \mathbb{R} \to E \subset L(F, G)$ be a function and let $Z \subset G^*$ be a norming space for G. For any $z \in Z$, consider the function $g_z : \mathbb{R} \to F^*$ defined by
$$\langle x, g_z(t) \rangle = \langle g(t) x, z \rangle, \text{ for } t \in \mathbb{R} \text{ and } x \in F.$$

In particular, considering the embedding $E = L(\mathbb{R}, E)$, for every $x^* \in E^*$ we have $g_{x^*}(t) = x^* g(t) = \langle g(t), x^* \rangle$ for $t \in \mathbb{R}$.

Considering $g_z : \mathbb{R} \to F^* \subset L(F, \mathbb{R})$, by Proposition 4 we have, for any interval $I \subset \mathbb{R}$,
$$svar_{F, \mathbb{R}}(g_z, I) = var(g_z, I).$$

The following proposition is the analog of Proposition 4.13.

7. Proposition. *For any interval $I \subset \mathbb{R}$ we have*
$$svar_{F, G}(g, I) = \sup_{z \in Z_1} var(g_z, I).$$

In particular, for every $t \in \mathbb{R}$ we have
$$svar_{F,G}(g, (-\infty, t]) = \sup_{z \in Z_1} var(g_z, (-\infty, t]).$$

The proof is similar to that of Proposition 4.13.

It is convenient to denote for every $t \in \mathbb{R}$,
$$\tilde{g}_{F,G}(t) = svar_{F,G}(g, (-\infty, t]).$$

The function $\tilde{g}_{F,G}$ is called the *semivariation function* of g with respect to (F, G).

If the pair (F, G) is understood, we shall write $\tilde{g}$ instead of $\tilde{g}_{F,G}$.

With these notations, from Properties 2–4 we deduce the following corresponding Properties $2'$–$4'$:

$2'$. If $t \leq t'$ then $\tilde{g}(t) \leq \tilde{g}(t')$.

$3'$. $\tilde{g}(t) \leq |g|(t)$.

$4'$. If $g : \mathbb{R} \to L(F, \mathbb{R})$ then $\tilde{g}_{F,\mathbb{R}}(t) = |g|(t)$.

In addition, if g has finite semivariation function $\tilde{g}$, we have the following property:

8. Proposition. *If $a < b$ then*
$$\tilde{g}(b) - \tilde{g}(a) \leq svar(g, [a, b]).$$

If, in addition, g is right continous, then
$$\tilde{g}(b) - \tilde{g}(a) \leq svar(g, (a, b]).$$

Proof. Let $z \in G_1^*$. By Property 18.6$''$ we have
$$|g_z|(b) - |g_z|(a) = var(g_z, [a, b]).$$

Using $|g_z|(a) \leq \tilde{g}(a)$ we deduce
$$|g_z|(b) - \tilde{g}(a) \leq |g_z|(b) - |g_z|(a) = var(g_z, [a, b]).$$

Taking the supremum for $z \in G_1^*$ we obtain the desired inequality.

The corresponding equality for g right continous is obtained from
$$|g_z|(b) - |g_z|(a) = var(g_z, (a, b]).$$

∎

The following proposition is similar to Proposition 4.15.

§20. FUNCTIONS WITH FINITE SEMIVARIATION AND THEIR STIELTJES INTEGRAL   247

9. Proposition. *Assume Z is a closed subspace of G^*, norming for G and let $I \subset \mathbb{R}$ be an interval. Then*

$$svar(g, I) < \infty \text{ iff } var(g_z, I) < \infty \text{ for every } z \in Z.$$

Proof. Let M be the set of all $\mathcal{R}$-step functions $s : \mathbb{R} \to F$ with $|s| \leq \varphi_I$. Consider the measure $m : \mathcal{R} \to E$ defined by $m(a, b] = g(b) - g(a)$. For each $z \in Z$ consider the measure $m_z : \mathcal{R} \to F^*$ defined by $m_z(a, b] = g_z(b) - g_z(a)$.

From Definition 1 we deduce that

$$svar(g, I) = \sup_{s \in M} |\int s\, dm|.$$

Using Proposition 4, for each $z \in Z$ we have

$$var(g_z, I) = svar(g_z, I) = \sup_{s \in M} |\int s\, dm_z|.$$

For each step function $s \in M$ define the linear functional $T_s \in Z^*$ by

$$T_s(z) = \int s\, dm_z, \text{ for } z \in Z.$$

Then $|T_s| = |\int s\, dm|$ and for each $z \in Z$ we have

$$\sup_{s \in M} |T_s(z)| = \sup_{s \in M} |\int s\, dm_z| = svar(g_z, I) = var(g_z, I).$$

Assume $var(g_z, I) < \infty$ for every $z \in Z$. Then $\sup_{s \in M} |T_s(z)| < \infty$ for $z \in Z$, hence, by the Banach–Steinhauss theorem, we have

$$svar(g, I) = \sup_{s \in M} |\int s\, dm| = \sup_{s \in M} |T_s| < \infty.$$

The converse implication is evident. ∎

C. The measure associated to a function

To the function $g : \mathbb{R} \to E \subset L(F, G)$ we associate the finitely additive measure $m_g : \mathcal{R} \to E$ defined by

$$m_g(a, b] = g(b) - g(a), \text{ for } a < b.$$

10. *For every $z \in G^*$ we have*

$$(m_g)_z = m_{(g_z)}.$$

In particular, considering the embedding $E = L(\mathbb{R}, E)$, for every $x^* \in E^*$ we have

$$x^* m_g = m_{x^* g}.$$

The following proposition is similar to Proposition 18.15.

11. Proposition. a) m_g *has finite (resp. bounded) semivariation relative to (F, G) iff g has finite (resp. bounded) semivariation relative to (F, G).*
b) *If $I \subset \mathbb{R}$ is an interval, then*

$$svar(g, I) \leq svar(m_g, I) \leq svar(g, \overline{I}).$$

c) *We have the equality*

$$svar(m_g, I) = svar(g, I)$$

if either $\inf I = -\infty$ *or* $\inf I \in I$.

Proof. For every $z \in G^*$ we have, by Proposition 4.13 and Proposition 7

$$sv(m_g, I) = \sup_{|z| \leq 1} v((m_g)_z, I)$$

and

$$sv(g, I) = \sup_{|z| \leq 1} v(g_z, I).$$

For each $z \in G^*$, we apply Propositions 18.15 to the function g_z:

$$v(g_z, I) \leq v(m_{g_z}, I) \leq v(g_z, \overline{I});$$

if $\inf I = -\infty$ or $\inf I \in I$, then

$$v(m_{g_z}, I) = v(g_z, I).$$

Taking the supremum for $|z| \leq 1$ and using Proposition 7, we obtain assertions b) and c). Then assertion a) follows from assertion b). ∎

12. Proposition. *Let $Z \subset G^*$ be a space norming for G and assume that for every $z \in Z$, the function g_z is right continuous.*
Then, for any interval $I \subset \mathbb{R}$ we have
$$svar(m_g, I) = svar(g, I).$$

Proof. Let $z \in Z$. Since the function g_z is right continuous, by Proposition 18.16, for every interval $I \subset \mathbb{R}$ we have

$$v(m_{g_z}, I) = v(g_z, I).$$

Taking the supremum for $z \in Z_1$, we obtain the desired equality. ∎

The problem is now to extend the measure m_g to a σ-additive measure with finite semivariation on the Borel σ-algebra $\mathcal{B}(\mathbb{R})$ or on the δ-ring $\mathcal{D}$ of bounded Borel sets.

13. Theorem. *Assume $c_0 \not\subset E$. Let $g : \mathbb{R} \to E$ be a function with finite (resp. bounded) semivariation relative to $(\mathbb{R}, E)$ and assume there is a space*

$Z \subset E^*$ norming for E such that x^*g is right continuous for every $x^* \in Z$. Then
a) g is right continuous;
b) The measure m_g can be extended uniquely to a σ-additive measure $m : \mathcal{D} \to E$ (resp. $m : \mathcal{B}(\mathbb{R}) \to E$) with finite (resp. bounded) semivariation $\tilde{m}_{\mathbb{R},E}$ and we have

$$\tilde{m}_{\mathbb{R},E} = (\tilde{m}_g)_{\mathbb{R},E}, \text{ on } \mathcal{R}.$$

c) If $E \subset L(F,G)$ and if g has finite (resp. bounded) semivariation relative to $(\mathbb{R}, E)$ and (F, G), then the extension m has finite (resp. bounded) semivariation $\tilde{m}_{F,G}$ and we have

$$\tilde{m}_{F,G} = (\tilde{m}_g)_{F,G}, \text{ on } \mathcal{R}.$$

Proof. Let $x^* \in Z$. The real-valued function x^*g is right continuous and has finite (resp. bounded) variation. By Theorem 18.19, the corresponding measure $x^*m_g = m_{x^*g}$ can be extended to a σ-additive measure with finite variation on $\mathcal{D}$ (resp. on $\mathcal{B}(\mathbb{R})$). In particular, x^*m_g is σ-additive on $\mathcal{R}$. Since m_g has finite (resp. bounded) semivariation relative to $(\mathbb{R}, E)$, m_g is locally bounded (resp. bounded), by Proposition 4.14.

We can now apply the extension Theorem 7.8 and deduce that m_g can be extended uniquely to a σ-additive measure $m : \mathcal{D} \to E$ (resp. $m : \mathcal{B}(\mathbb{R}) \to E$).

By Corollary 7.6, we have

$$\tilde{m}_{\mathbb{R},E} = (\tilde{m}_g)_{\mathbb{R},E}, \text{ on } \mathcal{R}$$

and this proves assertion b). Assertion c) also follows from Corollary 7.6.
Assertion a) follows from the σ-additivity of m_g on $\mathcal{R}$. ∎

The measure m is called the *Stieltjes measure associated to the function g with finite semivariation* and is still denoted by m_g.

Remark. If $E \subset L(F,G)$ isometrically, then Theorem 13 is valid if we assume that the semivariation of g relative to (F, G) is finite (resp. bounded), since, by Proposition 6, it follows that the semivariation of g relative to $(\mathbb{R}, E)$ is also finite (resp. bounded).

The following theorem evaluates the jumps of the semivariation $\tilde{g}_{F,G}$. It is an extension of Theorem 18.20 about the jumps of the variation $|g|$.

14. Theorem. *Let $g : \mathbb{R} \to E \subset L(F,G)$ be a right continuous function.*
a) *If $c_0 \not\subset E$ and g has finite semivariation $\tilde{g}_{\mathbb{R},E}$, then*

$$\Delta \tilde{g}_{\mathbb{R},E}(t) \leq |\Delta g(t)|, \text{ for } t \in \mathbb{R}.$$

b) *If $c_0 \not\subset E$, $c_0 \not\subset G$ and g has finite semivariations $\tilde{g}_{\mathbb{R},E}$ and $\tilde{g}_{F,G}$, then*

$$\Delta \tilde{g}_{F,G}(t) \leq |\Delta g(t)|, \text{ for } t \in \mathbb{R}.$$

Proof. Let $\mathcal{D}$ be the δ-ring of bounded Borel subsets of $\mathbb{R}$. In the hypothesis of assertion a), by Theorem 13 there is a σ-additive measure $m_g : \mathcal{D} \to E$ with finite semivariation $(\tilde{m}_g)_{\mathbb{R},E}$ satisfying $m_g(a,b] = g(b) - g(a)$, for $a < b$ in $\mathbb{R}$.

Assume now the hypothesis of assertion b) and prove that $\Delta \tilde{g}_{F,G}(t) \leq |\Delta g(t)|$, for $t \in \mathbb{R}$. Assertion a) will then follow as a particular case.

Let $t \in \mathbb{R}$ and (s_n) an increasing sequence in $\mathbb{R}$ with $s_n < t$ and $s_n \uparrow t$. Then $(s_n, t] \downarrow \{t\}$, hence

$$m_g(s_n, t] \to m_g\{t\}.$$

By Theorem 4.28, m_{F,G^*} is uniformly σ-additive; by Theorem 4.23 d) we have

$$(\tilde{m}_g)_{F,G}(s_n, t] \to (\tilde{m}_g)_{F,G}\{t\}.$$

A simple computation shows that

$$(\tilde{m}_g)_{F,G}\{t\} \leq |m_g\{t\}|,$$

with equality, if $E \subset L(F,G)$ isometrically.

At the same time, since g is cadlag (Theorem 13 a)), we have

$$g(s_n) \to g(t-)$$

and

$$\tilde{g}_{F,G}(s_n) \to \tilde{g}_{F,G}(t-).$$

We have also

$$m_g(s_n, t] = g(t) - g(s_n)$$

and (Propositions 8 and 11)

$$(\tilde{m}_g)_{F,G}(s_n, t] \geq svar(g, (s_n, t]) \geq \tilde{g}_{F,G}(t) - \tilde{g}_{F,G}(s_n).$$

Letting $n \to \infty$ we get

$$\Delta \tilde{g}_{F,G}(t) = \tilde{g}_{F,G}(t) - \tilde{g}_{F,G}(t-) \leq \lim (\tilde{m}_g)_{F,G}(s_n, t]$$
$$= (\tilde{m}_g)_{F,G}\{t\} \leq |m_g\{t\}|.$$

∎

D. The Stieltjes integral with respect to a function with finite semivariation

15. Let $g : \mathbb{R} \to E \subset L(F,G)$ be a right continuous function with finite semivariation relative to (F,G) and $m_g : \mathcal{R} \to E$ the finitely additive measure with finite semivariation $(\tilde{m}_g)_{F,G}$ associated to g.

§20. FUNCTIONS WITH FINITE SEMIVARIATION AND THEIR STIELTJES INTEGRAL

Assume m_g can be extended to a σ-additive measure still denoted by m_g on the δ-ring $\mathcal{D}$ of bounded Borel subsets of $\mathbb{R}$. This is the case, for example, if $c_0 \not\subset E$ and the semivariation $\tilde{g}_{\mathbb{R},E}$ is finite (Theorem 13). By Corollary 7.6, the extension has also finite semivariation $(\tilde{m}_g)_{F,G}$ on $\mathcal{D}$.

Consider the space $\mathcal{F}_{F,G}(m_g)$ defined in §5. The space $\mathcal{F}_{F,G}(m_g)$ is also denoted by $L^1_{F,G}(g)$.

For $f \in L^1_{F,G}(g)$ we define the *Stieltjes integral* $\int f dg$ by the equality

$$\int f dg = \int f dm_g.$$

We have $\int f dg \in G^{**}$,

$$\langle \int f dg, z \rangle = \int f dg_z, \text{ for } z \in G^*$$

and

$$|\int f dg| \leq (\tilde{m}_g)_{F,G}(f).$$

The following theorem gives instances when the integral $\int f dg$ belongs to G rather than G^{**}.

16. Theorem. *Assume $c_0 \not\subset E$ and let $g : \mathbb{R} \to E$ be a right continuous function with finite semivariation relative to $(\mathbb{R}, E)$.*
a) *For every function $f \in \mathcal{F}_{\mathbb{R},E}(m_g)$ we have $\int f dg \in E$.*
b) *If, in addition, $c_0 \not\subset G$ and g has finite semivariation relative to (F, G), then for every function $f \in \mathcal{F}_{F,G}(m_g)$ we have $\int f dg \in G$.*

Proof. We use Theorem 13 to deduce that m_g can be extended to a σ-additive measure on the δ-ring $\mathcal{D}$ generated by the intervals $(a, b]$. If g has bounded semivariation relative to $(\mathbb{R}, E)$, then m_g can be extended to $\mathcal{B}(\mathbb{R})$.

We notice that $S_n = [-n, n) \in \mathcal{D}$ and $\mathbb{R} = \bigcup S_n$. Then we can use Theorem 5.48 b) and b') to deduce that $\int f dm$ belongs to G or to E. ∎

17. Remark. Assume g is right continuous and has finite (resp. bounded) *variation*. Then g has also finite (resp. bounded) *semivariation* relative to any embedding $E \subset L(F, G)$. The corresponding measure m_g has finite (resp. bounded) semivariation $(\tilde{m}_g)_{F,G}$ on the δ-ring $\mathcal{D}$ (resp. on the σ-algebra $\mathcal{B}(\mathbb{R})$). Then $L^1_F(|m_g|) \subset \mathcal{F}_{F,G}(m_g)$ and

$$\tilde{m}_{F,G}(f) \leq \int |f| d|m_g|, \text{ for } f \in L^1_F(|m_g|).$$

For a function $f \in L^1_F(|m_g|)$, the Stieltjes integral $\int f dg$ is the same, whether we consider f in $L^1_F(|m_g|)$ or in $L^1_{F,G}(m_g)$.

From Proposition 5.46 we obtain the following property of the Stieltjes integral.

18. Proposition. *Let $g : \mathbb{R} \to E \subset L(F,G)$ be a right continuous function with finite variation (resp. finite semivariation relative to (F,G)) and let $x \in F$ and $z \in G^*$.*

Then the functions gx and $\langle gx, z \rangle$ have finite variation (resp. gx has finite semivariation relative to $(\mathbb{R}, G)$ and $\langle gx, z \rangle$ has finite variation).

Assume m_g can be extended to a σ-additive measure on the δ-ring $\mathcal{D}$ of bounded Borel sets, or on the σ-algebra $\mathcal{B}(\mathbb{R})$ of Borel sets (this is the case if $c_0 \not\subset E$ and g has finite semivariation relative to $(\mathbb{R}, E)$).

If $\varphi : \mathbb{R} \to \mathbb{R}$ is dg-integrable and if $\int \varphi dg \in E$, then φ is $d(gx)$-integrable and $d\langle gx, z \rangle$-integrable and we have

$$\left(\int \varphi dg\right)x = \int \varphi x \, dg = \int \varphi \, d(gx)$$

and

$$\left\langle \left(\int \varphi dg\right)x, z \right\rangle = \left\langle \int \varphi x \, dg, z \right\rangle = \int \varphi \, d\langle gx, z \rangle.$$

§21. PROCESSES WITH FINITE SEMIVARIATION

As we mentioned at the beginning of this chapter, in this paragraph we obtain the following results:

a) The summability of processes with integrable semivariation, if $c_0 \not\subset E$ and $c_0 \not\subset G$ (Theorem 12);

b) The stochastic integral can be computed as a Stieltjes integral, in the sense defined in §20 (Corollary 13);

c) To a process X with integrable semivariation we associate a σ-additive measure μ_X with finite semivariation, satisfying

$$\mu_X(M) = E(\int 1_M(s,\omega) dX_s(\omega)), \text{ for } M \in \mathcal{M},$$

where the integral is a Stieltjes integral in the sense of §20 (Theorem 8).

A. The semivariation process

1. If S is a set and $g : S \to E \subset L(F,G)$ is a function, then, for every $z \in G^*$ we define the function $g_z : S \to F^*$ by

$$\langle x, g_z(s) \rangle = \langle g(s)x, z \rangle, \text{ for } x \in F \text{ and } s \in S.$$

If $X : \mathbb{R}_+ \times \Omega \to E \subset L(F,G)$ is a process and $z \in G^*$ we define, as above, the process $X_z : \mathbb{R}_+ \times E \to F^*$ by

$$\langle x, (X_z)_t(\omega) \rangle = \langle X_t(\omega)x, z \rangle, \text{ for } x \in F, t \in \mathbb{R}_+ \text{ and } \omega \in \Omega.$$

For fixed $t \geq 0$, we can consider the function $X_t : \omega \mapsto X_t(\omega)$ from Ω into $E \subset L(F,G)$ and for $z \in G^*$ we can define $(X_t)_z : \Omega \to F^*$ satisfying

$$\langle x, (X_t)_z(\omega) \rangle = \langle X_t(\omega)x, z \rangle.$$

It follows that

$$(X_t)_z(\omega) = (X_z)_t(\omega), \text{ for } t \in \mathbb{R}_+ \text{ and } \omega \in \Omega.$$

2. Let $X : \mathbb{R}_+ \times \Omega \to E \subset L(F,G)$ be a process. We consider X extended to $\mathbb{R} \times \Omega$ with $X_t(\omega) = 0$ for $t < 0$ and any $\omega \in \Omega$.

The *semivariation* process of X relative to (F,G) is denoted by $\tilde{X}$ or $\tilde{X}_{F,G}$ and is defined by

$$\tilde{X}_t(\omega) = svar_{F,G}(X.(\omega), (-\infty, t]), \text{ for } t \in \mathbb{R} \text{ and } \omega \in \Omega.$$

It follows that

$$\tilde{X}_t(\omega) = 0, \text{ if } t < 0,$$
$$\tilde{X}_0(\omega) = |X_0(\omega)|,$$

and
$$\tilde{X}_t(\omega) = |X_0(\omega)| + svar_{F,G}(X.(\omega), [0,t]), \text{ if } t > 0.$$

We define $\tilde{X}_\infty : \Omega \to \overline{R}_+$ by
$$\tilde{X}_\infty(\omega) = \sup_{t \geq 0} \tilde{X}_t(\omega) = svar_{F,G}(X.(\omega), \mathbb{R}), \text{ for } \omega \in \Omega,$$

and we call it the *total semivariation* of X.

3. Definition. *We say that the process X has finite semivariation relative to (F,G), if for every $\omega \in \Omega$, the path $t \mapsto X_t(\omega)$ has finite semivariation relative to (F,G) on each interval $(-\infty, t]$.*

We say that X has p-integrable semivariation $\tilde{X}_{F,G}$ if the total semivariation $(\tilde{X}_{F,G})_\infty$ belongs to L^p.

The semivariation $\tilde{X}$ can be computed in terms of the variation of the processes X_z:
$$\tilde{X}_t(\omega) = \sup_{z \in G_1^*} |X_z|_t(\omega).$$

If X has finite semivariation $\tilde{X}$, then each X_z has finite variation $|X_z|$.

4. Let X be a right continuous process with finite semivariation relative to $(\mathbb{R}, E)$ and to (F, G). If $E \subset L(F, G)$ isometrically, it is enough to assume that the semivariation $\tilde{m}_{F,G}$ is finite (Proposition 20.6).

Throughout this paragraph we shall assume $c_0 \not\subset E$.

For each $\omega \in \Omega$, the path $X.(\omega)$ is right continuous and has finite semivariation relative to $(\mathbb{R}, E)$ and (F, G).

By Theorem 20.13 there is a σ-additive measure $m_{X(\omega)} : \mathcal{D} \to E \subset L(F, G)$ with finite semivariation $\tilde{m}_{X(\omega)} = (\tilde{m}_{X(\omega)})_{F,G}$ relative to (F, G), defined on the δ-ring $\mathcal{D}$ of bounded Borel sets, satisfying
$$m_{X(\omega)}(s, t] = X_t(\omega) - X_s(\omega), \text{ for } s < t \text{ in } \mathbb{R}.$$

We have
$$m_{X(\omega)}(s, 0] = X_0(\omega), \text{ if } s < 0,$$
and
$$m_{X(\omega)}[0, t] = X_t(\omega), \text{ if } t \geq 0.$$

For any measurable set $M \in \mathcal{M}$ and every $\omega \in \Omega$, we denote by $M(\omega)$ the section of M along ω. Then
$$\tilde{m}_{X(\omega)}(M(\omega)) = svar(X.(\omega), M(\omega)),$$
in case $M(\omega)$ is an interval; in particular,
$$\tilde{m}_{X(\omega)}(-\infty, t] = \tilde{X}_t(\omega).$$

If H is an F-valued, measurable process and $\omega \in \Omega$, then the path $H(\omega)$ is Borel measurable. We denote

$$\int H_s(\omega)dX_s(\omega) := \int H_s(\omega)dm_{X(\omega)},$$

the Stieltjes integral of $H.$ with respect to the function $X.$ with finite semivariation, in case the integral is defined. We denote also

$$\tilde{X}_{F,G}(H(\omega)) = \tilde{m}_{X(\omega)}(H(\omega)) = \sup_{z \in G_1^*} \int |H_s(\omega)|d|X_z|_s(\omega) \leq +\infty.$$

For processes X with integrable variation $|X|$, the integrals $\int |H_s|d|X|_s$ and $E(\int |H_s|d|X|_s)$ played an important role.

For processes X with integrable semivariation $\tilde{X}_{F,G}$, this role will be played by

$$E(\tilde{X}_{F,G}(H(\omega))) = E(\tilde{m}_{X(\omega)}(H(\omega))).$$

The following proposition states the measurability of the semivariation.

5. Proposition. *Let $X : \mathbb{R}_+ \times \Omega \to E \subset L(F,G)$ be a right continuous process with finite semivariation $\tilde{X}_{F,G}$ relative to (F,G).*
a) *For every $z \in G^*$ the process $X_z : \mathbb{R}_+ \times \Omega \to F^*$ is right continuous and has finite variation $|X_z|$.*
b) *If X is measurable (resp. optional, predictable), then so are X_z and $|X_z|$ for each $z \in G^*$.*
c) *Assume that either F is separable, or G is separable, or, more generally, there is a separable subspace $Z \subset G^*$ norming for G.*
  If X is measurable (resp. optional, predictable), then so is $\tilde{X}_{F,G}$.

Proof. Assume $\tilde{X}_{F,G}$ is finite. Let $z \in G^*$ and consider the process $X_z : \mathbb{R}_+ \times \Omega \to F^*$; for $x \in F$, $t \geq 0$ and $\omega \in \Omega$ we have

$$\langle x, (X_z)_t(\omega) \rangle = \langle X_t(\omega)x, z \rangle,$$

hence

$$|(X_z)_t(\omega)| \leq |X_t(\omega)|\,|z|.$$

For $s, t \in \mathbb{R}_+$ we have then

$$|(X_z)_s(\omega) - (X_z)_t(\omega)| = |(X_s - X_t)_z(\omega)|$$
$$\leq |X_s(\omega) - X_t(\omega)|\,|z|.$$

Since X is right continuous, it follows that X_z is also right continuous. From $|X_z| \leq \tilde{X}_{F,G}$ we deduce that X_z has finite variation $|X_z|$. Moreover, $|X_z|$ is also right continuous and assertion a) is proved. To prove assertion b), assume X is measurable and let (X^n) be a sequence of E-valued, $\mathcal{M}$-step processes such that $X^n \to X$ pointwise.

Let $z \in G^*$. Then each X_z^n is an $\mathcal{M}$-step process. In fact, if

$$X^n = \sum \varphi_{A_i} u_i,$$

with $A_i \in \mathcal{M}$ and $u_i \in E \subset L(F, G)$, then

$$X_z^n = \sum \varphi_{A_i} u_i^* z,$$

where $u_i^* : G^* \to F^*$ is the adjoint of $u_i : F \to G$. From the inequality

$$|(X_z^n)_t(\omega) - (X_z)_t(\omega)| \leq |X_t^n(\omega) - X_t(\omega)| \, |z|$$

we deduce that $X_z^n \to X_z$ pointwise, hence X_z is measurable. Since X_z is also right continuous, from Proposition 19.2 we deduce that its variation $|X_z|$ is measurable and this proves assertion b) concerning measurability.

The optional and the predictable parts are proved similarly, replacing $\mathcal{M}$ with $\mathcal{O}$ and $\mathcal{P}$ respectively.

To prove assertion c) assume there is a separable subspace $Z \subset G^*$ norming for G and X is measurable (resp. optional, predictable). We can choose a countable dense set $S_0 \subset Z$. Then

$$\tilde{X} = \sup\{|X_z| : z \in S_0, \, |z| \leq 1\}.$$

It follows that $\tilde{X}$ is measurable (resp. optional, predictable).

If G is itself separable, then there is a separable subspace $Z \subset G^*$ norming for G and by the above, assertion c) is true in this case.

Finally, if F is separable, then the set

$$\{X_t(\omega)x : (t, \omega) \in \mathbb{R}_+ \times \Omega, \, x \in F\}$$

is a separable subspace of G since the range of X is separable and we can assume G itself separable. ∎

The following theorem asserts the measurability of $\tilde{X}(H(\omega))$ and of the Stieltjes integral $\int H_s(\omega) dX_s(\omega)$.

6. Theorem. *Assume $c_0 \not\subset E$ and $c_0 \not\subset G$. Let $X : \mathbb{R}_+ \times \Omega \to E \subset L(F, G)$ be a right continuous, measurable process with finite semivariations $\tilde{X}_{\mathbb{R}, E}$ and $\tilde{X}_{F, G}$ and let $H : \mathbb{R}_+ \times \Omega \to F$ be a measurable process. Then:*

a) *The function $\omega \mapsto \tilde{X}_{F,G}(H(\omega)) \leq \infty$ is $\mathcal{F}$-measurable.*

a') *If F or G is separable and $\phi : \mathbb{R}_+ \times \Omega \to \mathbb{R}$ is measurable, then the function $\omega \mapsto \tilde{X}_{F,G}(\phi(\omega))$ is $\mathcal{F}$-measurable.*

b) *If $\tilde{X}_{F,G}(H(\omega)) < \infty$, a.s. on Ω, then the function $\omega \mapsto \int H_s(\omega) dX_s(\omega)$, defined a.s., is $\mathcal{F}$-measurable and has values in G.*

c) *If $E(\tilde{X}_{F,G}(H(\omega))) < \infty$, then the function $\omega \mapsto \int H_s(\omega) dX_s(\omega)$, defined a.s., is integrable and has values in G.*

d) If $E(\tilde{X}_{F,G}(\omega)) < \infty$ and H is bounded, then $\tilde{X}_{F,G}(H(\omega)) < \infty$, a.s. and the function $\omega \mapsto \int H_s(\omega) dX_s(\omega)$, defined a.s., is integrable and has values in G.

Proof. For each $\omega \in \Omega$, consider the measure $m_{X(\omega)}(a,b] = X_b(\omega) - X_a(\omega)$ for $a < b$. Since $c_0 \not\subset E$ and since $X.(\omega)$ is right continuous and has finite semivariation $\tilde{X}_{\mathbb{R},E}(\omega)$, we can extend $m_{X(\omega)}$ to a σ-additive measure $m_{X(\omega)} : \mathcal{D} \to E$ with finite semivariation $\tilde{m}_{X(\omega)}$ relative to $(\mathbb{R}, E)$ (Theorem 20.13).

Since X is measurable, its range is contained in a separable subspace $E_0 \subset E$ and we have $\tilde{X}_{\mathbb{R},E} = \tilde{X}_{\mathbb{R},E_0}$; therefore we can assume E separable. Since H is measurable, it has a separable range in F. Then the mapping $(t,\omega) \mapsto X_t(\omega) H_t(\omega)$ has separable range in G; therefore, we can assume G separable.

Let $Z \subset G^*$ be a separable space norming for G and let $Z_0 \subset Z$ be a countable dense set. For $z \in Z$, the function $X_z : \mathbb{R}_+ \times \Omega \to F^*$ has finite variation and is right continuous. By Theorem 19.3 a) the mapping $\omega \mapsto \int |H_s(\omega)| d|X_z|_s(\omega) \leq \infty$ is $\mathcal{F}$-measurable. From the equality

$$\tilde{X}_{F,G}(H(\omega)) = \sup_{\substack{|z|<1 \\ z \in Z_0}} \int |H_s(\omega)| d|X_z|_s(\omega)$$

we deduce that the mapping $\omega \mapsto \tilde{X}_{F,G}(H(\omega))$ is $\mathcal{F}$-measurable and assertion a) is proved.

Assertion a') is proved in the same way.

To prove assertion b), assume $\tilde{X}_{F,G}(H(\omega)) < \infty$ a.s. on Ω. Then the integral $\int H_s(\omega) dX_s(\omega)$ is defined a.s. and belongs to G, since $c_0 \not\subset G$ (Theorem 20.16 b). For each $z \in G^*$ we have $\int |H_s(\omega)| d|X_z|_s(\omega) < \infty$, a.s., hence, by Theorem 19.3 b), the function $\omega \mapsto \int H_s(\omega) d(X_z)_s(\omega)$, defined a.s., is $\mathcal{F}$-measurable. This means that the function $\omega \mapsto \int H_s(\omega) dX_s(\omega)$ is weakly $\mathcal{F}$-measurable. Since by the above, we can assume that G is separable, it follows that this function is $\mathcal{F}$-measurable and assertion b) is proved.

To prove assertion c), assume $E(\tilde{X}_{F,G}(H(\omega))) < \infty$. Then $\tilde{X}_{F,G}(H(\omega)) < \infty$ a.s. By assertion b), the function $\omega \mapsto \int H_s(\omega) dX_s(\omega)$, defined a.s., is $\mathcal{F}$-measurable. From the inequality

$$E(|\int H_s(\omega) dX_s(\omega)|) \leq E(\tilde{X}_{F,G}(H(\omega))) < \infty$$

we deduce that the function $\omega \to \int H_s(\omega) dX_s(\omega)$, defined a.s., is integrable and this is assertion c).

If $E((\tilde{X}_{F,G})_\infty(\omega)) < \infty$ and H bounded, then $E(\tilde{X}_{F,G}(H(\omega))) < \infty$. Then assertion d) follows from assertion c). ∎

The characterization in Theorem 19.4, of optional and predictable processes with integrable variation, can be extended for processes with integrable semivariation.

7. Theorem. *Assume $c_0 \not\subset E$. Let $X: \mathbb{R}_+ \times \Omega \to E$ be a right continuous, adapted process with integrable semivariation $\tilde{X}_{\mathbb{R},E}$. Then X is optional (resp. predictable) iff, for every real-valued, bounded, measurable process ϕ we have*

$$E(\int \phi_s dX_s) = E(\int {}^o\phi dX_s)(resp.\ E(\int {}^p\phi_s dX_s)).$$

Proof. Since ϕ is bounded and measurable, by Theorem 6 d), the integrals in the statement are defined. Let $z \in E^*$ and consider the right continuous, measurable process X_z with finite variation $|X_z|$ (Proposition 5). From the inequality $|X_z| \leq \tilde{X}_{\mathbb{R},E}$ we deduce that X_z has integrable variation. Since X is right continuous and adapted, it is measurable and separably valued; hence X is optional iff, for every $z \in E^*$, the process X_z is optional, iff (Theorem 19.4) for every real-valued, bounded, measurable process ϕ we have

$$E(\int \phi_s dX_z) = E(\int {}^o\phi_s dX_z),\ \text{for every } z \in G^*,$$

i.e.,

$$\langle E(\int \phi_s dX_s), z \rangle = \langle E(\int {}^o\phi dX_s) z \rangle,\ \text{for } z \in G^*,$$

i.e.,

$$E(\int \phi_s dX_s) = E({}^o\phi_s dX_s).$$

The part concerning predictable processes is proved similarly. ∎

B. The measure μ_X

We shall associate to a right continuous process X with integrable semivariation a measure μ_X on $\mathcal{M}$. For a process X with integrable variation, see Theorem 19.8.

We have to distinguish between the measure μ_X associated to X and the measure $m_{X(\omega)}$ associated to the function $X.(\omega)$, for $\omega \in \Omega$, by Theorem 20.13.

8. Theorem. *Assume $c_0 \not\subset E$. Let $X: \mathbb{R}_+ \times \Omega \to E$ be a right continuous, measurable process with integrable semivariation $\tilde{X}_{\mathbb{R},E}$. Then there is a σ-additive stochastic measure $\mu_X: \mathcal{M} \to E$ satisfying the following conditions:*
a) $\mu_X(M) = E(\int 1_M dX_s)$, *for $M \in \mathcal{M}$;*
b) *If $\phi: \mathbb{R}_+ \times \Omega \to \mathbb{R}$ is a measurable process with*

$$E(\tilde{X}_{\mathbb{R},E}(|\phi|)) = E((\tilde{m}_{X(\omega)})_{\mathbb{R},E}(\phi(\omega))) < \infty,$$

then $\phi \in \mathcal{F}_{\mathbb{R},E}(\mu_X)$. In this case the integrals $\int \phi d\mu_X$ and $E(\int \phi_s dX_s)$ are defined and we have

$$\int \phi d\mu_X = E(\int \phi_s dX_s) \in E$$

and
$$(\tilde{\mu}_X)_{\mathbb{R},E}(\phi) \le E(\tilde{X}_{\mathbb{R},E}(\phi(\omega))) = E((\tilde{m}_{X(\omega)})_{\mathbb{R},E}(\phi(\omega))).$$

c) *The measure μ_X is optional (resp. predictable) iff X is optional (resp. predictable).*

d) *Assume, in addition, that X has integrable semivariation $\tilde{X}_{F,G}$ and $c_0 \not\subset G$. Then μ_X has finite semivariation relative to (F,G). If $H : \mathbb{R}_+ \times \Omega \to F$ is a measurable process with*
$$E(\tilde{X}_{F,G}(|H|)) = E((\tilde{m}_{X(\omega)})_{F,G}(H(\omega))) < \infty,$$
then $H \in \mathcal{F}_{F,G}(\mu_X)$. In this case the integrals $\int H d\mu_X$ and $E(\int H_s dX_s)$ are defined and we have
$$\int H d\mu_X = E(\int H_s dX_s) \in G$$
and
$$(\tilde{\mu}_X)_{F,G}(H) \le E((\tilde{X}_{F,G})(H(\omega))) = E((\tilde{m}_{X(\omega)})_{F,G}(H(\omega))).$$

Proof. By Theorem 6 d) with $F = \mathbb{R}$ and $G = E$, for every set $M \in \mathcal{M}$, the integral $E(\int 1_M dX_s)$ is defined and belongs to E. Set
$$\mu_X(M) = E(\int 1_M dX_s), \text{ for } M \in \mathcal{M}.$$

It is clear that μ_X is additive on $\mathcal{M}$. To prove that μ_X is σ-additive let (M_n) be a sequence from $\mathcal{M}$ with $M_n \downarrow \phi$ and let $z \in E^*$. Then
$$\langle \mu_X(M_n), z \rangle = E(\langle \int 1_{M_n} dX_s, z \rangle) = E(\int 1_{M_n} d(X_z)_s).$$

Since X_z has finite variation, by the Monotone Convergence Theorem, the last term converges to 0. It follows that μ_X is a weakly σ-additive stochastic measure. By the Pettis Theorem 3.19, μ_X is σ-additive.

We shall prove now assertion d); assertion b) will then follow, as a particular case of d), with $F = \mathbb{R}$ and $G = E$.

Assume therefore that X has integrable semivariation $\tilde{X}_{F,G}$ (in addition to having integrable semivariation $\tilde{X}_{\mathbb{R},E}$) and $c_0 \not\subset G$ and let $H : \mathbb{R}_+ \times \Omega \to F$ be a measurable process such that
$$E(\tilde{X}_{F,G}(H(\omega))) < \infty.$$

By the first part of the proof of Theorem 6, we can assume G separable. By Theorem 6 a), the function $\omega \mapsto \tilde{X}_{F,G}(H(\omega))$ is $\mathcal{F}$-measurable, hence $E(\tilde{X}_{F,G}(H(\omega)))$ is defined.

By Theorem 6 c), the integral $\int H_s(\omega) dX_s(\omega)$ is defined a.s. and the function $\omega \mapsto \int H_s(\omega) dX_s(\omega)$ is $\mathcal{F}$-measurable and integrable; therefore the integral $E(\int H_s dX_s)$ is defined and belongs to G.

Let $z \in G^*$. Then the process $X_z : \mathbb{R}_+ \times \Omega \to F^*$ is right continuous, measurable and has integrable variation $|X_z| \leq \tilde{X}_{F,G}$.

By Theorem 19.8, the corresponding stochastic measure μ_{X_z} has finite variation $|\mu_{X_z}| = \mu_{|X_z|}$ and for every $M \in \mathcal{M}$ we have

$$\mu_{X_z}(M) = E(\int 1_M dX_z)$$

and

$$|\mu_{X_z}|(M) = E(\int 1_M d|X_z|).$$

We have also

$$\mu_{X_z} = (\mu_X)_z \text{ and } |\mu_{X_z}| = |(\mu_X)_z|.$$

In fact, for every $M \in \mathcal{M}$ and $x \in F$, we have, by Proposition 20.18,

$$\langle x, \mu_{X_z}(M) \rangle = E(\langle x, \int 1_M dX_z \rangle)$$
$$= E(\langle (\int 1_M dX)x, z \rangle) = \langle E(\int 1_M x dX), z \rangle$$
$$= \langle x E(\int 1_M dX), z \rangle = \langle x \mu_X(M), z \rangle = \langle x, (\mu_X)_z(M) \rangle,$$

which proves that $\mu_{X_z} = (\mu_X)_z$. Then

$$|(\mu_X)_z|(M) = |\mu_{X_z}|(M) = \mu_{|X_z|}(M)$$
$$= E(\int 1_M d|X_z|) = E(m_{|X_z(\omega)|}(M(\omega)))$$
$$= E(|(m_{X(\omega)})_z|(M(\omega)) \leq E((\tilde{m}_{X(\omega)})_{F,G}(M(\omega)))$$
$$\leq E((\tilde{m}_X)_{F,G}(\mathbb{R}_+ \times \Omega)) = E((\tilde{X}_{F,G})_\infty) < \infty,$$

therefore

$$(\tilde{\mu}_X)_{F,G}(M) = \sup_{|z| \leq 1} |(\mu_X)_z|(M) \leq E((\tilde{X}_{F,G})_\infty) < \infty,$$

hence μ_X has finite semivariation relative to (F, G).

By Theorem 19.8 we have also

$$\int |H| d|\mu_{X_z}| = E(\int |H| d|m_{X_z(\omega)}|)$$
$$\leq E((\tilde{m}_{X(\omega)})_{F,G}(H(\omega))) = E((\tilde{X}_{F,G})(H(\omega))) < \infty,$$

hence

$$(\tilde{\mu}_X)_{F,G}(H) = \sup_{|z| \leq 1} \int |H| d|\mu_{X_z}| \leq E((\tilde{X}_{F,G})(H(\omega))) < \infty,$$

§21. PROCESSES WITH FINITE SEMIVARIATION   261

therefore $H \in \mathcal{F}_{F,G}(\mu_X)$ and the integral $\int H d\mu_X$ is defined. This proves the inequality in assertion d). For every $z^* \in G^*$ we have $H \in L^1_F(\mu_{X_z})$. By Theorem 19.8, it follows that $E(\int |H| d|X_z|) < \infty$, hence $E(\int H dX_z)$ is defined and we have

$$\int H d\mu_{X_z} = E(\int H_s(\omega) d(X_z)_s(\omega)), \text{ for } z \in G^*,$$

therefore

$$\langle \int H d\mu_X, z \rangle = E(\langle \int H_s(\omega) dX_s(\omega), z \rangle)$$
$$= \langle E(\int H_s(\omega) dX_s(\omega)), z \rangle,$$

consequently

$$\int H d\mu_X = E(\int H_s(\omega) dX_s(\omega)),$$

and this proves the equality in assertion d).

Taking $F = \mathbb{R}$ and $H = \phi$, assertion b) follows. Assertion c) follows from Theorem 7 and Proposition 19.6: for every real-valued, bounded, measurable process ϕ we have

$$\int \phi d\mu_X = E(\int \phi_s dX_s) \text{ and } \int {}^o\phi d\mu_X = E(\int {}^o\phi_s dX_s).$$

We have

$$\int \phi d\mu_X = \int {}^o\phi d\mu_X$$

iff

$$E(\int \phi_s dX_s) = E(\int {}^o\phi_s dX_s),$$

that is, μ_X is optional iff X is optional. The proof for the predictable case is similar. ∎

Corollary 19.9 extends, with the same proof, for processes with integrable semivariation.

9. Corollary. *Assume $c_0 \not\subset E$. Let $A, B : \mathbb{R}_+ \times \Omega \to E$ be two adapted, right continuous processes with integrable semivariation relative to $(\mathbb{R}, E)$.*
a) *If for every stopping time T we have*

$$E(A_\infty - A_{T-} | \mathcal{F}_T) = E(B_\infty - B_{T-} | \mathcal{F}_T), \text{ a.s.}$$

(and $A_{0-} = B_{0-}$), then A and B are indistinguishable.
b) *If A and B are predictable, if $A_0 = B_0$ and*

$$E(A_\infty - A_t | \mathcal{F}_t) = E(B_\infty - B_t | \mathcal{F}_t), \text{ a.s. for every } t \geq 0,$$

then A and B are indistinguishable.

Corollary 19.10 also extends, with the same proof, for processes with integrable semivariation.

10. Corollary. *Assume $c_0 \not\subset E$. Let $A, B : \mathbb{R}_+ \to E$ be two adapted, right continuous processes with integrable semivariation relative to $(\mathbb{R}, E)$.*
a') *If for every stopping time T we have*
$$E(A_\infty - A_{T-}) = E(B_\infty - B_{T-}),$$
then A and B are indistinguishable.
b') *If A and B are predictable and if for every stopping time T we have*
$$E(A_\infty - A_T) = E(B_\infty - B_T),$$
then A and B are indistinguishable.

Finally, Corollary 19.11 extends for martingales with integrable semivariation.

11. Corollary. *Assume $c_0 \not\subset E$. If $A_t = E(A_\infty | \mathcal{F}_t)$ is a uniformly integrable, E-valued, cadlag, predictable martingale with $A_0 = 0$ and with integrable semivariation relative to $(\mathbb{R}, E)$, then A is evanescent.*

Proof. We take $B_t = 0$ for every $t \geq 0$ and we have $A_0 = 0 = B_0$ and
$$E(A_\infty - A_t | \mathcal{F}_t) = 0 = E(B_\infty - B_t | \mathcal{F}_t), \text{ for } t \geq 0.$$
By Corollary 9, A and B are indistinguishable. ∎

C. Summability of processes with integrable semivariation

This section contains the main results of this paragraph.

The following theorem asserts the summability of processes with integrable semivariation and the relationship between the measures μ_X and I_X. At the same time it states that the stochastic integral can be computed pathwise, as a Stieltjes integral (itself an integral with respect to a function with finite semivariation).

12. Theorem. *Assume $c_0 \not\subset E$ and $c_0 \not\subset G$. Let $X : \mathbb{R}_+ \times \Omega \to E \subset L(F, G)$ be a cadlag, adapted process with p-integrable semivariation relative to $(\mathbb{R}, E)$ and relative to (F, G). Then:*
a) *X is p-summable relative to (F, G)*
and
b) *$I_X(M)(\omega) = \int 1_M(s, \omega) dX_s(\omega) = m_{X(\omega)}(M(\omega))$, a.s., for $M \in \mathcal{P}$.*
c) *For every $M \in \mathcal{P}$ we have*
$$\mu_X(M) = E(I_X(M)).$$

d) For every $M \in \mathcal{P}$ we have
$$\tilde{\mu}_X(M) \leq (\tilde{I}_X)_{F, L_G^p}(M) \leq \|\tilde{X}_{F,G}(M(\omega))\|_p \leq \|(\tilde{X}_{F,G})_\infty\|_p.$$

Let $H : \mathbb{R}_+ \times \Omega \to F$ be a predictable process such that the function $\omega \mapsto \tilde{X}_{F,G}(H(\omega))$ belongs to L^p. Then:

a') $H \in \mathcal{F}_{F,G}(X) \cap \mathcal{F}_{F,G}(\mu_X)$;
b') $\int H dI_X \in L_G^p$, $\int H_s(\omega) dX_s(\omega) \in G$, a.s.

and
$$\left(\int H dI_X\right)(\omega) = \int H_s(\omega) dX_s(\omega), \quad a.s.$$

b'') We have $H \in L_{F,G}^1(X)$ iff the process $\left(\int_{[0,t]} H_s(\omega) dX_s(\omega)\right)_{t \geq 0}$ has a cadlag modification. In this case we have

$$(H \cdot X)_t(\omega) = \int_{[0,t]} H_s(\omega) dX_s(\omega), \quad a.s., \text{ for each } t \geq 0.$$

c') $\int H d\mu_X \in G$ and
$$\int H d\mu_X = E\left(\int H dI_X\right).$$

d') $(\tilde{\mu}_X)_{F,G}(H) \leq (\tilde{I}_X)_{F,L_G^p}(H) \leq \|\tilde{X}_{F,G}(H(\omega))\|_p$.

Proof. We shall divide the proof into several steps.
1) For every set $B \in \mathcal{R}$ we have
$$I_X(B)(\omega) = \int 1_B(s, \omega) dX_s(\omega) = m_{X(\omega)}(B(\omega)).$$

This follows from the definition of I_X, first for predictable rectangles: for $B = \{0\} \times A$ with $A \in \mathcal{F}_0$ we have
$$I_X(\{0\} \times A)(\omega) = 1_A(\omega) X_0(\omega) = \int 1_{\{0\} \times A}(s, \omega) dX_s(\omega)$$

and for $B = (s, t] \times A$ with $A \in \mathcal{F}_s$ we have
$$I_X((s, t] \times A)(\omega) = 1_A(\omega)(X_t(\omega) - X_s(\omega)) = \int 1_{(s,t] \times A}(s, \omega) dX_s(\omega).$$

Then, by additivity, the equality is valid for every $B \in \mathcal{R}$.

2) To prove that I_X has bounded semivariation relative to (F, L_G^p), let $B \in \mathcal{R}$, $(B_i)_{i \in I}$ a finite family of disjoint sets from $\mathcal{R}$ contained in B, and $(x_i)_{i \in I}$ a family of elements from F_1. Then

$$\|\sum I_X(B_i) x_i\|_p^p = E(|\sum I_X(B_i) x_i|^p)$$
$$= E(|\sum m_{X(\omega)}(B_i(\omega)) x_i|^p) \leq E(|(\tilde{m}_{X(\omega)})_{F,G}(B(\omega))|^p)$$
$$= \|(\tilde{m}_{X(\omega)})_{F,G}(B(\omega))\|_p^p = \|\tilde{X}_{F,G}(B(\omega))\|_p^p.$$

It follows that

$$(\tilde{I}_X)_{F,L_G^p}(B) \leq \|\tilde{X}_{F,G}(B(\omega))\|_{L^p} \leq \|(\tilde{X}_{F,G})_\infty\|_p.$$

3) Since X has p-integrable semivariation relative to (F,G), we have $(\tilde{X}_{F,G})_\infty \in L^p$, hence $(\tilde{X}_{F,G})_\infty(\omega) < \infty$, a.s. We redefine $X_t(\omega) = 0$ if $(\tilde{X}_{F,G})_\infty(\omega) = \infty$. Then, for every $\omega \in \Omega$, the path $X.(\omega)$ is right continuous and has bounded semivariation relative to (F,G). By Theorem 20.13, the measure $m_{X(\omega)}$ can be extended to a σ-additive measure $m_{X(\omega)} : \mathcal{B}(\mathbb{R}_+) \to E$ with bounded semivariation relative to (F,G). In particular, for every set $M \in \mathcal{P}$, we have $M(\omega) \in \mathcal{B}(\mathbb{R}_+)$, hence $m_{X(\omega)}(M(\omega))$ is defined.

4) We shall prove now that the function $\omega \mapsto m_{X(\omega)}(M(\omega))$ is measurable and belongs to L_E^p.

Denote by $\mathcal{P}_0$ the class of sets $M \in \mathcal{P}$ for which $\omega \mapsto m_{X(\omega)}(M(\omega))$ belongs to L_E^p. From step 1) we deduce that $\mathcal{P}_0$ contains $\mathcal{R}$. Let now (M_n) be a monotone sequence from $\mathcal{P}_0$ with limit M. For each $\omega \in \Omega$, the sequence $(M_n(\omega))$ is monotone in $\mathcal{B}(\mathbb{R}_+)$ and has limit $M(\omega)$. By hypothesis, for each n, the function $\omega \mapsto m_{X(\omega)}(M_n(\omega))$ belongs to L_E^p. We have also

$$|m_{X(\omega)}(M_n(\omega))| \leq \tilde{m}_{X(\omega)}(\mathbb{R}_+ \times \Omega) = \tilde{X}_\infty(\omega)$$

and $\tilde{X}_\infty \in L^p$ by hypothesis. By Lebesgue's Theorem we deduce that $m_{X(\cdot)}(M(\cdot)) \in L_E^p$, hence $M \in \mathcal{P}_0$ and, moreover,

$$m_{X(\cdot)}(M_n(\cdot)) \to m_{X(\cdot)}(M_n(\cdot)), \text{ in } L_E^p.$$

It follows that $\mathcal{P}_0 = \mathcal{P}$. It follows also that the mapping $M \mapsto m_{X(\cdot)}(M(\cdot))$ from $\mathcal{P}$ into L_E^p is σ-additive.

5) As we have seen in step 1), we have

$$I_X(B)(\cdot) = m_{X(\cdot)}(B(\cdot)), \text{ for } B \in \mathcal{R}.$$

We use this equality to extend I_X for every $M \in \mathcal{P}$ by

$$I_X(M)(\cdot) = m_{X(\cdot)}(M(\cdot)) \in L_E^p, \text{ for } M \in \mathcal{P}.$$

It is clear that I_X is additive on $\mathcal{P}$. If $M_n \downarrow \phi$ in $\mathcal{P}$, from the above proof we deduce that

$$I_X(M_n) - m_{X(\cdot)}(M_n(\cdot)) \to 0, \text{ in } L_E^p.$$

It follows that $I_X : \mathcal{P} \to L_E^p$ is σ-additive.

Since I_X has bounded semivariation on $\mathcal{R}$ relative to (F, L_G^p), by Corollary 7.6, I_X has bounded semivariation on $\mathcal{P}$ relative to (F, L_G^p).

We deduce that X is p-summable relative to (F,G) and this proves assertions a) and b).

6) The equality in assertion c) follows from the equality of assertion b), by taking expectations.

§21. PROCESSES WITH FINITE SEMIVARIATION

The inequalities of assertion d)

$$(\tilde{I}_X)_{F,L_G^p}(M) \leq \|\tilde{X}_{F,G}(M(\omega))\|_p \leq \|(\tilde{X}_{F,G})_\infty\|_p$$

were already proved in step 2) for sets $M \in \mathcal{R}$. For sets $M \in \mathcal{P}$ the proof is the same.

To prove the first inequality in assertion d), let $M \in \mathcal{P}$, let $(M_i)_{i \in I}$ be a finite family of disjoint sets from $\mathcal{P}$ contained in M, and $(x_i)_{i \in I}$ a family of elements from F_1. Then

$$|\sum_i \mu_X(M_i)x_i| = |\sum_i E(I_X(M_i)x_i|$$
$$\leq E(|\sum_i I_X(M_i)x_i|) \leq \|\sum_i I_X(M_i)x_i\|_p \leq (\tilde{I}_X)_{F,L_G^p}(M)$$

and the inequality

$$\tilde{\mu}_X(M) \leq (\tilde{I}_X)_{F,L_G^p}(M)$$

follows. This proves assertion d).

Let now $H : \mathbb{R}_+ \times \Omega \to \mathcal{F}$ be a predictable process such that the function $\tilde{X}_{F,G}(H(\omega))$ belongs to L^p and prove assertions b'), c') and d'). Assertion a') will follow from asserton d').

7) Let $H' : \mathbb{R}_+ \times \Omega \to F$ be a predictable simple process with $|H'| \leq |H|$, of the form

$$H' = \sum_i 1_{M_i} x_i, \text{ with } M_i \in \mathcal{P} \text{ and } x_i \in F.$$

From assertion b) we deduce that

$$(\int H' dI_X)(\omega) = \int H'_s(\omega) dX_s(\omega) = \int H'_s(\omega) dm_{X(\omega)}$$

and from assertion c) it follows that

$$\int H' d\mu_X = E(\int H' dI_X).$$

Then

$$|\int H' d\mu_X| = |E(\int H' dI_X)| \leq E(|\int H' dI_X|) = \|\int H' dI_X\|_1$$
$$\leq E(|\int H'_s(\omega) dm_{X(\omega)}|) \leq E(\tilde{m}_{X(\omega)}(H(\omega)))$$
$$= \|\tilde{X}_{F,G}(H(\omega))\|_1 \leq \|\tilde{X}_{F,G}(H(\omega))\|_p < \infty.$$

Taking the supremum for $|H'| \leq |H|$ we obtain assertion d'). Then assertion a') follows.

8) Since $c_0 \not\subset G$, we have $\int H d\mu_X \in G$ (Theorem 5.48 b). We have also $c_0 \not\subset L_G^p$ (Theorem 1.40); therefore $\int H dI_X \in L_G^p$ (Theorem 5.48 b). To prove

the equality in assertion b'), let (H^n) be a sequence of simple, predictable, F-valued processes such that $H^n \to H$ and $|H^n| \leq |H|$. Let $\omega \in \Omega$ and $z \in L^q_{G^*}$ with $\frac{1}{p} + \frac{1}{q} = 1$ and denote $y = z(\omega)$. Then $y \in G^*$. We have $H^n_s(\omega) \to H_s(\omega)$ and $|H^n_s(\omega)| \leq |H_s(\omega)|$ for each $s \geq 0$.

From $\|\tilde{X}_{F,G}(H(\omega))\|_p < \infty$ we deduce that

$$\tilde{m}_{X(\omega)}(H(\omega)) = \tilde{X}_{F,G}(H(\omega)) < \infty, \text{ a.s.}$$

If we modify H on an evanescent set, we can assume $\tilde{m}_{X(\omega)}(H(\omega)) < \infty$ for every $\omega \in \Omega$.

Then $H(\omega) \in \mathcal{F}_{F,G}(m_{X(\omega)}) \subset L^1_F((m_{X(\omega)})_y)$. We can apply Lebesgue's Theorem for the sequence $(H^n_s(\omega))$ in the space $L^1_F((m_{X(\omega)})_y)$ and deduce that

$$H(\omega) = \lim H^n(\omega), \text{ in } L^1_F((m_{X(\omega)})_y),$$

therefore

$$\int H(\omega)d(m_{X(\omega)})_y = \lim \int H^n(\omega)d(m_{X(\omega)})_y,$$

or

$$\langle \int H_s(\omega)dX_s(\omega), z(\omega) \rangle = \lim \langle \int H^n_s(\omega), z(\omega) \rangle.$$

9) Since

$$|\langle \int H_s(\omega)dX_s(\omega), z(\omega) \rangle| \leq \tilde{X}_{F,G}(H(\omega))|z(\omega)|$$

and the function $\omega \mapsto \tilde{X}_{F,G}(H(\omega))$ belongs to L^p, we can apply again Lebesgue's Theorem in the space L^1 and deduce that

$$\langle \int H_s(\cdot)dX_s(\cdot), z(\cdot) \rangle = \lim \langle \int H^n_s(\cdot)dX_s(\cdot), z(\cdot) \rangle, \text{ in } L^1.$$

10) At the same time, since $H \in \mathcal{F}_{F,G}(X) \subset L^1_F((I_X)_z)$, we can apply Lebesgue's Theorem in the space $L^1_F((I_X)_z)$ and deduce that

$$\int H d(I_X)_z = \lim \int H^n d(I_X)_z,$$

that is,

$$\langle \int H dI_X, z \rangle = \lim \langle \int H^n dI_X, z \rangle.$$

For each n we have

$$(\int H^n dI_X)(\omega) = \int H^n_s(\omega)dX_s(\omega),$$

hence

$$\langle (\int H^n dI_X)(\omega), z(\omega) \rangle = \langle \int H^n_s(\omega)dX_s(\omega), z(\omega) \rangle,$$

therefore, taking the expectations,

$$\langle \int H^n dI_X, z \rangle = \langle \int H^n_s(\cdot) dX_s(\cdot), z(\cdot) \rangle.$$

Taking the limit as $n \to \infty$ we obtain

$$\langle \int H dI_X, z \rangle = \langle \int H_s(\cdot) dX_s(\cdot), z(\cdot) \rangle.$$

Since $z \in L^q_{G^*}$ was arbitrary, it follows that

$$\int H dI_X = \int H_s(\cdot) dX_s(\cdot), \text{ in } L^p_G,$$

consequently

$$(\int H dI_X)(\omega) = \int H_s(\omega) dX_s(\omega), \text{ a.s.}$$

and this proves b').

11) Taking the expectations in the above equality we obtain assertion c'):

$$\int H d\mu_X = E(\int H_s(\cdot) dX_s(\cdot)) = E(\int H dI_X).$$

12) Replacing H with $1_{[0,t]} H$ we have

$$(\int_{[0,t]} H dI_X)(\omega) = \int_{[0,t]} H_s(\omega) dX_s(\omega), \text{ a.s.}$$

and assertion b'') follows. ∎

13. Corollary. *Assume $c_0 \not\subset E$ and $c_0 \not\subset G$. Let X be a cadlag, adapted, locally p-summable process relative to (F, G) and with finite semivariations $\tilde{X}_{\mathbb{R}, E}$ and $\tilde{X}_{F, G}$ and let H be an F-valued, predictable process.*

If $H \in L^1_{F,G}(X)_{loc}$ and if the function $\omega \mapsto \tilde{X}_{F,G}(H(\omega))$ belongs to L^p, then

$$(H \cdot X)_t(\omega) = \int_{[0,t]} H_s(\omega) dX_s(\omega), \text{ a.s., for each } t \geq 0.$$

D. The pathwise stochastic integral

In this section we shall assume that $c_0 \not\subset E$ and that $X : \mathbb{R}_+ \times \Omega \to E$ is a cadlag process with finite semivariations $\tilde{X}_{\mathbb{R}, E}$ and $\tilde{X}_{F, G}$. If the semivariation $(\tilde{X}_{F,G})_\infty$ is not integrable, then X is not necessarily summable and the stochastic integral $H \cdot X$ cannot be defined. But we can define a pathwise stochastic integral, in the same way we did in Section 19 F, for processes with finite variation, by using the Stieltjes integral.

14. Definition. Let $H : \mathbb{R}_+ \times \Omega \to F$.
a) We say that H is *pathwise integrable with respect to* X if for every $\omega \in \Omega$, the function $s \mapsto H_s(\omega)$ is Stieltjes integrable with respect to the function $s \mapsto X_s(\omega)$ with finite semivariation relative to (F, G).
b) We say H is *locally pathwise integrable with respect to* X if for every $\omega \in \Omega$ and $t \geq 0$, the function $s \mapsto H_s(\omega)$ is Stieltjes integrable with respect to the function $s \mapsto X_s(\omega)$ on the interval $[0, t]$.

15. Remark. If H is measurable, to say that H is locally pathwise integrable with respect to X means that for each $\omega \in \Omega$ and $t \geq 0$, we have

$$\tilde{X}_{F,G}(1_{[0,t]}H(\omega)) = (\tilde{m}_{X(\omega)})_{F,G}(1_{[0,t]}H(\omega)) < \infty,$$

that is, for each $z \in G^*$,

$$\int_{[0,t]} |H_s(\omega)| d|X_z|_s(\omega) = \int_{[0,t]} |H_s(\omega)| d|(m_{X(\omega)})_z| < \infty$$

(see Proposition 5.22).

16. Proposition. *Assume $c_0 \not\subset E$ and $c_0 \not\subset G$.*
If H is locally pathwise integrable with respect to X, then, for each $\omega \in \Omega$, the function $t \mapsto \int_{[0,t]} H_s(\omega) dX_s(\omega)$ is cadlag and has finite semivariation relative to $(\mathbb{R}, G)$, smaller than $\tilde{X}_{F,G}(1_{[0,t]}H(\omega))$.

Proof. We apply Proposition 5.48 b) with $\mathcal{D}$ the δ-ring of all bounded Borel subsets of $\mathbb{R}$ and deduce that for each $\omega \in \Omega$, we have $\int_B H_s(\omega) dX_s(\omega) \in G$ for each Borel set $B \subset \mathbb{R}_+$. Then, by Proposition 5.48 c), the mapping $B \mapsto \int_B H_s(\omega) dX_s(\omega)$ is σ-additive on $\mathcal{B}(\mathbb{R}_+)$. In particular, if $t_n \downarrow t$, then $[0, t_n] \downarrow [0, t]$, hence $\int_{[0,t_n]} H_s(\omega) dX_s(\omega) \to \int_{[0,t]} H_s(\omega) dX_s(\omega)$; if $t_n \uparrow t$ and $t_n \neq t$ then $[0, t_n] \uparrow [0, t)$, hence $\int_{[0,t_n]} H_s(\omega) dX_s(\omega) \to \int_{[0,t)} H_s(\omega) dX_s(\omega)$. It follows that the mapping $t \mapsto \int_{[0,t]} H_s(\omega) dX_s(\omega)$ is cadlag.

For each $\omega \in \Omega$, the measure μ corresponding to the function $t \mapsto \int_{[0,t]} H_s(\omega) dX_s(\omega)$ is determined by the equality

$$\mu[0,t] = \int_{[0,t]} H_s(\omega) dX_s(\omega) = \int_{[0,t]} H_s(\omega) dm_{X(\omega)}.$$

By Theorem 5.50 b) with $\varphi = 1_{[0,t]}$, we have

$$\tilde{\mu}_{\mathbb{R},G}(\varphi) \leq (\tilde{m}_{X(\omega)})_{F,G}(\varphi H(\omega)) = \tilde{X}_{F,G}(1_{[0,t]}H(\omega)) < \infty.$$

Since $\tilde{\mu}_{\mathbb{R},G}(\varphi)$ is equal to the semivariation relative to $(\mathbb{R}, G)$ of the function $t \mapsto \int_{[0,t]} H_s(\omega) dX_s(\omega)$ on the interval $[0, t]$, the assertion in Proposition 16 follows. ∎

Remark. Proposition 16 states that the process $(\int_{[0,t]} H_s(\omega) dX_s(\omega))_{t \geq 0}$ is cadlag and has finite semivariation relative to $(\mathbb{R}, G)$.

17. Proposition. *Assume $c_0 \not\subset E$ and $c_0 \not\subset G$. If X and H are adapted and if H is measurable and locally pathwise integrable with respect to X, then the process $(\int_{[0,t]} H_s(\omega) dX_s(\omega))_{t \geq 0}$ is adapted.*

Proof. We apply Theorem 6 b) to the processes $H' = 1_{[0,t]} H$ and $X' = 1_{[0,t]} X$. For fixed t, H' and X' are $\mathcal{B}(\mathbb{R}_+) \times \mathcal{F}_t$-measurable and H' is pathwise integrable with respect to X'. Then we deduce that the mapping $\omega \mapsto \int_{[0,t]} H_s(\omega) dX_s(\omega)$ is $\mathcal{F}_t$-measurable. ∎

Propositions 16 and 17 lead us to extend the definition of the stochastic integral for (not necessarily summable) processes X with finite semivariation.

18. Definition. *Assume $c_0 \not\subset E$ and $c_0 \not\subset G$. Let $X : \mathbb{R}_+ \times \Omega \to E \subset L(F,G)$ be a cadlag, adapted process with finite semivariations $\tilde{X}_{\mathbb{R},E}$ and $\tilde{X}_{F,G}$ and let $H : \mathbb{R}_+ \times \Omega \to F$ be a measurable, adapted (not necessarily predictable) process, locally pathwise integrable with respect to X. The pathwise stochastic integral of H with respect to X is the stochastic process denoted by $H \cdot X$ and defined by*

$$(H \cdot X)_t(\omega) = \int_{[0,t]} H_s(\omega) dX_s(\omega), \text{ for } (t,\omega) \in \mathbb{R}_t \times \Omega.$$

19. Proposition. *Assume $c_0 \not\subset E$ and $c_0 \not\subset G$ and let $X : \mathbb{R}_+ \times \Omega \to E \subset L(F,G)$ be a cadlag, adapted, locally p-summable process relative to (F,G) and with finite semivariations $\tilde{X}_{\mathbb{R},E}$ and $\tilde{X}_{F,G}$.*

If $H \in L^1_{F,G}(X)_{loc}$ and if H is locally pathwise integrable with respect to X, then the pathwise stochastic integral $H \cdot X$ of Definition 18 coincides with the stochastic integral of Definition 15.10.

We use Corollary 13.

In view of Proposition 19, the pathwise stochastic integral $H \cdot X$ will be also called the stochastic integral, if no confusion is possible.

In what follows, we assume that $c_0 \not\subset E$, $c_0 \not\subset G$ and that X is a cadlag, adapted process with *finite semivariations* $\tilde{X}_{\mathbb{R},E}$ and $\tilde{X}_{F,G}$ and $H : \mathbb{R}_+ \times \Omega \to F$ is a *measurable, adapted* process, *locally integrable* with respect to X.

20. Proposition. *The stochastic integral $H \cdot X$ is a cadlag, adapted process with finite semivariation relative to (F,G) satisfying*

$$(H \cdot X)^{\tilde{}}_{\mathbb{R},G}(1_{[0,t] \times \Omega}) \leq (\tilde{X}_{F,G})(1_{[0,t]} H(\omega)) < \infty.$$

We use Theorem 5.50.

21. Proposition. *If T is a stopping time, then X^T has finite semivariation relative to $(\mathbb{R}, E)$ and (F,G) and we have*

$$X^T = 1_{[0,T]} \cdot X \text{ and } X^{T-} = 1_{[0,T)} \cdot X.$$

Proof. For each $t \geq 0$ and $\omega \in \Omega$, we have

$$(1_{[0,T]} \cdot X)_t(\omega) = \int_{[0,t]} 1_{[0,T]}(s,\omega) dX_s(\omega)$$

$$= \int_{[0,t \wedge T(\omega)]} dX_s(\omega) = X_{t \wedge T(\omega)}(\omega) = X_t^T(\omega),$$

hence $1_{[0,T]} \cdot X = X^T$.

Similar proof for the second equality.

The semivariation of X^T relative to (F, G) satisfies

$$(X^T)\widetilde{_{F,G}}(\omega) = (\widetilde{X}_{F,G})_{T(\omega)}(\omega) < \infty.$$

■

22. Proposition. *Let T be a stopping time. Then H is locally integrable with respect to X^T (resp. X^{T-}) iff $1_{[0,T]}H$ (resp. $1_{[0,T)}H$) is locally integrable with respect to X. In this case we have*

$$H \cdot X^T = (1_{[0,T]}H) \cdot X = (H \cdot X)^T$$

(resp.

$$H \cdot X^{T-} = (1_{[0,T)}H) \cdot X = (H \cdot X)^{T-}).$$

Proof. H is locally integrable with respect to X^T iff H is locally integrable with respect to $(X^T)_z$ for each $z \in G^*$. Similarly, $1_{[0,T]}H$ is locally integrable with respect to X iff it is locally integrable with respect to X_z for each $z \in G^*$.

We have $(X^T)_z = (X_z)^T$ for $z \in G^*$. Since X_z has finite variation, by Proposition 19.24, H is locally integrable with respect to $(X_z)^T$ iff $1_{[0,T]}H$ is locally integrable with respect X_z. It follows that H is locally integrable with respect to X^T iff $1_{[0,T]}H$ is locally integrable with respect to X.

We use then Theorem 19.24 for X_z with $z \in G^*$ to deduce the equality in the statement of the property.

The proof for X^{T-} is done similarly. ■

23. Proposition. *If H is real-valued and K is an F-valued, measurable, adapted process, then K is locally integrable with respect to $H \cdot X$ iff KH is locally integrable with respect to X. In this case we have*

$$K \cdot (H \cdot X) = (KH) \cdot X.$$

Proof. Let $z \in G^*$. Then the process X_z has finite variation and, using Proposition 19.25, we deduce that K is locally integrable with respect to $H \cdot X_z$ iff KH is locally integrable with respect to X_z; in this case we have

$$K \cdot (H \cdot X_z) = (KH) \cdot X_z.$$

Now $H \cdot X_z = (H \cdot X)_z$, hence K is in locally integrable with respect to $H \cdot X_z$ iff it is locally integrable with respect to $(H \cdot X)_z$; in this case the above equality becomes
$$K \cdot (H \cdot X)_z = (KH) \cdot X_z.$$
Since $z \in G^*$ is arbitrary, we deduce that K is locally integrable with respect to $H \cdot X$ iff KH is locally integrable with respect to X, and in this case we have
$$K \cdot (H \cdot X) = (KH) \cdot X.$$

∎

24. Proposition. *If H is F-valued and K is a real-valued, measurable, adapted process such that KH is locally integrable with respect to X, then K is locally integrable with respect to $H \cdot X$ and we have*
$$K \cdot (H \cdot X) = (KH) \cdot X.$$

The proof is similar to that of Proposition 23, using Proposition 19.26.

25. Proposition. *We have*
$$\Delta(H \cdot X) = H \Delta X.$$

The proof is the same as that of Proposition 19.27.

26. Remark. We can extend Definition 19.28 of semilocal summability to processes Z of the form $Z = X + Y$, where X is locally p-summable and Y is cadlag, adapted and with finite semivariation; and we can extend the definition of the stochastic integral: $H \cdot X = H \cdot X + H \cdot Y$.

Theorem 19.30 also extends for this situation.

§22. DUAL PROJECTIONS

In Sections A and B we define and study the dual projections of measures and of processes with integrable *semivariation*. The dual projections for measures or of processes with integrable *variation* follow as a particular case.

The existence of dual projections is proved for processes with integrable *variation* (Theorems 8 and 13).

In Section E some examples of processes with locally integrable variation are given. Finally, in Section F, the decomposition theorem of a local martingale as a sum of a local martingale with jumps ≤ 1 and a local martingale with locally integrable variation is proved in Theorem 19.

As a consequence, every local martingale is locally summable (Corollary 20).

A. Dual projection of measures

Let $m : \mathcal{M} \to E$ be a σ-additive measure. Then m has finite semivariation $\tilde{m}_{\mathbb{R},E}$.

For every set $A \in \mathcal{M}$ we consider the optional projection $^o(1_A)$ and the predictable projection $^p(1_A)$ of 1_A and set

$$m^o(A) = \int {}^o(1_A) dm$$

and

$$m^p(A) = \int {}^p(1_A) dm.$$

It is clear that $m^o, m^p : \mathcal{M} \to E$ are additive.

1. Theorem. *The set functions $m^o, m^p : \mathcal{M} \to E$ are σ-additive measures.*

If m has finite semivariation relative to (F, G), then m^o and m^p have finite semivariation relative to (F, G).

If m has finite variation $|m|$, then m^o and m^p have finite variation satisfying

$$|m^o| \leq |m|^o$$

and

$$|m^p| \leq |m|^p.$$

Proof. To prove that m^o is σ-additive, let (A_n) be a decreasing sequence from $\mathcal{M}$ with $A_n \downarrow \phi$. Then $^o(1_{A_n}) \downarrow \phi$. For each $z \in E^*$, the measure $m_z : \mathcal{M} \to \mathbb{R}$ is σ-additive and has finite variation $|m_z|$. Then $\int {}^o(1_{A_n}) dm_z \to 0$. From

$$\langle m^o(A_n), z \rangle = \langle \int {}^o(1_{A_n}) dm, z \rangle = \int {}^o(1_{A_n}) dm_z$$

we deduce that m^o is weakly σ-additive. By the Pettis Theorem 3.20, m^o is σ-additive.

Assume now that m has finite variation $|m|$ and consider the σ-additive measure $|m|^o$ defined by

$$|m|^o(A) = \int {}^o(1_{A_n}) d|m|, \text{ for } A \in \mathcal{M}.$$

For $A \in \mathcal{M}$ we have

$$|m^o(A)| = |\int {}^o(1_A) dm| \le \int {}^o(1_A) d|m| = |m|^o(A).$$

It follows that m^o has finite variation satisfying $|m^o| \le |m|^o$.

If m has finite semivariation $\tilde{m}_{F,G}$, using Proposition 5.46 we deduce that $(m^o)_z = (m_z)^o$ for each $z \in G^*$. Then

$$\tilde{m}^o(A) = \sup_{z \in G_1^*} |(m^o)_z|(A) = \sup_{z \in G_1^*} |(m_z)^o|(A)$$
$$\le \sup_{z \in G_1^*} \int {}^o(1_A) d|m_z| = \tilde{m}_{F,G}({}^o(1_A)) \le \tilde{m}_{F,G}(\mathbb{R}_+ \times \Omega) < \infty.$$

The proof for the predictable part is similar. ∎

2. Definition. *The measures m^o and m^p are called the optional dual projection and the predictable dual projection of m, respectively.*

From the Definition 19.5 it follows that m is optional (resp predictable) iff $m = m^o$ (resp. $m = m^p$).

The characterization of dual projections can be done in terms of functions ϕ:

3. Proposition. *Let $m : \mathcal{M} \to E$ be a σ-additive measure. Then, for every real-valued, measurable, bounded process ϕ we have*

$$\int \phi \, dm^o = \int {}^o\phi \, dm$$

and

$$\int \phi \, dm^p = \int {}^p\phi \, dm.$$

Proof. From the definition of m^o, for every real-valued $\mathcal{M}$-step process ϕ we have

$$\int \phi \, dm^o = \int {}^o\phi \, dm.$$

If ϕ is real-valued, bounded, and measurable, there is a sequence (ϕ^n) of real-valued, $\mathcal{M}$-step processes such that $\phi^n \to \phi$ uniformly and $|\phi^n| \le |\phi|$. It follows that ${}^o\phi^n \to {}^o\phi$ uniformly, hence (Theorem 5.35),

$$\int \phi^n \, dm^o \to \int \phi \, dm^o \text{ and } \int {}^o\phi^n \, dm \to \int {}^o\phi \, dm.$$

Since for each n we have
$$\int \phi^n dm^o = \int {}^o\phi^n dm$$
we deduce that
$$\int \phi dm^o = \int {}^o\phi dm.$$
The proof for m^p is similar. ∎

B. Dual projections of processes

In all the definitions and theorems below, involving a right continuous process X with finite semivariation $\tilde{X}_{\mathbb{R},E}$, the condition $c_0 \not\subset E$ will be automatically assumed, in order to ensure the existence of the Stieltjes integral $\int \phi dX_s$ (Theorem 20.13). However, if X has finite variation $|X|$, we shall not assume that $c_0 \not\subset E$, since, in this case, the Stieltjes integral exists without imposing this condition.

4. Definition. *Let $X : \mathbb{R}_+ \times \Omega \to E$ be a right continuous, measurable process.*

Assume that X has integrable variation $|X|$ (resp. $c_0 \not\subset E$ and X has integrable semivariation $\tilde{X}_{\mathbb{R},E}$).

A right continuous, optional process $Y : \mathbb{R}_+ \times \Omega \to E$ with integrable variation $|Y|$ (resp. with integrable semivariation $\tilde{Y}_{\mathbb{R},E}$) is called the optional dual projection with integrable variation (resp. with integrable semivariation) of X, if for every real-valued, bounded, measurable process ϕ we have

$$E(\int \phi_s dY_s) = E(\int {}^o\phi_s dX_s).$$

A right continuous, predictable process $Z : \mathbb{R}_+ \times \Omega \to E$ with integrable variation $|Z|$ (resp. with integrable semivariation $\tilde{Z}_{\mathbb{R},E}$) is called the predictable dual projection with integrable variation (resp. with integrable semivariation) of X, if, for every real-valued, bounded, measurable process ϕ we have

$$E(\int \phi_s dZ_s) = E(\int {}^p\phi_s dX_s).$$

We denote $Y = X^o$ and $Z = X^p$. With these notations the above equalities can be written as
$$E(\int \phi_s dX_s^o) = E(\int {}^o\phi_s dX_s)$$
and
$$E(\int \phi_s dX_s^p) = E(\int {}^p\phi_s dX_s).$$

Remarks. a) When we say that a right continuous, measurable process $X : \mathbb{R}_+ \times \Omega \to E$ has an optional dual projection or a predictable dual projection, this will imply that either X has integrable variation $|X|$ or $c_0 \not\subset E$ and X has integrable semivariation $\tilde{X}_{\mathbb{R},E}$.
b) The above definition does not exclude the possibility that a process with finite *variation* has dual projections with finite *semivariation* (but not necessarily with finite variation); or that a process with finite *semivariation* (but not necessarily with finite variation) has dual projections with finite *variation*.

5. Proposition. *If the optional (resp. predictable) dual projection in Definition 4 exists, then it is unique, outside an evanescent set.*

Proof. With the notations of Definition 4, assume X has two optional dual projections A and B. We assume that both A and B have integrable variation, or $c_0 \not\subset E$ and A and B have integrable semivariation. Then for every real-valued, bounded, measurable process ϕ we have

$$E(\int \phi_s dA_s) = E(\int \phi_s dB_s).$$

Let T be a stopping time and take $\phi = 1_{[T,\infty)}$. Then the above equality is written

$$E((A_\infty - A_{T-})) = E((B_\infty - B_{T-})).$$

By Corollaries 19.10 and 21.10, A and B are undistinguishable.
The proof for the predictable dual projection is similar. ■

The following theorem gives a characterization of processes having the same dual projection.

6. Theorem. *Let $A, B : \mathbb{R}_+ \times \Omega \to E \subset L(F, G)$ be two right continuous, measurable processes.*
a) *Assume A and B have optional dual projections. Then $A^o = B^o$ iff*

$$E(A_\infty - A_{T-}|\mathcal{F}_T) = E(B_\infty - B_{T-}|\mathcal{F}_T), \quad a.s.,$$

for any stopping time T (where $A_\infty = A_{\infty-}$ and $B_\infty = B_{\infty-}$).
a') *If $A^o = B^o$, then for any stopping time T we have*

$$E(A_\infty - A_T|\mathcal{F}_T) = E(B_\infty - B_T|\mathcal{F}_T), \quad a.s.$$

b) *Assume A and B have predictable dual projections. Then $A^p = B^p$ iff*

$$E(A_0|\mathcal{F}_0) = E(B_0|\mathcal{F}_0), \quad a.s.$$

and

$$E(A_\infty - A_t|\mathcal{F}_t) = E(B_\infty - B_t|\mathcal{F}_t), \quad a.s.,$$

for every $t \geq 0$.

b′) If $A^p = B^p$, then for any predictable stopping time T we have

$$E(A_\infty - A_{T-}|\mathcal{F}_{T-}) = E(B_\infty - B_{T-}|\mathcal{F}_{T-}), \text{ a.s.}$$

and

$$E(A_\infty - A_T|\mathcal{F}_{T-}) = E(B_\infty - B_T|\mathcal{F}_{T-}), \text{ a.s.}$$

Proof. Assume $A^o = B^o$ and let T be a stopping time. Since $\phi = 1_{[T,\infty)}$ is optional, we have $^o\phi = 1_{[T,\infty)}$, hence, from Definition 4 we deduce

$$E(\int_{[T,\infty)} dA_s) = E(\int_{[T,\infty)} dB_s),$$

that is,

$$E(A_\infty - A_{T-}) = E(B_\infty - B_{T-}).$$

Replacing T with T_H for $H \in \mathcal{F}_T$ we get

$$E(A_\infty - A_{T-}|\mathcal{F}_T) = E(B_\infty - B_{T-}|\mathcal{F}_T), \text{ a.s.}$$

This proves the first implication of assertion a). Taking $\phi = 1_{(T,\infty)}$, ϕ is still optional and we get, as above

$$E(A_\infty - A_T|\mathcal{F}_T) = E(B_\infty - B_T|\mathcal{F}_T), \text{a.s.}$$

which is assertion a′).

Assertion b′) is proved similarly, taking T predictable and then replacing T with T_H, for $H \in \mathcal{F}_{T-}$.

To prove the other implication of assertion a), assume that for every stopping time T we have

$$E(A_\infty - A_{T-}|\mathcal{F}_T) = E(B_\infty - B_{T-}|\mathcal{F}_T), \text{ a.s.}$$

Let μ_A and μ_B be the σ-additive measures associated to A and B by Theorems 19.8 and 21.8. Then

$$\mu_A([T,\infty)) = E(\int 1_{[T,\infty)} dA_s) = E(A_\infty - A_{T-})$$

and

$$\mu_B([T,\infty)) = E(A_\infty - A_{T-}).$$

It follows that μ_A and μ_B are equal on the class of sets $[T,\infty)$ which generates the optional σ-algebra $\mathcal{O}$; therefore $\mu_A = \mu_B$ on $\mathcal{O}$. Then, for every real-valued, measurable process ϕ we have

$$\int \phi d\mu_A^o = \int {}^o\phi d\mu_A = \int {}^o\phi d\mu_B = \int \phi d\mu_B^o,$$

that is, $\mu_A^o = \mu_B^o$ on $\mathcal{M}$.

But we have $\mu_A^o = \mu_{A^o}$ and $\mu_B^o = \mu_{B^o}$; therefore $\mu_{A^o} = \mu_{B^o}$. It follows that for every stopping time T we have

$$\mu_{A^o}([T,\infty)) = \mu_{B^o}([T,\infty)),$$

that is,

$$E(A_\infty^o - A_{T-}^o) = E(B_\infty^o - B_{T-}^o).$$

By Corollaries 19.10 and 21.10 it follows that $A^o = B^o$ outside an evanescent set. Assertion a) is proved.

The proof of the predictable case is similar, using the fact that the predictable σ-algebra $\mathcal{P}$ is generated by the sets $\{0\} \times A$ with $A \in \mathcal{F}_0$ and $(s,t] \times A$ with $A \in \mathcal{F}_t$. ∎

The following theorem gives a relationship between the jumps of a process and of its dual projections.

7. Theorem. *Let $X : \mathbb{R}_+ \times \Omega \to E$ be a right continuous, measurable process. Assume X has an optional (resp. predictable) dual projection Y. Then the jump ΔY_T of Y at an optional (reps. predictable) stopping time T satisfies*

$$\Delta Y_T = E(\Delta X_T | \mathcal{F}_T), \text{ a.s.}$$

(resp.

$$\Delta Y_T = E(\Delta X_T | \mathcal{F}_{T-}), \text{ a.s.}),$$

with the convention $\Delta X_T = \Delta Y_T = 0$ on $\{T = \infty\}$.

Proof. We consider the optional case. The predictable case is proved similarly.

Since X and Y have the same optional dual projection, by Theorem 6, for every stopping time T we have

$$E(X_\infty - X_{T-} | \mathcal{F}_T) = E(Y_\infty - Y_{T-} | \mathcal{F}_T), \text{ a.s.}$$

and

$$E(X_\infty - X_T | \mathcal{F}_T) = E(Y_\infty - Y_T | \mathcal{F}_T), \text{ a.s.}$$

Taking the difference we get

$$E(\Delta X_T | \mathcal{F}_T) = E(\Delta Y_T | \mathcal{F}_T), \text{ a.s.}$$

The process Y is optional and right continuous, hence Y_- is left continuous. It follows that Y_T and Y_{T-} are $\mathcal{F}_T$-measurable; therefore ΔY_T is $\mathcal{F}_T$-measurable; hence $E(\Delta Y_T | \mathcal{F}_T) = \Delta Y_T$; consequently

$$\Delta Y_T = E(\Delta X_T | \mathcal{F}_T), \text{ a.s.}$$

∎

C. Existence of dual projections

The existence of dual projections is proved only for processes with integrable variation.

8. Theorem. *Let $X : \mathbb{R}_+ \times \Omega \to E$ be a right continuous, measurable process with integrable variation $|X|$. Assume E has the Radon–Nikodym Property (for example E is a separable dual space, or a reflexive space, in particular a Hilbert space).*

Then X has optional dual projection with integrable variation and predictable dual projection with integrable variation.

Proof. Since X is measurable, it is separably valued; therefore we can assume E itself is separable.

We apply first Theorem 19.8 and obtain a σ-additive measure $\mu_X : \mathcal{M} \to E$ with finite variation $|\mu_X| = \mu_{|X|}$, such that

$$\int \phi \, d\mu_X = E(\int \phi_s \, dX_s), \text{ for } \phi \in L^1_{\mathbb{R}}(\mu_X).$$

Consider the optional projection μ_X^o of μ_X. Then $\mu_X^o : \mathcal{M} \to E$ is σ-additive, has finite variation and satisfies

$$\int \phi \, d\mu_X^o = \int {}^o\phi \, d\mu_X = E(\int {}^o\phi_s \, dX_s)$$

for every real-valued, bounded, measurable process ϕ. Consider now the embedding $E = L(\mathbb{R}, E)$ and take a separable space $Z \subset E^*$, norming for E. By Theorem 19.12 d), there is a right continuous, measurable process $Y : \mathbb{R}_+ \times \Omega \to E$, with integrable variation, such that

$$\int \phi \, d\mu_X^o = E(\int \phi \, dY_s)$$

for any real-valued, bounded measurable process ϕ. Moreover, by Theorem 19.12 b), since μ_X^o is optional, the process Y_s is optional. It follows that

$$E(\int \phi \, dY_s) = E(\int {}^o\phi \, dX_s)$$

for any real-valued, bounded, measurable process ϕ; therefore $Y = X^o$.

The existence of the dual predictable projection X^p is proved similarly. ∎

Remark. If X is right continuous and has integrable *semivariation* but not necessarily finite variation, we are not able to prove the existence of dual projections, since we do not have a correspondent of Theorem 19.12, for measures with finite semivariation.

D. Processes with locally integrable variation or semivariation

9. Definition. *Let $X : \mathbb{R}_+ \times \Omega \to E \subset L(F, G)$ be a right continuous, measurable process with finite variation (resp. finite semivariation $\tilde{X} = \tilde{X}_{F,G}$). Assume G is separable.*

If $X_0 = 0$, we say X has locally integrable variation (resp. locally integrable semivariation $\tilde{X}_{F,G}$), if there is an increasing sequence (T_n) of stopping times with $T_n \uparrow \infty$ such that, for each n, the stopped process X^{T_n} has integrable variation $|X^{T_n}|$ (resp. integrable semivariation $(X^{T_n})\tilde{}_{F,G}$).

If $X_0 \neq 0$, we say X has locally integrable variation $|X|$ (resp. locally integrable semivariation $\tilde{X}_{F,G}$), if $E(|X_0| \,|\, \mathcal{F}_0) < \infty$, a.s. and $X - X_0$ has locally integrable variation $|X - X_0|$ (resp. locally integrable semivariation $(X - X_0)\tilde{}_{F,G}$).

It follows that if X has locally integrable variation, then it has also locally integrable semivariation. If X has finite variation $|X|$ (resp. finite semivariation $\tilde{X}_{F,G}$) and T is a stopping time, then X^T has finite variation $|X^T|$ (resp. finite semivariation $(X^T)\tilde{}_{F,G}$) and we have

$$|X^T| = |X|^T \text{ and } (X^T)\tilde{}_{F,G} = (\tilde{X}_{F,G})^T.$$

To say that X has locally integrable variation $|X|$ (resp. locally integrable semivariation $\tilde{X}_{F,G}$) means that $E(|X|_{T_n}) < \infty$ (resp. $E((\tilde{X}_{T_n})_{F,G}) < \infty$) for each n, in the above definition.

The condition that G is separable is imposed in order to ensure that the semivariation $\tilde{X}_{F,G}$ is measurable (Proposition 21.5 c).

The semivariation $\tilde{X}_{\mathbb{R},E}$ is measurable, since X is measurable, hence separably valued; therefore we can assume E separable.

10. Remarks. a) If (T_n) is a (not necessarily increasing) sequence of stopping times with $\sup T_n = \infty$ and if $E(|X|_{T_n}) < \infty$ (resp. $E((\tilde{X}_{T_n})_{F,G}) < \infty$) for each n, then X has locally integrable variation $|X|$ (resp. locally integrable semivariation $\tilde{X}_{F,G}$). In fact, let $S_n = \sup\{T_1, \ldots, T_n\}$. Then (S_n) is an increasing sequence of stopping times with $S_n \uparrow \infty$ and for each n we have

$$E(|X|_{S_n}) < \infty \text{ (resp } E((\tilde{X}_{S_n})_{F,G}) < \infty).$$

b) A process with finite variation (resp. finite semivariation) is not necessarily with locally integrable variation (resp. locally integrable semivariation). However, this is true if X is predictable, as will be proved in Theorem 14 below.

Processes with locally integrable variation or semivariation are locally summable:

11. Theorem. *Let $X : \mathbb{R}_+ \times \Omega \to E \subset L(F, G)$ be a right continuous, adapted process. Assume that either X has locally integrable variation $|X|$ or $c_0 \not\subset E$,*

$c_0 \not\subset G$, G is separable and X has locally integrable semivariations $\tilde{X}_{\mathbb{R},E}$ and $\tilde{X}_{F,G}$.

Then X is locally summable relative to (F, G).

Proof. If X has locally integrable variation we use Theorem 19.13. If X has locally integrable semivariations we use Theorem 21.12. ∎

The definition of dual projections can be extended for processes with locally integrable variation or semivariation.

12. Definition. *Let $X : \mathbb{R}_+ \times \Omega \to E$ be a right continuous, measurable process.*

Assume that X has locally integrable variation $|X|$ (resp. $c_0 \not\subset E$ and X has locally integrable semivariation $\tilde{X}_{\mathbb{R},E}$).

We say that a right continuous, optional process $Y : \mathbb{R}_+ \times \Omega \to E$ with locally integrable variation (resp. locally integrable semivariation $\tilde{Y}_{\mathbb{R},E}$) is the optional dual projection of X, if there is an increasing sequence (T_n) of stopping times with $T_n \uparrow \infty$, such that for each n, X^{T_n} and Y^{T_n} have integrable variation (resp. integrable semivariation relative to $(\mathbb{R}, E)$) and Y^{T_n} is the optional dual projection of X^{T_n}.

A similar definition is given for the predictable dual projection of X.

The optional dual projection of X is denoted by X^o and the predictable dual projection of X is denoted by X^p.

With these notations Definition 12 states that

$$(X^o)^{T_n} = (X^{T_n})^o \text{ and } (X^p)^{T_n} = (X^{T_n})^p.$$

From Proposition 5 we deduce that the optional and the predictable dual projections of X, if they exist, are unique up to an evanescent process.

The existence of the dual projections stated in Theorem 8 can be extended to processes with locally integrable variations.

13. Theorem. *Let $X : \mathbb{R}_+ \times \Omega \to E$ be a right continuous, measurable process with locally integrable variation. Assume E has the Radon–Nikodym Property. Then X has optional and predictable dual projections with locally integrable variation.*

Proof. Let (T_n) be an increasing sequence of stopping times with $T_n \uparrow \infty$ such that X^{T_n} has integrable variation, for each n.

By Theorem 8, X^{T_n} has an optional projection Y^n with integrable variation, which is unique up to an evanescent process.

We have $Y^{n+1} = Y^n$ on $[0, T_n]$, up to an evanescent process. In fact, for $k \leq n$ and for any real-valued, bounded, measurable process ϕ we have

$$E(\int \phi d(1_{[0,T_k]} Y^n)) = E(\int \phi 1_{[0,T_k]} dY^n)$$
$$= E(\int {}^o(\phi 1_{[0,T_k]}) dX^{T_n})$$
$$= E(\int {}^o\phi d(1_{[0,T_k]} X^{T_n})) = E(\int {}^o\phi d(1_{[0,T_k]} X^{T_k}));$$

in particular, for $k = n$,

$$E(\int \phi d(1_{[0,T_k]} Y^k)) = E(\int {}^o\phi d(1_{[0,T_k]} X^{T_k})).$$

It follows that for $k \leq n$ we have

$$E(\int \phi d(1_{[0,T_k]} Y^n)) = E(\int \phi d(1_{[0,T_k]} Y^k)).$$

Taking $\phi = 1_{[T,\infty)}$ with T any stopping time we get

$$E(1_{[0,T_k]}(Y^n_\infty - Y^n_{T-})) = E(1_{[0,T_k]}(Y^k_\infty - Y^k_{T-})).$$

By Corollary 19.10, we have $1_{[0,T_k]} Y^n = 1_{[0,T_k]} Y^k$, up to an evanescent process. In particular, $Y^{n+1} = Y^n$ on $[0, T_n]$ up to an evanescent process.

We set $Y = \lim Y^n$ if the limit exists and $Y = 0$ otherwise. Then Y is optional and for each n we have $Y^{T_n} = Y^n = (X^{T_n})^o$; that is, Y is the optional dual projection of X.

The existence of the predictable dual projection is proved similarly. ∎

E. Examples of processes with locally integrable variation or semivariation

A first example of a locally integrable, increasing process is given in the following theorem:

14. Theorem. *If $M : \mathbb{R}_+ \times \Omega \to E$ is a local martingale, then the increasing process M^* defined by*

$$M^*_t(\omega) = \sup |M_t(\omega)|, \text{ for } \omega \in \Omega \text{ and } t \geq 0,$$

is locally integrable.

Proof. We have $M^*_0 = |M_0|$ and $E(M^*_0 | \mathcal{F}_0) = |M_0| < \infty$. From the inequality $M^*_t - M^*_0 \leq (M - M_0)^*$ it follows that if $(M - M_0)^*$ is locally integrable, then so is $M^* - M^*_0$; therefore, we can assume $M_0 = 0$.

Let (R_n) be a fundamental sequence for M, that is, (R_n) is an increasing sequence of stopping times with $R_n \uparrow \infty$ and for each n, $M^{R_n} I_{\{R_n > 0\}}$ is a uniformly integrable martingale. Then M^{R_n} is a uniformly integrable martingale, since $M_0 = 0$. We can assume $R_n < \infty$ everywhere for each n; otherwise we replace R_n with $R_n \wedge n$. Let

$$S_n = R_n \wedge \inf\{t : |M_t| \geq n\}.$$

Then $S_n \uparrow \infty$. Since $S_n \leq R_n$ and M^{R_n} is a uniformly integrable martingale, we have $E(M_{S_n}) < \infty$. If $t < S_n$ then $|M_t| < n$; therefore $M^*_{S_n} \leq n \vee |M_{S_n}|$, hence $E(M^*_{S_n}) < \infty$. It follows that M^* is locally integrable. ∎

Another example of processes with locally integrable semivariation is that of predictable processes with finite semivariation.

However, optional processes with finite semivariation do not necessarily have locally integrable semivariation.

15. Theorem. *Let $X : \mathbb{R}_+ \times \Omega \to E \subset L(F,G)$ be a right continuous, predictable process with finite variation $|X|$ (resp. finite semivariation $\tilde{X}_{F,G}$; in this case assume G is separable).*

Then X has locally integrable variation $|X|$ (resp. locally integrable semivariation $\tilde{X}_{F,G}$).

Proof. Assume $X_0 = 0$. Since X is predictable, if G is separable, its semivariation $\tilde{X} = \tilde{X}_{F,G}$ is also predictable (Proposition 21.5) For each n set

$$S_n = \inf\{t : \tilde{X}_t \geq n\}.$$

Then S_n is a predictable stopping time and $S_n \uparrow \infty$, since $\tilde{X}$ is finite. Moreover $S_n > 0$, since $\tilde{X}_0 = 0$. For each n there is an increasing sequence $(S_{nk})_k$ such that $S_{nk} < S_n$ for each k and $\lim_k S_{nk} = S_n$. For each n and k we have $\tilde{X}_{S_{nk}} < n$, hence $E(\tilde{X}_{S_{nk}}) < \infty$.

Since $\sup_{nk} S_{nk} = \infty$, we deduce, by Remark 10 a) that $\tilde{X}$ is locally integrable.

If $X_0 \neq 0$, then $X - X_0$ is predictable and has finite semivariation $(X - X_0)^\sim \leq \tilde{X} + |X_0|$. By the first part of the proof, the semivariation $(X - X_0)^\sim$ is locally integrable. Since $\tilde{X}$ is predictable, we have $E(\tilde{X}_0|\mathcal{F}_0) = \tilde{X}_0 < \infty$ a.s. It follows that X has locally integrable semivariation.

The case of finite variation is proved similarly, without assuming G separable. ∎

We give another example of processes with locally integrable variation or semivariation.

16. Theorem. *Let $M : \mathbb{R}_+ \times \Omega \to E \subset L(F,G)$ be a local martingale.*
a) *If M has finite variation $|M|$, then M has locally integrable variation $|M|$.*

b) *Assume $c_0 \not\subset E$ and $c_0 \not\subset G$. If M has finite semivariations $\tilde{M}_{\mathbb{R},E}$ and $\tilde{M}_{F,G}$, then M has locally integrable semivariations $\tilde{M}_{\mathbb{R},E}$ and $\tilde{M}_{F,G}$.*

Proof. Denote $\tilde{M} = \tilde{M}_{F,G}$. Assume $M_0 = 0$. The semivariation $\tilde{M}$ is an increasing process with $\tilde{M}_0 = 0$. We have (Theorems 18.20 and 20.14)

$$\Delta \tilde{M}_s \leq \Delta |M|_s = |\Delta M_s| \leq 2M_s^*, \text{ for each } s \geq 0.$$

Since, by Theorem 14, M^* is locally integrable, there is an increasing sequence (T_n) of stopping times with $T_n \uparrow \infty$ such that $E(M_{T_n}^*) < \infty$ for each n. Let

$$S_n = T_n \wedge \inf\{t : \tilde{M}_t \geq n\}.$$

Then S_n is a stopping time and $S_n \uparrow \infty$. We have $\tilde{M}_{S_n-} \leq n$, hence

$$\tilde{M}_{S_n} = \tilde{M}_{S_n-} + \Delta \tilde{M}_{S_n} \leq n + 2M_{S_n}^*,$$

therefore $E(\tilde{M}_{S_n}) < \infty$.

In case M has finite variation, the proof is similar. ∎

Corollary 21.11 extends for local martingales:

17. Theorem. *Let $M : \mathbb{R}_+ \times \Omega \to E$ be a predictable local martingale with $M_0 = 0$. Assume that either M has finite variation, or M has finite semivariation $\tilde{M}_{\mathbb{R},E}$ and $c_0 \not\subset E$.*
Then M is evanescent.

Proof. By Theorem 15, the variation $|M|$, respectively the semivariation $\tilde{M}_{\mathbb{R},E}$, is locally integrable. Let (T_n) be an increasing sequence of stopping times with $T_n \uparrow \infty$ such that, for each n, $M^{T_n} I_{\{T_n > 0\}}$ is a uniformly integrable martingale with integrable semivariation relative to $(\mathbb{R}, E)$, vanishing at 0. By Corollaries 19.11 and 21.11, $M^{T_n} I_{\{T_n > 0\}}$ is evanescent, hence M itself is evanescent. ∎

The following theorem gives a characterization of the dual predictable projection of an optional process in terms of local martingales.

18. Theorem. *Let $A : \mathbb{R}_+ \times \Omega \to E$ be a right continuous, optional process.*
a) *Assume A has finite variation $|A|$ (resp. $c_0 \not\subset E$ and A has finite semivariation $\tilde{A} = \tilde{A}_{\mathbb{R},E}$).*
If there is a right continuous, predictable process $B : \mathbb{R}_+ \times \Omega \to E$ with finite variation $|B|$ (resp. finite semivariation $\tilde{B} = \tilde{B}_{\mathbb{R},E}$) such that $A - B$ is a local martingale vanishing at 0, then A and B have locally integrable variation (resp. locally integrable semivariation relative to $(\mathbb{R}, E)$) and $B = A^p$.
b) *Assume A has locally integrable variation $|A|$ (resp. $c_0 \not\subset E$ and A has locally integrable semivariation $\tilde{A}_{\mathbb{R},E}$).*
If A has a predictable dual projection B, then $A - B$ is a local martingale vanishing at 0.

Proof. Assume first there is a right continuous, predictable process B with finite semivariation $\tilde{B}$ such that $M = A - B$ is a local martingale vanishing at 0. Then $A_0 = B_0$ and we can assume $A_0 = 0$ and $B_0 = 0$; otherwise we replace A and B by $A - A_0$ and $B - A_0$ respectively. Since B has separable range, we can assume E separable.

Let (U_n) be an increasing sequence of stopping times with $U_n \uparrow \infty$, reducing the local martingale M, that is, such that for each n, $M^{U_n} I_{\{U_n > 0\}}$ is a uniformly integrable martingale vanishing at 0.

By Theorem 15, B has locally integrable semivariation $\tilde{B}$ relative to $(\mathbb{R}, E)$. Let (V_n) be an increasing sequence of stopping times with $V_n \uparrow \infty$ such that $E(\tilde{B}_{V_n}) < \infty$ for every n.

For each n let $W_n = \inf\{t : \tilde{A}_t \geq n\}$ and $T_n = U_n \wedge V_n \wedge W_n$. Since $\tilde{A}$ is optional and finite (Theorem 21.5 c), W_n is a stopping time and $W_n \uparrow \infty$. Then $T_n \uparrow \infty$ and $\tilde{A}_{T_n-} \leq n$. We have

$$|A_{T_n}| \leq |B_{T_n}| + |M_{T_n}| \leq \tilde{B}_{T_n} + |M_{T_n}|$$

and

$$E(|A_{T_n}|) \leq E(\tilde{B}_{T_n}) + E(|M_{T_n}|) < \infty.$$

Using Theorem 20.14, from the inequality

$$\Delta \tilde{A}_{T_n} \leq |\Delta A_{T_n}| \leq |A_{T_n}| + |A_{T_n-}| \leq |A_{T_n}| + n$$

it follows

$$E(\Delta \tilde{A}_{T_n}) \leq E(|A_{T_n}|) + n < \infty.$$

We have

$$\tilde{A}_{T_n} = \Delta \tilde{A}_{T_n} + \tilde{A}_{T_n-} \leq \Delta \tilde{A}_{T_n} + n,$$

hence $\tilde{A}_{T_n}$ is integrable.

For each n we shall prove that $B^{T_n} = (A^{T_n})^p$.

Since M^{T_n} is a uniformly integrable martingale vanishing at 0, for every stopping time T we have

$$E(M_\infty^{T_n} - M_{T-}^{T_n}) = 0,$$

therefore

$$E(A_\infty^{T_n} - A_{T-}^{T_n}) = E(B_\infty^{T_n} - B_{T-}^{T_n}),$$

that is,

$$E(\int 1_{[T,\infty)} dA^{T_n}) = E(\int 1_{[T,\infty)} dB^{T_n}).$$

Using the measures $\mu_n = \mu_{A^{T_n}}$ and $\nu_n = \mu_{B^{T_n}}$ associated to A^{T_n} and B^{T_n} by Theorem 21.8, this means that

$$\mu_n([T,\infty)) = \nu_n([T,\infty)).$$

Since μ_n and ν_n are σ-additive on $\mathcal{M}$ and the intervals $[T,\infty)$ generate the optional σ-algebra $\mathcal{O}$, it follows that $\mu_n = \nu_n$ on $\mathcal{O}$; in particular, $\mu_n = \nu_n$ on $\mathcal{P}$. Let ϕ be a real-valued, bounded, measurable process and $^p\phi$ its predictable projection. Then
$$\int {}^p\phi\, d\mu_n = \int {}^p\phi\, d\nu_n$$
hence
$$E(\int {}^p\phi\, dA^{T_n}) = E(\int {}^p\phi\, dB^{T_n}).$$
Since B^{T_n} is predictable we have also
$$E(\int {}^p\phi\, dB^{T_n}) = E(\int \phi\, dB^{T_n}),$$
therefore
$$E(\int \phi\, dB^{T_n}) = E(\int {}^p\phi\, dA^{T_n}).$$

It follows that $B^{T_n} = (A^{T_n})^p$; consequently, $B = A^p$. This proves assertion a) and also the uniqueness of B (Proposition 5). The proof for the finite variation case is similar, using Theorems 18.20 and 19.8 instead of Theorems 20.14 and 21.8, respectively.

To prove assertion b), assume $c_0 \not\subset E$, that A has locally integrable semi-variation $\tilde{A}$ relative to $(\mathbb{R}, E)$ and has a dual predictable projection B. Let (T_n) be an increasing sequence of finite stopping times with $T_n \uparrow \infty$ such that, for each n, A^{T_n} and B^{T_n} have integrable semivariation relative to $(\mathbb{R}, E)$ and $B^{T_n} = (A^{T_n})^p$. Then, by Theorem 6 b), we have
$$E(A_0^{T_n}|\mathcal{F}_0) = E(B_0^{T_n}|\mathcal{F}_0)$$
and
$$E(A_\infty^{T_n} - A_t^{T_n}|\mathcal{F}_t) = E(B_\infty^{T_n} - B_t^{T_n}|\mathcal{F}_t),$$
therefore $A_0 = B_0$ and
$$A_t^{T_n} - B_t^{T_n} = E(A_{T_n} - B_{T_n}|\mathcal{F}_t).$$

It follows that $A - B$ is a local martingale vanishing at 0.

If A has locally integrable variation, statement b) follows as a particular case. ∎

F. Decomposition of local martingales

We can now prove the decomposition theorem of local martingales, which will ensure the local summability of local martingales.

19. Theorem. *Assume E has the Radon–Nikodym Property and let $M : \mathbb{R}_+ \times \Omega \to E$ be a local martingale. Then M can be written as $M = U + V$,* where $U : \mathbb{R}_+ \times \Omega \to E$ is a local martingale with $U_0 = 0$ and jumps $|\Delta U_t| \leq 1$ and $V : \mathbb{R}_+ \times \Omega \to E$ is a local martingale with locally integrable variation.

Proof. We can assume $M_0 = 0$; otherwise we replace M by $M - M_0$. Set

$$A_t = \sum_{s \leq t} \Delta M_s 1_{\{|\Delta M_s| > \frac{1}{2}\}}, \text{ for } t \geq 0.$$

Since M is cadlag, it has only a finite number of points s with $|\Delta M_s| > \frac{1}{2}$, in any compact interval. It follows that for each fixed $\omega \in \Omega$, the sum defining $A_t(\omega)$ has only a finite number of terms $\neq 0$. The process A_t is adapted, right continuous and with finite variation. For each $t \geq 0$ we have

$$\Delta A_t = A_t - A_{t-}$$
$$= \sum_{s \leq t} \Delta M_s 1_{\{|\Delta M_s| > \frac{1}{2}\}} - \sum_{s < t} \Delta M_s 1_{\{|\Delta M_s| > \frac{1}{2}\}} = \Delta M_t 1_{\{|\Delta M_t| > \frac{1}{2}|\}};$$

therefore,

$$|\Delta A_t| \leq |\Delta M_t| \leq 2 M_t^*.$$

The variation of A is

$$|A|_t = \sum_{s \leq t} |\Delta M_s| 1_{\{|\Delta M_s| > \frac{1}{2}\}};$$

hence,

$$\Delta |A|_t = |\Delta A_t| \leq 2 M_t^*.$$

Since M is a local martingale, the process M^* is locally integrable (Theorem 14). Let (R_n) be an increasing sequence of stopping times with $R_n \uparrow \infty$ and $E(M_{R_n}^*) < \infty$.

For each n set

$$S_n = R_n \wedge \inf\{s : |A|_s \geq n\}.$$

Since $|A|$ is right continuous and increasing, it is cadlag. $|A|$ is also adapted; therefore S_n is a stopping time and $S_n \leq R_n$, hence $E(M_{S_n}^*) < \infty$.

For $s < S_n$ we have $|A|_s < n$; therefore $|A|_{S_n-} \leq n$. Then

$$0 \leq |A|_{S_n} = \Delta |A|_{S_n} + |A|_{S_n-} \leq 2 M_{S_n}^* + n \in L^1,$$

hence $|A|$ is locally integrable; that is, $A : \mathbb{R}_+ \times \Omega \to E$ has locally integrable variation.

By Theorem 13, A has a predictable dual projection $B : \mathbb{R}_+ \times \Omega \to E$, which has locally integrable variation. Let

$$V = A - B.$$

Then V is a local martingale with locally integrable variation.

The process $U = M - V$ is a local martingale and we have $M = U + V$. It remains to prove that $|\Delta U_t| \leq 1$ for every $t \geq 0$.

We have
$$|\Delta U_t| \leq |\Delta M_t - \Delta A_t| + |\Delta B_t|.$$

The first term is smaller than $\frac{1}{2}$:
$$|\Delta M_t - \Delta A_t| = |\Delta M_t - \Delta M_t 1_{\{|\Delta M_t| > \frac{1}{2}\}}|$$
$$= |\Delta M_t| 1_{\{|\Delta M_t| \leq \frac{1}{2}\}} \leq \frac{1}{2}.$$

We have to prove $|\Delta B_t| \leq \frac{1}{2}$. Let S be a finite stopping time such that:
α) $M^S I_{\{S>0\}}$ is a uniformly integrable martingale,
β) A^S has integrable variation,
and
γ) $B^S = (A^S)^p$, the predictable dual projection of A^S.

Let T be a predictable stopping time. Then
$$E(M_T^S I_{\{T<\infty\}} | \mathcal{F}_{T-}) = M_{T-}^S I_{\{T<\infty\}},$$

therefore
$$E(\Delta M_T^S I_{\{T<\infty\}} | \mathcal{F}_{T-}) = 0.$$

From the equality $\Delta A_t = \Delta M_t 1_{\{|\Delta M_t| > \frac{1}{2}\}}$ we deduce that
$$\Delta A_T^S = \Delta M_T^S 1_{\{|\Delta M_T^S| > \frac{1}{2}\}}.$$

By Theorem 7 we have
$$\Delta B_T^S = E(\Delta A_T^S | \mathcal{F}_{T-}).$$

Then
$$|\Delta B_T^S| = |E(\Delta A_T^S | \mathcal{F}_{T-})|$$
$$= |E(\Delta M_T^S 1_{\{|\Delta M_T^S| > \frac{1}{2}\}} | \mathcal{F}_{T-})|$$
$$= |E(\Delta M_T^S 1_{\{T<\infty\}} - \Delta M_T^S 1_{\{|\Delta M_T^S| > \frac{1}{2}\}} | \mathcal{F}_{T-})|$$
$$= |E(\Delta M_T^S 1_{\{|\Delta M_T^S| \leq \frac{1}{2}\}} | \mathcal{F}_{T-})|$$
$$\leq E(|\Delta M_T^S| 1_{\{|\Delta M_T^S| \leq \frac{1}{2}\}} | \mathcal{F}_{T-}) \leq \frac{1}{2}, \text{ a.s.}$$

Let now (T_n) be an increasing sequence of finite stopping times with $T_n \uparrow \infty$, such that for each n, conditions α), β), and γ) are satisfied for $S = T_n$. Then $|\Delta B_T^{T_n}| \leq \frac{1}{2}$, a.s. for each n. Letting $n \to \infty$ we get $|\Delta B_T| 1_{\{T<\infty\}} \leq \frac{1}{2}$, a.s. Since ΔB_t is predictable, the set $\{\Delta B_t \neq 0\}$ is contained in the union of graphs of a sequence of predictable stopping times ([D–M], IV. 88 B). It follows that any jump ΔB_t of B is of the form ΔB_T for some predictable stopping time T, hence $|\Delta B_t| \leq \frac{1}{2}$. We deduce that $|\Delta U_t| \leq 1$ for every $t \geq 0$. ∎

20. Corollary. *Assume E and G are Hilbert spaces and let $M : \mathbb{R}_+ \times \Omega \to E \subset L(F,G)$ be a local martingale. Then M is 1-locally summable with respect to (F,G).*

Proof. By Theorem 19 we can write $M = U + V$, where U is a local martingale with $U_0 = 0$ and $|\Delta U_t| \leq 1$ and V is a local martingale with locally integrable variation. By Theorem 11, V is locally summable relative to (F, G).

Let (T_n) be an increasing sequence of stopping times with $T_n \uparrow \infty$, such that for each n, $U^{T_n} 1_{\{T_n > 0\}}$ is a uniformly integrable martingale, with jumps ≤ 1.

For each n let
$$S_n = \inf\{t : |U_t| \geq n\} \wedge T_n$$

Then S_n is a stopping time and $S_n \uparrow \infty$. The process $U^{S_n} 1_{\{S_n > 0\}}$ is still a uniformly integrable martingale and
$$|U_{S_n}| \leq |U_{S_n-}| + |\Delta U_{S_n}| \leq n+1;$$

therefore $U^{S_n} I_{\{S_n > 0\}}$ is bounded; consequently it is a square integrable martingale. By Theorem 17.7, a square integrable martingale is 2–summable, hence it is 1–summable. It follows that U is locally summable; consequently $M = U + V$ is locally summable. ∎

Chapter 6
The Itô Formula

§23. THE ITÔ FORMULA

In this paragraph we establish the Itô formula for locally semi summable processes with values in Banach spaces. For real-valued semimartingales we refer to [D–M. VIII. 27].

A. Preliminary results

In the sequel we shall state and prove the results concerning only local summability. The results concerning local p-summability will then follow as corollaries.

Let $X : \mathbb{R}_+ \times \Omega \to E$ be a cadlag, adapted process.

1. We shall reserve the notation $(v(n,k))_{k \geq 0}$, $n = 1, 2 \ldots$ for a family of stopping times satisfying the following three conditions:

(i) For each n, we have $v(n,0) = 0$ and $v(n,k) \uparrow \infty$, as $k \to \infty$;

(ii) $\lim_n \sup_k (v(n, k+1) - v(n,k)) = 0$;

(iii) There is a sequence $a_n \downarrow 0$ such that for $t \in [v(n,k), v(n,k+1))$, we have

$$|X_t - X_{v(n,k)}| \leq a_n.$$

Given $a_n \downarrow 0$, an example of such a family of stopping times is:

$$v(n,0) = 0,$$

and

$$v(n, k+1) = \inf\{t > v(n,k) : |X_t - X_{v(n,k)}| > a_n\} \wedge (v(n,k) + a_n).$$

These stopping times will play an important role in the development of Itô's formula.

For each n, we denote

$$X^n = \sum_{k \geq 0} X_{v(n,k)} 1_{(v(n,k), v(n,k+1)]}.$$

From condition (ii) we deduce that $X^n \to X_-$ pointwise. From condition (iii), we deduce that

$$|X_{t-} - X_{v(n,k)}| \leq a_n, \text{ for } t \in (v(n,k), v(n, k+1)],$$

and this shows that $X^n \to X_-$ uniformly on $\mathbb{R}_+ \times \Omega$.

We use the standard convention that $Y_{0-} = 0$, for any process Y.

2. Theorem. *Assume that X is locally summable relative to (F, G) and let $H : R_+ \times \Omega \to F$ be a cadlag, adapted process; therefore, H_- is locally integrable with respect to X. Assume further that either*
a) the set of measures $(I_X)_{F, L_{G^*}^\infty}$ is locally uniformly σ-additive (this is the case, for example, if $F = \mathbb{R}$);
or
b) there is a sequence $b_n \downarrow 0$ such that for $t \in [v(n,k), v(n,k+1))$ we have

$$|H_t - H_{v(n,k)}| \leq b_n.$$

Then for every $t \geq 0$, we have

$$\int_{[0,t]} H_{s-} dX_s = \lim \text{prob}_n \sum_{k \geq 0} H_{v(n,k) \wedge t}(X_{v(n,k+1) \wedge t} - X_{v(n,k) \wedge t})$$

$$= \lim \text{prob}_n \sum_{k \geq 0} H_{v(n,k)}(X_{v(n,k+1)} - X_{v(n,k)}) 1_{\{v(n,k) \leq t\}}.$$

Moreover, there is a subsequence (r_n) such that a.s., as $n \to \infty$, the first limit is uniform on compact time intervals and the second limit is pointwise.

Proof. For each n, define

$$H^n = \sum_{k \geq 0} H_{v(n,k)} 1_{(v(n,k), v(n,k+1)]}$$

$$= \sum_{k \geq 0} H_{v(n,k) \wedge t} 1_{(v(n,k), v(n,k+1)]}.$$

The process H^n is σ-elementary, hence locally integrable with respect to X and
$$\int_{[0,t]} H_s^n dX_s = \sum_{k \geq 0} H_{v(n,k) \wedge t}(X_{v(n,k+1) \wedge t} - X_{v(n,k) \wedge t}).$$

Condition (ii) of the stopping times $v(n,k)$ implies that $H^n \to H_-$ pointwise. If, in addition, H satisfies condition (b) above, then $H^n \to H_-$ uniformly on $\mathbb{R}_+ \times \Omega$.

Let $M_i \uparrow \infty$ and for each i, define the stopping time
$$T_i = \inf\{t : |H_t| > M_i\}.$$

Then $T_i \uparrow \infty$ and on $[0, T_i)$, we have $|H_t| \leq M_i$; therefore $|H_t^n| \leq M_i$, for each n. Since H^n is left continuous, we deduce that $|H_t^n| \leq M_i$ on $[0, T_i]$ for each n. The process
$$\Phi = \sum_{i \geq 0} M_{i+1} 1_{(T_i, T_{i+1}]}$$

is σ-elementary; if we choose $x \in F$, with $|x| = 1$, then $H^0 = \Phi x$ is locally integrable with respect to X and for each n we have $|H^n| \leq |H^0|$. Under either of the conditions a) or b) above, we can apply the Lebesgue Theorem 15.15 and deduce that
$$\int_{[0,t]} H_{s-} dX_s = \lim \text{prob}_n \int_{[0,t]} H_s^n dX_s$$
$$= \lim \text{prob}_n \sum_{k \geq 0} H_{v(n,k) \wedge t}(X_{v(n,k+1) \wedge t} - X_{v(n,k) \wedge t}),$$

and that there exists a subsequence (r_n) such that a.s., the convergence is uniform on any compact time interval.

The equality involving the second limit follows from the fact that if $t \in [v(n,j), v(n,j+1))$, then
$$\lim_n \sum_{k \geq 0} [H_{v(n,k)}(X_{v(n,k+1)} - X_{v(n,k)}) 1_{\{v(n,k) \leq t\}} -$$
$$- H_{v(n,k) \wedge t}(X_{v(n,k+1) \wedge t} - X_{v(n,k) \wedge t})]$$
$$= \lim_n [H_{v(n,j)}(X_{v(n,j+1)} - X_{v(n,j)}) - H_{v(n,j)}(X_t - X_{v(n,j)})] = 0.$$

∎

3. Remarks. a) In the proof of Theorem 2, we did not use condition (iii) of the stopping times.
b) If there are no conditions imposed on the family $v(n,k)$ of stopping times, other than conditions (i) and (ii), we can always choose $v(n,k)$ to satisfy conditions (i)–(iii) and condition (b) above. In fact, we can take $v(n,0) = 0$

and

$$v(n, k+1) = \inf\{t > v(n,k) : |X_t - X_{v(n,k)}| > a_n \text{ or } |H_t - H_{v(n,k)}| > b_n\} \wedge$$
$$\wedge (v(n,k) + a_n + b_n).$$

But, in some cases, for example, when the existence of the quadratic variation $[X]$ is assumed (see Section C), then, in general, the stopping times $v(n, k)$ cannot be replaced. They can be replaced if E is a Hilbert space and X is summable relative to the inner product (see Section C).

c) Since $H_{0-} = 0$, we have $\int_{(0,t]} H_{s-} dX_s = \int_{[0,t]} H_{s-} dX_s$.

The following theorem is the analog of Theorem 2 for processes of finite variation.

4. Theorem. *Assume that X has finite variation and let $H : \mathbb{R}_+ \times \Omega \to F$ be a cadlag, adapted process. Then H_- is pathwise integrable with respect to X and for every $t \geq 0$ we have*

$$\int_{[0,t]} H_{s-} dX_s = \lim_n \sum_{k \geq 0} H_{v(n,k) \wedge t}(X_{v(n,k+1) \wedge t} - X_{v(n,k) \wedge t})$$
$$= \lim_n \sum_{k \geq 0} H_{v(n,k)}(X_{v(n,k+1)} - X_{v(n,k)}) 1_{\{v(n,k) \leq t\}}.$$

pointwise. The first limit is uniform on compact time intervals if H satisfies the following condition:

b(ω)) *For each $\omega \in \Omega$, there is a sequence $b_n \downarrow 0$ (depending on ω) such that for $t \in [v(n,k), v(n,k+1))$ we have $|H_t - H_{v(n,k)}| \leq b_n$.*

Proof. Using the notations in the proof of Theorem 2, we have $H^n \to H_-$ pointwise. Let $t \geq 0$. Since H is cadlag, it is pathwise bounded on $[0, t]$ by a constant M (depending on ω and t); also $|H^n| \leq M$ on $[0, t]$ for each n. We can then apply the Lebesgue theorem pathwise for the Stieltjes integral and deduce that, pathwise,

$$\int_{[0,t]} H_{s-} dX_s = \lim_n \int_{[0,t]} H^n_s dX_s.$$

If condition b(ω) is satisfied, then $H^n \to H_-$ uniformly for each ω. Let $[0, a]$ be a compact interval. If $t \in [0, a]$, then

$$\left| \int_{[0,t]} H^n_s dX_s - \int_{[0,t]} H_{s-} dX_s \right| \leq \sup_{s \in [0,a]} |H^n_s - H_{s-}| \, |X|_a \to 0,$$

which implies that

$$\int_{[0,t]} H_{s-} dX_s = \lim_n \int_{[0,t]} H^n_s dX_s$$

uniformly on $[0, a]$. This proves the conclusion concerning the first limit in the statement. The part concerning the second limit is proved as in the proof of Theorem 2. ■

5. Theorem. *Assume that X is locally summable relative to (F, G) and suppose $h : E \to F$ is a continuous function. Then $h(X_-)$ is locally integrable with respect to X. Assume further that either*
a) *The set of measures $(I_X)_{F, L_{G^*}^\infty}$ is uniformly σ-additive (for example, when $F = \mathbb{R}$);*
or
b') *h is uniformly continuous and bounded on bounded subsets of E.*
Then for every $t > 0$, we have

$$\int_{(0,t]} h(X_{s-}) dX_s = \lim \operatorname{prob}_n \sum_{k \geq 0} h(X_{v(n,k) \wedge t})(X_{v(n,k+1) \wedge t} - X_{v(n,k) \wedge t})$$

$$= \lim \operatorname{prob}_n \sum_{k \geq 0} h(X_{v(n,k)})(X_{v(n,k+1)} - X_{v(n,k)}) 1_{\{v(n,k) \leq t\}}$$

and there is a subsequence (r_n) such that, a.s., the first limit is uniform on any compact time interval and the second limit is pointwise.

Proof. The process $H_t = h(X_t)$ is F-valued, cadlag and adapted. We note that $H_{t-} = h(X_{t-})$ for $t > 0$, but not necessarily for $t = 0$. Condition a) above is identical to condition a) of Theorem 2; therefore the conclusion follows in this case.

Now assume condition b').
Denote

$$X^n = \sum_{k \geq 0} X_{v(n,k)} 1_{(v(n,k), v(n,k+1)]}$$

and

$$H^n = \sum_{k \geq 0} H_{v(n,k)} 1_{(v(n,k), v(n,k+1)]}.$$

We notice that for $t > 0$ we have $H_t^n = h(X_t^n)$. By condition (iii) of the stopping times $v(n, k)$, we have $X_t^n \to X_{t-}$ uniformly on $\mathbb{R}_+ \times \Omega$. Since h is continuous, we have $H_t^n \to H_{t-}$ pointwise. We shall prove that $H_t^n \to H_{t-}$ locally uniformly. Let $M_i \uparrow \infty$ and for each i, define the stopping time

$$T_i = \inf\{t : |H_t| > M_i\}.$$

Then $T_i \uparrow \infty$ and $1_{[0,T_i)}|X| \leq M_i$; thus $1_{[0,T_i]}|X_-| \leq M_i$ for each n and i. Since $1_{[0,T_i]} X^n \to 1_{[0,T_i]} X_-$ uniformly and since h is uniformly continuous on the bounded set $B_i = \{|x| \leq M_i\}$, we have

$$1_{[0,T_i]} h(X^n) \to 1_{[0,T_i]} h(X_-), \text{ uniformly.}$$

Since $H_0^n = H_{0-}^n = 0$, we deduce that $1_{[0,T_i]} H^n \to 1_{[0,T_i]} H_-$ uniformly.

Note that h is bounded on B_i; set $N_i = \sup\{|h(x)| : |x| \leq M_i\}$. Then $1_{[0,T_i]}|H^n| \leq N_i$ for all n. The σ-elementary process $H^0 = \sum_{i \geq 1} xN_i 1_{(T_i, T_{i+1}]}$, where $x \in F$ and $|x| = 1$, is locally integrable with respect to X, and we have $|H^n| \leq |H^0|$ for all n. By the Lebesgue Theorem 15.15, we deduce that (since $H_0^n = H_{0-} = 0$),

$$\int_{[0,t]} H^n dX \to \int_{[0,t]} H_- dX$$

in probability and uniformly on compact time intervals for a certain subsequence (r_n). The conclusion follows by using the equalities

$$1_{[0,t]} H^n = 1_{(0,t]} h(X^n) \text{ and } 1_{[0,t]} H_- = 1_{(0,t]} h(X_-).$$

This proves the part of the theorem concerning the first limit. The part concerning the second limit follows from the fact that the difference of the sums involved in the two limits converges to zero a.s., as a consequence of condition (ii) of the stopping times $v(n,k)$ and the cadlag property of H. ∎

Remark. Under assumption a), we do not need condition (iii) of the stopping times $v(n,k)$.

6. Corollary. *Assume that $F = L(E, G)$; hence $E \subset L(F, G)$. Assume X is locally summable relative to (F, G). Let $f : E \to G$ be a function of class C^2. Then $f' : E \to L(E, G) = F$ is continuous and $f'(X_-)$ is locally integrable with respect to X.*

Assume further that either

a) the set of measures $(I_X)_{F, L_{G^*}^\infty}$ is uniformly σ-additive (for example, if $F = \mathbb{R}$);

or

b'') the second derivative $f'' : E \to L(E \hat{\otimes}_\pi E, G)$ is bounded on bounded subsets of E.

Then, for every $t > 0$, we have

$$\int_{(0,t]} f'(X_{t-}) dX_s = \lim_n \sum_{k \geq 0} f'(X_{v(n,k) \wedge t})(X_{v(n,k+1) \wedge t} - X_{v(n,k) \wedge t})$$

$$= \lim_n \sum_{k \geq 0} f'(X_{v(n,k)})(X_{v(n,k+1)} - X_{v(n,k)}) 1_{\{v(n,k) \leq t\}}$$

in probability and there exists a subsequence (r_n) such that, a.s., the first limit is uniform on compact time intervals and the second limit is pointwise.

Proof. Let $B \subset E$ be a bounded set and let $S = \{|x| < r\}$ be an open sphere containing B. Let $M = \sup\{|f''(x)| : |x| < r\} < \infty$. We can apply the mean value theorem and deduce that

$$|f'(x) - f'(y)| \leq M|x-y|, \text{ for } x, y \in S.$$

It follows that f' is uniformly continuous on S. Also, we deduce that f' is bounded on S. Since the function $h = f'$ satisfies condition b') of Theorem 5, the conclusion follows. ∎

Remark. Under assumption a) we do not use condition (iii) of the stopping times.

The following theorem and corollary are the analog of Theorem 5 and Corollary 6 respectively, for processes of finite variation.

7. Theorem. *Assume X has finite variation and let $h : E \to F$ be a continuous function. Then $h(X_-)$ is pathwise locally integrable with respect to X and for every $t > 0$ we have*

$$\int_{(0,t]} H(X_{s-})dX_s = \lim_n \sum_{k \geq 0} h(X_{v(n,k) \wedge t})(X_{v(n,k) \wedge t} - X_{v(n,k) \wedge t})$$

$$\lim_n \sum_{k \geq 0} h(X_{v(n,k)})(X_{v(n,k+1)} - X_{v(n,k)}) 1_{\{v(n,k) \geq t\}}$$

pointwise. If h is uniformly continuous and bounded on bounded subsets of E, then the first limit is uniform on bounded time intervals.

Proof. If we set $H_t = h(X_t)$, then H is cadlag, adapted, and the first part of the theorem follows from Theorem 4, using $1_{[0,t]}H_- = 1_{(0,t]}h(X_-)$. Assume now that h is uniformly continuous and bounded on bounded subsets of E. Fix $\omega \in \Omega$ and $a > 0$; we shall prove that the first limit is uniform for $t \in (0, a]$. We shall use the notations of Theorem 5. For a given sequence $M_i \uparrow \infty$, we can find a sequence of stopping times $T_i \uparrow \infty$ such that $1_{[0,T_i]}|X^n| \leq M_i$ and

$$1_{[0,T_i]}X_t^n \to 1_{[0,T_i]}X_{t-}, \text{ uniformly.}$$

Since h is uniformly continuous on the set $B_i = \{|x| \leq M_i\}$, we deduce that

$$1_{[0,T_i]}h(X_t^n) \to 1_{[0,T_i]}h(X_{t-}), \text{ uniformly.}$$

Taking i such that $a < T_i(\omega)$, we deduce that $H_t^n \to H_{t-}$ uniformly on $[0, a]$. Let

$$N = \sup\{|h(x)| : |x| \leq M_i\} < \infty.$$

Then $|H_t^n| \leq N$ for all n. We can then apply the Lebesgue Theorem for the Stieltjes integral and conclude that for every $t \in (0, a]$ we have

$$\left| \int_{(0,t]} h(X_{s-})dX_s - \int_{(0,t]} h(X_s^n)dX_s \right| \leq \sup_{s \in [0,a]} |H_{s-} - H_s^n| \, |X|_a \to 0$$

as $n \to \infty$, which proves the second conclusion of the theorem. ∎

8. Corollary. *Assume X has finite variation and let $f : E \to G$ be a function of class C^2 such that the second derivative $f'' : E \to L(E \hat{\otimes}_\pi E, G)$ is bounded* on bounded subsets of E. Then $f'(X_-)$ is locally integrable with respect to X on $[0,t]$ for every $t > 0$ and we have

$$\int_{(0,t]} f'(X_{s-})dX_s = \lim_n \sum_{k \geq 0} f'(X_{v(n,k) \wedge t})(X_{v(n,k+1) \wedge t} - X_{v(n,k) \wedge t})$$

$$= \lim_n \sum_{k \geq 0} f'(X_{v(n,k)})(X_{v(n,k+1)} - X_{v(n,k)}) 1_{\{v(n,k) \leq t\}}.$$

The proof is the same as that of Corollary 6.

B. The vector quadratic variation

9. In this section, E and D are Banach spaces and $B : E \times E \to D$ is a continuous bilinear mapping, denoted by $B(x,y) = xy$, such that

$$|x| = \sup\{|xy| : |y| \leq 1\} \text{ for every } x \in E.$$

We denote by $B' : E \times E \to D$ the bilinear mapping defined by $B'(x,y) = B(y,x) = yx$, for $x, y \in E$. We write $x^2 = x \cdot x$. Then $|x^2| \leq |x|^2$.

Important examples of such bilinear mapping are:
1) The tensor product $B(x,y) = x \otimes y$ from $E \times E$ into $E \hat{\otimes}_\pi E$. We write $x^{\otimes 2} = x \otimes x$. In this case $|x^{\otimes 2}| = |x|^2$ for $x \in E$.
2) The inner product $B(x,y) = \langle x, y \rangle$, of $E \times E$ into $\mathbb{R}$ if E is a Hilbert space. In this case $x^2 = \langle x, x \rangle$ for $x \in E$.

If $B : E \times E \to D$ is a bilinear mapping as above, we can embed isometrically $E \subset L(E, D)$. We say a process $X : \mathbb{R}_+ \times \Omega \to E$ is locally summable or semilocally summable relative to the bilinear mapping B, if we regard X as taking values in $L(E, D)$ and X is locally summable or semilocally summable relative to (E, D).

If X is semilocally summable relative to both bilinear mappings B and B', then we can integrate E-valued processes H and the stochastic integrals relative to B and B' are denoted, respectively, $\int H \cdot dX$ and $\int dX \cdot H$ and have values in D. In particular, if X is semilocally summable relative to both B and B', then X_- is locally integrable with respect to X for both B and B' and the stochastic integrals $\int_{[0,t]} X_{s-} \cdot dX_s$ and $\int_{[0,t]} dX_s \cdot X_{s-}$ are defined and have values in D.

We remark that if B is the inner product, then local summability relative to B implies local summability relative to B', since in this case $B = B'$

We now define a process which is fundamental to the establishment of Itô's formula.

10. Definition. *Assume that* $X : \mathbb{R}_+ \times \Omega \to E$ *is semilocally summable relative to B and B'. The D-valued, cadlag, adapted process* $[[X]]^B$ *defined by*

$$[[X]]_t^B = X_t^2 - \int_{[0,t]} (X_{s-} \cdot dX_s + dX_s \cdot X_{s-})$$

is called the *vector quadratic variation (or vector square bracket)* of the process X with respect to B.

We note that $[[X]]_0^B = X_0^2$.

If $B(x,y) = x \otimes y$ and $D = E \hat{\otimes}_\pi E$, we denote $[[X]] = [[X]]^B$ and call $[[X]]$ the *tensor quadratic variation* of X. Hence, if X is semilocally summable relative to the tensor product mapping $(x,y) \to x \otimes y$, then

$$[[X]]_t = X_t^{\otimes 2} - \int_{[0,t]} (X_{s-} \otimes dX_s + dX_s \otimes X_{s-}).$$

If E is a Hilbert space and X is semilocally summable relative to the inner product B, we denote $[X] = [[X]]^B$ and call $[X]$ the *scalar quadratic variation* of X. In this case we have

$$[X]_t = |X_t|^2 - 2\int_{[0,t]} X_{s-} dX_s.$$

We shall present, in the next section, an extension of $[X]$, when E is a general Banach space.

Remark. If E is a Hilbert space and X is a semimartingale, then both $[[X]]$ and $[X]$ can be defined since in this case X is semilocally summable relative to the corresponding bilinear maps.

11. Theorem. *Assume $X : \mathbb{R}_+ \times \Omega \to E$ is semilocally summable relative to B and B'. For each $t \geq 0$, we have*

$$[[X]]_t^B = X_0^2 + \lim \mathrm{prob}_n \sum_{k \geq 0} \left(X_{v(n,k+1) \wedge t} - X_{v(n,k) \wedge t}\right)^2$$

$$= X_0^2 + \lim \mathrm{prob}_n \sum_{k \geq 0} \left(X_{v(n,k+1)} - X_{v(n,k)}\right)^2 1_{\{v(n,k) \leq t\}}.$$

If X is locally summable, then there is a subsequence such that the first limit is a.s., uniform on compact time intervals and the second limit exists pointwise a.s.

If X has finite variation, the first limit is a.s. uniform on compact time intervals and the second limit is pointwise a.s.

Proof. We use Theorems 2 and 4 with H and H^n replaced by X and X^n, defined by

$$X^n = \sum_{k \geq 0} X_{v(n,k)} 1_{(v(n,k), v(n,k+1)]}.$$

We have $X^n \to X_-$ pointwise uniformly, by condition (iii) of the stopping times $v(n,k)$. The processes X^n and X_- are locally integrable with respect to X. If X is locally summable, then, by Theorem 2

$$\int_{[0,t]} X_s^n dX_s \to \int_{[0,t]} X_{s-} dX_s \text{ and } \int_{[0,t]} dX_s \cdot X_s^n \to \int_{[0,t]} dX_s \cdot X_{s-}$$

in probability and uniformly on compact time intervals for a certain subsequence. For each $t \geq 0$, we have

$$\sum_{k \geq 0} \left(X_{v(n,k+1) \wedge t} - X_{v(n,k) \wedge t}\right)^2 = \sum_{k \geq 0} \left(X^2_{v(n,k+1) \wedge t} - X^2_{v(n,k) \wedge t}\right)$$
$$- \sum_{k \geq 0} X_{v(n,k) \wedge t} \cdot \left(X_{v(n,k+1) \wedge t} - X_{v(n,k) \wedge t}\right)$$
$$- \sum_{k \geq 0} \left(X_{v(n,k+1) \wedge t} - X_{v(n,k) \wedge t}\right) \cdot X_{v(n,k) \wedge t}$$

and this converges in probability, as $n \to \infty$, to

$$X_t^2 - X_0^2 - \int_{[0,t]} (X_{s-} \cdot dX_s + dX_s \cdot X_{s-}) = [[X]]_t^B - X_0^2.$$

If X has finite variation, we apply Theorem 4, with H replaced by X, using the fact that condition (iii) of the stopping times $v(n,k)$ implies condition $b(\omega)$ of Theorem 4. ∎

12. Remarks. a) As a rule, either of the sums involved in the convergence to the vector quadratic variation $[[X]]$ will be denoted by W^n:

$$[[X]]_t = X_0^2 + \lim \text{prob} W_t^n$$

b) If the set of measures $(I_X)_{E, L^\infty_{D^*}}$ is uniformly σ-additive, we can require that the stopping times $v(n,k)$ satisfy only conditions (i) and (ii) of Section A.

c) We shall see in the following sections that if X is semilocally summable relative to B and B' and has "finite quadratic variation," then $[[X]]^B$ has finite variation.

d) If E is a Hilbert space and X is semilocally summable relative to the inner product, then $[X]$ is an increasing, positive process and we have

$$[X]_t = X_0^2 + \lim \text{prob}_n \sum_{k \geq 0} \left|X_{v(n,k+1) \wedge t} - X_{v(n,k) \wedge t}\right|^2.$$

C. The quadratic variation

In this section $X : \mathbb{R}_+ \times \Omega \to E$ is a cadlag, adapted process and $B(x,y) = xy$ is a bilinear mapping of $E \times E$ into D, with $|x| = \sup\{|xy| : |y| \leq 1\}$, for $x \in E$. In order to ensure that $[[X]]$ has paths with finite variation, we need to examine the quadratic variation $[X]$.

13. Definition. *We say that X has finite quadratic variation if there exists a family of stopping times $(v(n,k))_{n,k}$ (cf. Section A), such that*

$$\lim_n \sum_{k \geq 0} \left|X_{v(n,k+1) \wedge t} - X_{v(n,k) \wedge t}\right|^2$$

exists and is finite a.s. for every $t \geq 0$. The process

$$[X]_t = |X_0|^2 + \lim_n \sum_{k \geq 0} |X_{v(n,k+1) \wedge t} - X_{v(n,k) \wedge t}|^2$$

is called the quadratic variation of X.

Note that $[X_0] = |X_0|^2$ and $[X]$ is positive, increasing and adapted. Moreover,

$$[X]_t = |X_0|^2 + \lim_n \sum_{k \geq 0} |X_{v(n,k+1)} - X_{v(n,k)}|^2 \, 1_{\{v(n,k) \leq t\}}.$$

In fact, for $v(n,k) \leq t < v(n,k+1)$, the difference between the sum in the definition and the above sum is $|X_t - X_{v(n,k)}|^2 - |X_{v(n,k+1)} - X_{v(n,k)}|^2$ and since X is right continuous, this has limit zero, as $n \to \infty$. In the above equality one can replace $1_{\{v(n,k) \leq t\}}$ by $1_{\{v(n,k) < t\}}$ if $t > 0$.

14. Remarks. a) As a rule, either of the sums involved in the definition of the quadratic variation $[X]$ will be denoted by V^n, hence

$$[X]_t = |X_0|^2 + \lim_n V_t^n.$$

b) If E is a Hilbert space and X is a semimartingale, then X has finite quadratic variation $[X]$, which is equal to the vector quadratic variation $[[X]]^B$, relative to the inner product B. This follows from Theorem 11 and the equality $x^2 = |x|^2$ for $x \in E$.

Next we state the following important property concerning the variation of $[[X]]^B$.

15. Proposition. *Assume that X is semilocally summable relative to B and B' and suppose that X has finite quadratic variation. Then $[[X]]^B$ has finite variation and*

$$\left| [[X]]_t^B - [[X]]_s^B \right| \leq [X]_t - [X]_s, \ for \ s \leq t.$$

Proof. For $0 \leq s < t$, we have

$$\left| \sum_{k \geq 0} \left(X_{v(n,k+1)} - X_{v(n,k)} \right)^2 1_{\{s < v(n,k) \leq t\}} \right|$$

$$\leq \sum_{k \geq 0} |X_{v(n,k+1)} - X_{v(n,k)}|^2 1_{\{s < v(n,k) \leq t\}}.$$

If we pass to the limit along a convenient subsequence, we obtain the desired inequality. ∎

16. Remark. When $[[X]]^B$ has finite variation and takes values in D, one can integrate pathwise, with respect to $[[X]]^B$, processes H with values in F if

$F \subset L(D, G)$. Although $[[X]]_t^B$ has an integral representation, we cannot use this formula in evaluating $\int H_s d[[X]]_s^B$, since we do not know if the processes $\int X_{s-} \cdot dX_s$ and $\int dX_x \cdot X_{s-}$ have local finite variation or if they are locally summable relative to the bilinear mapping $D \times F \to G$.

We shall use the following convergence result in studying integration with respect to $[[X]]^B$ in Theorem 18.

17. Proposition. *Let $(x, y) \mapsto xy$ be a continuous bilinear mapping from $E \times F$ into G. Suppose $h_n, h : [a, b] \to E$ are functions with equally bounded variations such that $\lim_n h_n = h$ pointwise. Let $g : [a, b] \to F$ be a caglad function. Then*

$$\int_{[a,s]} gdh^n \to \int_{[a,s]} gdh, \ for \ s \in [a, b].$$

Moreover, the convergence is uniform if $h_n \to h$ uniformly.

Proof. Choose a number V larger than $\sup_n var(h_n, [a, b])$ and $var(h, [a, b])$. Let $\epsilon > 0$. Obtain a division $a = t_0 < t_1 < \cdots < t_{p+1} = b$ such that the oscillation of g on each interval $(t_i, t_{i+1}], 0 \leq i \leq p$, is smaller than $\epsilon/4V$. To do this, set $g(t) = g(a)$ for $t < a$ and $g(t) = g(b)$ for $t > b$; for each $t \in [a, b]$, obtain an open interval (a_t, b_t) containing t such that g has oscillation less than $\epsilon/4V$ on $(a_t, t]$, by left continuity of g and oscillation less than $\epsilon/4V$ on $(t, b_t]$, since right limits exist. Use the compactness of $[a, b]$ to obtain a finite covering $(a_{t_i}, b_{t_i}), i = 0, 1, \ldots, p+1$ of $[a, b]$. We then intersect $(a_{t_i}, t_i] \cap (a, t_i]$ and $(t_i, b_{t_i}] \cap (t_i, b]$.

We adopt the convention that $h^n(a-) = h(a-) = 0$. If $s \in [a, b]$, then

$$I_n := \int_{[a,s]} gd(h^n - h)$$
$$= \int_{\{a\}} gd(h^n - h) + \sum_{0 \leq i \leq p} \int_{(t_i, t_{i+1}] \cap [a,s]} [g(t) - g(t_i)] dh^n(t)$$
$$- \sum_{0 \leq i \leq p} \int_{(t_i, t_{i+1}] \cap [a,s]} [g(t) - g(t_i)] dh(t)$$
$$+ \sum_{0 \leq i \leq p} \int_{(t_i, t_{i+1}] \cap [a,s]} g(t_i) d(h^n - h)(t).$$

Then

$$\left| \int_{\{a\}} gd(h^n - h) \right| \leq |g(a)||h^n(a) - h(a)| < \epsilon/4$$

for $n \geq n_0$, for a suitable n_0.

Also,

$$\sum_{0\leq i\leq p}\left|\int_{(t_i,t_{i+1}]\cap[a,s]}[g(t)-g(t_i)]dh^n\right|$$
$$\leq \sum_{0\leq i\leq p}\int_{(t_i,t_{i+1}]}|g(t)-g(t_i)|d|h^n| < \epsilon/4$$

and similarly

$$\sum_{0\leq i\leq p}\left|\int_{(t_i,t_{i+1}]\cap[a,s]}[g(t)-g(t_i)]dh\right| < \epsilon/4.$$

Finally,

$$\left|\int_{(t_i,t_{i+1}]\cap[a,s]} g(t_i)d(h^n-h)\right|$$
$$\leq \|g\|_\infty |h^n(t_{i+1}\wedge s) - h(t_{i+1}\wedge s) - (h^n(t_i\wedge s) - h(t_i\wedge s))|$$
$$< \epsilon/4(p+1),$$

for n sufficiently large, for $0 \leq i \leq p$. Note that $\|g\|_\infty < \infty$, since g is caglad.

Thus, for n large enough, $|I_n| < \epsilon$, that is, $|I_n| \to 0$, which proves the first conclusion. If $h_n \to h$ uniformly, then $I_n \to 0$ uniformly on $[a,b]$. ∎

We apply the above results to study the integral with respect to $[[X]]^B$.

18. Theorem. *Assume that X is semilocally summable relative to the bilinear maps B and B' and suppose X has finite quadratic variation $[X]$.*

Let $F = L(D,G)$ and let H be an F-valued cadlag, adapted process. Then $[[X]]^B$ has finite variation, hence the Stieltjes integral $\int_{[0,t]} H_{s-} d[[X]]_s^B$ is defined pathwise. Let $(v(m,k))$ be a family of stopping times used to define the quadratic variation $[X]$. Assume that there is a sequence $b_m \downarrow 0$ such that for $t \in [v(m,k), v(m,k+1))$, we have

b)
$$|H_t - H_{v(m,k)}| \leq b_m.$$

Then there is a subsequence $\{n\}$ of $\{m\}$ such that a.s.

$$\int_{[0,t]} H_{s-} d[[X]]_s^B = \lim_n \sum_{k\geq 0} H_{v(n,k)\wedge t}\left(X_{v(n,k+1)\wedge t} - X_{v(n,k)\wedge t}\right)^2$$
$$= \lim_n \sum_{k\geq 0} H_{v(n,k)}\left(X_{v(n,k+1)} - X_{v(n,k)}\right)^2 1_{\{v(n,k)\leq t\}}.$$

Proof. Denote

$$H^m = \sum_{k\geq 0} H_{v(m,k)} 1_{(v(m,k),v(m,k+1)]}.$$

Then $H^m \to H_-$ pathwise, uniformly on compact time intervals by condition b) above.

Denote
$$W_t^m = \sum_{k \geq 0} \left(X_{v(m,k+1)} - X_{v(m,k)}\right)^2 1_{\{v(m,k) \leq t\}}.$$

This process is simple, constant on $[v(m,k), v(m,k+1))$ and has jumps at $v(m,k)$ equal to $(X_{v(m,k+1)} - X_{v(m,k)})^2$. Let $|W^m|$ denote the variation of the process W^m defined by

$$|W^m|_t = |W_0^m| + var\ (W^m, [0,t]).$$

Then $|W^m|_0 = |W_0^m|$ and for every t we have

$$\begin{aligned}|W^m|_t &= \sum_{k \geq 0} \left|\Delta W_{v(m,k)}^m\right| 1_{\{v(m,k) \leq t\}} \\ &= \sum_{k \geq 0} |\left(X_{v(m,k+1)} - X_{v(m,k)}\right)^2| 1_{\{v,(m,k) \leq t\}} \\ &\leq \sum_{k \geq 0} \left|X_{v(m,k+1)} - X_{v(m,k)}\right|^2 1_{\{v,(m,k) \leq t\}}.\end{aligned}$$

Denote the last term by V_t^m. Now let $m \to \infty$. We have $\lim_m V_t^m = [X]_t - |X_0|^2$ and $\lim\ \mathrm{prob}_m W_t^m = [[X]]_t^B - X_0^2$. Choose a subsequence $\{n\}$ of $\{m\}$ such that $\lim_n W_t^n = [[X]]_t^B - X_0^2$ a.s. by Theorem 11. Since the sequence (V^n) is bounded a.s. on $[0,t]$, the sequence of variations $(|W^n|)$ is also bounded a.s. on $[0,t]$. By Proposition 17, we deduce that a.s.,

$$\int_{[0,t]} H_{s-} dW_s^n \to \int_{[0,t]} H_{s-} d([[X]]_s^B - X_0^2) = \int_{[0,t]} H_{s-} d[[X]]_s^B.$$

On the other hand

$$\int_{[0,t]} (H_s^n - H_{s-}) dW_s^n \to 0\ a.s.$$

In fact, let $\epsilon > 0$. Then there is an n_0 such that for $n \geq n_0$ and $s \in [0,t]$, we have $|H_s^n(\omega) - H_{s-}(\omega)| < \epsilon$; therefore

$$\left|\int_{[0,t]} (H_s^n(\omega) - H_{s-}(\omega)) dW_s^n(\omega)\right| \leq \epsilon |W^n|_t(\omega) \leq \epsilon V_t^n(\omega).$$

It follows that

$$\limsup_n \left|\int_{[0,t]} (H_s^n(\omega) - H_{s-}(\omega)) dW_s^n(\omega)\right| \leq \epsilon [X]_t(\omega),\ a.s.$$

Thus
$$\lim_n \int_{[0,t]} (H_s^n(\omega) - H_{s-}(\omega))dW_s^n(\omega) = 0, \text{ a.s.}$$

Now observe that
$$\int_{[0,t]} H_s^n dW_s^n = \sum_{k \geq 0} H_{v(n,k)} \Delta W_{v(n,k)}^n 1_{\{v(n,k) \leq t\}}$$
$$= \sum_{k \geq 0} H_{v(n,k) \wedge t} \left(X_{v(n,k+1)} - X_{v(n,k)}\right)^2 1_{\{v(n,k) \leq t\}}$$

and
$$\int_{[0,t]} H_s^n dW_s^n - \sum_{k \geq 0} H_{v(n,k)} \left(X_{v(n,k+1) \wedge t} - X_{v(n,k) \wedge t}\right)^2$$
$$= H_{v(n,k)} \left[\left(X_{v(n,k+1)} - X_{v(n,k)}\right)^2 - \left(X_t - X_{v(n,k)}\right)^2\right] 1_{[v(n,k), v(n,k+1))}(t),$$

and this last term approaches zero as $n \to \infty$. This completes the proof. ∎

19. Remark. Assume E is a Hilbert space and the bilinear mapping B is the inner product. In this case $[X] = [[X]]^B$ and by Theorem 11 any family of stopping times satisfying conditions (i)–(iii) of Section A can be used in the definition of $[X]$. Then we can take $v(n,k)$ to satisfy condition b) of Theorem 2, which implies condition b) of the above theorem.

20. Corollary. *Assume X is semilocally summable relative to the tensor product and has finite quadratic variation $[X]$ and let $f : E \to G$ be a function of class C^2, such that f'' is uniformly continuous on bounded subsets of E. Then the Stieltjes integral $\int_{(0,t]} f''(X_{s-})d[[X]]_s$ is defined pathwise and there is a family of stopping times $v(n,k)$ (which can be used in the definition of $[X]$) such that, a.s.,*

$$\int_{(0,t]} f''(X_{s-})d[[X]]_s = \lim_n \sum_{k \geq 0} f''(X_{v(n,k) \wedge t})(X_{v(n,k+1) \wedge t} - X_{v(n,k) \wedge t})^{\otimes 2}$$
$$= \lim_n \sum_{k \geq 0} f''(X_{v(n,k)})(X_{v(n,k+1)} - X_{v(n,k)})^{\otimes 2} 1_{\{v(n,k) \leq t\}}.$$

Proof. If we set $F = L(E \hat{\otimes}_\pi E, G)$, then $f'' : E \to F$. Let $H_t = f''(X_t)$, for $t \geq 0$. Then H satisfies condition b) of Theorem 18. In fact, there is a sequence $T_i \uparrow \infty$ of stopping times such that $|1_{[0,T_i)} X| \leq i$. Let $\omega \in \Omega$ and $t_0 \geq 0$. Let i be such that $t_0 < T_i(\omega)$. Let $b_n \downarrow 0$. Since f'' is uniformly continuous on the bounded set $\{x : |x| \leq i\} \subset E$, there is a $\delta > 0$ such that for $x, y \in E$, each bounded by i, with $|x - y| < \delta$, we have $|f''(x) - f''(y)| < b_n$. Since $a_n \downarrow 0$, there is a $j(n)$, such that $a_{j(n)} < \delta$. For any $t \leq t_0$, let k be such that
$$t \in [v(j(n), k), v(j(n), k+1)).$$

Then by condition (iii) of Section A, we have
$$|X_t - X_{v(j(n),k)}| \le a_{j(n)} < \delta;$$
therefore
$$|f''(X_t(\omega)) - f''\left(X_{v(j(n),k)}(\omega)\right)| < b_n.$$
Hence H_t satisfies condition b) of Theorem 18 for the family of stopping times $v(j(n),k)$. We can then apply Theorem 18 and deduce the corollary, using the fact that $1_{[0,t]}H_- = 1_{(0,t]}f''(X_-)$. ∎

D. The process of jumps

In this section we shall develop further results about the jumps of X which will enable us to prove Itô's formula.

The first is a simple but useful lemma.

21. Lemma. *Let $X : \mathbb{R}_+ \times \Omega \to E$ be cadlag and adapted. Let $(v(n,k))$ be the usual family of stopping times defined in Section A satisfying conditions (i)–(iii). Then for every n and ω we have*
$$\{s : |\Delta X_s| > 2a_n\} \subset \{v(n,k) : k \ge 0\}.$$

Proof. Assume $|\Delta X_s| > 2a_n$. If $s = 0$, then $s = v(n,0)$. If $s > 0$ and $v(n,k) < s \le v(n,k+1)$, then $s = v(n,k+1)$, since, otherwise, $|\Delta X_s - X_{v(n,k)}| \le a_n$ and, hence $|X_s - X_{s'}| \le 2a_n$ for $v(n,k) < s' < s$; therefore, letting $s' \uparrow s$, we obtain $|\Delta X_s| \le 2a_n$, a contradiction. ∎

22. Theorem. *Assume that X is E-valued, cadlag, adapted and has finite quadratic variation $[X]$ and let $(v(n,k))$ be a family of stopping times defining $[X]$. Then:*

a) *The family $\{|\Delta X_s|^2 : s \le t\}$ is summable for each t;*

b) *The process of jumps defined for any $t \ge 0$ by*
$$J_t := \sum_{s \le t} |\Delta X_s|^2$$
is increasing, cadlag, adapted and satisfies
$$\sum_{s \le t} |\Delta X_s|^2 \le [X]_t;$$

c) *We have*
$$\sum_{s \le t} |\Delta X_s|^2 = \lim_n \sum_{k \ge 0} |\Delta X_{v(n,k)}|^2 1_{\{v(n,k) \le t\}}$$
$$= \lim_n \sum_{k \ge 0} |\Delta X_{v(n,k)}|^2 1_{\{|\Delta X_{v(n,k)}| > 2a_n\}} 1_{\{v(n,k) \le t\}}$$

uniformly for t on any compact time interval.

d) *Let $V_t^n = \sum_{k \geq 0} |X_{v(n,k+1) \wedge t} - X_{v(n,k) \wedge t}|^2$ and assume that for every ω there exists a subsequence (n_i), which may depend on ω, such that $V_t^{n_i}(\omega)$ converges uniformly to $[X]_t(\omega) - |X_0(\omega)|^2$ on compact time intervals. Then $[X]$ is cadlag and the process*

$$[X]_t^c(\omega) = [X]_t(\omega) - \sum_{s \leq t} |\Delta X_s|^2(\omega)$$

is continuous.

Remark. The assumption in d) is satisfied, for example, if E is a Hilbert space and X is locally summable relative to the inner product, by Theorem 11.

Proof. X is cadlag and by using condition (ii) for stopping times $v(n,k)$, we deduce that for $s > 0$, we have

$$\Delta X_s = \lim_n \sum_{k \geq 0} \left(X_{v(n,k+1)} - X_{v(n,k)} \right) 1_{[v(n,k),v(n,k+1))}(s).$$

Since, for each s and ω, the above sum reduces to one term, we deduce that

$$|\Delta X_s|^2 = \lim_n \sum_{k \geq 0} \left| X_{v(n,k+1)} - X_{v(n,k)} \right|^2 1_{[v(n,k),v(n,k+1))}(s)$$

Let $0 < s_1 < s_2 < \cdots, s_j \leq t$ and choose $(v(n,k))$ to be a defining sequence for $[X]$. Then

$$\sum_{1 \leq i \leq j} |\Delta X_{s_i}|^2$$
$$= \lim_n \sum_{1 \leq i \leq j} \sum_{k \geq 0} \left| X_{v(n,k+1)} - X_{v(n,k)} \right|^2 1_{[v(n,k),v(n,k+1))}(s_i)$$
$$\leq \lim_n \sum_{k \geq 0} \left| X_{v(n,k+1)} - X_{v(n,k)} \right|^2 1_{\{v(n,k) \leq t\}} = [X]_t - |X_0|^2.$$

It follows that $\sum_{0 < s \leq t} |\Delta X_s|^2 \leq [X]_t - |X_0|^2$, hence $\sum_{0 \leq s \leq t} |\Delta X_{s_i}|^2 \leq [X]_t$ which proves a).

It is obvious that the process of jumps J is increasing and cadlag. The fact that J is adapted will follow from c).

To prove c), we note that, by the above Lemma 21, we have

$$\sum_{s \leq t} |\Delta X_s|^2 - \sum_{k \geq 0} |\Delta X_{v(n,k)}|^2 1_{\{v(n,k) \leq t\}}$$
$$\leq \sum_{s \leq t} |\Delta X_s|^2 - \sum_{k \geq 0} |\Delta X_{v(n,k)}|^2 1_{\{|\Delta X_{v(n,k)}| > 2a_n\}} 1_{\{v(n,k) \leq t\}}$$
$$\leq \sum_{s \leq t} |\Delta X_s|^2 - \sum_{s \leq t} |\Delta X_s|^2 1_{\{|\Delta X_s| > 2a_n\}} = \sum_{s \leq t} |\Delta X_s|^2 1_{\{|\Delta X_s| \leq 2a_n\}}$$

and the last term converges to zero, as $n \to \infty$. To prove this let $\epsilon > 0$. Then there is a finite family of numbers $0 \le s(i) \le t$ with $i = 1, 2, ..., N$ such that

$$\sum_{s \le t} |\Delta X_s|^2 < \sum_{1 \le i \le N} |\Delta X_{s(i)}|^2 + \epsilon.$$

Since $a_n \downarrow 0$, there is an n_0 such that for $n \ge n_0$ we have $4Na_n^2 < \epsilon$. Let $n \ge n_0$. Then

$$\sum_{i \le N} |\Delta X_{s(i)}|^2 1_{\{|\Delta X_{s(i)}| \le 2a_n\}} \le N4a_n^2 < \epsilon,$$

therefore

$$\sum_{1 \le i \le N} |\Delta X_{s(i)}|^2 \le \sum_{1 \le i \le N} |\Delta X_{s(i)}|^2 1_{\{|\Delta X_{s(i)}| > 2a_n\}} + \epsilon$$

$$\le \sum_{s \le t} |\Delta X_s|^2 1_{\{|\Delta X_s| > 2a_n\}} + \epsilon.$$

It follows that

$$\sum_{s \le t} |\Delta X_s|^2 - \sum_{s \le t} |\Delta X_s|^2 1_{\{|\Delta X_s| > 2a_n\}} < 2\epsilon;$$

consequently

$$\lim_n \sum_{s \le t} |\Delta X_s|^2 1_{\{|\Delta X_s| \le 2a_n\}} = 0.$$

But $\sum_{s \le t} |\Delta X_s|^2 1_{\{|\Delta X_s| \le 2a_n\}}$ is increasing as a function of t; therefore, it converges to zero, as $n \to \infty$, uniformly on compact time intervals.

Now we shall prove d). For $t = 0$ we have $[X]_0 - J_0 = |X_0|^2 - |\Delta X_0|^2 = 0$, hence $[X] - J$ is continuous at 0. Now fix ω; we shall show that $[X] - J$ is continuous on $(0, +\infty)$. By the assumption in d), there is a subsequence (n_i) such that $V_t^{n_i}(\omega) \to [X]_t(\omega) - |X_0(\omega)|^2$ uniformly on compact time intervals. To simplify the notation, assume $n_i = n$. Each V^n is cadlag, since for fixed t, V_t^n is a finite sum; as a consequence, $[X]$ is cadlag. For $t \in (v(n,j), v(n,j+1)]$, the jump of V^n at t is

$$\Delta V_t^n = |X_t - X_{v(n,j)}|^2 - |X_{t-} - X_{v(n,j)}|^2,$$

therefore $|\Delta V_t^n| \le 2a_n^2$. Denote

$$J_t^n = \sum_{k \ge 0} |\Delta X_{v(n,k)}|^2 1_{\{v(n,k) \le t\}}.$$

Then by c), $J_t^n \to J_t$, uniformly on compact time intervals. The jumps of J_t^n occur at $t = v(n, j+1)$ and

$$\Delta J_{v(n,j+1)}^n = |\Delta X_{v(n,j+1)}|^2.$$

For $t = v(n, j+1)$ we have

$$|\Delta(V_t^n - J_t^n)|$$
$$= \left| \left|X_{v(n,j+1)} - X_{v(n,j)}\right|^2 - \left|X_{v(n,j+1)-} - X_{v(n,j)}\right|^2 - \left|\Delta X_{v(n,j+1)}\right|^2 \right|$$
$$\leq 2\left|\Delta X_{v(n,j+1)}\right| \left|X_{v(n,j+1)-} - X_{v(n,j)}\right|$$
$$\leq 2 \sup_{s \leq t} |\Delta X_s| \, a_n.$$

Since $\sum_{s \leq t} |\Delta X_s|^2 < \infty$, we have $\sup_{s \leq t} |\Delta X_s| < \infty$; therefore $\lim_n |\Delta(V_t^n - J_t^n)| = 0$, uniformly on compact time intervals.

If $t \neq v(n, j+1)$, then for any n and j, we have $\Delta J_t^n = 0$, hence

$$|\Delta(V_t^n - J_t^n)| = |\Delta V_t^n| \leq 2a_n^2.$$

As a result, in either case, for $t > 0$, we have $\lim_n |\Delta(V_t^n - J_t^n)| = 0$ uniformly on bounded time intervals. Since $\lim_n J_t^n = J_t$ uniformly on compact time intervals, we can interchange limits and conclude that for $t > 0$,

$$\Delta([X]_t - J_t) = \lim_{s \uparrow t} \{([X]_t - [X]_s) - (J_t - J_s)\}$$
$$= \lim_n \lim_{s \uparrow t} \{(V_t^n - V_s^n) - (J_t^n - J_s^n)\}$$
$$= \lim_n \Delta(V_t^n - J_t^n) = 0.$$

Thus $[X] - J$ is continuous at $t > 0$. This completes the proof. ∎

We now turn to the regularity of $[[X]]^B$.

23. Theorem. *Let $B: E \times E \to D$ be a continuous bilinear map, denoted $B(x, y) = xy$, such that $|x| = \sup\{|xy| : |y| \leq 1\}$ for each $x \in E$. Assume X is E-valued, cadlag, adapted, and has finite quadratic variation $[X]$. Let $(v(n, k))$ be a family of stopping times defining $[X]$. Then:*

a) *The family $\{\Delta(X_s)^2; s \leq t\}$ is summable for each t;*
b) *The process of vector-valued jumps*

$$vJ_t := \sum_{s \leq t} (\Delta X_s)^2$$

is cadlag, adapted, and has finite variation $|vJ|$; in addition, we have

$$|vJ|_t \leq \sum_{s \leq t} |\Delta X_s|^2;$$

c) *We have*

$$\sum_{s \leq t} (\Delta X_s)^2 = \lim_n \sum_{k \geq 0} (\Delta X_{v(n,k)})^2 1_{\{v(n,k) \leq t\}}$$
$$= \lim_n \sum_{k \geq 0} (\Delta X_{v(n,k)})^2 1_{\{|\Delta X_{v(n,k)}| > 2a_n\}} 1_{\{v(n,k) \leq t\}}$$

uniformly for t in any compact time interval.
d) Assume that X is semilocally summable relative to B and B'. Then the process

$$[[X]]_t^c = [[X]]_t^B - \sum_{s \leq t} (\Delta X_s)^2$$

is continuous.
e) Let $H : \mathbb{R}_+ \times \Omega \to L(D,G)$ be a cadlag, adapted process. The family $\{H_{s-}(\Delta X_s)^2 : s \leq t\}$ is summable and we have

$$\begin{aligned} J_t(H,X) &:= \sum_{0 < s \leq t} H_{s-}(\Delta X_s)^2 \\ &= \lim_n \sum_{k>0} H_{v(n,k)-}(\Delta X_{v(n,k)})^2 1_{\{v(n,k) \leq t\}} \\ &= \lim_n \sum_{k>0} H_{v(n,k)-}(\Delta X_{v(n,k)})^2 1_{\{|\Delta X_{v(n,k)}| > 2a_n\}} 1_{\{v(n,k) \leq t\}} \end{aligned}$$

uniformly on compact time intervals. The process $J_t(H,X)$ is adapted and right continuous.

Proof. a) Since $|x^2| \leq |x|^2$ for $x \in E$, by Theorem 22 it follows that $\{(X_s)^2; s \leq t\}$ is absolutely summable.
b) The process of jumps $J_t = \sum_{s \leq t} |\Delta X_s|^2$ is increasing and finite. For $t' < t$ we have

$$\begin{aligned} |vJ_t - vJ_{t'}| &= \left| \sum_{t' < s \leq t} (\Delta X_s)^2 \right| \\ &\leq \sum_{t' < s \leq t} |\Delta X_s|^2 = J_t - J_{t'}. \end{aligned}$$

It follows that the process vJ has finite variation

$$|vJ|_t := |vJ_0| + var(vJ, [0,t]);$$

and $|vJ|_t \leq J_t$. The fact that vJ is cadlag and adapted follows from c) and from the fact that the limits in c) of the cadlag processes are uniform on compact time intervals.
c) Note that, by Lemma 21,

$$\left| \sum_{s \leq t} (\Delta X_s)^2 - \sum_{k \geq 0} (\Delta X_{v(n,k)})^2 1_{\{v(n,k) \leq t\}} \right|$$
$$\leq \sum_{s \leq t} |\Delta X_s|^2 1_{\{|\Delta X_s| \leq 2a_n\}},$$

and

$$\left| \sum_{s \leq t}(\Delta X_s)^2 - \sum_{k \geq 0}(\Delta X_{v(n,k)})^2 1_{\{|\Delta X_s|>2a_n\}} 1_{\{v(n,k)\leq t\}} \right|$$

$$\leq \sum_{s \leq t} |\Delta X_s|^2 1_{\{|\Delta X_s|\leq 2a_n\}}.$$

Since the last term converges to zero, as $n \to \infty$, uniformly on any compact time interval, assertion c) follows.

d) Denote

$$W_t^n = \sum_{k \geq 0}(X_{v(n,k+1)\wedge t} - X_{v(n,k)\wedge t})^2,$$

and

$$vJ_t^n = \sum_{k \geq 0}(\Delta X_{v(n,k)})^2 1_{\{v(n,k)\leq t\}}.$$

Note that if $t \in (v(n,j), v(n,j+1)]$, then

$$\Delta W_t^n = (X_t - X_{v(n,j)})^2 - (X_{t-} - X_{v(n,j)})^2.$$

If $t = 0$, then $[[X]]_0^B - vJ_0 = X_0^2 - (\Delta X_0)^2 = 0$, hence $[[X]]^B - vJ$ is continuous at 0. If $t > 0$ and $t \in (v(n,j), v(n,j+1)]$, the jumps of vJ_t^n occur at $t = v(n, j+1)$ and $\Delta vJ_t^n = (\Delta X_{v(n,j+1)})^2$; for this value of t, we have

$$|\Delta(W_t^n - vJ_t^n)| = \Big|(X_{v(n,j+1)} - X_{v(n,j)})^2 - (X_{v(n,j+1)-} - X_{v(n,j)})^2$$
$$-(\Delta X_{v(n,j+1)})^2\Big|$$
$$= \Big|\Delta X_{v(n,j+1)}(X_{v(n,j+1)-} - X_{v(n,j)})$$
$$+(X_{v(n,j+1)-} - X_{v(n,j)})\Delta X_{v(n,j+1)}\Big|$$
$$\leq 2\big|\Delta X_{v(n,j+1)}\big|\big|(X_{v(n,j+1)-} - X_{v(n,j)})\big|$$
$$\leq 2\sup_{s \leq t}|\Delta X_s|a_n,$$

where $a_n \downarrow 0$ is the sequence used in condition (iii) of the stopping times $v(n, k)$. In this case, $\Delta(W_t^n - vJ_t^n) \to 0$, as $n \to \infty$, uniformly on compact time intervals. If $t \in (v(n,j), v(n,j+1))$, then $\Delta vJ_t^n = 0$ and

$$|\Delta(W_t^n - vJ_t^n)| \leq |X_t - X_{v(n,j)}|^2 + |X_{t-} - X_{v(n,j)}| \leq 2a_n^2.$$

Again, $\Delta(W_t^n - vJ_t^n) \to 0$, as $n \to \infty$, uniformly on compact time intervals.

By Theorem 11 and assertion c), there exists a subsequence (n_r) such that $\lim_r (W_t^{n_r} - vJ_t^{n_r}) = [[X]]_t^B - \sum_{s \leq t}(\Delta X_s)^2$ uniformly on compact time intervals. If we interchange limits, it follows that $[[X]]_t^B - \sum_{s \leq t}(\Delta X_s)^2$ is a continuous process.

e) Since H is cadlag, for fixed t and ω, the function $s \mapsto H_s(\omega)$ is bounded on the interval $[0, t]$ by a constant $M_t(\omega)$.

If we denote
$$J_t = \sum_{s \leq t} |\Delta X_s|^2,$$
then
$$\sum_{0 < s \leq t} |H_{s-}(\Delta X_s)^2| \leq M_t J_t,$$
hence $J_t(H, X)$ is defined. By Lemma 21 we have
$$\left| \sum_{0 < s \leq t} H_{s-}(\Delta X_s)^2 - \sum_{k > 0} H_{v(n,k)-}(\Delta X_{v(n,k)})^2 1_{\{v(n,k) \leq t\}} \right|$$
$$\leq M_t \sum_{s \leq t} |\Delta X_s|^2 1_{\{|\Delta X_s| \leq 2a_n\}}$$

and

$$\left| \sum_{0 < s \leq t} H_{s-}(\Delta X_s)^2 - \sum_{k > 0} H_{v(n,k)-}(\Delta X_{v(n,k)})^2 1_{\{|\Delta X_s| > 2a_n\}} 1_{\{v(n,k) \leq t\}} \right|$$
$$\leq M_t \sum_{s \leq t} |\Delta X_s|^2 1_{\{|\Delta X_s| \leq 2a_n\}},$$

and the last term converges to 0, as $n \to \infty$, uniformly on compact time intervals (see the proof of Theorem 22). Assertion e) follows.

From the equalities of assertion e) we also deduce that $J_t(H, X)$ is adapted. This process is cadlag since J_t is cadlag and for s, t we have
$$|J_t(H, X) - J_s(H, X)| \leq M_t(J_t - J_s).$$
This completes the proof. ∎

We shall use the following corollary of assertion e) in establishing Itô's formula.

24. Corollary. *Assume X is E-valued, cadlag, adapted and has finite quadratic variation $[X]$. Let $(v(n, k))$ be a family of stopping times defining $[X]$. Let $f : E \to G$ be a function of class C^2; hence $f'' : E \to L(E \hat{\otimes}_\pi E, G)$. Then the family $\{f''(X_{s-})(\Delta X_s)^{\otimes 2} : s \leq t\}$ is summable and*
$$\sum_{0 < s \leq t} f''(X_{s-})(\Delta X_s)^{\otimes 2} = \lim_n \sum_{k > 0} f''(X_{v(n,k)-})(\Delta X_{v(n,k)})^{\otimes 2} 1_{\{v(n,k) \leq t\}}$$
uniformly on compact time intervals.

Proof. We take $H_t = f''(X_t)$ and $D = E \hat{\otimes}_\pi E$ in assertion e) of Theorem 23. We notice that the equalities in assertion e) of Theorem 23 and in Corollary

24 remain true if we allow $0 \leq s \leq t$ in the summation of the left-hand side and $k \geq 0$ in the summation of the right-hand side. ∎

We state Taylor's formula for reference.

25. Proposition. *Let $f : E \to G$ be a function of class C^2, with f'' uniformly continuous on bounded subsets of E. Then there exists a function $R : E \times E \to L(E \hat{\otimes}_\pi E, G)$, with $R(x, x) = 0$ for $x \in E$, such that $\lim_{x \to y} R(y, x) = 0$ uniformly with respect to x, belonging to any given bounded set and such that for $x, y \in E$, we have*

$$f(y) = f(x) + f'(x)(y - x) + \frac{1}{2} f''(x)(y - x)^{\otimes 2} + R(y, x)(y - x)^{\otimes 2}.$$

For the proof, we use the Taylor formula

$$f(y) = f(x) + f'(x)(y - x) + \int_{[0,1]} (1 - s) f''(x + s(y - x)) \, ds (y - x)^{\otimes 2}$$

and let $R(y, x) = \int_{[0,1]} (1 - s)(f''(x + s(y - x)) - f''(x)) \, ds$. ∎

26. Proposition. *Suppose X is E-valued and has finite quadratic variation $[X]$. Let $f : E \to G$ be a function of class C^2 such that f'' is bounded on bounded subsets of E.*

Then, for every t, we have

$$\sum_{s \leq t} |f(X_s) - f(X_{s-}) - f'(X_{s-}) \Delta X_s| < \infty.$$

The process

$$Q_t := \sum_{s \leq t} (f(X_s) - f(X_{s-}) - f'(X_{s-}) \Delta X_s)$$

is adapted, cadlag and has finite variation.

Proof. Since X is cadlag, for fixed ω and t, $X_s(\omega)$ is bounded in norm on $[0, t]$ by a constant a depending on ω and t; hence f'' is bounded in norm by a constant c on the set $\{x : |x| \leq a\}$. We then have

$$|X_{t-} + u(X_t - X_{t-})| = |X_{t-}(1 - u) + X_t u| \leq a,$$

therefore

$$|f''(X_{t-} + u(X_t - X_{t-}))| \leq c,$$

consequently

$$\left| \int_{[0,1]} (1 - u) f''(X_{t-} + u(X_t - X_{t-})) du \right|$$
$$\leq \int_{[0,1]} (1 - u) |f''(X_{t-} + u(X_t - X_{t-}))| \, du \leq \frac{c}{2}.$$

If we use Taylor's formula for f, it follows that

$$|f(X_t) - f(X_{t-}) - f'(X_{t-})(X_t - X_{t-})|$$
$$\leq \left(\frac{c}{2}\right)|(X_t - X_{t-})^{\otimes 2}| = \left(\frac{c}{2}\right)|\Delta X_t|^2.$$

Since X has finite quadratic variation, the summability follows, hence Q_t is defined and we have

$$|Q_t - Q_{t'}| \leq \left(\frac{c}{2}\right)(J_t - J_{t'}),$$

for $t' < t$. This proves that Q_t is cadlag and of finite variation.

To show that Q_t is adapted, note that the jumps of $f(X_t)$ and X are contained in a countable union of disjoint graphs of stopping times T_n. Hence

$$Q_t = \sum_{1 \leq n} f(X_{T_n \wedge t}) - f(X_{(T_n \wedge t)-}) - f'(X_{(T_n \wedge t)-})\Delta X_{T_n \wedge t},$$

and it follows that Q_t is adapted. ∎

E. Itô's formula

27. Theorem. *Let $X : \mathbb{R}_+ \times \Omega \to E$ be semilocally summable with respect to the bilinear mappings $B_1(x, y) = x \otimes y$ and $B'_1(x, y) = y \otimes x$ of $E \times E$ into $E \hat{\otimes}_\pi E$ and with respect to $B_2(x, y) = y(x)$ of $E \times L(E, G)$ into G (that is, we regard X, in this instance, as taking values in $L(L(E, G), G))$. Assume that X has finite quadratic variation $[X]$.*

Let $f : E \to G$ be a function of class C^2 such that $f'' : E \to L(E\hat{\otimes}_\pi E, G)$ is uniformly continuous and bounded on bounded subsets of E.

Then for every $t > 0$, we have, a.s.,

$$f(X_t) = f(X_0) + \int_{(0,t]} f'(X_{s-})\, dX_s + \frac{1}{2}\int_{(0,t]} f''(X_{s-})\, d[[X]]_s$$
$$+ \sum_{0 < s \leq t} [f(X_s) - f(X_{s-}) - f'(X_{s-})\Delta X_s - \frac{1}{2}f''(X_{s-})(\Delta X_s)^{\otimes 2}]$$
$$= f(X_0) + \int_{(0,t]} f'(X_{s-})\, dX_s + \frac{1}{2}\int_{(0,t]} f''(X_{s-})\, d[[X]]_s^c$$
$$+ \sum_{0 < s \leq t} [f(X_s) - f(X_{s-}) - f'(X_{s-})\Delta X_s].$$

Proof. (cf [Me]) All the processes in the formula are defined. In fact, $f' : E \to L(E, G)$ is continuous, hence $f'(X_{s-})$ is caglad; therefore it is locally integrable with respect to X, which is semilocally summable relative to B_2; consequently, the first integral is defined. Next, $f''(X_{s-})$ is caglad;

therefore pathwise integrable, on every compact time interval, with respect to the process $[[X]]$, which is of finite variation; consequently the second integral is also defined. The sum in the formula is defined and is cadlag and adapted by Corollary 24 and Proposition 26. Now we have to establish the equality.

Choose a sequence $a_n \downarrow 0$ and the corresponding family $(v(n,k))$ of stopping times defining the quadratic variation $[X]$ (see Definition 13).

Fix $t > 0$. By Taylor's formula, for each n we have

$$f(X_t) - f(X_0) = \sum_{k \geq 0} [f(X_{v(n,k+1) \wedge t}) - f(X_{v(n,k) \wedge t})]$$

(1) $$= \sum_{k \geq 0} f'(X_{v(n,k) \wedge t})(X_{v(n,k+1) \wedge t} - X_{v(n,k) \wedge t})$$

(2) $$+ \frac{1}{2} \sum_{k \geq 0} f''(X_{v(n,k) \wedge t})(X_{v(n,k+1) \wedge t} - X_{v(n,k) \wedge t})^{\otimes 2}$$

(3) $$+ \sum_{k \geq 0} R_{n,k}(X_{v(n,k+1) \wedge t} - X_{v(n,k) \wedge t})^{\otimes 2}$$

where we set $R_{n,k} = R(X_{v(n,k+1) \wedge t}, X_{v(n,k) \wedge t})$. Note that $R_{n,k} = R_{n,k} 1_{\{v(n,k) < t\}}$. By Corollaries 6 and 8, taking a subsequence if necessary, we deduce that the sum (1) converges pointwise to $\int_{(0,t]} f'(X_{s-})\, dX_s$. By Corollary 20, again by taking a subsequence if necessary, we deduce that the sum (2) converges pointwise to $\frac{1}{2} \int_{(0,t]} f''(X_{s-})\, d[[X]]_s$.

We must prove that $\lim_n \sum_{k \geq 0} R_{n,k}(X_{v(n,k+1) \wedge t} - X_{v(n,k) \wedge t})^{\otimes 2}$ is equal, pointwise, to the sum

$$\sum_{0 < s \leq t} [f(X_s) - f(X_{s-}) - f'(X_{s-})\Delta X_s - \frac{1}{2} f''(X_{s-})(\Delta X_s)^{\otimes 2}].$$

The above sum can be written as

$$\sum_{0 < s \leq t} R(X_s, X_{s-})(\Delta X_s)^{\otimes 2}.$$

We shall assume, at first, that X is bounded and let $B = \{x : |x| < a\}$ be an open ball in E containing the range of X.

For $v(n,k) < t$, we can write

$$R_{n,k}(X_{v(n,k+1) \wedge t} - X_{v(n,k) \wedge t})^{\otimes 2} = R_{n,k}(X_{v(n,k+1) \wedge t} - X_{v(n,k)})^{\otimes 2}$$
$$= f(X_{v(n,k+1) \wedge t}) - f(X_{v(n,k)}) - f'(X_{v(n,k)})(X_{v(n,k+1) \wedge t} - X_{v(n,k)})$$
$$- \frac{1}{2} f''(X_{v(n,k)})(X_{v(n,k+1) \wedge t} - X_{v(n,k)})^{\otimes 2}$$
$$= v_{n,k} + w_{n,k} + y_{n,k},$$

where

$$v_{n,k} = f(X_{v(n,k+1)\wedge t}) - f(X_{v(n,k+1)\wedge t-}) - f'(X_{v(n,k)})\Delta X_{v(n,k+1)\wedge t}$$
$$- \frac{1}{2}f''(X_{v(n,k)})(\Delta X_{v(n,k+1)\wedge t})^{\otimes 2};$$

$$w_{n,k} = f(X_{v(n,k+1)\wedge t-})$$
$$- f(X_{v(n,k)}) - f'(X_{v(n,k)})(X_{v(n,k+1)\wedge t-} - X_{v(n,k)})$$
$$- \frac{1}{2}f''(X_{v(n,k)})(X_{v(n,k+1)\wedge t-} - X_{v(n,k)})^{\otimes 2}$$
$$= R(X_{v(n,k+1)\wedge t-}, X_{v(n,k)})(X_{v(n,k+1)\wedge t-} - X_{v(n,k)})^{\otimes 2};$$

$$y_{n,k} = -\frac{1}{2}f''(X_{v(n,k)})[(\Delta X_{v(n,k+1)\wedge t}) \otimes (X_{v(n,k+1)\wedge t-} - X_{v(n,k)})$$
$$+ (X_{v(n,k+1)\wedge t-} - X_{v(n,k)}) \otimes \Delta X_{v(n,k+1)\wedge t}].$$

We shall first prove

(a) $\lim_n \sum_{k\geq 0} |w_{n,k}| 1_{\{v(n,k)<t\}} = 0$ uniformly on compact time intervals.

By Proposition 25, given an $\epsilon > 0$, there is a $\delta > 0$ such that $|R(y,x)| < \epsilon$ whenever $|x - y| < \delta$ and $x, y \in B$. Choose n_0 such that $a_n < \delta$ for $n \geq n_0$. Then by condition (iii) of the stopping times $v(n,k)$, we have for $v(n,k) < t$, $|X_{v(n,k+1)\wedge t-} - X_{v(n,k)}| < \delta$, hence

$$|R(X_{v(n,k+1)\wedge t-}, X_{v(n,k)})| < \epsilon,$$

for $n \geq n_0$. Thus for any time $t > 0$ and $n \geq n_0$, we have

$$\sum_{k\geq 0} |w_{n,k}| 1_{\{v(n,k)<t\}} \leq \epsilon \sum_{k\geq 0} |X_{v(n,k+1)\wedge t} - X_{v(n,k)}|^2 1_{\{v(n,k)<t\}}$$

$$\leq 2\epsilon \left(\sum_{k\geq 0} |\Delta X_{v(n,k+1)\wedge t}|^2 1_{\{v(n,k)<t\}} + \sum_{k\geq 0} |X_{v(n,k+1)\wedge t} - X_{v(n,k)}|^2 \right).$$

As a result, for any $t > 0$, we have

$$\limsup_n \sum_{k\geq 0} |w_{n,k}| 1_{\{v(n,k)<t\}}$$

$$\leq 2\epsilon \left(\sum_{s\leq t} |\Delta X_s|^2 + [X]_t - [X]_0 \right),$$

and the last term converges to zero, as $\epsilon \downarrow 0$, uniformly on compact time intervals. This proves (a).

We choose a sequence $b_n \downarrow 0$ such that $a_n \leq b_n^2$ and $2a_n \leq b_n$ (e.g. $b_n = 2\sqrt{a_n}$). Define the following subsets of Ω:

$$A_{n,k} = \{|\Delta X_{v(n,k+1)}| > b_n\}$$

and
$$B_{n,k} = \{|\Delta X_{v(n,k+1)}| \leq b_n\} = \Omega - A_{n,k}.$$

Next we prove

(b) $$\lim_n \sum_{k \geq 0} |y_{n,k}| 1_{A_{n,k}} 1_{\{v(n,k)<t\}} = 0,$$

uniformly on compact time intervals.

To see this, let $|f''|$ have bound M on B. Note that on $A_{n,k}$, we have, for $t > v(n,k)$,

$$|X_{v(n,k+1)\wedge t-} - X_{v(n,k)}| \leq a_n \leq b_n^2 < b_n |\Delta X_{v(n,k+1)}|;$$

therefore

$$|\Delta X_{v(n,k+1)\wedge t} \otimes (X_{v(n,k+1)\wedge t-} - X_{v(n,k)})| \leq b_n |\Delta X_{v(n,k+1)}|^2$$

and

$$|(X_{v(n,k+1)\wedge t-} - X_{v(n,k)}) \otimes \Delta X_{v(n,k+1)\wedge t}| \leq b_n |\Delta X_{v(n,k+1)}|^2.$$

It follows that

$$\sum_{k \geq 0} |y_{n,k}| 1_{A_{n,k}} 1_{\{v(n,k)<t\}} \leq M b_n |\Delta X_{v(n,k+1)}|^2$$
$$\leq M b_n \sum_{s \leq t} |\Delta X_s|^2,$$

and the last term tends to 0, as $n \to \infty$, uniformly on compact time intervals, which proves (b).

Next we prove that for $t \geq 0$, we have

(c) $$\lim_n \sum_{k \geq 0} R(X_{v(n,k+1)\wedge t}, X_{v(n,k)\wedge t})(X_{v(n,k+1)\wedge t} - X_{v(n,k)\wedge t})^{\otimes 2} 1_{B_{n,k}} = 0$$

and

(d) $$\lim_n \sum_{k \geq 0} R(X_{v(n,k+1)}, X_{v(n,k)})(X_{v(n,k+1)} - X_{v(n,k)})^{\otimes 2} 1_{B_{n,k}} 1_{\{v(n,k)\leq t\}} = 0.$$

To prove this, let $\epsilon > 0$ and as before, choose a $\delta > 0$ such that for x and y in B, with $|x - y| < \delta$, we have $|R(y,x)| < \epsilon$. Choose n_0 so that $a_{n_0} + b_{n_0} < \delta$. Then on $B_{n,k}$, with $n \geq n_0$, we have

$$|X_{v(n,k+1)} - X_{v(n,k)}| \leq b_n + a_n < \delta,$$

which implies that

$$|R(X_{v(n,k+1)}, X_{v(n,k)})| 1_{B_{n,k}} \leq \epsilon.$$

Denote

$$M_n = \left| \sum_{k \geq 0} R(X_{v(n,k+1)}, X_{v(n,k)})(X_{v(n,k+1)} - X_{v(n,k)})^{\otimes 2} 1_{B_{n,k}} 1_{\{v(n,k) \leq t\}} \right|,$$

and

$$V_t^n = \sum_{k \geq 0} |X_{v(n,k+1)} - X_{v(n,k)}|^2 1_{\{v(n,k) \leq t\}}.$$

Then $M_n \leq \epsilon V_t^n$. Since $V_t^n \to [X]_t - [X]_0$, the sequence (V_t^n) is bounded. We deduce $\lim_n M_n = 0$. This proves (d).

To prove (c), fix $t \geq 0$. Then for every n, there is a j such that $v(n,j) \leq t < v(n,j+1)$. The difference between the two sums in (c) and (d) is equal to

$$A_n = R(X_t, X_{v(n,j)})(X_t - X_{v(n,j)})^{\otimes 2} \\ - R(X_{v(n,j+1)}, X_{v(n,j)})(X_{v(n,j+1)} - X_{v(n,j)})^{\otimes 2}$$

on $B_{n.j}$. As $n \to \infty$, we have $v(n,j) \uparrow t$ and $v(n,j+1) \downarrow t$; therefore $\lim A_n = 0$, which means that the two limits in (c) and (d) are equal. This proves (c).

We shall now prove that for every $t > 0$, we have

(e)
$$\lim_n \sum_{k \geq 0} v_{n,k} 1_{A_{n,k}} 1_{\{v(n,k)<t\}}$$
$$= \lim_n \sum_{k \geq 0} [\, f(X_{v(n,k+1)\wedge t}) - f(X_{v(n,k+1)\wedge t-}) - f'(X_{v(n,k)\wedge t})\Delta X_{v(n,k+1)\wedge t}$$
$$- \frac{1}{2} f''(X_{v(n,k)})(\Delta X_{v(n,k+1)\wedge t})^{\otimes 2} \,] 1_{A_{n,k}} 1_{\{v(n,k)<t\}}$$
$$= \sum_{0<s\leq t} R(X_s, X_{s-})(\Delta X_s)^{\otimes 2}$$
$$= \sum_{0<s\leq t} \left[f(X_s) - f(X_{s-}) - f'(X_{s-})\Delta X_s - \frac{1}{2}f''(X_{s-})(\Delta X_s)^{\otimes 2} \right],$$

and

(f) $$\lim_n \sum_{k \geq 0} v_{n,k} 1_{\{v(n,k)<t\}} = \sum_{0<s\leq t} R(X_s, X_{s-})(\Delta X_s)^{\otimes 2}.$$

To prove this, we define the following functions, for $t > 0$:
$$v'_{n,k} = f(X_{v(n,k+1)\wedge t}) - f(X_{v(n,k+1)\wedge t-})$$
$$- f'(X_{v(n,k+1)\wedge t-})\Delta X_{v(n,k+1)\wedge t}$$
$$- \frac{1}{2}f''(X_{v(n,k+1)\wedge t-})(\Delta X_{v(n,k+1)\wedge t})^{\otimes 2}$$
$$= R(X_{v(n,k+1)\wedge t}, X_{v(n,k+1)\wedge t-})(\Delta X_{v(n,k+1)\wedge t})^{\otimes 2};$$
$$v''_{n,k} = [f'(X_{v(n,k+1)\wedge t-}) - f'(X_{v(n,k)})]\Delta X_{v(n,k+1)\wedge t};$$
$$v'''_{n,k} = \frac{1}{2}[f''(X_{v(n,k+1)\wedge t-}) - f''(X_{v(n,k)})](\Delta X_{v(n,k+1)\wedge t})^{\otimes 2}.$$

Thus $v_{n,k} = v'_{n,k} + v''_{n,k} + v'''_{n,k}$.

In order to prove (e) and (f), we shall prove first (g), (h) and (i) below:

(g)
$$\lim_n \sum_{k\geq 0} v'_{n,k} 1_{A_{n,k}} 1_{\{v(n,k)<t\}} = \sum_{0<s\leq t} R(X_s, X_{s-})(\Delta X_s)^{\otimes 2}$$
$$= \lim_n \sum_{k\geq 0} R(X_{v(n,k+1)\wedge t}, X_{v(n,k+1)\wedge t-})(\Delta X_{v(n,k+1)\wedge t})^{\otimes 2} 1_{A_{n,k}} 1_{\{v(n,k)<t\}};$$

(h)
$$\lim_n \sum_{k\geq 0} v''_{n,k} 1_{A_{n,k}} 1_{\{v(n,k)<t\}} = 0;$$

and

(i)
$$\lim_n \sum_{k\geq 0} v'''_{n,k} 1_{A_{n,k}} 1_{\{v(n,k)<t\}} = \lim_n \sum_{k\geq 0} v'''_{n,k} 1_{\{v(n,k)<t\}} = 0$$

uniformly on compact time intervals.

To prove (g), note that, since $2a_n < b_n$, by Lemma 21, we have
$$\{s > 0 : |\Delta X_s| > b_n\} \subset \{s > 0 : |\Delta X_s| > 2a_n\} \subset \{v(n,k) : k \geq 0\}.$$

Then
$$\left| \sum_{0<s\leq t} R(X_s, X_{s-})(\Delta X_s)^{\otimes 2} - \sum_{k\geq 0} v'_{n,k} 1_{A_{n,k}} 1_{\{v(n,k)<t\}} \right|$$
$$= \left| \sum_{0<s\leq t} R(X_s, X_{s-})(\Delta X_s)^{\otimes 2} - \right.$$
$$\left. - \sum_{k\geq 0} R(X_{v(n,k+1)\wedge t}, X_{v(n,k+1)\wedge t-})(\Delta X_{v(n,k+1)\wedge t})^{\otimes 2} 1_{A_{n,k}} \right|$$
$$\leq \sum_{0<s\leq t} |R(X_s, X_{s-})| \, |\Delta X_s|^2 1_{B_{n,k}}$$

and

$$\left| \sum_{0<s\leq t} R(X_s, X_{s-})(\Delta X_s)^{\otimes 2} - \right.$$
$$\left. - \sum_{k\geq 0} R(X_{v(n,k+1)\wedge t}, X_{v(n,k+1)\wedge t-})(\Delta X_{v(n,k+1)\wedge t})^{\otimes 2} 1_{A_{n,k}} 1_{\{v(n,k)<t\}} \right|$$
$$\leq \sum_{s\leq t} |R(X_s, X_{s-})| \, |\Delta X_s|^2 1_{B_{n,k}}.$$

Let $\epsilon > 0$ and take $\delta > 0$ such that if $x, y \in B$, with $|x - y| < \delta$, then $|R(y,x)| < \epsilon$. Choose n_0 so that $b_{n_0} < \delta$.

For $n \geq n_0$ and $s > 0$, we have, on $B_{n,k}$,

$$|X_s - X_{s-}| = |\Delta X_s| \leq b_n < \delta,$$

therefore $|R(X_s, X_{s-})| < \epsilon$; consequently

$$\sum_{0<s\leq t} |R(X_s, X_{s-})| |\Delta X_s|^2 1_{B_{n,k}} \leq \epsilon \sum_{0<s\leq t} |\Delta X_s|^2 1_{B_{n,k}},$$

and the last term tends to 0 as $\epsilon \downarrow 0$.

It follows that

$$\lim_n \sum_{k\geq 0} v'_{n,k} 1_{A_{n,k}} 1_{\{v(n,k)<t\}} = \sum_{0<s\leq t} R(X_s, X_{s-})(\Delta X_s)^{\otimes 2}$$

uniformly on compact time intervals, that is, the first equality of (g) holds.

In a similar fashion, one can show that

$$\lim_n \sum_{k\geq 0} R(X_{v(n,k+1)\wedge t}, X_{v(n,k+1)\wedge t-})(\Delta X_{v(n,k+1)\wedge t})^{\otimes 2} 1_{A_{n,k}} 1_{\{v(n,k)<t\}}$$
$$= \sum_{0<s\leq t} R(X_s, X_{s-})(\Delta X_s)^{\otimes 2}$$

uniformly on compact time intervals. This establishes (g).

§23. THE ITÔ FORMULA

To prove (h), let M again be the bound of $|f''|$ on B. Use the mean value theorem to deduce

$$\sum_{k \geq 0} v''_{n,k} 1_{A_{n,k}} 1_{\{v(n,k)<t\}}$$
$$\leq M|X_{v(n,k+1)\wedge t-} - X_{v(n,k)}| \, |\Delta X_{v(n,k+1)\wedge t}| 1_{A_{n,k}} 1_{\{v(n,k)<t\}}$$
$$\leq M a_n \sum_{k \geq 0} |\Delta X_{v(n,k+1)\wedge t}| 1_{A_{n,k}} 1_{\{v(n,k)<t\}}$$
$$\leq M b_n \sum_{k \geq 0} b_n |\Delta X_{v(n,k+1)\wedge t}| 1_{A_{n,k}} 1_{\{v(n,k)<t\}}$$
$$\leq M b_n \sum_{s \leq t} |\Delta X_s|^2,$$

and the last sum tends to 0, as $n \to \infty$, uniform on compact time intervals. This proves (h).

To prove (i), we use the uniform continuity of f'' on B. Let $\epsilon > 0$ and take $\delta > 0$ such that $x, y \in B$, with $|x - y| < \delta$, implies $|f''(x) - f''(y)| < \epsilon$. Take n_0 such that for $n \geq n_0$ we have $a_n < \delta$; therefore, if $v(n,k) < t$, then

$$|X_{v(n,k+1)\wedge t-} - X_{v(n,k)}| \leq a_n < \delta;$$

consequently

$$|f''(X_{v(n,k+1)\wedge t-}) - f''(X_{v(n,k)})| < \epsilon.$$

It follows that for $n \geq n_0$, we have

$$\sum_{k \geq 0} v'''_{n,k} 1_{A_{n,k}} 1_{\{v(n,k)<t\}}$$
$$\leq \sum_{k \geq 0} |f''(X_{v(n,k+1)\wedge t-}) - f''(X_{v(n,k)})| |\Delta X_{v(n,k+1)\wedge t}|^2 1_{A_{n,k}} 1_{\{v(n,k)<t\}}$$
$$\leq \epsilon \sum_{k \geq 0} |\Delta X_{v(n,k+1)\wedge t}|^2 1_{A_{n,k}} 1_{\{v(n,k)<t\}} \leq \epsilon \sum_{s \leq t} |\Delta X_s|^2,$$

and similarly

$$\sum_{k \geq 0} v'''_{n,k} 1_{\{v(n,k)<t\}} \leq \epsilon \sum_{s \leq t} |\Delta X_s|^2.$$

This proves (i).

Since $v_{n,k} = v'_{n,k} + v''_{n,k} + v'''_{n,k}$, from (g), (h) and (i), we deduce that

$$\lim_n \sum_{k \geq 0} v_{n,k} 1_{A_{n,k}} 1_{\{v(n,k)<t\}} = \sum_{0 < s \leq t} R(X_s, X_{s-})(\Delta X_s)^{\otimes 2},$$

which proves (e). From (a), (b) and (e), we deduce that

$$\lim_n \sum_{k\geq 0} R_{n,k}(X_{v(n,k+1)\wedge t} - X_{v(n,k)\wedge t})^{\otimes 2} 1_{A_{n,k}}$$
$$= \lim_n \sum_{k\geq 0} R_{n,k}(X_{v(n,k+1)\wedge t} - X_{v(n,k)\wedge t})^{\otimes 2} 1_{A_{n,k}} 1_{\{v(n,k)<t\}}$$
$$= \lim_n \sum_{k\geq 0} (v_{n,k} + w_{n,k} + y_{n,k}) 1_{A_{n,k}} 1_{\{v(n,k)<t\}}$$
$$= \sum_{0<s\leq t} R(X_s, X_{s-})(\Delta X_s)^{\otimes 2}.$$

Since, by (c), we have

$$\lim_n \sum_{k\geq 0} R_{n,k}(X_{v(n,k+1)\wedge t} - X_{v(n,k)\wedge t})^{\otimes 2} 1_{B_{n,k}} = 0,$$

we deduce finally that

$$\lim_n \sum_{k\geq 0} R_{n,k}(X_{v(n,k+1)\wedge t} - X_{v(n,k)\wedge t})^{\otimes 2} = \sum_{0<s\leq t} R(X_s, X_{s-})(\Delta X_s)^{\otimes 2}$$

and this proves the theorem when X is bounded. We note that to establish the above equality, we only used the fact that X has finite quadratic variation.

For the case X is unbounded, obtain a sequence $T_i \uparrow \infty$ of stopping times such that X^{T_i-} still has finite quadratic variation and $[X^{T_i-}]_t = [X]_t^{T_i-}$; therefore, for each X^{T_i-}, the above equality is true. For a given ω and $t > 0$, choose i such that $T_i(\omega) > t$. Then for every $0 < s \leq t$, we have $X_s^{T_i-} = X_s$ and the equality is valid for $X_t(\omega)$. This concludes the proof. ∎

Chapter 7

Stochastic Integration in the Plane

In this chapter we study the summability of two-parameter processes and the properties of their stochastic integral. The approach is measure-theoretic.

This study is similar to that of Chapter 2 for one parameter processes.

§24. PRELIMINARIES

A. Order relation in $\mathbb{R}^2$

1. The points in $\overline{\mathbb{R}}^2 = [-\infty, \infty]^2$ are denoted by $z = (s, t)$.

The order relation $z = (s, t) \leq z' = (s', t')$ is defined by $s \leq s'$ and $t \leq t'$. We write $z < z'$ if $s < s'$ and $t < t'$. We have $z \wedge z' = (s \wedge s', t \wedge t')$.

The point $(-\infty, -\infty)$ will be denoted by $-\infty$; similarly, $(+\infty, +\infty)$ will be denoted by ∞. Also $(0,0)$ will be denoted by 0.

For $-\infty \leq z \leq z' \leq \infty$ we define the rectangle

$$(z, z'] = \{u \in \overline{\mathbb{R}}^2 : z < u \leq z'\}.$$

A similar definition is given for $[z, z']$, $[z, z')$ and (z, z'). Other rectangles of interest are $(s, s'] \times (t, t')$ and $(s, s') \times (t, t']$.

For $z \in \overline{\mathbb{R}}^2$ we denote $R_z = (-\infty, z]$. The first quadrant of the plane $\mathbb{R}^2$ is denoted by $\mathbb{R}_+^2$ and consists of the points $z = (s, t) \geq 0$ of $\mathbb{R}^2$.

B. The increment $\Delta_{zz'} g$

2. Let $g : \mathbb{R}^2 \to E$ be a function and $R = (z, z']$ a rectangle with $z = (s, t)$ and $z' = (s', t')$.

The *increment* of g on the rectangle R is denoted by $g(R)$, or $\Delta_R g$, or $\Delta_R(g)$, or $\Delta_{(z,z']} g$, or even $\Delta_{z,z'} g$ or $\Delta_{zz'} g$ and is defined by

$$\Delta_{zz'} g = g(s', t') + g(s, t) - g(s', t) - g(s, t').$$

If $s = s'$ or $t = t'$, then $\Delta_{zz'} g = 0$.

If $g(z) = 0$ outside $\mathbb{R}_+^2$ and if $u < 0 < z$, then $\Delta_{uz} g = g(z)$. Then we define $\Delta_{[0,z]} g$ by the equality

$$\Delta_{[0,z]} g = \lim_{u \uparrow 0} \Delta_{(u,z]} g = g(z).$$

It is convenient to extend g with $g(s, t) = 0$ if $s = -\infty$ or $t = -\infty$. If g vanishes outside $\mathbb{R}_+^2$, then, extending the above definition of the increment, we have

$$\Delta_{R_z} g = g(z), \text{ for } z \geq -\infty.$$

C. Right continuity

3. Let $g : \mathbb{R}^2 \to E$ be a function.

We say that g is *right continuous*, if for every $z \in \mathbb{R}^2$ we have

$$\lim_{\substack{z' \to z \\ z' \geq z}} g(z') = g(z).$$

The above limit is also denoted by $\lim_{z' \downarrow z} g(z')$. We say g is cadlag if it is right continuous and at every point $z \in \mathbb{R}^2$ it has left limit

$$\lim_{\substack{z' \to z \\ z' < z}} g(z').$$

We say g is *incrementally right continuous* if for every $z \in \mathbb{R}^2$ we have

$$\lim_{z' \downarrow z} \Delta_{zz'} g = 0.$$

If g is right continuous then it is also incrementally right continuous. The converse is not true, in general.

Example. Let $f : \mathbb{R} \to E$ be a function and define $g : \mathbb{R}^2 \to E$ by $g(s, t) = f(t)$, for $(s, t) \in \mathbb{R}^2$. For every $z \leq z'$ we have $\Delta_{zz'} g = 0$, hence g is incrementally right continuous. But if f is not right continuous, then g is not right continuous on $\mathbb{R}^2$.

D. The filtration

4. $(\Omega, \mathcal{F}, \mathcal{P})$ is a probability space and $(\mathcal{F}_s^1)_{s \in \mathbb{R}}$ and $(\mathcal{F}_t^2)_{t \in \mathbb{R}}$ are two filtrations satisfying the usual conditions. We set $\mathcal{F}_\infty^1 = \mathcal{F}_\infty^2 = \mathcal{F}$.

For $z = (s, t)$ with $-\infty < s \leq +\infty$ and $-\infty < t \leq +\infty$ we define

$$\mathcal{F}_z = \mathcal{F}_{st} = \mathcal{F}_s^1 \cap \mathcal{F}_t^2.$$

Then the filtration $(\mathcal{F}_z)_{z \in \mathbb{R}^2}$ satisfies the usual conditions. We have

$$\mathcal{F}_{s\infty} = \mathcal{F}_s^1 \text{ and } \mathcal{F}_{\infty t} = \mathcal{F}_t^2.$$

We assume that the *commutation* condition is satisfied, that is, the conditional expectations $E(\cdot | \mathcal{F}_s^1)$ and $E(\cdot | \mathcal{F}_t^2)$ commute. Then

$$E(\cdot | \mathcal{F}_{st}) = E(\cdot | \mathcal{F}_s | \mathcal{F}_t).$$

E. The predictable σ-algebra

5. $\mathcal{R}$ is the ring generated by the subsets of $\mathbb{R}^2 \times \Omega$ of the form $(z, z'] \times A$ with $z, z' \in \mathbb{R}^2$ and $A \in \mathcal{F}_z$. The σ-algebra $\mathcal{P}$ of subsets of $\mathbb{R}^2 \times \Omega$ generated by $\mathcal{R}$ is called the σ-algebra of *predictable* subsets of $\mathbb{R}^2 \times \Omega$.

We define also the σ-algebra $\mathcal{P}(\infty)$ of the predictable subsets of $\overline{\mathbb{R}}^2 \times \Omega$, generated by the sets of the form $(z, z'] \times A$ with $z, z' \in \overline{\mathbb{R}}^2$ and $A \in \mathcal{F}_z$. Then $\mathcal{P}$ is the trace of $\mathcal{P}(\infty)$ on $\mathbb{R}^2 \times \Omega$.

We can give an alternative description of $\mathcal{P}(\infty)$. Consider the σ-algebra $\overline{\mathcal{P}}^1$ of predictable subsets of $\overline{\mathbb{R}} \times \Omega$, relative to the filtration $(\mathcal{F}_s^1)_{s \in \overline{\mathbb{R}}}$ and denote by $\mathcal{P}^1(\infty)$ the σ-algebra of the sets of the form $\{(s, \infty, \omega) : (s, \omega) \in A\}$ with $A \in \overline{\mathcal{P}}^1$. Then $\mathcal{P}^1(\infty)$ is the σ-algebra of the predictable subsets of $\overline{\mathbb{R}} \times \{\infty\} \times \Omega$, relative to the filtration $(\mathcal{F}_{s\infty})_{s \in \overline{\mathbb{R}}}$.

Consider also the σ-algebra $\overline{\mathcal{P}}^2$ of predictable subsets of $\overline{\mathbb{R}} \times \Omega$, relative to the filtration $(\mathcal{F}_t^2)_{t \in \overline{\mathbb{R}}}$ and the σ-algebra $\mathcal{P}^2(\infty)$ of the sets of the form $\{(\infty, t, \omega) : (t, \omega) \in A\}$ with $A \in \overline{\mathcal{P}}^2$. Then $\mathcal{P}^2(\infty)$ is the σ-algebra of the predictable subsets of $\{\infty\} \times \overline{\mathbb{R}} \times \Omega$, relative to the filtration $(\mathcal{F}_{\infty t})_{t \in \overline{\mathbb{R}}}$.

Then $\mathcal{P}(\infty)$ is the class of the sets of the form $A \cup A^1 \cup A^2$ with $A \in \mathcal{P}$, $A^1 \in \mathcal{P}^1(\infty)$ and $A^2 \in \mathcal{P}^2(\infty)$.

F. Stopping times

6. A *stopping time* is a mapping $Z : \Omega \to (-\infty, +\infty]^2$ such that for every $a \in \mathbb{R}^2$ we have $\{Z \leq a\} \in \mathcal{F}_z$. A stopping time is called also a *stopping point*. If Z is a stopping time and if $Z = (S, T)$, with $S, T : \Omega \to (-\infty, +\infty]$, then S is a stopping time for $(\mathcal{F}_t^1)_{t \in \mathbb{R}}$ and T is a stopping time for $(\mathcal{F}_t^2)_{t \in \mathbb{R}}$. The converse is not true.

(S, T) is a stopping point iff S is an $\mathcal{F}^1$ stopping time and $\mathcal{F}_T^2$-measurable and T is an $\mathcal{F}^2$ stopping time and $\mathcal{F}_S^1$-measurable.

The two dimensional stopping times do not satisfy the needs of the general theory of stochastic processes. There is no cross-section theorem for stopping points.

If U and V are stopping points then $U \wedge V$ is not necessarily a stopping point.

Example. If S is an $\mathcal{F}^1$ stopping time and T is an $\mathcal{F}^2$ stopping time, then (S, ∞) and (∞, T) are stopping points and we have $(S,T) = (S, \infty) \wedge (\infty, T)$, but (S,T) is not necessarily a stopping point.

If $Z = (S,T)$ is a stopping point and $X : \mathbb{R}^2 \times \Omega \to \mathbb{R}$ is a progressive process, then the process $X_{S \wedge s, T \wedge t}$ is not adapted. If Z is a stopping point, the stochastic interval (Z, ∞) is a predictable set; but the stochastic interval $(-\infty, Z]$ is not even adapted.

As a consequence of the above, the stopping points cannot be used for stopping.

It follows that the theorems about one-parameter processes that use stopping times in the proof do not necessarily have extensions for two-parameter processes.

One such important example is the summability Theorem 14.3. There is no similar summability criterion for two-parameter processes.

For a more detailed account the reader is referred to [M].

G. Stochastic processes

7. We shall consider processes $X : \mathbb{R}^2 \times \Omega \to E$. Such a process is said to be *right continuous* if $\lim_{u \downarrow z} X_u(\omega) = X_z(\omega)$ for every $z \in \mathbb{R}^2$ and $\omega \in \Omega$; X is said to be *cadlag* if it is right continuous and has *left limit* $\lim_{\substack{u \uparrow z \\ u < z}} X_u(\omega)$, for every $z \in \mathbb{R}^2$ and $\omega \in \Omega$.

We say that X is *adapted* (to the filtration $(\mathcal{F}_z)$), if X_z is $\mathcal{F}_z$-measurable for every $z \in \mathbb{R}^2$.

Remark. We shall see that if X is p-summable (see definition 25.3) then the left limits X_{z-}, $X_{s-,t}$ and $X_{s,t-}$ exist in L_E^p for each $z \in \mathbb{R}^{-2}$ (Theorem 26.4). We can then extend X to the points (s, ∞) or (∞, t) or ∞, such that the extended process is left continuous in L_E^p at these points.

8. For a process $X : \mathbb{R}^2 \times \Omega \to E$ and a rectangle $R = (z, z']$ with $z < z'$ in $\mathbb{R}^2$ the *increment* of X is denoted by $X(R)$, $\Delta_R(X)$, $\Delta_{[z,z']}X$, $\Delta_{z,z'}X$ or $\Delta_{zz'}X$ and is defined by

$$\Delta_{zz'}X = X_{s't'} + X_{st} - X_{s't} - X_{s,t'}$$

if $z = (s,t)$ and $z' = (s',t')$.

We define also

$$X(R_z) = \Delta_{-\infty, z}X = X_z.$$

If we extend X with $X_z = 0$ if $z = (s, \infty)$ or $z = (\infty, t)$ or $z = \infty$, then the above equality follows from the general definition of the increment.

If $X_z = 0$ outside $\mathbb{R}_+^2 \times \Omega$ and if $u < 0 < z$, then $\Delta_{uz} X = X_z$. Then we can define
$$\Delta_{[0,z]} X = \lim_{u \uparrow 0} \Delta_{(u,z]} = X_z.$$

H. Extension of processes from $\mathbb{R}_+^2 \times \Omega$ to $\mathbb{R}^2 \times \Omega$

9. Let $X : \mathbb{R}_+^2 \times \Omega \to E$ be a process, $(\mathcal{F}_s^1)_{s \in \mathbb{R}_+}$ and $(\mathcal{F}_t^2)_{t \in \mathbb{R}_+}$ two filtrations satisfying the usual conditions; set $\mathcal{F}_{st} = \mathcal{F}_s^1 \cap \mathcal{F}_t^2$.

It is convenient to extend the process X and the filtration $(\mathcal{F}_{st})$ to the whole space $\mathbb{R}^2 \times \Omega$. If, in addition X is adapted to the filtration $(\mathcal{F}_{st})$ we want the extension of X to be adapted to the extended filtration. Moreover, if X is a martingale we want the extension to be a martingale (See Definition 27.1, for the definition of a martingale).

There are several extensions possible. The reader can use any one of them, according to his or her needs.
a) Set $X_z = 0$ if $z \notin \mathbb{R}_+^2$, $X_{s,-\infty} = X_{-\infty,t} = X_{-\infty} = 0$. This extension of X is adapted to any extension of the filtration.

If X is a martingale, this extension is no longer a martingale, in general.
b) Set $X_z = X_0$ and $\mathcal{F}_z = \mathcal{F}_0$ for $z \notin \mathbb{R}_+^2$. If X is adapted, then this extension is still adapted to the extended filtration. This extension is not too interesting, unless X_z vanishes along the coordinate axes.
c) Set

$$X_{s,t} = X_{0,t}, \text{ for } s < 0 \text{ and } t \geq 0$$
$$X_{s,t} = X_{s,0}, \text{ for } t < 0 \text{ and } s \geq 0$$
$$X_{s,t} = X_{0,0}, \text{ for } s < 0 \text{ and } t < 0.$$

Similarly, set $\mathcal{F}_s^1 = \mathcal{F}_0^1$ if $x < 0$, $\mathcal{F}_t^2 = \mathcal{F}_0^2$ if $t < 0$ and $\mathcal{F}_{st} = \mathcal{F}_s^1 \cap \mathcal{F}_t^2$ for $(s,t) \in \mathbb{R}^2$. Then

$$\mathcal{F}_{st} = \mathcal{F}_0^1 \cap \mathcal{F}_t^2 = \mathcal{F}_{0,t}, \text{ if } s < 0 \text{ and } t \geq 0$$
$$\mathcal{F}_{st} = \mathcal{F}_s^1 \cap \mathcal{F}_0^2 = \mathcal{F}_{s0}, \text{ if } t < 0 \text{ and } s \geq 0$$
$$\mathcal{F}_{st} = \mathcal{F}_0^1 \cap \mathcal{F}_0^2 = \mathcal{F}_{00}, \text{ if } s < 0 \text{ and } t < 0.$$

The extension $(\mathcal{F}_{st})$ still satisfies the usual conditions on $\mathbb{R}^2 \times \Omega$. If X is adapted, then the above extension of X is adapted to the extended filtration. If X is a martingale on $\mathbb{R}_+^2 \times \Omega$, then this extension is a martingale on $\mathbb{R}^2 \times \Omega$.
d) Set

$$X_{s,t} = E(X_{0,t}), \text{ if } s < 0 \text{ and } t \geq 0$$
$$X_{s,t} = E(X_{s,0}), \text{ if } t < 0 \text{ and } s \geq 0$$
$$X_{s,t} = E(X_{0,0}), \text{ if } s < 0 \text{ and } t < 0.$$

For each $z \notin \mathbb{R}_+^2$, take $\mathcal{F}_z$ the trivial σ-algebra, generated by the negligible sets.

If X is adapted, or a martingale, then so is this extension.

§25. SUMMABLE PROCESSES

In this paragraph we sketch the construction of the stochastic integral for two-parameter processes, using the measure-theoretic approach that was used in Chapter 2 for one-parameter processes.

A. The measure I_X

1. Let $X : \mathbb{R}^2 \times \Omega \to E \subset L(F, G)$ be a two-parameter, cadlag, adapted process, such that $X_z \in L_E^p$ for each $z \in \mathbb{R}^2$. We define the additive measure

$$I_X : \mathcal{R} \to L_E^p \subset L(F, L_G^p)$$

by

$$I_X((z, z'] \times A) = 1_A \Delta_{zz'} X, \text{ for } A \in \mathcal{F}_z$$

and then extended by additivity to the ring $\mathcal{R}$.

If the process X is understood we shall write I instead of I_X.

2. Since $L_E^p \subset L(F, L_G^p)$, we can consider the semivariation of I_X relative to the pair (F, L_G^p)

We shall write $(\tilde{I}_X)_{F,G}$ instead of $(\tilde{I}_X)_{F,L_G^p}$, and we shall call it the semivariation of I_X relative to (F, G). If the process X is understood, we shall write simply, $\tilde{I}_{F,L_G^p}$ or $\tilde{I}_{F,G}$:

$$\tilde{I}_{F,G}(A) = \sup \| \sum I_X(A_i) x_i \|_p, \text{ for } A \in \mathcal{R},$$

where the supremum is taken for all finite families $(A_i)_{i \in I}$ of disjoint sets from $\mathcal{R}$ contained in A and all finite families $(x_i)_{i \in I}$ of elements from F_1.

B. Summable processes

3. Definition. *We say that a cadlag, adapted process*

$$X : \mathbb{R}^2 \times \Omega \to E \subset L(F, G)$$

is p-summable relative to the pair (F, G), if I_X has a σ-additive extension $I_X : \mathcal{P} \to L_E^p$ with finite semivariation relative to (F, G).

Examples of summable two-parameter processes are:
a) Square integrable martingales with values in a Hilbert space (Theorem 28.9)
b) Processes with integrable variation (Theorem 30.6)
c) Processes with integrable semivariation, provided that $c_0 \not\subset E$ and $c_o \not\subset G$ (Theorem 32.6).

Remark. For one-parameter processes $(X_z)_{z \in \mathbb{R}_+}$ we have a simple criterion of extension of I_X: if $c_0 \not\subset E$ and if I_X is bounded in L_E^p on $\mathcal{R}$, then I_X can be extended to a σ-additive measure $I_X : \mathcal{P} \to L_E^p$ (Theorem 14.3).

But for two-parameter processes we do not have such an extension theorem, since the proof involves stopping times.

An alternative extension method is to find a positive, σ-additive measure μ on $\mathcal{R}$ such that $I_X \ll \mu$ and then use the general extension Theorem 7.3.

We shall use this method for two-parameter square integrable martingales (Theorem 28.5).

C. The seminorm $\tilde{I}_X$ and the space $\mathcal{F}_{F,G}(X)$

4. Assume X is p-summable relative to (F, G) and consider I_X extended to $\mathcal{P}$. Then we can apply the general integration theory of §5.

Let $U \subset L_{G^*}^q$, $\frac{1}{p} + \frac{1}{q} = 1$, be a norming spaces for L_G^p. For any Banach space D we denote by $\mathcal{F}_D((I_X)_{F,L_G^p})$ or $\mathcal{F}_D((I_X)_{F,G})$ or $\mathcal{F}_D(I_{F,G})$ the set of predictable processes $H : \mathbb{R}^2 \times \Omega \to D$ such that

$$\tilde{I}_X(H) = (\tilde{I}_X)_{F,L_G^p} := \sup\left\{ \int |H| d|(I_X)_u| : u \in U_1 \right\} < \infty,$$

where $|(I_X)_u|$ is the variation of the σ-additive measure $(I_X)_u : \mathcal{P} \to F^*$, defined by

$$\langle x, (I_X)_u(A) \rangle = \langle I_X(A)x, u \rangle, \text{ for } x \in F \text{ and } A \in \mathcal{P}.$$

Then $\mathcal{F}_D(I_{F,G})$ is a complete vector space for the seminorm $\tilde{I}_X(H)$ (Corollary 5.25).

D. The integral $\int H dI_X$

5. Assume X is p-summable relative to (F, G). If $D = F$, the space $\mathcal{F}_F((I_X)_{F,G})$ will be denoted by $\mathcal{F}_{F,G}(I_X)$ or $\mathcal{F}_{F,G}(X)$. We have

$$\mathcal{F}_{F,G}(I_X) \subset \bigcap_{u \in U} L_F^1((I_X)_u)$$

with equality, if U is closed in $(L_G^p)^*$ (Corollary 5.23).

If $H \in \mathcal{F}_{F,G}(I_X)$ we can define the integral $\int H dI_X$: for each $u \in U$ we have $H \in L_F^1((I_X)_u)$, hence the integral $\int H d(I_X)_u$ is defined and is a scalar. The mapping $u \mapsto \int H d(I_X)_u$ is a continuous linear functional on U, denoted $\int H dI_X$, and called the integral of H with respect to I_X. We have then,

$$\int H dI_X \in U^*,$$

$$\left\langle \int H dI_X, u \right\rangle = \int H d(I_X)_u, \text{ for } u \in U$$

and

$$\left| \int H dI_X \right| \leq (\tilde{I}_X)_{F,G}(H).$$

If $H \in \mathcal{F}_{F,G}(I_X)$ and $A \in \mathcal{P}$, then $1_A H \in \mathcal{F}_{F,G}(I_X)$. We denote

$$\int_A H dI_X = \int 1_A H dI_X.$$

Then

$$\int H dI_X = \int_{\mathbb{R}^2 \times \Omega} H dI_X.$$

Remark. If we take $U = L_{G^*}^q$, then $\int H dI_X \in (L_{G^*}^q)^*$; and if $U = (L_G^p)^*$, then $\int H dI_X \in (L_G^p)^{**}$.

E. The stochastic integral $H \cdot X$

Assume X is p-summable relative to (F, G) and also that $\int_{(-\infty, z]} H dI_X \in L_G^p$ for every $z \in \mathbb{R}^2$. We denote by the same symbol the equivalence class $\int_{(-\infty, z]} H dI_X$ in L_G^p, as well as any representative of this class.

If we choose a representative in each equivalence class, we obtain a process $(\int_{(-\infty, z]} H dI_X)_{z \in \mathbb{R}^2}$ with values in G. This process is adapted:

6. Theorem. *If $H \in \mathcal{F}_{F,G}(X)$ and if $\int_{(-\infty, z]} H dI_X \in L_G^p$ for every $z \in \mathbb{R}^2$, then the process $(\int_{(-\infty, z]} H dI_X)_{z \in \mathbb{R}^2}$ is adapted.*

The proof is the same as that of Theorem 10.4.

The process $(\int_{(-\infty, z]} H dI_X)_{z \in \mathbb{R}^2}$ is not necessarily cadlag. This leads to the following definition:

7. Definition. *We denote by $L_{F,G}^1(X)$ the set of processes $H \in \mathcal{F}_{F,G}(X)$ satisfying the following two conditions:*
a) $\int_{(-\infty, z]} H dI_X \in L_G^p$, *for every* $z \in \mathbb{R}^2$;
b) *The process* $(\int_{(-\infty, z]} H dI_X)_{z \in \mathbb{R}^2}$ *has a cadlag modification.*

If $H \in L_{F,G}^1(X)$, then any cadlag modification of the process $(\int_{(-\infty, z]} H dI_X)_{z \in \mathbb{R}^2}$ is called the *stochastic integral* of H with respect to X and is denoted by $H \cdot X$ or $\int H dX$:

$$(H \cdot X)_z(\omega) = (\int_{[-\infty, z]} H dI_X)(\omega), \text{ a.s.}$$

The stochastic integral is defined up to an evanescent process.

8. Remark. For a p-summable, one-parameter process X, the space $L_{F,G}^1(X)$ is complete for the seminorm $\tilde{I}_X$ (Corollary 12.2). In the proof of the completeness we used the convergence Theorem 12.1, which itself used stopping times.

For a two-parameter p-summable process, an analog of Theorem 12.1 cannot be proved. It follows that we do not know whether or not the space $L_{F,G}^1(X)$ is complete, in general.

However, if X is a p-summable martingale with $p > 1$ and G is reflexive, then $L^1_{F,G}(X) = \mathcal{F}_{F,G}(X)$, hence $L^1_{F,G}(X)$ is complete (Theorem 27.5). Also, if X has integrable variation, the $L^1_{F,G}(X)$ is complete (Theorem 30.7).

§26. PROPERTIES OF THE STOCHASTIC INTEGRAL

A. Convergence theorems

Let X be an E-valued, p-summable process relative to (F, G).

The following convergence theorem is the analog for two-parameter processes of Theorem 10.3. The proof is the same as that of Theorem 10.3.

1. Theorem. *Let $(H^n)_{0 \leq n < \infty}$ be a sequence from $\mathcal{F}_{F,G}(X)$ such that $|H^n| \leq |H^0|$ for each n and $H^n \to H^0$ pointwise. Assume that*
(i) $\int H^n dI_X \in L_G^p$ for $n \geq 1$
and
(ii) *the sequence $(\int H^n dI_X)_n$ converges pointwise on Ω, weakly in G.*
Then:
a) $\int H dI_X \in L_G^p$
and
b) $\int H^n dI_X \to \int H dI_X$ *in the $\sigma(L_G^p, L_{G^*}^q)$ topology of L_G^p, as well as pointwise, weakly in G.*
c) *If $(\int H^n dI_X)_n$ converges pointwise in Ω, strongly in G, then $\int H^n dI_X \to \int H dI_X$, strongly in L_G^1.*

The Vitali and the Lebesgue theorems below are weaker than for one-parameter processes (Theorems 12.5 and 12.6) in that uniform convergence on the bounded sets of $\mathbb{R}^2$ is not asserted in the conclusion. The proof follows from the general Vitali and Lebesgue convergence theorems (Theorems 5.36 and 5.37).

2. Theorem. (Vitali) *Let (H^n) be a sequence from $\mathcal{F}_{F,G}(X)$ and let H be an F-valued predictable process. Assume that*
(i) $\tilde{I}_{F,G}(H^n 1_A) \to 0$ *as $\tilde{I}_{F,G}(A) \to 0$, uniformly in n;*
and that either one of the conditions (ii) *or* (iii) *below is true:*
(ii) $H^n \to H$ *in $\tilde{I}_{F,G}$-measure;*
(iii) $H^n \to H$ *pointwise and $I_{F, L_{G^*}^q}$ is uniformly σ-additive.*
Then $H \in \mathcal{F}_{F,G}(X)$ and $H^n \to H$ in $\mathcal{F}_{F,G}(X)$.
Conversely, if $H^n, H \in \mathcal{F}_{F,G}(X)$ and $H^n \to H$ is $\mathcal{F}_{F,G}(X)$, then conditions (i) *and* (ii) *are satisfied.*

3. Theorem. (Lebesgue) *Let (H^n) be a sequence from $\mathcal{F}_{F,G}(X)$ and let H be an F-valued, predictable process. Assume that:*
(i) *There is a process $\phi \in \mathcal{F}_\mathbb{R}(\mathcal{B}, I_{F,G})$ such that $|H^n| \leq \phi$ for every n;*
and that either one of the conditions (ii) *or* (iii) *below is true:*
(ii) $H^n \to H$ *in $\tilde{I}_{F,G}$-measure;*
(iii) $H^n \to H$ *pointwise and $I_{F, L_{G^*}^q}$ is uniformly σ-additive.*
Then $H^n \in \mathcal{F}_F(\mathcal{B}, X)$ and $H^n \to H$ in $\mathcal{F}_{F,G}(X)$.

B. Extension of I_X to $\mathcal{P}(\infty)$

Assume X is p-summable relative to (F, G) and consider the σ-additive measure $I_X : \mathcal{P} \to L_E^p$ with finite semivariation $(\tilde{I}_X)_{F, L_G^p}$.

Although X is not defined on the whole space $\overline{\mathbb{R}}^2 \times \Omega$, we can extend I_X to the σ-algebra $\mathcal{P}(\infty)$ of predictable subsets of $\overline{\mathbb{R}}^2 \times \Omega$, by the equality

$$I_X(B) = I_X(B \cap (\mathbb{R}^2 \times \Omega)), \text{ for } B \in \mathcal{P}(\infty).$$

It follows that I_X is 0 on the sub algebras $\mathcal{P}^1(\infty)$ and $\mathcal{P}^2(\infty)$. The extended I_X is still σ-additive in L_E^p on $\mathcal{P}(\infty)$ and has finite semivariation relative to (F, L_G^p):

$$(\tilde{I}_X)_{F, L_G^p}(B) = (\tilde{I}_X)_{F, L_G^p}(B \cap (\mathbb{R}^2 \times \Omega)), \text{ for } B \in \mathcal{P}(\infty).$$

One can define the space $\mathcal{F}_{F, L_G^p}(\overline{\mathbb{R}}^2 \times \Omega, I_X)$ and we have

$$\int_{\overline{\mathbb{R}}^2 \times \Omega} H dI_X = \int_{\mathbb{R}^2 \times \Omega} H dI_X, \text{ for } H \in \mathcal{F}_{F, L_G^p}(\overline{\mathbb{R}}^2 \times \Omega, I_X).$$

C. Existence of left limits of X in L_E^p

Assume X is p-summable relative to (F, G). The following theorem states the existence, in the topology of L_E^p, of the left limits of X at the points (z, ω) for $z \in \overline{\mathbb{R}}^2$.

4. Theorem. *For every point $z = (s, t) \in \overline{\mathbb{R}}^2$ we have*

$$\lim_{v \uparrow z} X_v = I_X((-\infty, z) \times \Omega),$$

$$\lim_{\alpha \uparrow s} X_{\alpha t} = I_X((-\infty, s) \times (-\infty, t] \times \Omega),$$

$$\lim_{\beta \uparrow t} X_{s\beta} = I_X((-\infty, s] \times (-\infty, t) \times \Omega),$$

in the topology of L_E^p.

If the limit X_{z-} (respectively $X_{s-,t}, X_{s,t-}$) exists pointwise on Ω in E, then

$$I_X((-\infty, z) \times \Omega) = X_{z-}, \text{ a.s.,}$$

(respectively

$$I_X((-\infty, s) \times (-\infty, t] \times \Omega) = X_{s-,t}, \text{ a.s.,}$$
$$I_X((-\infty, s] \times (-\infty, t) \times \Omega) = X_{s,t-}, \text{ a.s.}).$$

Proof. Let $z = (s, t) \in \overline{\mathbb{R}}^2$ and let (v_n) be a sequence from $\mathbb{R}^2$ with $v_n \uparrow z$. Since I_X is σ-additive on $\mathcal{P}(\infty)$ we have

$$I_X((-\infty, z) \times \Omega) = \lim I_X((-\infty, v_n] \times \Omega)$$
$$= \lim \Delta_{-\infty v_n} X = \lim X_{v_n}$$

in L_E^p. It follows that
$$\lim_{v \uparrow z} X_v = I_X((-\infty, z) \times \Omega), \text{ in } L_E^p.$$

In case the left limit X_{z-} exists pointwise, we have $\lim X_{v_n} = X_{z-}$; therefore the two limits are equal a.s.:
$$I_X((-\infty, z) \times \Omega) = X_{z-}, \text{ a.s.}$$

The other assertions in the theorem are proved similarly. ∎

The above theorem leads to the following definition:

5. Definition. *Let $z = (s,t)$ be any point in $\overline{\mathbb{R}}^2 \times \Omega$. If one of the left limits $X_{z-}, X_{s-,t}, X_{s,t-}$ does not exist pointwise, we set*
$$X_{z-} = I_X((-\infty, z) \times \Omega),$$
$$X_{s-,t} = I_X((-\infty, s) \times (-\infty, t] \times \Omega)$$

and
$$X_{s,t-} = I_X((-\infty, s] \times (-\infty, t) \times \Omega).$$

For each point of the form $z = (s, \infty)$ or $z = (\infty, t)$ or $z = \infty$, we set
$$X_z = X_{z-} = I_X((-\infty, z) \times \Omega).$$

D. Some properties of the integral $\int H dI_X$

Let X be a p-summable process relative to (F, G).

In this section we prove equalities of the form $\int hH dI_X = h \int H dI_X$, where h is a random variable.

The first property is an extension of the equality by which the measure I_X was defined.

6. Proposition. *Let $z \leq z'$ in $\mathbb{R}^2$ and $h : \Omega \to F$ be an $\mathcal{F}_z$-measurable, bounded random variable. Then*
$$\int h 1_{(z,z']} dI_X = h \Delta_{zz'} X, \text{ a.s.}$$

Proof. If $h = 1_A y$ with $A \in \mathcal{F}_z$ and $y \in F$, then
$$\int h 1_{(z,z']} dI_X = \int y 1_A 1_{(z,z']} dI_X = 1_A y \Delta_{zz'} X$$
$$= h \Delta_{zz'} X.$$

Then the equality remains valid for $\mathcal{F}_z$-step functions h.

Let now h be as in the statement of the theorem and let (h_n) be a sequence of $\mathcal{F}_z$-step functions such that $h_n \to h$ and $|h_n| \leq |h|$. We can apply Theorem 1 for $H^n = h_n 1_{(z,z']}$ and $H^0 = H = h 1_{(z,z']}$ and deduce that $\int h 1_{(z,z']} dI_X \in L^p_G$ and

$$\int h_n 1_{(z,z']} dI_X \to \int h 1_{(z,z']} dI_X,$$

pointwise, weakly in G. At the same time

$$h_n \Delta_{zz'} X \to h \Delta_{zz'} X, \text{ strongly in } G.$$

For each n we have

$$\int h_n 1_{(z,z']} dI_X = h_n \Delta_{zz'} X.$$

Then the two limits are equal,

$$\int h 1_{(z,z]} dI_X = h \Delta_{zz'} X, \text{ a.s.}$$

∎

Remark. If $z' = (s', \infty)$ or $z' = (\infty, t')$, we have

$$\int h 1_{(z,z')} dI_X = h \Delta_{z,z'-} X, \text{ a.s.}$$

The following two theorems are extensions of the preceding one. In the first theorem, h is scalar-valued; in the second theorem, h is F-valued.

7. Theorem. *Let $z \leq z'$ in $\mathbb{R}^2$, $h : \Omega \to \mathbb{R}$ be a bounded, $\mathcal{F}_z$-measurable function and $H \in \mathcal{F}_{F,G}(X)$. Then*
a) *If $\int 1_{(z,z']} H dI_X \in L^p_G$ we have*

$$\int h 1_{(z,z']} H dI_X = h \int 1_{(z,z']} H dI_X.$$

b) *If $H \in L^1_{F,G}(X)$, then $1_{(z,z']} H$ and $h 1_{(z,z']} H$ belong to $L^1_{F,G}(X)$ and we have*

$$(h 1_{(z,z']} H) \cdot X = h[(1_{(z,z']} H) \cdot X].$$

Proof. To prove assertion a), we assume first $H = 1_{(u,u'] \times A} y$ with $A \in \mathcal{F}_u$ and $y \in F$.

If $(z, z'] \cap (u, u'] = \phi$, then $1_{(z,z']} H = 0$ and the equality is satisfied.

If $(z, z']$ and $(u, u']$ are not disjoint, then $(z, z'] \cap (u, u'] = (v, v']$, where $v = \sup(z, u)$. Then $1_A y$ and $h 1_A y$ are $\mathcal{F}_v$-measurable and we can apply Proposition 6:

$$\int h 1_{(z,z']} H dI_X = \int h 1_A y 1_{(v,v']} dI_X = h 1_A y \Delta_{vv'} X$$

$$= h \int 1_A y 1_{(v,v']} dI_X = h \int 1_{(z,z']} H dI_X.$$

Then the equality remains valid for $H = 1_B y$ with $B \in \mathcal{R}$ and $y \in F$:
$$\int h 1_{(z,z']} 1_B y dI_X = h \int 1_{(z,z']} 1_B y dI_X.$$

For $u \in L^q_{G^*}$ with $\frac{1}{p} + \frac{1}{q} = 1$, we have $hu \in L^q_{G^*}$ and
$$\int h 1_{(z,z']} 1_B y (dI_X) u = \int 1_{(z,z']} 1_B y d(I_X) hu.$$

In fact,
$$\langle \int h 1_{(z,z']} 1_B y dI_X, u \rangle = \langle h \int 1_{(z,z']} 1_B y dI_X, u \rangle$$
$$= \langle \int 1_{(z,z']} 1_B y dI_X, hu \rangle,$$

where the bracket means the duality between L^p_G and $L^q_{G^*}$.

By a monotone class argument, the above equality remains valid for $B \in \mathcal{P}$ and $u \in L^q_{G^*}$. Then, for any predictable step process H and $u \in L^q_{G^*}$ we have
$$\int h 1_{(z,z']} H d(I_X) u = \int 1_{(z,z']} H d(I_X) hu.$$

If $H \in \mathcal{F}_{F,G}(X)$ we approximate H by a sequence (H^n) of simple predictable processes with $|H^n| \le |H|$ and apply Lebesgue's theorem in the spaces $L^1_F((I_X)_u)$ and $L^1_F((I_X)_{hu})$ and deduce that
$$\int h 1_{(z,z']} H d(I_X)_u = \int 1_{(z,z']} H d(I_X)_{hu}, \text{ for } u \in L^q_{G^*}.$$

Assume now that $\int 1_{(z,z']} H dI_X \in L^p_G$. Then $h \int 1_{(z,z]} H dI_X \in L^p_G$ and
$$\langle h \int 1_{(z,z']} H dI_X, u \rangle = \langle \int 1_{(z,z']} H dI_X, hu \rangle$$
$$= \int 1_{(z,z']} H d(I_X)_{hu} = \int h 1_{(z,z']} H d(I_X)_u$$
$$= \langle \int h 1_{(z,z']} dI_X, u \rangle.$$

Since $L^q_{G^*}$ is norming both for L^p_G and $(L^q_{G^*})^*$, we deduce that
$$\int h 1_{(z,z']} H dI_X = h \int 1_{(z,z']} H dI_X$$

and assertion a) is proved.

Assume now $H \in L^1_{F,G}(X)$. Then $\int_{(-\infty,v]} H dI_X \in L^p_G$ for every $v \in \mathbb{R}^2$ and the process $(\int_{(-\infty,v]} H dI_X)_{v \in \mathbb{R}^2}$ is chosen to be cadlag. To prove that $1_{(z,z]} H \in L^1_{F,G}(X)$ we remark first that
$$1_{(z,z']} = 1_{(-\infty,z']} + 1_{(-\infty,z]} - 1_{(-\infty,(s',t)]} - 1_{(-\infty,(s,t')]},$$

where $z = (s, t)$ and $z' = (s', t')$. Now

$$\int_{(-\infty, v]} 1_{(-\infty, z']} H dI_X = \int_{(-\infty, v \wedge z']} H dI_X \in L_G^p$$

and the mapping $v \mapsto \int_{(-\infty, v \wedge z']} H dI_X$ is cadlag, therefore $1_{(-\infty, z]} H \in L_{F,G}^1(X)$.

Similarly, all the processes $1_{(-\infty, z]} H$, $1_{(-\infty, (s, t')]} H$, $1_{(-\infty, (s', t)]} H$ belong to $L_{F,G}^1(X)$; therefore $1_{(z, z']} H \in L_{F,G}^1(X)$. Then from assertion a) we deduce that $h 1_{(z, z']} H \in L_{F,G}^1(X)$ as well as the equality in assertion b). ∎

In the following theorem the function h is F-valued.

8. Theorem. *Let $z \leq z'$ in $\mathbb{R}^2$, $h : \Omega \to F$ a bounded, $\mathcal{F}_z$-measurable function and $H \in \mathcal{F}_{\mathbb{R}}((I_X)_{F,G}) \cap \mathcal{F}_{\mathbb{R}}((I_X)_{\mathbb{R}, E})$. Then*
a) *If $\int 1_{(z, z']} H dI_X \in L_E^p$ we have*

$$\int h 1_{(z, z']} H dI_X = h \int 1_{(z, z']} H dI_X .$$

b) *If $H \in L_{\mathbb{R}, E}^1(X)$, then $1_{(z, z']} H \in L_{\mathbb{R}, E}^1(X)$, $h 1_{(z, z']} H \in L_{F,G}^1(X)$ and we have*

$$(h 1_{(z, z']} H) \cdot X = h[(1_{(z, z']} H) \cdot X] .$$

Proof. Assume first that $h = h' y$ with $y \in F$ and h' scalar-valued, bounded and $\mathcal{F}_z$-measurable. Then, using Proposition 5.46, we have

$$\int 1_{(z, z']} H y dI_X = y \int 1_{(z, z']} H dI_X \in L_G^p .$$

By Theorem 7 we have then

$$\int h 1_{(z, z']} H dI_X = h' \int 1_{(z, z']} H y dI_X = h \int 1_{(z, z']} H dI_X .$$

This equality holds then for any $\mathcal{F}_z$-step function h. Assume now h is as in the statement and let (h_n) be a sequence of $\mathcal{F}_z$-step functions with $h_n \to h$ and $|h_n| \leq h$. Then we can apply the convergence Theorem 1 to the sequence $H^n = h_n 1_{(z, z']} H$ converging pointwise to $h 1_{(z, z']} H$.

We have $\int H^n dI_X \in L_G^p$ and

$$\int H^n dI_X = \int h_n 1_{(z, z']} H dI_X = h_n \int 1_{(z, z']} H dI_X,$$

which converges pointwise in G to $h \int 1_{(z, z']} H dI_X$. Then

$$\int h_n 1_{(z, z']} H dI_X \longrightarrow \int h 1_{(z, z']} H dI_X \text{ pointwise.}$$

From the equality
$$\int h_n 1_{(z,z']} H dI_X = h_n \int 1_{(z,z']} H dI_X$$
we deduce that
$$\int h 1_{(z,z']} H dI_X = h \int 1_{(z,z']} H dI_X$$
and this proves assertion a).

To prove assertion b), assume $H \in L^1_{\mathbb{R},E}(X)$. Then we prove that $1_{(z,z']}H \in L^1_{\mathbb{R},E}(X)$ the same way we did for assertion b) in Theorem 7. From the equality in assertion a) we deduce that $h1_{(z,z']}H \in L^1_{F,G}(X)$, as well as the equality in assertion b). ∎

The following theorem gives the left limit $(H \cdot X)_{z-}$ of the stochastic integral in terms of the integral with respect to I_X.

9. Theorem. *Let $H \in L^1_{F,G}(X)$ and $z \in \mathbb{R}^2$. Then $(H \cdot X)_{z-} \in L^p_G$ and*
$$(H \cdot X)_{z-} = \int_{(-\infty,z)} H dI_X.$$

Proof. Let $z_n \uparrow z$. Then $1_{(-\infty,z_n]}H \to 1_{(-\infty,z)}H$ pointwise and for each n we have $|1_{(-\infty,z_n]}H| \leq |H|$ and
$$\int 1_{(-\infty,z_n]} H dI_X = (H \cdot X)_{z_n} \in L^p_G.$$

We have also $(H \cdot X)_{z_n} \to (H \cdot X)_{z-}$ in G. We can apply Theorem 1 with $H^n = 1_{(-\infty,z_n]}H$ and we get $\int_{(-\infty,z)} H dI_X \in L^p_G$ and
$$\int 1_{(-\infty,z_n]} H dI_X \to \int 1_{(-\infty,z)} H dI_X,$$
pointwise and in L^1_G, hence
$$(H \cdot X)_{z-} = \int 1_{(-\infty,z]} H dI_X \in L^p_G.$$
∎

10. Corollary. *The stochastic integral is cadlag in L^p_G.*

Remark. If $H \cdot X$ is itself p-summable and $z = (s, \infty)$ or $z = (\infty, t)$, then, by Theorem 4, $(H \cdot X)_{z-}$ exists in L^p_G and the above proof remains valid.

E. Summability of stopped processes

We consider now the p-summability of the stopped process X^z defined by $X_v^z = X_{z \wedge v}$ for $v \in \mathbb{R}^2$. The following theorem is a partial extension for two-parameter processes of the following equality for one-parameter processes: $(H \cdot X)^T = (1_{[0,T]} H) \cdot X$, where T is a stopping time. (Theorem 11.6). Assume X is p-summable relative to (F, G).

11. Theorem. *Let $H \in L^1_{F,G}(X)$ and $z \in \mathbb{R}^2$. Then $1_{(-\infty,z]} H \in L^1_{F,G}(X)$ and*

$$(H \cdot X)^z = (1_{(-\infty,z]}) \cdot X.$$

Proof. Since $H \in L^1_{F,G}(X)$, we can choose the process $\left(\int_{(-\infty,z]} H \, dI_X \right)_{z \in \mathbb{R}^2}$ with values in G and cadlag. Then, for $z \in \mathbb{R}^2$ we have

$$\int_{(-\infty,v]} 1_{(-\infty,z]} H \, dI_X = \int_{(-\infty,v \wedge z]} H \, dI_X,$$

therefore $1_{(-\infty,z]} H \in L^1_{F,G}(X)$ and this equality can be written

$$\left[1_{(-\infty,z]} H \cdot X \right]_v = (H \cdot X)_{v \wedge z}, \text{ a.s.}$$

∎

The following theorem also extends for two-parameter processes, properties of one-parameter processes involving stopping times (Theorem 11.9).

12. Theorem. *Let $z \in \mathbb{R}^2$. Then*
a) *X^z is p-summable relative to (F, G) and we have*

$$X^z = 1_{(-\infty,z]} \cdot X$$

and

$$(I_{X^z}(A) = I_X(((-\infty, z] \times \Omega) \cap A), \text{ for } A \in \mathcal{P}.$$

b) *For every F-valued, predictable process H we have*

$$\text{svar}_{F, L^p_G} I_{X^z}(H) = \text{svar}_{F, L^p_G} I_X(1_{(-\infty,z]} H).$$

c) *We have $H \in \mathcal{F}_{F,G}(X^z)$ iff $1_{(-\infty,z]} H \in \mathcal{F}_{F,G}(X)$ and in this case we have*

$$\int H \, dI_{X^z} = \int 1_{(-\infty,z]} H \, dI_X.$$

d) *We have $H \in L^1_{F,G}(X^z)$ iff $1_{(-\infty,z]} H \in L^1_{F,G}(X)$ and in this case we have*

$$H \cdot X^z = (1_{(-\infty,z]} H) \cdot X$$

§26. PROPERTIES OF THE STOCHASTIC INTEGRAL

d') If $H \in L^1_{F,G}(X)$, then $1_{(-\infty,z]}H \in L^1_{F,G}(X)$ and $H \in L^1_{F,G}(X^z)$ and we have
$$(H \cdot X)^z = H \cdot X^z = (1_{(-\infty,z]}H) \cdot X.$$

e) If the set of measures $(I_X)_{F,L^q_{G*}}$ is uniformly σ-additive, then so is $(I_{X^z})_{F,L^q_{G*}}$.

Proof. For $u < v$ in $\mathbb{R}^2$ and B in $\mathcal{F}_u$ we have
$$\begin{aligned} I_{X^z}((u,v] \times B) &= 1_B \Delta_{uv} X^z = 1_B \Delta_{u \wedge z, v \wedge z} X \\ &= I_X(((u,v] \cap (-\infty, z]) \times B) \\ &= I_X(((-\infty, z] \times \Omega) \cap ((u,v] \times B)). \end{aligned}$$

It follows that
$$I_{X^z}(A) = I_X(((-\infty, z] \times \Omega) \cap A) \text{ for } A \in \mathcal{R}.$$

Since $I_X : \mathcal{P} \to L^p_G$ is σ-additive, the above equality extends I_{X^z} to a σ-additive measure $I_{X^z} : \mathcal{P} \to L^p_G$. We have further
$$\operatorname{svar}_{F,L^p_G} I_{X^z}(A) = \operatorname{svar}_{F,L^p_G} I_X(((-\infty, z] \times \Omega) \cap A), \text{ for } A \in \mathcal{P}$$

which can be easily proved. Since I_X has finite semivariation relative to (F,G), it follows that I_{X^z} has finite semivariation relative to (F,G), hence X^z is p-summable relative to (F,G). To prove assertion a) it remains to prove the equality $X^z = 1_{(-\infty,z]} \cdot X$. This equality will follow from assertion d) with $H \equiv 1$.

To prove assertion b) let $u \in L^q_{G*}$, $\frac{1}{p} + \frac{1}{q} = 1$ with $\|u\|_q \leq 1$. Then, from assertion a) we deduce
$$(I_{X^z})_u(A) = (I_X)_u(((-\infty, z] \times \Omega) \cap A), \text{ for } A \in \mathcal{P}.$$

Then we have equality for the variations:
$$|(I_{X^z})_u|(A) = |(I_X)_u|((-\infty, z] \times \Omega) \cap A), \text{ for } A \in \mathcal{P}.$$

It follows that for any F-valued predictable process H we have
$$\int |H| d|(I_{X^z})_u| = \int 1_{(-\infty,z]} |H| d|(I_X)_u|.$$

Taking the supremum for $\|u\|_q \leq 1$ we obtain the equality in assertion b).

From assertion b) we deduce that
$$H \in \mathcal{F}_{F,G}(X^z) \text{ iff } 1_{(-\infty,z]} H \in \mathcal{F}_{F,G}(X).$$

From the equality
$$(I_{X^z})_u(A) = (I_X)_u(((-\infty, z] \times \Omega) \cap A), \text{ for } A \in \mathcal{P}$$

proved above, we deduce that if $H \in \mathcal{F}_{F,G}(X^z)$, then

$$\int H\, dI_{X^z} = \int 1_{(-\infty,z]} H\, dI_X$$

and this proves assertion c).

If we replace H with $1_{(-\infty,v]}H$ in assertion c) we deduce that $1_{(-\infty,v]}H \in \mathcal{F}_{F,G}(X^z)$ iff $1_{(-\infty,v]}1_{(-\infty,z]}H \in \mathcal{F}_{F,G}(X)$ and in this case we have

$$\int_{(-\infty,v]} H\, dI_{X^z} = \int_{(-\infty,v]} 1_{(-\infty,z]} H\, dI_X \, .$$

It follows that $H \in L^1_{F,G}(X^z)$ iff $1_{(-\infty,z]}H \in L^1_{F,G}(X)$ and in this case we have

$$(H \cdot X^z)_v = ((1_{(-\infty,z]}H) \cdot X)_v$$

and this is assertion d).

If now $H \in L^1_{F,G}(X)$, then, by Theorem 11, we deduce that $1_{(-\infty,z]}H \in L^1_{F,G}(X)$ and

$$(1_{(-\infty,z]}H) \cdot X = (H \cdot X)^z .$$

Then assertion d') follows. Assertion e) follows from the equality

$$|(I_{X^z})_u|(A) = |(I_X)_u|(((-\infty, z] \times \Omega) \cap A)\, , \text{ for } A \in \mathcal{P}.$$

∎

F. Summability of the stochastic integral

Let $X : \mathbb{R}^2 \times \Omega \to E \subset L(F,G)$ be a p-summable process relative to (F,G).

For $H \in L^1_{F,G}(X)$, the stochastic integral $H \cdot X$ is again a cadlag, adapted process with values in G and we can ask whether $H \cdot X$ is itself p-summable. We give a partial answer in the following two theorems, the first one for H scalar-valued and the second one for H with values in F.

13. Theorem. *Let* $H \in \mathcal{F}_\mathbb{R}((I_X)_{F,G}) \cap \mathcal{F}_\mathbb{R}((I_X)_{\mathbb{R},E})$ *and assume that* $H \in L^1_{\mathbb{R},E}(X)$ *and* $\int_A H\, dI_X \in L^p_E$ *for every* $A \in \mathcal{P}$. *Then*
a) $H \cdot X$ *is p-summable relative to (F,G) and*

$$dI_{H \cdot X} = d(HI_X)$$

where HI_X is the measure defined by

$$(HI_X)(A) = \int_A H\, dI_X \, , \text{ for } A \in \mathcal{P}.$$

b) *For any predictable process $K \geq 0$ we have*

$$(\tilde{I}_{H \cdot X})_{F,G}(K) = (\tilde{I}_X)_{F,G}(KH)\, .$$

c) $K \in L^1_{F,G}(H \cdot X)$ iff $KH \in L^1_{F,G}(X)$ and in this case we have

$$K \cdot (H \cdot X) = (KH) \cdot X.$$

d) Assume $(I_X)_{F,L^q_{G^*}} = \{|(I_X)_u| : u \in L^q_{G^*}, \|u\|_q \leq 1\}$ is a uniformly σ-additive family of measures with values in F^*.
Then $(I_{H \cdot X})_{F,L^q_{G^*}}$ is uniformly σ-additive iff $H \in \mathcal{F}_\mathbb{R}(\mathcal{B}, (I_X)_{F,G}) =$ the closure in $\mathcal{F}_\mathbb{R}((I_X)_{F,G})$ of the set of bounded processes.

The proof is the same as that of Theorem 13.1 for one-parameter processes.

14. Theorem. Let $H \in L^1_{F,G}(X)$ and assume $\int_A H \, dI_X \in L^p_G$ for every $A \in \mathcal{P}$. Then:
a) $H \cdot X$ is p-summable relative to $(\mathbb{R}, G)$ and

$$dI_{H \cdot X} = d(HI_X).$$

b) For every predictable process $K \geq 0$ we have

$$(\tilde{I}_{H \cdot X})_{\mathbb{R},G}(K) \leq (\tilde{I}_X)_{F,G}(KH) = (|K|I_X)\widetilde{_{F,G}}(H).$$

c) If K is a real-valued, predictable process such that $KH \in L^1_{F,G}(X)$, then $K \in L^1_{\mathbb{R},G}(H \cdot X)$ and we have

$$K \cdot (H \cdot X) = K \cdot (H \cdot X).$$

The proof is similar to that of Theorem 13.1 for one-parameter processes, replacing the equality

$$(\tilde{I}_H \cdot X)_{F,G}(A) = (\tilde{I}_X)_{F,G}(1_A H)$$

with the inequality

$$(\tilde{I}_H \cdot X)_{\mathbb{R},G}(A) \leq (\tilde{I}_X)_{F,G}(1_A H)$$

(since $|(I_{H \cdot X})_u|(A) \leq \int_A |H| d|(I_X)_u|$).

Chapter 8
Two-Parameter Martingales

This chapter contains the main results concerning the stochastic integral of a two-parameter process, in case the process is a martingale.

If M is a square integrable martingale with values in a Hilbert space E, then M is 2-summable (Theorem 28.9) and we have $L^1_{\mathbb{R}, L^2_E(M)} = \mathcal{F}_{\mathbb{R}, L^2_E}(M)$ (Theorem 28.14).

§27. MARTINGALES

There are two main results in this paragraph:
1. A two-parameter martingale with values in a Banach space E and with paths in L^p_E, with $p > 1$, has a cadlag modification (Theorem 3).
2. If $X : \mathbb{R}^2 \times \Omega \to E \subset L(F, G)$ is a p-summable martingale with $p > 1$ and if G is reflexive, then the stochastic integral $H \cdot X$ is again p-summable (Theorem 4) and $L^1_{F, L^p_G}(X)$ is complete (Corollary 5).

We start with the definition of the martingale:

1. Definition. *Let $M : \mathbb{R}^2 \times \Omega \to E$ be an adapted process such that $M_z \in L^1_E$ for every $z \in \mathbb{R}^2$.*
a) *M is called a martingale if*
$$E(M_{z'}|\mathcal{F}_z) = M_z, \text{ for } z \leq z' \text{ in } \mathbb{R}^2.$$
b) *M is called a weak martingale if*
$$E(\Delta_{zz'}M|\mathcal{F}_z) = 0, \text{ for } z \leq z' \text{ in } \mathbb{R}^2.$$

c) *M* is called a **strong martingale** if

$$E(M_{z'} - M_z | \mathcal{F}_s^1 \vee \mathcal{F}_t^2) = 0, \text{ for } z = (s,t) \leq z' \text{ in } \mathbb{R}^2.$$

d) *M* is a **1-martingale**, if for each $t \in \mathbb{R}$, the one-parameter process $(M_{st})_{s \in \mathbb{R}}$ is a martingale with respect to the filtration $(\mathcal{F}_{st})_{s \in \mathbb{R}}$. A **2-martingale** is defined similarly.

We state without proof some properties of the martingales.

If $(M_{st})_{s \in \mathbb{R}}$ is a martingale for the filtration $(\mathcal{F}_{st})_{s \in \mathbb{R}}$, then it is also a martingale for the filtration $(\mathcal{F}_{s\infty})_{s \in \mathbb{R}}$.

Any strong martingale is a martingale.

A process *M* is a martingale iff it is a 1-martingale and a 2-martingale.

Any 1-martingale (or 2-martingale) is a weak martingale (see [C-W], Proposition 1.1 and [M], p. 9).

2. Theorem. (Doob's inequality) *Let M be an E-valued, right continuous martingale and denote $M^* = \sup_z |M_z|$. Then*

$$\lambda P\{M^* > \lambda\} \leq C(1 + \sup_z E[|M_\infty| \log^+ |M_\infty|])$$

where C is a constant.

For $1 < p < \infty$ we have

$$E(M^{*p}) \leq \left(\frac{p}{p-1}\right)^{2p} \sup_z E(|M_z|^p).$$

For the proof see [C.1] and [M] (Theorem 1.1).

The following theorem proves the existence of cadlag modifications of a martingale with values in a Banach space.

For scalar-valued martingales, Bakry [Ba] (see also [M], Theorems 9.2 and 9.3) proved that a martingale with paths in $L \log^+ L$ has cadlag modifications. We mention that $L^p \subset L \log^+ L$ if $p > 1$.

3. Theorem. *Let $M : \mathbb{R}^2 \times \Omega \to E$ be a martingale with $M_z \in L_E^p$ for every $z \in \mathbb{R}^2$. If $p > 1$, then M has a cadlag modification.*

Proof. The theorem is true if $M = \sum \phi^i x_i$, finite sum, with $x_i \in E$ and ϕ^i scalar martingales with $\phi_z^i \in L^p$ for $z \in \mathbb{R}^2$, since, by [B], ϕ^i has a cadlag modification.

Assume now *M* is a martingale with $M_z \in L_E^p$ for each $z \in \mathbb{R}^2$. Let $a \in \mathbb{R}^2$ and consider the restricted martingale $(M_z)_{z \in R_a}$, where $R_a = (-\infty, a]$. Then $M_z = E(M_a | \mathcal{F}_z)$ for $z \in R_a$. Let (Y_a^n) be a sequence of step functions from L_E^p converging to M_a in L_E^p. For each *n*, the martingale $E(Y_a^n | \mathcal{F}_z)_{z \in R_a}$ has a cadlag modification $(Y_z^n)_{z \in R_a}$ and for each $z \in R_a$ we have $Y_z^n \to M_z$ in L_E^p, hence (Y_z^n) is a Cauchy sequence in L_E^p; therefore in L_E^1:

$$E(|Y_z^n - Y_z^m|) = E(|E(Y_a^n - Y_a^m | \mathcal{F}_z)|)$$
$$\leq E((E(|Y_a^n - Y_a^m| | \mathcal{F}_z))) \leq E(|Y_a^n - Y_a^m|) \to 0;$$

it follows that
$$\sup_{z \in R_a} E(|Y_z^n - Y_z^m|) \leq E(|Y_a^n - Y_a^m|) \to 0, \text{ as } n, m \to \infty.$$

From Doob's inequality we deduce that (Y_z^n) is a Cauchy sequence in probability, uniformly for $z \in R_a$. There is a subsequence $(n_k)_{k \in \mathbb{N}}$ such that $(X_t^{n_k})_k$ is Cauchy a.s., uniformly for $z \in R_a$, converging to a limit Y_z a.s. and also to M_z in the mean. Hence $Y_z = M_z$ a.s. It follows that $(Y_z)_{z \in R_a}$ is a cadlag modification of $(M_z)_{z \in R_a}$.

Let now $a = (n, n)$. By the above, for each n, there is a cadlag modification $(Y_z^n)_{z \leq (n,n)}$ of $(M_z)_{z \leq (n,n)}$. We have
$$Y_z^{n+1} = Y_z^n = M_z, \quad \text{a.s. for } z \leq (n, n).$$

Let N be a negligible set such that outside N we have
$$Y_z^n = Y_z^m = M_z, \text{ for every } n \text{ and } m \text{ with } n < m$$

and for $z \leq (n, n)$ with rational coordinates. Since $(Y_z^n)_{z \leq (n,n)}$ is cadlag, for each n, we deduce that, outside N, we have
$$Y_z^n = Y_z^m, \text{ for every } z \in \mathbb{R}^2 \text{ with } z \leq (n,n) \text{ and } n < m.$$

It follows that $Y_z = \lim_{n \to \infty} Y_z^n$ exists outside N and
$$Y_z = M_z \text{ outside } N \text{ for every } z \in \mathbb{R}^2.$$

If we set $Y_z = 0$ on N, we obtain a cadlag modification Y of M, and the theorem is proved. ∎

We can prove now that the stochastic integral is again a martingale.

4. Theorem. *Let $M : \mathbb{R}^2 \times \Omega \to E \subset L(F, G)$ be a p-summable martingale and let $H \in \mathcal{F}_{F,L_G^p}(I_M)$ be such that $\int_{(-\infty, z]} H dI_M \in L_G^p$ for every $z \in \mathbb{R}^2$. Then*
1) *$(\int_{(-\infty, z]} H dI_M)_{z \in \mathbb{R}^2}$ is a uniformly integrable martingale, bounded in L_E^p.*
2) *If $p > 1$, then $H \in L_{F,G}^1(M)$.*
3) *If $p > 1$ and G is reflexive, then $H \cdot M$ is p-summable relative to $(\mathbb{R}, G)$.*

Proof. Let $z \in \mathbb{R}^2$ and prove that
$$(*) \qquad E(\int_{(-\infty,\infty)} H dI_M | \mathcal{F}_z) = \int_{(-\infty, z]} H dI_M.$$

Assume $H = 1_{(z', z''] \times B} x$ with $B \in \mathcal{F}_{z'}$ and $x \in F$. Let $z = (s, t)$, $z' = (s', t')$ and $z'' = (s'', t'')$. Then $\mathcal{F}_z = \mathcal{F}_{st} = \mathcal{F}_s^1 \cap \mathcal{F}_t^2$ and $B \in \mathcal{F}_{z'} = \mathcal{F}_{s'}^1 \cap \mathcal{F}_{t'}^2$. We have

$$\int_{(-\infty,\infty)} H dI_M = \int 1_{(z', z''] \times B} x dI_M$$
$$= 1_B x \Delta_{z' z''} M = 1_B x (M_{s't'} + M_{s''t''} - M_{s't''} - M_{s''t'}).$$

There are several cases.
a) $s \leq s' \leq s''$. Then $\mathcal{F}_s^1 \subset \mathcal{F}_{s'}^1 \subset \mathcal{F}_{s''}^1$, therefore

$$E(\int_{(-\infty,\infty)} H dI_M | \mathcal{F}_z)$$
$$= E(1_B x(M_{s't'} + M_{s''t''} - M_{s't''} - M_{s''t'}|\mathcal{F}_s^1|\mathcal{F}_t^2))$$
$$= E(1_B x(M_{st'} + M_{st''} - M_{st''} - M_{st'}|\mathcal{F}_t^2) = 0,$$

and $(-\infty, z] \cap (z', z''] = \phi$; therefore

$$\int_{(-\infty,z]} H dI_M = \int_{(-\infty,z]} 1_{(z',z'']\times B} x dI_M = 0,$$

and equality $(*)$ is satisfied.

By symmetry, equality $(*)$ is satisfied also if $t \leq t' \leq t''$.

b) Assume $s' \leq s \leq s''$ and $t' \leq t \leq t''$. Then $z' \leq z \leq z''$ and $B \in \mathcal{F}_{s'}^1 \subset \mathcal{F}_s^1$ and also $B \in \mathcal{F}_{t'}^2 \subset \mathcal{F}_t^2$; therefore, using the fact that M is a 1-martingale and a 2-martingale, we get

$$E(\int_{(-\infty,\infty)} H dI_M | \mathcal{F}_z)$$
$$= E(1_B x(M_{s't'} + M_{s''t''} - M_{s't''} - M_{s''t'}|\mathcal{F}_s^1|\mathcal{F}_t^2))$$
$$= E(1_B x(M_{s't'} + M_{st''} - M_{s't''} - M_{st'}|\mathcal{F}_t^2)$$
$$= 1_B x(M_{s't'} + M_{st} - M_{s't} - M_{st'}),$$

and $(-\infty, z] \cap (z', z''] = (z', z]$; therefore

$$\int_{(-\infty,z]} H dI_M = 1_B x \Delta_{z'z} M = 1_B x(M_{s't'} + M_{st} - M_{s't} - M_{st'}).$$

Consequently, equality $(*)$ is satisfied.

c) Assume now $s' \leq s \leq s''$ and $t' \leq t'' \leq t$. Then, again $B \in \mathcal{F}_{s'}^1 \subset \mathcal{F}_s^1$ and $B \in \mathcal{F}_{t'}^2 \subset \mathcal{F}_t^2$; therefore,

$$E(\int_{(-\infty,\infty)} H dI_M | \mathcal{F}_z)$$
$$= E(1_B x(M_{s't'} + M_{s''t''} - M_{s't''} - M_{s''t'}|\mathcal{F}_s^1|\mathcal{F}_t^2))$$
$$= E(1_B x(M_{s't'} + M_{st''} - M_{s't''} - M_{st'}|\mathcal{F}_t^2)$$
$$= 1_B x(M_{s't'} + M_{st''} - M_{s't''} - M_{st'}),$$

and $(-\infty, z) \cap (z', z''] = (z', (s, t'')]$; therefore,

$$\int_{(-\infty,z]} H d_M = \int_{(-\infty,z]} 1_{(z',z'']\times B} x dI_M$$
$$= \int 1_{(z',(s,t'')]\times B} x dI_M = 1_B x(M_{s't'} + M_{st''} - M_{s't''} - M_{st'}),$$

so that equality $(*)$ is satisfied.

By symmetry, equality $(*)$ is satisfied if $s' \leq s'' \leq s$ and $t' \leq t \leq t''$.

d) Assume that $s' \leq s'' \leq s$ and $t' \leq t'' \leq t$, i.e., $z' \leq z'' \leq z$. Then $(s', t'') \leq z$ and $(s'', t') \leq z$; therefore,

$$E(\int_{(-\infty,\infty)} H dI_M | \mathcal{F}_z) = E(1_B x \Delta_{z'z''} M | \mathcal{F}_z)$$
$$= 1_B x \Delta_{z'z''} M$$

and $(-\infty, z] \cap (z', z''] = (z', z'']$; therefore

$$\int_{(-\infty,z]} H dI_M = \int 1_{(z',z''] \times B} x dI_M = 1_B x \Delta_{z'z''} M,$$

consequently, equality $(*)$ is satisfied.

It follows that equality $(*)$ is satisfied if H is a predictable simple process over the ring $\mathcal{R}$.

Assume now that H is a predictable, F-valued process and let $y^* \in G^*$ and $A \in \mathcal{F}_z$. Set $u = 1_A y^* \in L^q_{G^*}$ with $\frac{1}{p} + \frac{1}{q} = 1$. Consider the measure $(I_M)_u : \mathcal{P} \to F^*$. The $\mathcal{R}$-simple processes are dense in $L^1_F((I_M)_u)$. There is a sequence (H^n) of $\mathcal{R}$-simple processes such that $H^n \to H$ in $L^1_F((I_M)_u)$. Then

$$\int_{(-\infty,\infty)} H^n d(I_M)_u \to \int_{(-\infty,\infty)} H d(I_M)_u,$$

that is,

$$\langle \int_{(-\infty,\infty)} H^n dI_M, u \rangle \to \langle \int_{(-\infty,\infty)} H dI_M, u \rangle,$$

the bracket meaning the duality between L^p_G and $L^q_{G^*}$, hence,

$$E(\langle \int_{(-\infty,\infty)} H^n dI_M, 1_A y^* \rangle) \to E(\langle \int_{(-\infty,\infty)} H dI_M, 1_A y^* \rangle),$$

where the bracket, this time, means the duality between G and G^*. Then

$$\langle E(1_A \int_{(-\infty,\infty)} H^n dI_M), y^* \rangle \to \langle E(1_A \int_{(-\infty,\infty)} H dI_M), y^* \rangle.$$

Since $1_{(-\infty,z]} H^n \to 1_{(-\infty,z]} H$ in $L^1_F((I_M)_u)$, we deduce similarly, that

$$\langle E(1_A \int_{(-\infty,z]} H^n dI_M), y^* \rangle \to \langle E(1_A \int_{(-\infty,z]} H dI_M), y^* \rangle.$$

For each n, we have, by equality $(*)$,

$$\langle E(1_A \int_{(-\infty,\infty)} H^n dI_M), y^* \rangle = \langle E(1_A \int_{(-\infty,z]} H^n dI_M), y^* \rangle.$$

It follows that the two sequences have the same limit:

$$(**) \quad \langle E(1_A \int_{(-\infty,\infty)} H dI_M), y^* \rangle = \langle E(1_A \int_{(-\infty,z]} H dI_M), y^* \rangle.$$

Since, by hypothesis, $\int_{(-\infty,\infty)} H dI_M \in L_G^p$ and $\int_{(-\infty,z]} H dI_M \in L_G^p$, it follows that

$$E(1_A \int_{(-\infty,\infty)} H dI_M) \in G \text{ and } E(1_A \int_{(-\infty,z]} H dI_M) \in G.$$

Since $y^* \in G^*$ was arbitrary, from equality $(**)$ we deduce that

$$E(1_A \int_{(-\infty,\infty)} H dI_M) = E(1_A \int_{(-\infty,z]} H dI_M).$$

By Theorem 25.6, the process of representatives $(\int_{(-\infty,z]} H dI_M)_{z \in \mathbb{R}^2}$ is adapted. Since $A \in \mathcal{F}_z$ was arbitrary and since $\int_{(-\infty,z]} H dI_M$ is $\mathcal{F}_z$-measurable, it follows that

$$E(\int_{(-\infty,\infty)} H dI_M | \mathcal{F}_z) = \int_{(-\infty,z]} H dI_M.$$

Consequently, $(\int_{(-\infty,z]} H dI_M)_{z \in \mathbb{R}^2}$ is a uniformly integrable martingale, which is bounded in L_G^p. This proves assertion 1).

If $p > 1$, then the martingale $(\int_{(-\infty,z]} H dI_M)_{z \in \mathbb{R}^2}$ has a cadlag modification, by Theorem 3; therefore $H \in L_{F,G}^1(M)$ and this proves assertion 2).

Assertion 3) follows from assertion 2) and Theorem 13.2, since, G being reflexive, L_G^p is also reflexive; therefore $\int_A H dI_M \in L_G^p$ for every $A \in \mathcal{P}$. ∎

5. Corollary. *Let $M : \mathbb{R}^2 \times \Omega \to E \subset L(F,G)$ be a p-summable martingale with $p > 1$. If G is reflexive, then*

$$L_{F,L_G^p}^1(M) = \mathcal{F}_{F,L_G^p}(M)$$

and $L_{F,L_G^p}^1(M)$ is complete.

This follows from assertion 2) of the preceding theorem and the fact that $\mathcal{F}_{F,L_G^p}$ is complete.

§28. SQUARE INTEGRABLE MARTINGALES

A martingale $M : \mathbb{R}^2 \times \Omega \to E$ is called a *square integrable martingale* if there is a random variable $M_\infty \in L^2_E$ such that

$$M_z = E(M_\infty | \mathcal{F}_z), \text{ for every } z \in \mathbb{R}^2.$$

The main result of this section is that if E and G are Hilbert spaces and $E \subset L(F,G)$, then an E-valued square integrable martingale is 2-summable relative to (F,G).

A. A decomposition theorem

We shall need the following Doob–Mayer decomposition theorem:

1. Theorem. *If M is an E-valued, right continuous square integrable martingale, then there is a unique predictable, right continuous, incrementally increasing, positive process $\langle M \rangle$ such that $|M|^2 - \langle M \rangle$ is a weak martingale.*

For the proof (in case M is real-valued), see [C-W] (Theorem 1.5) and [M] (Theorem 3.1 and Section 10).

The process $\langle M \rangle$ is called the *sharp bracket* of M.

To prove that I_M is σ-additive on $\mathcal{R}$ and can be extended to a σ-additive measure on $\mathcal{P}$, we use the Doléans measure $\mu_{\langle M \rangle}$ and prove that $I_M \ll \mu_{\langle M \rangle}$.

For the rest of the paragraph, E and G are Hilbert spaces and $M : \mathbb{R}_+ \times \Omega \to E \subset L(F,G)$ is a square integrable martingale.

B. The measures $\|I_M(\cdot)\|^2_{L^2_E}$ and $\mu_{\langle M \rangle}$

We prove first that for disjoint sets $A, B \in \mathcal{R}$, $I_M(A)$ and $I_M(B)$ are orthogonal in L^2_E. This property will be extended in Theorem 6 for sets $A, B \in \mathcal{P}$.

2. Lemma. *Let $M, N : \mathbb{R}^2 \times \Omega \to E \subset L(F,G)$ be two square integrable martingales, A, B two disjoint sets from $\mathcal{R}$, and $x, y \in F$. Then*

$$I_M(A) \perp I_N(B) \text{ in } L^2_E \text{ and } I_M(A)x \perp I_N(B)y \text{ in } L^2_G.$$

Proof. We denote by $\langle \cdot, \cdot \rangle_E$ the inner product in E and by $\langle \cdot, \cdot \rangle_{L^2_E} = E(\langle \cdot, \cdot \rangle_E)$ the inner product in L^2_E.

Assume $A = (z, z'] \times C$ with $C \in \mathcal{F}_z$ and $B = (u, u'] \times D$ with $D \in \mathcal{F}_u$. Denote $z = (s,t)$, $z' = (s',t')$, $u = (\alpha, \beta)$ and $u' = (\alpha', \beta')$.

Since $A \cap B = \phi$ we have either $C \cap D = \phi$ or $(z, z'] \cap (u, u'] = \phi$.

If $C \cap D = \phi$, then

$$E(\langle I_M(A), I_N(B) \rangle_E) = E(1_C 1_D \langle \Delta_{zz'} M, \Delta_{uu'} N \rangle_E) = 0,$$

hence $I_M(A) \perp I_N(B)$ in L^2_E.

Assume now that $(z,z'] \cap (u,u'] = \phi$. Then either $(s,s'] \cap (\alpha,\alpha'] = \phi$, or $(t,t'] \cap (\beta,\beta'] = \phi$. We shall assume that $(s,s'] \cap (\alpha,\alpha'] = \phi$ (the other case is proved similarly). To make a choice, we assume that $s < s' \leq \alpha < \alpha'$. Since $(\alpha,\beta) \leq (\alpha',\beta)$ we have

$$E(N_{\alpha'\beta} - N_{\alpha\beta}|\mathcal{F}_{\alpha\beta}) = 0;$$

Since $(\alpha,\beta) \leq (\alpha,\beta') \leq (\alpha',\beta')$ we have also

$$E(N_{\alpha'\beta'} - N_{\alpha\beta'}|\mathcal{F}_{\alpha\beta}) = 0.$$

Since $(N_{v\beta})_v$ is a martingale for the filtration $(\mathcal{F}_{v\beta})_v$, it is also a martingale for the filtration $(\mathcal{F}_{v\infty})_v$; hence,

$$E(N_{\alpha'\beta} - N_{\alpha\beta}|\mathcal{F}_{\alpha\infty}) = 0.$$

Similarly, $(N_{v\beta'})_v$ is a martingale for $(\mathcal{F}_{v\infty})_v$, hence,

$$E(N_{\alpha'\beta'} - N_{\alpha\beta'}|\mathcal{F}_{\alpha\infty}) = 0.$$

It follows that

$$E(I_N(B)|\mathcal{F}_{\alpha\infty}) = 1_D E(\Delta_{uu'} N|\mathcal{F}_{\alpha\infty})$$
$$= 1_D E((N_{\alpha'\beta'} - N_{\alpha\beta'}) - (N_{\alpha'\beta} - N_{\alpha\beta})|\mathcal{F}_{\alpha\infty}) = 0.$$

Since $I_M(A) = 1_C \Delta_{zz'} M$ is $\mathcal{F}_{\alpha\infty}$-measurable, it follows that

$$E(\langle I_M(A), I_N(B) \rangle_E | \mathcal{F}_{\alpha\infty}) = \langle I_M(A), E(I_N(B)|\mathcal{F}_{\alpha,\infty}) \rangle_E = 0,$$

therefore, taking expectations,

$$E(\langle I_M(A), I_N(B) \rangle_E) = 0,$$

that is, $I_M(A) \perp I_N(B)$ in L_E^2.

If $A, B \in \mathcal{R}$, then A and B are finite disjoint unions of sets of the preceding form:

$$A = \bigcup_{i \in I} A_i \text{ and } B = \bigcup_{j \in J} B_j.$$

Since $A \cap B = \phi$, we have $A_i \cap B_j = \phi$ for every i and j, hence $I_M(A_i) \perp I_N(B_j)$ in L_E^2. Then

$$I_M = \sum_i I_M(A_i) \perp \sum_j I_N(B_j) = I_N(B), \text{ in } L_E^2.$$

If now, $x, y \in F$, then Mx and Ny are square integrable martingales with values in G; therefore, by the above,

$$I_{Mx}(A) \perp I_{Ny}(B), \text{ in } L_G^2, \text{ if } A, B \in \mathcal{R} \text{ are disjoint.}$$

But $I_{Mx}(A) = I_M(A)x$, which can be seen by considering first $A = (z, z'] \times C$; and similarly $I_{Ny}(B) = I_N(B)y$. Therefore

$$I_N(A)x \perp I_N(B)y, \text{ in } L_G^2, \text{ if } A, B \in \mathcal{R} \text{ are disjoint.}$$

∎

Remark. The property stated in Lemma 2 will be extended in Theorem 6 for any disjoint sets A, B in $\mathcal{P}$.

An immediate consequence of Lemma 2 is the following lemma:

3. Lemma. *If $M : \mathbb{R}^2 \times \Omega \to E$ is a square integrable martingale, then the mapping $A \mapsto \|I_M(A)\|_{L_E^2}^2$ is finitely additive on $\mathcal{R}$.*

Proof. If $(A_i)_{i \in I}$ is a finite family of disjoint sets from $\mathcal{R}$, then, by Lemma 8.1, the family $(I_M(A_i))_{i \in I}$ is orthogonal in L_E^2, hence

$$\|I_M(\bigcup_i A_i)\|_{L_E^2}^2 = \|\sum_i I_M(A_i)\|_{L_E^2}^2 = \sum_i \|I_M(A_i)\|_{L_E^2}^2.$$

∎

Remark. In Theorem 7 we shall prove that the mapping $A \mapsto \|I_M(A)\|_{L_E^2}^2$ is σ-additive on $\mathcal{P}$.

In the next theorem we shall use the sharp bracket $\langle M \rangle$ associated to a square integrable martingale $M : \mathbb{R}^2 \times \Omega \to E$ in Theorem 1; $\langle M \rangle$ is a right continuous, incrementally increasing, positive process, such that $|M|^2 - \langle M \rangle$ is a weak martingale.

The process $\langle M \rangle$ has integrable variation $|\langle M \rangle|$ (not to be confused with the absolute value of $\langle M \rangle$) and we have $|\langle M \rangle|_z = \langle M \rangle_z$ for $z \in \mathbb{R}^2$. In Theorem 30.4 we shall prove that there is a positive, σ-additive measure $\mu_{\langle M \rangle}$ on $\mathcal{B}(\mathbb{R}^2) \times \mathcal{F}$ (the Doléans measure of $\langle M \rangle$), satisfying

$$\mu_{\langle M \rangle}(A) = E(\int 1_A(z, \omega) d\langle M \rangle_z(\omega)), \text{ for } A \in \mathcal{B}(\mathbb{R}^2) \times \mathcal{F}.$$

The measure $\mu_{\langle M \rangle}$ will be used in Theorem 5 to prove that I_M can be extended to a σ-additive measure on $\mathcal{P}$. It follows that in a logical order, §29 and §30 should be presented before the present paragraph.

4. Lemma. *If $M : \mathbb{R}^2 \times \Omega \to E$ is a square integrable martingale, then*

$$\|I_M(A)\|_{L_E^2}^2 = \mu_{\langle M \rangle}(A), \text{ for } A \in \mathcal{R}.$$

Proof. Let $A = (z, z'] \times C$ with $C \in \mathcal{F}_z$. Then

$$\|I_M(A)\|_{L_E^2}^2 = E(|I_M(A)|^2) = E(1_C |\Delta_{zz'} M|^2)$$
$$= E(1_C \Delta_{zz'} \langle M \rangle) = E(\int 1_{(z,z'] \times C} d\langle M \rangle)$$
$$= \mu_{\langle M \rangle}(A).$$

Since both mappings $A \mapsto \|I_M(A)\|_{L_E^2}^2$ and $\mu_{\langle M \rangle}$ are finitely additive on $\mathcal{R}$ and coincide on the sets $(z, z'] \times C$ with $C \in \mathcal{F}_z$, these mappings coincide on $\mathcal{R}$. ∎

Remark. The equality in Lemma 4 will be extended in Theorem 7 for every $A \in \mathcal{P}$.

5. Theorem. *If $M : \mathbb{R}^2 \times \Omega \to E$ is a square integrable martingale, then I_M can be extended to a σ-additive measure $I_M : \mathcal{P} \to L_E^2$.*

Proof. From Lemma 4 we deduce that $I_M \ll \mu_{\langle M \rangle}$ on $\mathcal{R}$. Since $\mu_{\langle M \rangle}$ is positive and σ-additive on $\mathcal{P}$, by Theorem 7.3, I_M can be extended to a σ-additive measure on $\mathcal{P}$ and we still have $I_M \ll \mu_{\langle M \rangle}$ on $\mathcal{P}$. ∎

The following theorem extends Lemma 2 to sets from $\mathcal{P}$.

6. Theorem. *Let $M, N : \mathbb{R}^2 \times \Omega \to E \subset L(F, G)$ be two square integrable martingales. Then for every disjoint sets $A, B \in \mathcal{P}$ and $x, y \in F$ we have*

$$I_M(A) \perp I_N(B) \text{ in } L_E^2 \text{ and } I_M(A)x \perp I_N(A)y \text{ in } L_G^2.$$

Proof. This property has been proved in Lemma 2 for $A, B \in \mathcal{R}$. Let now $A \in \mathcal{R}$ and denote by $\sum_A$ the class of sets $B \in \mathcal{P}$ such that $I_M(A) \perp I_N(B - A)$ in L_E^2. Then $\sum_A$ is a monotone class containing $\mathcal{R}$, hence $\sum_A = \mathcal{P}$. If $B \in \mathcal{P}$, let $\sum'_B$ be the class of sets $A \in \mathcal{P}$ such that $I_M(A) \perp I_M(B - A)$ in L_E^2. Again, $\sum'_B$ is a monotone class containing $\mathcal{R}$, hence $\sum'_B = \mathcal{P}$. It follows that if $A, B \in \mathcal{P}$ are disjoint, then $I_M(A) \perp I_M(B)$ in L_E^2. The second property follows from the first one, since Mx and Ny are square integrable martingales with values in G. ∎

The following theorem extends Lemma 4 to sets from $\mathcal{P}$.

7. Theorem. *If $M : \mathbb{R}^2 \times \Omega \to E$ is a square integrable martingale, then*

$$\|I_M(A)\|_{L_E^2}^2 = \mu_{\langle M \rangle}(A), \text{ for } A \in \mathcal{P}.$$

Proof. From Theorem 6, we deduce that the mapping $A \mapsto \|I_M(A)\|_{L_E^2}^2$ is finitely additive on $\mathcal{P}$.

Let $A_n \downarrow \phi$ in $\mathcal{P}$. Since I_M is σ-additive on $\mathcal{P}$, by Theorem 5, we have $I_M(A_n) \to 0$; therefore $\|I_M(A_n)\|_{L_E^2}^2 \to 0$; consequently $A \to \|I_M(A)\|_{L_E^2}^2$ is σ-additive on $\mathcal{P}$. Since $\mu_{\langle M \rangle}$ is also σ-additive on $\mathcal{P}$ and since, by Lemma 4, the two σ-additive measures coincide on $\mathcal{R}$, it follows that they are equal on $\mathcal{P}$. ∎

C. Summability of the square integrable martingales in Hilbert spaces

We shall show now that I_M has finite semivariation relative to the embedding $E \subset L(F, G)$.

8. Theorem. *Let $M : \mathbb{R}^2 \times \Omega \to E$ be a square integrable martingale.*
a) *For any embedding $E \subset L(F, G)$, with G a Hilbert space, I_M has finite semivariation relative to (F, L_G^2) and we have*

$$(\tilde{I}_M)_{F, L_G^2}(A) \leq \|I_M(A)\|_{L_E^2} \leq \sup_{z \in \mathbb{R}^2} \|M_z\|_{L_E^2} < \infty, \text{ for } A \in \mathcal{P}.$$

b) *For the particular embedding $E = L(\mathbb{R}, E)$, we have equality:*

$$(\tilde{I}_M)_{\mathbb{R}, L_E^2}(A) = \|I_M(A)\|_{L_E^2}, \text{ for } A \in \mathcal{P}.$$

c) *If M is scalar-valued and D is any Hilbert space, then for the embedding $\mathbb{R} \subset L(D, D)$ we have the equality*

$$(\tilde{I}_M)_{D, L_D^2}(A) = \|I_M(A)\|_{L_\mathbb{R}^2}, \text{ for } A \in \mathcal{P}.$$

Proof. Let $A \in \mathcal{P}$, let $(A_i)_{i \in I}$ be a finite family of disjoint sets from $\mathcal{P}$ with union A, and $(x_i)_{i \in I}$ a family of elements from F with $|x_i| \leq 1$. For $i \neq j$ we have $I_M(A_i) \perp I_M(A_j)$ in L_E^2 and $I_M(A_i)x_i \perp I_M(A_j)x_j$ in L_G^2, by Theorem 6. Then

$$\|\sum I_M(A_i)x_i\|_{L_G^2}^2 = \sum \|I_M(A_i)x_i\|_{L_G^2}^2 \leq \sum \|I_M(A_i)\|_{L_E^2}^2$$
$$= \|\sum I_M(A_i)\|_{L_E^2}^2 = \|I_M(A)\|_{L_E^2}^2.$$

It follows that

$$\|\sum I_M(A_i)x_i\|_{L_G^2} \leq \|I_M(A)\|_{L_E^2}.$$

Taking the supremum we obtain

$$(\tilde{I}_M)_{F, L_G^2}(A) \leq \|I_M(A)\|_{L_E^2}.$$

Since $A \mapsto \|I_M(A)\|_{L_E^2}^2$ is a positive, σ-additive measure we have

$$\|I_M(A)\|_{L_E^2}^2 \leq \|I_M(\mathbb{R}^2 \times \Omega)\|_{L_E^2}^2$$
$$= \|\Delta_{-\infty, \infty} M\|_{L_E^2}^2 = \sup_z \|\Delta_{-\infty, z} M\|_{L_E^2}^2 = \sup_z \|M_z\|_{L_E^2}^2,$$

therefore $\|I_M(A)\|_{L_E^2} \leq \sup_{z \in \mathbb{R}^2} \|M_z\|_{L_E^2}$ and this proves assertion a).

To prove assertion b), consider $E = L(\mathbb{R}, E)$ and let $A \in \mathcal{P}$. Then for any $\alpha \in \mathbb{R}$ with $|\alpha| = 1$ we have

$$\|I_M(A)\|_{L_E^2} = \|I_M(A)\alpha\|_{L_E^2} \leq (\tilde{I}_M)_{\mathbb{R}, L_E^2}(A),$$

and the equality follows. Similar computation in case M is scalar valued and we consider $\mathbb{R} \subset L(D,D)$: for any $A \in \mathcal{P}$ and any $x \in D$ with $|x| \leq 1$ we have

$$\|I_M(A)\|_{L^2_{\mathbb{R}}} = \|I_M(A)x\|_{L^2_D} \leq (\tilde{I}_M)_{D,L^2_D}(A),$$

and the equality follows. In particular, taking $D = \mathbb{R}$, we get

$$\|I_M(A)\|_{L^2_{\mathbb{R}}}(A) = (\tilde{I}_M)_{\mathbb{R},L^2_{\mathbb{R}}}(A), \text{ for } A \in \mathcal{P}.$$

∎

9. Theorem. *A square integrable martingale $M : \mathbb{R} \times \Omega \to E$ is 2-summable relative to any embedding $E \subset L(F,G)$, with E and G Hilbert spaces.*

A real-valued square integrable martingale is 2-summable relative to any embedding $\mathbb{R} \subset L(D,D)$, with D Hilbert space.

Proof. This follows from Theorems 5 and 8. ∎

10. Corollary. *A square integrable martingale $M : \mathbb{R} \times \Omega \to E$ is 1-summable relative to any embedding $E \subset L(F,G)$ with G Hilbert space and we have*

$$(\tilde{I}_M)_{F,L^1_G} \leq (\tilde{I}_M)_{F,L^2_G}.$$

A real-valued square integrable martingale M is 1-summable relative to any embedding $\mathbb{R} \subset L(D,D)$ with D Hilbert space and we have

$$(\tilde{I}_M)_{\mathbb{R},L^1_{\mathbb{R}}} \leq (\tilde{I}_M)_{D,L^1_D} \leq (\tilde{I}_M)_{D,L^2_D} = \|\tilde{I}_M\|_{\mathbb{R},L^2_{\mathbb{R}}}.$$

Proof. In fact, the inequality $(\tilde{I}_M)_{\mathbb{R},L^1_{\mathbb{R}}} \leq (\tilde{I}_M)_{D,L^1_D}$ is true since the embedding $L^1_{\mathbb{R}} \subset L(D,L^1_D)$ is an isometry. ∎

11. Corollary. *If $M : \mathbb{R}^2 \times \Omega \to E \subset L(F,G)$ is a square integrable martingale, then*

$$(\tilde{I}_M)_{F,L^2_G}(A) \leq (\tilde{I}_M)_{\mathbb{R},L^2_E}(A), \text{ for } A \in \mathcal{P}.$$

If M is scalar-valued, then we consider $\mathbb{R} \subset L(D,D)$ with D a Hilbert space and then

$$(\tilde{I}_M)_{D,L^2_D}(A) = (\tilde{I}_M)_{\mathbb{R},L^2_{\mathbb{R}}}(A), \text{ for } A \in \mathcal{P}.$$

Proof. The first inequality follows from Theorem 8 a) and b); and the second equality follows from Theorem 8 b) and c). ∎

Remark. If the embedding $L^2_E \subset L(F,L^2_G)$ is an isometry, then we have the inequality

$$(\tilde{I}_M)_{\mathbb{R},L^2_E} \leq (\tilde{I}_M)_{F,L^2_G}.$$

However, in general, the embedding $L^2_E \subset L(F,L^2_G)$ is not an isometry, even if the embedding $E \subset L(F,G)$ is an isometry. For a square integrable martingale the above corollary states that we have the converse inequality. If M is not a square integrable martingale, but, say, a summable process, then we may have $(\tilde{I}_M)_{F,L^2_G}(A) \leq (\tilde{I}_M)_{\mathbb{R},L^2_E}(A)$ for some sets A, and the opposite inequality for some other sets.

12. Corollary. *If $M : \mathbb{R}^2 \times \Omega \to E \subset L(F, G)$ is a square integrable martingale, then the set of measures*

$$(I_M)_{F,L^2_G} = \{(I_M)_u : u \in L^2_G, \|u\|_2 \leq 1\}$$

is uniformly integrable.

Proof. In fact, for $u \in L^2_G$ with $\|u\|_2 \leq 1$ we have

$$|(I_M)_u|(A) \leq (\tilde{I}_M)_{F,L^2_G}(A) \leq \|I_M(A)\|_{L^2_E}, \text{ for } A \in \mathcal{P};$$

if $A_n \downarrow \phi$, then $I_M(A_n) \to 0$ in L^2_E, hence $(\tilde{I}_M)_{F,L^2_G}(A_n) \to 0$ and $|(I_M)_u|(A_n) \to 0$, uniformly for $u \in L^2_G$ with $\|u\|_2 \leq 1$. ∎

D. The space $\mathcal{F}_{F,G}(I_M)$

The following theorem is an extension of Corollary 11, for processes, rather than sets.

13. Theorem. *Let $M : \mathbb{R}^2 \times \Omega \to E \subset L(F, G)$ be a square integrable martingale. Then*
a)
$$\mathcal{F}_{\mathbb{R}}((I_M)_{\mathbb{R},L^2_E}) \subset \mathcal{F}_{\mathbb{R}}((I_M)_{F,L^2_G})$$

and for every $H \in \mathcal{F}_{\mathbb{R}}((I_M)_{\mathbb{R},L^2_E})$ we have

$$(\tilde{I}_M)_{F,L^2_G}(H) \leq (\tilde{I}_M)_{\mathbb{R},L^2_E}(H).$$

b) *If M is scalar-valued and we consider $\mathbb{R} \subset L(D, D)$ with D a Hilbert space, then*

$$\mathcal{F}_{\mathbb{R}}((I_M)_{\mathbb{R},L^2_\mathbb{R}}) = \mathcal{F}_{\mathbb{R}}((I_M)_{D,L^2_D})$$

and for every $H \in \mathcal{F}_{\mathbb{R}}((I_M)_{\mathbb{R},L^2_\mathbb{R}})$ we have

$$(\tilde{I}_M)_{D,L^2_D}(H) = (\tilde{I}_M)_{\mathbb{R},L^2_\mathbb{R}}(H).$$

Proof. Let $H \in \mathcal{F}_{\mathbb{R}}((I_M)_{\mathbb{R},L^2_E})$ and let $K : \mathbb{R}^2 \times \Omega \to F$ be a $\mathcal{P}$-simple process, $K = \sum 1_{A_i} x_i$, finite sum, with $A_i \in \mathcal{P}$ mutually disjoint and $x_i \in F$. Assume $|K| \leq |H|$. The family $(I_M(A_i))_i$ is orthogonal in L^2_E and the family $(I_M(A_i) x_i)_i$ is orthogonal in L^2_G. Then

$$\|\int K dI_M\|^2_{L^2_G} = \|\sum I_M(A_i) x_i\|^2_{L^2_G}$$
$$= \sum \|I_M(A_i) x_i\|^2_{L^2_G} \leq \sum \|I_M(A_i)\|^2_{L^2_E} |x_i|^2$$
$$= \sum \||I_M(A_i)| x_i|\|^2_{L^2_E} = \|\sum |I_M(A_i)| |x_i|\|^2_{L^2_E}$$
$$= \|\int |K| d I_M\|^2_{L^2_E},$$

therefore,
$$\left\| \int K dI_M \right\|_{L_G^2} \le \left\| \int |K| dI_M \right\|_{L_E^2} \le (\tilde{I}_M)_{\mathbb{R}, L_E^2}(H).$$

Taking the supremum for $|K| \le |H|$ we obtain
$$(\tilde{I}_M)_{F, L_G^2}(H) \le (\tilde{I}_M)_{\mathbb{R}, L_E^2}(H) < \infty,$$

hence $H \in \mathcal{F}_{\mathbb{R}}((I_M)_{F, L_G^2})$ and this proves assertion a).

To prove assertion b), assume M is scalar-valued and consider $\mathbb{R} \subset L(D, D)$, with D a Hilbert space. Let $H \in \mathcal{F}_{\mathbb{R}}((I_M)_{D, L_D^2})$ and let $K = \sum 1_{A_i} \alpha_i$, finite sum, with $A_i \in \mathcal{P}$ mutually disjoint, $\alpha_i \in \mathbb{R}$ and $|K| \le |H|$. Let also $x \in D$ with $|x| = 1$. Then

$$\left\| \int K dI_M \right\|_{L_{\mathbb{R}}^2}^2 = \left\| \sum I_M(A_i) \alpha_i \right\|_{L_{\mathbb{R}}^2}^2$$
$$= \sum \| I_M(A_i) \alpha_i \|_{L_{\mathbb{R}}^2}^2 = \sum \| I_M(A_i) \alpha_i x \|_{L_D^2}^2$$
$$= \left\| \sum I_M(A_i) \alpha_i x \right\|_{L_D^2}^2 = \left\| \int K x dI_M \right\|_{L_D^2}^2,$$

therefore
$$\left\| \int K dI_M \right\|_{L_{\mathbb{R}}^2} = \left\| \int K x dI_M \right\|_{L_D^2}^2 \le (\tilde{I}_M)_{D, L_D^2}(H).$$

Taking the supremum, we get
$$(\tilde{I}_M)_{\mathbb{R}, L_{\mathbb{R}}^2}(H) \le (\tilde{I}_M)_{D, L_D^2}(H) < \infty.$$

Using assertion a), we get the equality
$$(\tilde{I}_M)_{D, L_D^2}(H) = (\tilde{I}_M)_{\mathbb{R}, L_{\mathbb{R}}^2}(H),$$

and assertion b) is proved. ∎

14. Theorem. a) If $M : \mathbb{R}^2 \times \Omega \to E \subset L(\mathbb{R}, E)$ is a square integrable martingale, then
$$L_{\mathbb{R}, L_E^2}^1(M) = \mathcal{F}_{\mathbb{R}, L_E^2}(I_M)$$

and the $\mathcal{R}$-simple, real-valued processes are dense in $\mathcal{F}_{\mathbb{R}, L_E^2}(I_M)$.

b) If M is a scalar-valued, square integrable martingale and D is a Hilbert space, then
$$L_{D, L_D^2}^1(M) = \mathcal{F}_{D, L_D^2}(I_M)$$

and the D-valued, $\mathcal{R}$-simple processes are dense in $\mathcal{F}_{D, L_D^2}(I_M)$.

Proof. a) Let $H \in \mathcal{F}_{\mathbb{R}, L_E^2}(M)$ and prove that H can be approximated in $\mathcal{F}_{\mathbb{R}, L_E^2}(M)$ by bounded processes.

We have $|H| \in \mathcal{F}_{\mathbb{R}, L_E^2}(M) = L_{\mathbb{R}, L_E^2}^1(M)$, by Corollary 7.3.

By Theorem 27.4, $|H| \cdot M$ is an E-valued square integrable martingale; therefore, by theorem 8 b),

$$(\tilde{I}_{|H| \cdot M})_{\mathbb{R}, L_E^2}(A) = \|I_{|H| \cdot M}(A)\|_{L_E^2}, \text{ for } A \in \mathcal{P}.$$

By Theorem 6.1 b) we have

$$(\tilde{I}_{|H| \cdot M})_{\mathbb{R}, L_E^2}(A) = (\tilde{I}_M)_{\mathbb{R}, L_E^2}(1_A |H|), \text{ for } A \in \mathcal{P}.$$

It follows that for $A \in \mathcal{P}$ we have

$$(\tilde{I}_M)_{\mathbb{R}, L_E^2}(1_A H) = (\tilde{I}_M)_{\mathbb{R}, L_E^2}(1_A |H|)$$
$$= (\tilde{I}_{|H| \cdot M})_{\mathbb{R}, L_E^2}(A) = \|I_{|H| \cdot M}(A)\|_{L_E^2}.$$

Since $I_{|H| \cdot M}$ is σ-additive in L_E^2 on $\mathcal{P}$, it follows that if $A_n \downarrow \phi$, then $I_{|H| \cdot M}(A_n) \to 0$ in L_E^2; therefore $(\tilde{I}_M)_{\mathbb{R}, L_E^2}(1_{A_n} H) \to 0$. By Proposition 5.40, H can be approximated by bounded processes in $\mathcal{F}_{\mathbb{R}, L_E^2}(M)$. By Proposition 5.44 and Corollary 12, the bounded processes can be approximated by $\mathcal{R}$-simple processes in $\mathcal{F}_{\mathbb{R}, L_E^2}(M)$. Therefore, H can be approximated by $\mathcal{R}$-simple processes in $\mathcal{F}_{\mathbb{R}, L_E^2}(M)$.

b) Assume M is a scalar-valued, square integrable martingale. Let $H \in \mathcal{F}_{D, L_D^2}(M)$ and prove H can be approximated in $\mathcal{F}_{D, L_D^2}(M)$ by bounded processes. We have

$$(\tilde{I}_M)_{D, L_D^2}(|H|) = (\tilde{I}_M)_{D, L_D^2}(H) < \infty,$$

therefore $|H| \in \mathcal{F}_{\mathbb{R}}((I_M)_{D, L_D^2})$. From Theorem 13 we deduce that

$$(\tilde{I}_M)_{D, L_D^2}(|H|) = (\tilde{I}_M)_{\mathbb{R}, L_{\mathbb{R}}^2}(|H|),$$

hence $|H| \in \mathcal{F}_{\mathbb{R}}((I_M)_{\mathbb{R}, L_{\mathbb{R}}^2}) = \mathcal{F}_{\mathbb{R}, L_{\mathbb{R}}^2}(M) = L_{\mathbb{R}, L_{\mathbb{R}}^2}^1(M)$, by assertion a) of the theorem. If $A_n \downarrow \phi$ in $\mathcal{P}$, by the same argument as in the proof of assertion a) we deduce that $(\tilde{I}_M)_{\mathbb{R}, L_{\mathbb{R}}^2}(H 1_{A_n}) \to 0$. Then also $(\tilde{I}_M)_{D, L_D^2}(H 1_{A_n}) \to 0$. By Proposition 5.40 we deduce that H can be approximated in $\mathcal{F}_{D, L_D^2}(M)$ by bounded processes. Finally, since, by Corollary 12, $(I_M)_{D, L_D^2}$ is uniformly σ-additive, the bounded processes can be approximated in $\mathcal{F}_{D, L_D^2}(M)$ by $\mathcal{R}$-simple processes (Proposition 5.44). This proves the theorem. ∎

E. Isometric isomorphism of $L_{F,G}^1(M)$ and $L_F^2(\mu_{\langle M \rangle})$

Let M be an E-valued square integrable martingale. Then $|M|^2 - \langle M \rangle$ is a weak martingale, hence

$$E(\Delta_{zz'}|M|^2) = E(\Delta_{zz'}\langle M \rangle), \text{ for } z < z' \text{ in } \mathbb{R}^2.$$

Consider the Doléans measure $\mu_{|M|^2}$ defined by

$$\mu_{|M|^2}(A) = E(I_{|M|^2}(A)), \text{ for } A \in \mathcal{R},$$

and the Doléans measure $\mu_{\langle M\rangle}$ defined by

$$\mu_{\langle M\rangle}(A) = E(I_{\langle M\rangle}(A)) = E(\int 1_A d\langle M\rangle),$$

for $A \in \mathcal{B}(\mathbb{R}^2) \times \mathcal{F}$. The relationship between these two measures is

$$\mu_{|M|^2}(A) = \mu_{\langle M\rangle}(A), \text{ for } A \in \mathcal{R}.$$

It follows that $\mu_{|M|^2}$ can be extended uniquely to a σ-additive measure on $\mathcal{P}$ and the extension of $\mu_{|M|^2}$ is equal to $\mu_{\langle M\rangle}$ on $\mathcal{P}$.

We introduce a third σ-additive measure. Since $M_\infty \in L_E^2$, we take $u = M_\infty \in (L_E^2)^*$ and consider the σ-additive measure $(I_M)_u$ denoted by $\langle I_M, M_\infty\rangle$.

15. Proposition. *We have*

$$\mu_{|M|^2} = \mu_{\langle M\rangle} = \langle I_M, M_\infty\rangle, \text{ on } \mathcal{P}.$$

Proof. It is enough to prove the equality for sets $A = (z, z'] \times C$ with $C \in \mathcal{F}_z$. We have

$$\langle I_M(A), M_\infty\rangle_{L_E^2} = E(\langle I_M(A), M_\infty\rangle_E)$$
$$= E(1_C \langle \Delta_{zz'} M, M_\infty\rangle_E)$$
$$= E[1_C(\langle M_z, M_\infty\rangle + \langle M_{z'}, M_\infty\rangle - \langle M_{st'}, M_\infty\rangle - \langle M_{s't}, M_\infty\rangle)]$$
$$= E[1_C(\langle M_z, M_z\rangle + \langle M_{z'}, M_{z'}\rangle - \langle M_{st'}, M_{st'}\rangle - \langle M_{s't}, M_{s't}\rangle)]$$
$$= E[1_C(|M_z|^2 + |M_{z'}|^2 - |M_{s't}|^2 - |M_{st'}|^2)]$$
$$= E(1_C \Delta_{zz'} |M|^2) = E(1_C \Delta_{zz'} \langle M\rangle)$$
$$= E(\int 1_A d\langle M\rangle) = \mu_{\langle M\rangle}(A).$$

Since all three measures are σ-additive on $\mathcal{P}$ and coincide on $\mathcal{R}$, they are equal on $\mathcal{P}$. ∎

The relationship between all these measures and the seminorm $(\tilde{I}_M)_{F,G}$ is given in the following theorem.

16. Theorem. *Let $M : \mathbb{R}^2 \times \Omega \to E \subset L(F,G)$ be a square integrable martingale. Then*
a) $L_F^2(\langle I_M, M_\infty\rangle) \subset \mathcal{F}_{F, L_G^2}(I_M) = L_{F,G}^1(M)$,
and for $H \in L_F^2(\langle I_M, M_\infty\rangle)$ we have

$$\|\int H dI_M\|_{L_G^2} \leq (\tilde{I}_M)_{F, L_G^2} \leq \|H\|_{L_F^2(\langle I_M, M_\infty\rangle)}.$$

b) *For the particular embedding $E = L(\mathbb{R}, E)$ we have*
$$L_{\mathbb{R}}^2(\langle I_M, M_\infty \rangle) = \mathcal{F}_{\mathbb{R}, L_E^2}(I_M) = L_{\mathbb{R}, L_E^2}^1(M)$$
and for $H \in \mathcal{F}_{\mathbb{R}, L_D^2}(I_M)$ we have
$$\| \int H dI_M \|_{L_E^2} = (\tilde{I}_M)_{\mathbb{R}, L_E^2}(H) = \|H\|_{L_{\mathbb{R}}^2(\mu_{\langle M \rangle})} = \|H\|_{L_{\mathbb{R}}^2(\langle I_M, M_\infty \rangle)}.$$

c) *If M is scalar-valued and D is a Hilbert space, we have*
$$L_D^2(\langle I_M, M_\infty \rangle) = \mathcal{F}_{D, L_D^2}(I_M) = L_{D, L_D^2}^1(M)$$
and for $H \in \mathcal{F}_{D, L_D^2}(I_M)$ we have
$$\| \int H dI_M \|_{L_D^2}^2 = (\tilde{I}_M)_{D, L_D^2} = \|H\|_{L_D^2(\mu_{\langle M \rangle})}^2 = \|H\|_{L_D^2(\langle I_M, M_\infty \rangle)}.$$

Proof. To prove assertion a), assume first $H = \sum 1_{A_i} x_i$, finite sum, with $A_i \in \mathcal{P}$ mutually disjoint and $x_i \in F$. By Theorem 6, the family $(I_M(A_i) x_i)_i$ is orthogonal in L_G^2 and we have

$$\| \int H dI_M \|_{L_G^2}^2 = \| \sum I_M(A_i) x_i \|_{L_G^2}^2 = \sum \| I_M(A_i) x_i \|_{L_G^2}^2$$
$$\leq \sum \| I_M(A_i) \|_{L_E^2}^2 |x_i|^2 = \sum E(|I_M(A_i)|^2) |x_i|^2$$
$$= \sum E(I_{|M|^2}(A_i)) |x_i|^2 = E(\int |H|^2 dI_{|M|^2})$$
$$= \int |H|^2 d\mu_{|M|^2} = \int |H|^2 d\mu_{\langle M \rangle} = \int |H|^2 d\langle I_M, M_\infty \rangle.$$

Let now $H \in L_F^2(\langle I_M, M_\infty \rangle)$. For each F-valued $\mathcal{P}$-step process K with $|K| \leq |H|$, we have, by the above,
$$\| \int K dI_M \|_{L_G^2} \leq (\int |K|^2 d\langle I_M, M_\infty \rangle)^{\frac{1}{2}} \leq \|H\|_{L_F^2(\langle I_M, M_\infty \rangle)} < \infty;$$
hence, taking the supremum for $|K| \leq |H|$, we get
$$(\tilde{I}_M)_{F, L_G^2}(H) \leq \|H\|_{L_F^2(\langle I_M, M_\infty \rangle)},$$
therefore $H \in \mathcal{F}_{F, L_G^2}(I_M)$. The equality $\mathcal{F}_{F, L_G^2}(I_M) = L_{F, G}^1(M)$ follows from Corollary 7.3 and the inequality $\| \int H dI_M \|_{L_G^2} \leq (\tilde{I}_M)_{F, L_G^2}$ follows from the definition of $\int H dI_M$. This proves assertion a).

To prove assertion b), consider a simple process $H = \sum 1_{A_i} \alpha_i$ with $A_i \in \mathcal{P}$ mutually disjoint and $\alpha_i \in \mathbb{R}$. The computation in the previous case yields this time an equality:
$$\| \int H dI_M \|_{L_E^2}^2 = \int |H|^2 d\mu_{|M|^2} = \int |H|^2 d\mu_{\langle M \rangle}$$
$$= \int |H|^2 d\langle I_M, M_\infty \rangle.$$

Now let $H \in \mathcal{F}_{\mathbb{R},L_E^2}(I_M)$.

By Theorem 14 a), the $\mathcal{P}$-simple processes are dense in $\mathcal{F}_{\mathbb{R},L_E^2}(I_M)$. Then there is a sequence (H^n) of $\mathcal{P}$-simple processes converging to H in $\mathcal{F}_{\mathbb{R},L_E^2}$. Since the integral $\int H dI_M$ is continuous, we deduce that

$$\int H^n dI_M \to \int H dI_M, \text{ in } L_E^2.$$

Since $\mathcal{F}_{\mathbb{R},L_E^2}(I_M) \subset L_{\mathbb{R}}^1(\langle I_M, M_\infty \rangle)$, we have $H^n \to H$ in $L_{\mathbb{R}}^1(\langle I_M, M_\infty \rangle)$; therefore, there is a subsequence converging pointwise, $\langle I_M, M_\infty \rangle$-a.s. We can assume that $H^n \to H$ pointwise, $\langle I_M, M_\infty \rangle$-a.s. The sequence (H^n) is Cauchy in $L_{\mathbb{R}}^2(\langle I_M, M_\infty \rangle)$. In fact,

$$\|H^n - H^m\|_{L_{\mathbb{R}}^2(\langle I_M, M_\infty \rangle)} = \|\int (H^n - H^m) dI_M\|_{L_E^2} \to 0$$

as $n, m \to \infty$. It follows that $H \in L_{\mathbb{R}}^2(\langle I_M, M_\infty \rangle)$ and $H^n \to H$ in $L_{\mathbb{R}}^2(\langle I_M, M_\infty \rangle)$. We deduce then that

$$\mathcal{F}_{\mathbb{R},L_E^2}(I_M) \subset L_{\mathbb{R}}^2(\langle I_M, M_\infty \rangle)$$

and

$$\int |H^n|^2 d\langle I_M, M_\infty \rangle \to \int |H|^2 d\langle I_M, M_\infty \rangle.$$

From assertion a) it follows that

$$L_{\mathbb{R}}^2(\langle I_M, M_\infty \rangle) = \mathcal{F}_{\mathbb{R},L_E^2}(I_M).$$

For each n we have, by the above,

$$\|\int H^n dI_M\|_{L_E^2} = \|H^n\|_{L_{\mathbb{R}}^2(\langle I_M, M_\infty \rangle)}.$$

Taking the limits, we get

$$\|\int H dI_M\|_{L_E^2} = \|H\|_{L_{\mathbb{R}}^2(\langle I_M, M_\infty \rangle)}.$$

From the inequalities of assertion a) we deduce the equalities

$$(\tilde{I}_M)_{\mathbb{R},L_E^2}(H) = \|\int H dI_M\|_{L_E^2} = \|H\|_{L_{\mathbb{R}}^2(\langle I_M, M_\infty \rangle)}.$$

The other equalities follow from Proposition 15.

The proof of assertion c) is similar to that of assertion b). ∎

17. Corollary. *Let $M : \mathbb{R}^2 \times \Omega \to E$ be a square integrable martingale.*
a) *The spaces $L_{\mathbb{R},L_E^2}^1(M)$ and $L_{\mathbb{R}}^2(\mu_{\langle M \rangle})$ contain the same predictable processes and are isometrically isomorphic.*

b) *If M is scalar-valued and D is a Hilbert space, then the spaces $L^1_{D,L^2_D}(M)$ and $L^2_D(\mu_{\langle M \rangle})$ contain the same predictable processes and are isometrically isomorphic.*

Remark. The classical approach to the stochastic integral with respect to a real-valued, two-parameter, square integrable martingale M, is to prove the isometry $H \mapsto \int H dI_M$ for $\mathcal{R}$-simple processes H from $L^2(\mu_{\langle M \rangle})$ into $L^2(P)$ and to extend this isometry to the whole space $L^2(\mu_{\langle M \rangle})$ (see [C-W] and [M]).

In our approach, we obtain this isometry from Theorem 16 b) or c).

Chapter 9
Two-Parameter Processes with Finite Variation

In this chapter we extend for two-parameter processes, the results of § 19 concerning one-parameter processes with finite variation.

We study first, in § 29, functions with finite variation in the plane and define their Stieltjes integral.

§29. FUNCTIONS WITH FINITE VARIATION IN THE PLANE

In this section we shall study the functions of two variables with finite variation, which will be used in §30 in the study of the two-parameter stochastic processes with integrable variation.

We shall extend for functions of two variables with finite variation, the familiar properties of the functions of one variable with finite variation, such as the existence of lateral limits (Theorem 37), the right continuity of the variation (which follows from Theorem 16), and the Jordan decomposition (Theorem 34). Because the order relation on $\mathbb{R}^2$ is partial, the proofs of the properties in $\mathbb{R}^2$ are more laborious than in $\mathbb{R}$.

A. Monotone functions

1. Let $g : \mathbb{R}^2 \to \mathbb{R}$ be a *real-valued* function. We say that g is *increasing* if $z \leq z'$ implies $g(z) \leq g(z')$. We say that g is *incrementally increasing* if $z \leq z'$ implies $\Delta_{zz'} g \geq 0$.

A function g can be increasing without being incrementally increasing and vice versa.

Example. Let $g : \mathbb{R}^2 \to \mathbb{R}$ be the function defined by

$$\begin{aligned} g(s,t) &= t + s(1-t), & &\text{if } 0 \leq (s,t) \leq 1, \\ g(s,t) &= 0, & &\text{if } s \leq 0 \text{ or } t \leq 0 \\ g(s,t) &= 1, & &\text{if } s \geq 1 \text{ and } t \geq 0 \text{ or if } s \geq 0 \text{ and } t \geq 1. \end{aligned}$$

For $0 \leq (s,t) \leq (s',t') \leq (1,1)$ we have

$$g(s',t') - g(s,t) = (1-s')(t'-t) + (1-t)(s'-s) \geq 0,$$

hence g is increasing on the unit square. Then g is increasing on the whole plane. But

$$\Delta_{(s,t),(s',t')} g = (s'-s)(t-t') \leq 0,$$

hence g is not incrementally increasing.

The function $-g$ is incrementally increasing but not increasing.

B. Partitions

2. A *partition* is a finite family $P = (R_i)_{i \in I}$ of bounded rectangles of the form $R_i = (z_i, z_i']$ such that $R_i \cap R_j = \phi$ if $i \neq j$. Some of the rectangles may be empty.

If all the rectangles R_i of a partition P are contained in a subset $T \subset \mathbb{R}^2$ we say that P is a *partition in* T.

If the union $\bigcup_{i \in I} R_i$ of the rectangles of a partition P is equal to a set $T \subset \mathbb{R}^2$, we say that P is a *partition of* T.

We remark that the union of a partition is not necessarily a rectangle. But if it is a rectangle, it is of the form $(z, z']$.

A *grid* is a particular partition of the form $Q = (R_{ij})$ with

$$R_{ij} = ((s_i, t_j), (s_{i+1}, t_{j+1})] = (s_i, s_{i+1}] \times (t_j, t_{j+1}]$$

where $\sigma : s_0 \leq s_1 \leq \cdots \leq s_n$ and $\tau : t_0 \leq t_1 \leq \cdots \leq t_m$ are *divisions* in $\mathbb{R}$. We shall write also $Q = \sigma \times \tau$.

We say that a partition $P = (R_i)_{i \in I}$ is *finer than* (or a *refinement of*) a partition $P' = (R_j')_{j \in J}$ if every rectangle R_j' is the union of a subfamily $(R_i)_{i \in I(j)}$ with $I(j) \subset I$.

3. Proposition. *For any partition $P = (R_i)_{i \in I}$ there is a grid $Q = \sigma \times \tau$ finer than P. Each rectangle R_i is the union of a subgrid $Q_i = \sigma_i \times \tau_i$.*

Proof. Let $P = (R_i)_{i \in I}$ be a partition with $R_i = (s_i, s_i'] \times (t_i, t_i']$. We arrange all the points s_i and s_i' in increasing order and obtain a division σ. Similarly, we obtain a division τ by arranging in increasing order all the points t_i and t_i'. Then the grid $Q = \sigma \times \tau$ is a refinement of the partition P and each rectangle R_i is the union of a subfamily Q_i of Q, which is a grid $\sigma_i \times \tau_i$, where σ_i consists of the points of σ between s_i and s'_i and τ_i consists of the points of τ between t_i and t'_i.

4. Proposition. *For any partitions P and P', there is a partition P'' finer than both P and P'.*

Proof. Let $\sigma \times \tau$ be a grid finer than P and $\sigma' \times \tau'$ a grid finer than P'. Let σ'' be the division obtained by arranging in increasing order all the points of the divisions σ and σ'; let τ'' be the division obtained similarly from τ and τ'. Then the grid $\sigma'' \times \tau''$ is finer than both $\sigma \times \tau$ and $\sigma' \times \tau'$; therefore $\sigma'' \times \tau''$ is finer than both P and P'. ∎

It follows that the set of partitions is directed for the order relation defined by refinement.

5. Proposition. *If $P = (R_i)_{i \in I}$ is a partition with union a rectangle R, then*

$$\Delta_R g = \sum_{i \in I} \Delta_{R_i} g.$$

Proof. The proposition is true if P is a grid, $P = \sigma \times \tau$ with

$$\sigma : s_0 < s_1 < \cdots < s_n \text{ and } \tau : t_0 < t_1 < \cdots < t_m,$$

since at each point (s_i, t_j) different from $(s_0, t_0), (s_0, t_m), (s_n, t_0), (s_n, t_m)$, the terms $g(s_i, t_j)$ come in the sum $\sum_{i,j} \Delta_{R_{ij}} g$ with opposite signs and cancel each other. Let now $P = (R_i)_{i \in I}$ be a partition with union R. By Proposition 3 there is a grid $Q = \sigma \times \tau$ finer than P and with union R. Each rectangle R_k is the union of a subgrid Q_k of Q. By the first part of the proof we have

$$\Delta_{R_k} g = \sum_{R_{ij} \in Q_k} \Delta_{R_{ij}} g,$$

hence

$$\sum_{k \in I} \Delta_{R_k} g = \sum_{k \in I} \sum_{R_{ij} \in Q_k} \Delta_{R_{ij}} g$$
$$= \sum_{R_{ij} \in Q} \Delta_{R_{ij}} g = \Delta_R g.$$

∎

Remark. This property will be used to associate to g a finitely additive measure m_g in Section K.

C. Variation corresponding to a partition

Let $g : \mathbb{R}^2 \to E$ be a function.

6. Definition. *For any partition $P = (R_i)_{i \in I}$, the variation of g corresponding to the partition P is a number denoted by $var(g, P)$ or $v(g, P)$ and defined by*

$$var(g, P) = \sum_{i \in I} |\Delta_{R_i} g|.$$

We list some of the properties of the variation $v(g, P)$.

7. *If P and P' are two partitions and $P \subset P'$, then*

$$var(g, P) \leq var(g, P').$$

8. *If P and P' are disjoint partitions, then $P \cup P'$ is a partition and*

$$var(g, P \cup P') = var(g, P) + var(g, P').$$

9. *Assume g is right continuous and let $Q = \sigma \times \tau$ be a grid with*

$$\sigma : s_0 < s_1 < \cdots < s_n \text{ and } \tau : t_0 < t_1 < \cdots < t_m.$$

Then for every $\varepsilon > 0$ there is a grid $Q' = \sigma' \times \tau'$ with

$$\sigma' : s'_0 < s'_1 < \cdots < s'_{n-1} < s_n \text{ and } \tau' : t'_0 < t'_1 < \cdots < t'_{m-1} < t_m,$$

such that $s_i < s'_i$ for $0 \leq i < n$, $t_j < t'_j$ for $0 \leq j < m$, the points s'_i and t'_j are rational and

$$|var(g, Q) - var(g, Q')| < \varepsilon.$$

D. Variation of a function on a rectangle

Let $g : \mathbb{R}^2 \to E$ be a function.

10. Definition. *For any rectangle R (not necessarily bounded and not necessarily of the form $(z, z']$) the variation of g on R is a number denoted by $var(g, R)$ or $v(g, R)$ and defined by the equality*

$$var(g, R) = \sup_P var(g, P) = \sup_P \sum_{R_i \in P} |\Delta_{R_i} g|,$$

where the supremum is taken for all the partitions $P = (R_i)_{i \in I}$ consisting of rectangles R_i with vertices in R.

We use the same type of notation for the variation corresponding to a partition P, namely $v(g, P)$ and the variation on a rectangle R, namely $v(g, R)$. If there is any danger of confusion we shall specify that P is a partition and R is a rectangle.

We say that g has *finite variation* if $v(g, R) < \infty$ for any bounded rectangle R.

We say that g has *bounded variation* if $v(g, \mathbb{R}^2) < \infty$.

There is an important intermediary situation, when $v(g, (-\infty, z]) < \infty$ for every $z \in \mathbb{R}^2$. In this case we say that g has *finite variation function* $|g|$. The variation function $|g|$ is defined for every $z \in \mathbb{R}^2$ by

$$|g|(z) = v(g, (-\infty, z]).$$

(See 18 infra).

We list some properties of the variation $v(g, R)$.

11. *If R and R' are rectangles and if $R \subset R'$, then*

$$v(g, R) \leq v(g, R').$$

12. *If $-\infty \leq z < z' \leq +\infty$, then*

$$var(g, (z, z')) = \sup_{u<z'} var(g, (z, u]) = \sup_{u<z'} var(g, (z, u));$$

if $z' < \infty$, then

$$var(g, (z, z']) = \sup_{u>z} var(g, (u, z']) = \sup_{u>z} var(g, [u, z']).$$

In particular,

$$var(g, (-\infty, z]) = \sup_{u<z} var(g, (u, z]) = \sup_{u<z} var(g, [u, z]).$$

For every $z \in \mathbb{R}^2$ we have

$$var(g, (z, +\infty)) = \sup_{u>z} var(g, (z, u]) = \sup_{u>z} var(g, (z, u)),$$

and

$$var(g, \mathbb{R}^2) = \sup_{z<z'} var(g, (z, z']).$$

13. Proposition. a) *For any rectangle R we have*

$$var(g, R) = \sup_Q var(g, Q)$$

where the supremum is taken for all the grids Q consisting of rectangles with vertices in R.
b) *If g is right continuous, the supremum can be taken using only grids $Q = \sigma \times \tau$ with $\sigma : s_0 < s_1 < \cdots < s_n$ and $\tau : t_0 < t_1 < \cdots < t_m$ such that $s_0, \ldots s_{n-1}$ and $t_0, \ldots t_{m-1}$ are rational.*

Proof. For the proof of assertion a), use Property 7. For the proof of assertion b) use Property 9. ■

14. Proposition. *Let $(R_i)_{i \in I}$ be a family of (bounded or unbounded) rectangles R_i with union a rectangle R.*

Assume that for $i \neq j$, the rectangles R_i and R_j contain the intersection of their boundaries, but have no common interior points.

Then
$$var(g, R) = \sum_{i \in I} var(g, R_i)$$

Proof. For each $i \in I$ let P_i be a partition of rectangles with vertices in R_i. For $i \neq j$, the partitions P_i and P_j are disjoint and the union $P = \bigcup_{i \in I} P_i$ is a partition of rectangles with vertices in R. Then
$$\sum_{i \in I} var(g, P_i) = var(g, P) \leq var(g, R),$$
therefore
$$\sum_{i \in I} var(g, R_i) \leq var(g, R).$$

If $var(g, R_i) = \infty$, then $var(g, R) = \infty$ and we have equality.

Assume $var(g, R_i) < \infty$ for every $i \in I$ and prove the converse inequality. Let $Q = \sigma \times \tau$ be a grid of rectangles with vertices in R with
$$\sigma : s_0 < s_1 < \cdots < s_n \text{ and } \tau : t_0 < t_1 < \cdots < t_m.$$

For each $i \in I$ denote by (a_i, a_i') and (b_i, b_i') the vertices of the main diagonal of R_i. For each $i \in I$, if (a_i, a_i') belongs to R, we add a_i to the division σ and a_i' to the division τ. Similarly, if (b_i, b_i') belongs to R, we add b_i to σ and b_i' to τ. We obtain a division σ' finer than σ and a division τ' finer than τ.

The grid $Q' = \sigma' \times \tau'$ is finer than Q and its rectangles have vertices in R. For each $i \in I$, those rectangles of Q' contained in R_i form a grid Q_i of rectangles with vertices in R_i and for $i \neq j$, the grids Q_i and Q_j are disjoint. Then, by Property 8 we have
$$var(g, Q) \leq var(g, Q') = \sum_{i \in I} var(g, Q_i') \leq \sum_{i \in I} var(g, R_i),$$
therefore
$$var(g, R) \leq \sum_{i \in I} var(g, R_i).$$

∎

15. Proposition. *Assume g is right continuous. If $(R_i)_{i \in I}$ is a partition with union a rectangle R, then*
$$var(g, R) = \sum_{i \in I} var(g, R_i).$$

Proof. The inequality

$$\sum_{i \in I} var(g, R_i) \le var(g, R)$$

is proved the same way it was done for Proposition 14.

To prove the converse inequality, let $Q = \sigma \times \tau$ be a grid of rectangles with vertices in R. Again as in the proof of Proposition 14, there is a grid Q' finer than Q and containing those diagonal points of the rectangles R_i that belong to R. For each $i \in I$ denote by Q'_i the subgrid of Q' consisting of rectangles of Q' contained in R_i. Let $\varepsilon > 0$. Since g is right continuous, we can find, by Property 9, a grid Q''_i with vertices in R_i such that

$$|var(g, Q'_i) - var(g, Q''_i)| < \varepsilon/\text{card}I \,.$$

For $i \ne j$, the grids Q''_i and Q''_j are disjoint and their union is a partition Q'' with vertices in R. Then, by property 8,

$$\begin{aligned} var(g, Q) &\le var(g, Q') = \sum_{i \in I} var(g, Q'_i) \\ &\le \sum_{i \in I} var(g, Q''_i) + \varepsilon \\ &\le \sum_{i \in I} var(g, R_i) + \varepsilon, \end{aligned}$$

therefore

$$var(g, R) \le \sum_{i \in I} var(g, R_i) \,.$$

∎

E. Limits of the variation

One of the most important properties of the variation is that it inherits the right continuity from the function g.

16. Theorem. *Assume g has finite variation and is right continuous. Then for any points $z_0 \le z$ in $\mathbb{R}^2$ we have*

$$\lim_{\substack{u \to z \\ u \ge z}} var(g, (z_0, u]) = var(g, (z_0, z])$$

and

$$\lim_{\substack{u \to z \\ u \ge z}} var(g, [z_0, u]) = var(g, [z_0, z]) \,.$$

If, in addition, g has finite variation function $|g|$, then for any $z \in \mathbb{R}^2$ we have

$$\lim_{\substack{u \to z \\ u \ge z}} var(g, (-\infty, u]) = var(g, (-\infty, z]) \,.$$

Proof. (Cf. [L]) We remark first that by Property 11 the functions $u \mapsto var(g, [z_0, u])$ and $u \mapsto var(g, (z_0, u])$ are increasing; therefore the limits in the statement of the theorem exist.

For $-\infty \leq z_0 \leq z \leq u$ with $z_0 = (s_0, t_0)$, $z = (s, t)$ and $u = (p, r)$ we have the following union of rectangles:

$$(z_0, u] = (z_0, z] \cup [z, u] \cup [s, p] \times (t_0, t] \cup (s_0, s] \times [t, r],$$

and if $z_0 \in \mathbb{R}$,

$$[z_0, u] = [z_0, z] \cup [z, u] \cup [s, p] \times [t_0, t] \cup [s_0, s] \times [t, r].$$

By Proposition 14 we have corresponding equalities for the variations

$$var(g, (z_0, u]) = var(g, (z_0, z]) + var(g, [z, u])$$
$$+ var(g, [s, p] \times (t_0, t]) + var(g, (s_0, s] \times [t, r]),$$

and if $z_0 \in \mathbb{R}$,

$$var(g, [z_0, u]) = var(g, [z_0, z]) + var(g, [z, u])$$
$$+ var(g, [s, p] \times [t_0, t]) + var(g, [s_0, s] \times [t, r]).$$

We shall prove that the last three terms in these inequalities have the right limit equal to 0 as $u \downarrow z$.

a) $\lim_{u \downarrow z} var(g, [z, u]) = 0$.

Assume the contrary: there is an $\alpha > 0$ such that $var(g, [z, u]) > \alpha$ for every $u > z$. We shall construct by induction a sequence $u_0 > u_1 > u_2 > \cdots > z$ with $u_n \to z$ and for each n, a partition P_n contained in $[z, u_n] - [z, u_{n+1}]$, such that

$$var(g, P_n) > \alpha - \varepsilon/2^n.$$

In fact, let $u_0 = (p_0, r_0) > z$ and $\varepsilon > 0$. Since $var(g, [z, u_0]) > \alpha$, there is a grid $\sigma_0 \times \tau_0$ with

$$\sigma_0 : s = s_{00} < s_{01} < \cdots < s_{0n} = p_0 \text{ and } \tau_0 : t = t_{00} < t_{01} < \cdots < t_{0m} = r_0$$

such that

$$var(g, \sigma_0 \times \tau_0) > \alpha.$$

Since g is incrementally right continuous, there is a $u_1 > z$ such that for every z' with $z < z' \leq u_1$ we have $|\Delta_{zz'} g| < \varepsilon/2$. In particular,

$$|\Delta_{zu_1} g| < \varepsilon/2.$$

We can choose $u_1 = (p_1, r_1) \leq (s_{01}, t_{01})$. Then we can add the point p_1 to the division σ_0 and the point r_1 to the division τ_0 and change the numbering in σ_0 and τ_0 such that $p_1 = s_{01}$ and $r_1 = t_{01}$.

With this addition we still have $var(g, \sigma_0 \times \tau_0) > \alpha$. Denote by P_0 the partition $\sigma_0 \times \tau_0 - \{(z, u_1]\}$. Then

$$var(g, P_0) = var(g, \sigma_0 \times \tau_0) - |\Delta_{zu_1} g| > \alpha - \varepsilon/2.$$

Starting now with the point $u_1 = (p_1, r_1)$ and reasoning as above, we can find a grid $\sigma_1 \times \tau_1$ with

$$\sigma_1 : s = s_{10} < s_{11} < \cdots < s_{1n} = p_1 \text{ and } \tau_1 : t = t_{10} < t_{11} < \cdots < t_{1m} = r_1$$

such that

$$var(g, \sigma_1 \times \tau_1) > \alpha,$$

and a point $u_2 = (p_2, r_2)$ such that $z < u_2 < u_1$ and

$$|\Delta_{zu_2} g| < \varepsilon/2^2.$$

We can add the point p_2 to the division σ_1 and the point r_2 to the division τ_1 and then change the notation to have $p_2 = s_{11}$ and $r_2 = t_{11}$. Then we still have $var(g, \sigma_1 \times \tau_1) > \alpha$.

Denoting by P_1 the partition $\sigma_1 \times \tau_1 - \{(z, u_2]\}$, we get

$$var(g, P_1) = var(g, \sigma_1 \times \tau_1) - |\Delta_{[z, u_2]} g| > \alpha - \varepsilon/2^2.$$

Continuing this way we obtain the desired sequence (u_n) and the partitions P_n contained in $[z, u_n] - [z, u_{n+1}]$, such that

$$var(g, P_n) > \alpha - \varepsilon/2^n.$$

Then, for each n, the union $\bigcup_{0 \leq i \leq n} P_i$ is a partition contained in $[z, u_0]$, hence

$$var(g, [z, u_0]) \geq var(g, \bigcup_{0 \leq i \leq n} P_i)$$
$$= \sum_{0 \leq i \leq n} var(g, P_i) \geq \sum_{0 \leq i \leq n} (\alpha - \varepsilon/2^i) \geq n\alpha - \varepsilon.$$

It follows that $var(g, [z, u_0]) = \infty$, which contradicts the hypothesis.

b) $\lim_{\substack{p \to s \\ p > s}} var(g, [s, p] \times (t_0, t]) = 0$.

Assume the contrary: there is an $\alpha > 0$ such that for every $p > s$ we have $var(g, [s, p] \times (t_0, t]) > \alpha$. We shall construct by induction a sequence $p_0 > p_1 > \ldots s$ with $p_n \to s$ and for each n a partition P_n contained in $[s, p_0] \times (t_0, t]$ such that for $i \neq j$, the rectangles of P_i are disjoint from the rectangles of P_j and such that

$$var(g, P_n) > \alpha - \varepsilon/2^n.$$

In fact, let $p_0 > s$ and $\varepsilon > 0$. Since $var(g, [s, p_0] \times (t_0, t]) > \alpha$, there is a grid $\sigma_0 \times \tau_0$ with

$$\sigma_0 : s = s_{00} < s_{01} < \cdots < s_{0n} = p_0 \text{ and } \tau_0 : t = t_{00} < t_{01} < \cdots < t_{0m} = t,$$

with t_{00} finite and $t_{00} \geq t_0$, such that

$$var(g, \sigma_0 \times \tau_0) > \alpha.$$

Since g is right continuous, it is right continuous with respect to the first variable, hence, for each $j \leq m$ we have

$$\lim_{p \downarrow s} \Delta_{(s,p] \times (t_{0,j-1}, t_{0j}]} g = 0.$$

Then there is an r_j such that $s < r_j < s_{01}$ and

$$|\Delta_{(s,r_j] \times (t_{0,j-1}, t_{0j}]} g| < \varepsilon/2^{j+1}.$$

Moreover, we can choose $r_1 > r_2 > \cdots > r_m > s$. If we add the points $r_1, r_2, \ldots, r_m$ to the division σ_0 we obtain a division σ_0' and a grid $Q_0 = \sigma_0' \times \tau_0$ finer than $\sigma_0 \times \tau_0$. It follows that

$$var(g, Q_0) > \alpha.$$

We have also

$$\sum_{1 \leq j \leq m} |\Delta_{(s,r_j] \times (t_{0,j-1}, t_{0j}]} g| < \sum_{1 \leq j \leq m} \varepsilon/2^{j+1} < \varepsilon/2.$$

It follows that

$$var(g, Q_0) - \sum_{1 \leq j \leq m} |\Delta_{(s,r_j] \times (t_{0,j-1}, t_{0j}]} g| > \alpha - \varepsilon/2.$$

Denoting by P_1 the partition obtained by subtracting from the grid Q_0 all the rectangles $(s, r_j] \times (t_{0,j-1}, t_{0j}]$, we obtain

$$var(g, P_1) = var(g, Q_0) - \sum_{1 \leq j \leq m} |\Delta_{(s,r_j] \times (t_{0,j-1}, t_{0j}]} g| > \alpha - \varepsilon/2.$$

Let $p_1 = r_m$. Then $var(g, [s, p_1] \times (t_0, t]) > \alpha$. We can repeat the above reasoning and obtain a grid $\sigma_1 \times \tau_1$ contained in $[s, p_1] \times (t_0, t]$ with

$$\sigma_1 : s = s_{10} < s_{11} < \cdots < s_{1n} = p_1 \text{ and } \tau_1 : t = t_{10} < t_{11} < \cdots < t_{1m} = t$$

such that t_{10} is finite and $t_{10} \geq t_0$ and such that

$$var(g, \sigma_1 \times \tau_1) > \alpha.$$

The indices n and m in this second step are different from the indices n and m in the first step, but we used the same letters so that the notation does not become even more complicated. Since g is right continuous in the first variable, for each $j \leq m$ there is a point r'_j with $s < r'_j < s_{11}$ such that $r'_1 > r'_2 > \cdots > r'_m > s$ and

$$\sum_{1 \leq j \leq m} |\Delta_{(s, r'_j] \times (t_{1,j-1}, t_{1j}]} g| < \varepsilon/2^2.$$

If we add the points r'_j to the division σ_1 we obtain a division σ'_1 and a grid $Q_1 = \sigma'_1 \times \tau_1$ finer than $\sigma_1 \times \tau_1$, hence

$$var(g, Q_1) > \alpha.$$

Denoting by P_2 the partition obtained by subtracting from the grid Q_1 all the rectangles $(s, r'_j] \times (t_{1,j-1}, t_{1j}]$, we have

$$var(g, P_2) = var(g - Q_1) - \sum_{1 \leq j \leq m} |\Delta_{(s, r'_j] \times (t_{1,j-1}, t_{1j}]} g| > \alpha - \varepsilon/2^2.$$

We remark that the rectangles of the partition P_1 are disjoint from the rectangles of the partition P_2. We continue this way, by induction and obtain the desired sequences (p_n) and (P_n).

For each n, the union $\bigcup_{i \leq n} P_i$ is a partition contained in $[s, p_0] \times (t_0, t]$ and

$$var(g, \bigcup_{i \leq n} P_i) = \sum_{i \leq n} var(g, P_i) \geq \sum_{i \leq n} (\alpha - \varepsilon/2^i)$$

$$= n\alpha - \sum_{i \leq n} \varepsilon/2^i > n\alpha - \varepsilon.$$

It follows that

$$var(g, [s, p_0] \times (t_0, t]) \geq n\alpha - \varepsilon, \text{ for each } n,$$

hence

$$var(g, [s, p_0] \times (t_0, t]) = \infty,$$

which contradicts the hypothesis.

c) If t_0 is finite, exactly the same proof shows that

$$\lim_{\substack{p \to s \\ p > s}} var(g, [s, p] \times [t_0, t]) = 0.$$

d) We can prove, similarly, that

$$\lim_{\substack{r \to t \\ r > t}} var(g, (s_0, s] \times [t, r]) = 0$$

and

$$\lim_{\substack{r \to t \\ r > t}} var(g, [s_0, s] \times [t, r]) = 0.$$

e) Using the preceding steps we deduce the equalities in the statement of the theorem. ∎

17. Remark. The converse of Theorem 16 is not true in general. If

$$\lim_{\substack{u \to z \\ u \geq z}} var(g, [z_0, u]) = var(g, [z_0, z]),$$

we can only deduce that g is incrementally right continuous. In fact, we deduce first that

$$\lim_{\substack{u \to z \\ u \geq z}} var(g, [z, u]) = 0$$

and from the inequality $|\Delta_{zu} g| \leq var(g, (z, u])$ we deduce that

$$\lim_{\substack{u \to z \\ u > z}} \Delta_{zu} g = 0.$$

The fact that g is not necessarily right continuous follows from the following example:

Let $f : \mathbb{R} \to \mathbb{R}$ be a function which is not right continuous and define $g(s,t) = f(s)$, for $(s,t) \in \mathbb{R}^2$. Then g is not right continuous. But for any $u < v$ in $\mathbb{R}^2$ we have $\Delta_{uv} g = 0$; therefore, for any $z_0 < z \leq u$ in $\mathbb{R}^2$ and for any partition $P = (R_i)$ in $(z_0, u]$ we have $\sum_i |\Delta_{R_i} g| = 0$; therefore $var(g, (z_0, u]) = 0$; consequently

$$\lim_{\substack{u \to z \\ u > z}} var(g, (z_0, u]) = var(g, (z_0, z]) = 0.$$

F. The variation function $|g|$

18. Let $g : \mathbb{R}^2 \to E$ be a function and $z_0 \in \mathbb{R}^2$. We define the *variation function* $z \mapsto |g|_{z_0}(z)$ on the rectangle (z_0, ∞) by

$$|g|_{z_0}(z) = var(g, (z_0, z]) \leq \infty, \text{ for } z \geq z_0,$$

and on the rectangle $R_{z_0} = (-\infty, z_0)$ by

$$|g|_{z_0}(z) = -var(g, (z, z_0]), \text{ for } z \leq z_0.$$

However, if z and z_0 are not comparable, then $|g|_{z_0}(z)$ is not defined.

We define also the *variation function* $|g| = |g|_{-\infty}$ on the whole extended plane $\overline{\mathbb{R}}^2$ by

$$|g|(z) = var(g, (-\infty, z]) \leq \infty, \text{ for } z \in \mathbb{R}^2,$$
$$|g|(-\infty) = 0,$$
$$|g|(+\infty) = var(g, \mathbb{R}^2) \leq \infty.$$

The advantage of $|g|$ over $|g|_{z_0}$ is that $|g|$ is defined on the whole plane. The disadvantage of $|g|$ is that it may be infinite even if $|g|_{z_0}$ is finite. We list some properties of $|g|$ and $|g|_{z_0}$.

19. *If* $-\infty \leq u < z_0 \leq z$ *with* $u = (p, r)$, $z_0 = (s_0, t_0)$, *and* $z = (s, t)$, *then*

$$v(g, (u, z]) = v(g, (u, z_0]) + v(g, [z_0, z])$$
$$+ v(g, [s_0, s] \times (r, t_0]) + v(g, (p, s_0] \times [t_0, t]),$$
$$v(g, [u, z]) = v(g, [u, z_0]) + v(g, [z_0, z])$$
$$+ v(g, [s_0, s] \times [r, t_0]) + v(g, [p, s_0] \times [t_0, t]),$$

and

$$|g|_u(z) = |g|_u(z_0) + v(g, [z_0, z])$$
$$+ v(g, [s_0, s] \times (r, t_0]) + v(g, (p, s_0] \times [t_0, t]).$$

We use Proposition 14.

20. *For* $z_0 < z$ *we have*

$$v(g, (-\infty, z]) = v(g, (-\infty, z_0]) + v(g, [z_0, z])$$
$$+ v(g, [s_0, s] \times (-\infty, t_0]) + v(g, (-\infty, s_0] \times [t_0, t]),$$

that is,

$$|g|(z) = |g|(z_0) + v(g, [z_0, z])$$
$$+ v(g, [s_0, s] \times (-\infty, t_0]) + v(g, (-\infty, s_0] \times [t_0, t]).$$

We take the supremum for $u > -\infty$ in the first equality of Property 19.

G. Functions with finite variation

Functions with finite variation have additional properties. Let $g : \mathbb{R}^2 \to E$ be a function with *finite variation*.

21. *If* $u < z_0 < z$ *in* $\mathbb{R}^2$, *then*

$$\Delta_{z_0 z} |g|_u = var(g, [z_0, z]).$$

If g *has finite variation function* $|g|$, *then*

$$\Delta_{z_0 z} |g| = var(g, [z_0, z]).$$

Use Properties 19 and 20.

22. *If g has finite variation function $|g|$, then $|g|$ is increasing and incrementally increasing.*

Use Properties 11 and 21.

23. *If $u < z_0 < z$ in $\mathbb{R}^2$, then*

$$|\Delta_{z_0 z} g| \leq \Delta_{z_0 z} |g|_u .$$

In fact, $|\Delta_{z_0 z} g| \leq var(g, [z_0, z])$; we use then Property 21.

24. *If g has finite variation function $|g|$ and if $|g|$ is incrementally right continuous, then so is g.*

25. *If g has finite variation function $|g|$ and if g is right continuous, then $|g|$ is also right continuous.*

This follows from Theorem 16.

Remark. The converse is also true, provided g vanishes outside a quadrant (Proposition 31).

H. Functions vanishing outside a quadrant

We say that a function $g : \mathbb{R}^2 \to E$ vanishes outside the first quadrant around a point $z_0 = (s_0, t_0) \in \mathbb{R}^2$ if $g(s,t) = 0$ for $s < s_0$ or $t < t_0$. This is the case of stochastic processes which are usually defined on the first quadrant $\mathbb{R}_+^2$ and are extended with 0 elsewhere.

We remark that if g vanishes outside the first quadrant around z_0 and if $z_1 \leq z_0$, the g vanishes outside the first quadrant around z_1. For this reason we say sometimes that g vanishes outside a first quadrant, without mentioning the point z_0.

Let $g : \mathbb{R}^2 \to E$ be a function vanishing outside the first quadrant around a point $z_0 = (s_0, t_0) \in \mathbb{R}^2$. We translate some of the preceding properties for this situation.

26. *If $-\infty \leq u < z_0 \leq z$, then*

$$\Delta_{uz} g = g(z).$$

27. *If $-\infty \leq u < z_0 \leq z$ with $u = (p, r)$, $z_0 = (s_0, t_0)$ and $z = (s,t)$, then*

$$\begin{aligned} |g|_u(z) &= |g(z_0)| + var(g, [z_0, z]) \\ &\quad + var(g(\cdot, t_0), [s_0, s]) + var(g(s_0, \cdot), [t_0, t]). \end{aligned}$$

This follows from Property 19. In fact,

$$var(g, (u, z_0]) = |g(z_0)|$$
$$var(g, [s_0, s] \times (r, t_0]) = var(g(\cdot, t_0), [s_0, s])$$

and
$$var(g, (p, s_0] \times [t_0, t]) = var(g(s_0, \cdot), [t_0, t])\,.$$

28. *If $u < z_0 \leq z$ with $z_0 = (s_0, t_0)$ and $z = (s, t)$, then*
$$|g|(z) = |g|_u(z)\,.$$

In fact, the right-hand side term does not depend on u.

29. *If g has finite variation, then g has finite variation function $|g|$.*

Use Property 28.

30. Proposition. *If g has finite variation function $|g|$, then, for $z < z'$ we have*
$$|g(z') - g(z)| \leq |g|(z') - |g|(z)\,.$$

Proof. Let $z < z'$. Since g vanishes outside the first quadrant around z_0, it vanishes outside the first quadrant around $z \wedge z_0$; therefore, replacing z_0 with $z \wedge z_0$, if necessary, we can assume that $z_0 \leq z < z'$. Denote $z_0 = (s_0, t_0)$, $z = (s, t)$ and $z' = (s', t')$. Using property 27 with $u = -\infty$, we have
$$var(g, [z_0, z']) = |g|(z') - |g|(z_0)| - var(g(\cdot, t_0), [s_0, s']) - var(g(s_0, \cdot), [t_0, t'])$$
and
$$var(g, [z_0, z]) = |g|(z) - |g|(z_0)| - var(g(\cdot, t_0), [s_0, s]) - var(g(s_0, \cdot), [t_0, t]),$$
whence
$$var(g, [z_0, z']) - var(g, [z_0, z]) = |g|(z') - |g|(z) - var(g(\cdot, t_0), [s, s']) - var(g(s_0, \cdot), [t, t'])\,.$$
Consider now the rectangles $R = [s_0, s'] \times [t, t']$ and $P = [s, s'] \times [t_0, t]$. Then
$$\begin{aligned}
|g(z') - g(z)| &\leq |g(s', t') - g(s', t)| + |g(s', t) - g(s, t)| \\
&= |\Delta_R g + g(s_0, t') - g(s_0, t)| + |\Delta_P g + g(s', t_0) - g(s, t_0)| \\
&\leq |\Delta_R g| + |g(s_0, t') - g(s_0, t)| + |\Delta_P g| + |g(s', t_0) - g(s, t_0)| \\
&\leq var(g, R) + var(g, P) + var(g(s_0, \cdot), [t, t']) + var(g(\cdot, t_0), [s, s']) \\
&= var(g, [z_0, z']) - var(g, [z_0, z]) + var(g(s_0, \cdot), [t, t']) + var(g, (\cdot, t_0), [s, s']) \\
&= |g|(z') - |g|(z).
\end{aligned}$$
∎

31. Proposition. *Assume g has finite variation and vanishes outside a quadrant. Then g is right continuous iff $|g|$ is right continuous.*

It follows from Property 25 and Proposition 30.

378 Ch.9 TWO-PARAMETER PROCESSES WITH FINITE VARIATION

I. Variation of real-valued functions

Let $g : \mathbb{R}^2 \to \mathbb{R}$ be a real-valued function.

32. Proposition. *If g is incrementally increasing, then g has finite variation and for $z_0 < z$ we have*

$$var(g, [z_0, z]) = \Delta_{z_0 z} g\,.$$

If, in addition, g has finite variation function $|g|$, then

$$\Delta_{z_0 z}|g| = \Delta_{z_0 z} g$$

Proof. If g is incrementally increasing, then $\Delta_R g \geq 0$ for any bounded rectangle $R = (u, u']$. If $P = (R_i)_{i \in I}$ is a partition with union $R = (z_0, z]$, then, by Property 5 we have

$$var(g, P) = \sum_{i \in I} |\Delta_{R_i} g| = \sum_{i \in I} \Delta_{R_i} g = \Delta_{z_0 z} g\,.$$

Then

$$var(g, [z_0, z]) = \Delta_{z_0 z} g\,.$$

If the variation function $|g|$ is finite, we use Property 21: $\Delta_{z_0 z}|g| = var(g, [z_0, z])$ to deduce $\Delta_{z_0 z}|g| = \Delta_{z_0 z} g$. ∎

33. Proposition. *If g is incrementally increasing and vanishes outside the first quadrant around a point z_0, then g has finite variation function $|g|$ and $|g| = g$.*

Proof. By Proposition 32, g has finite variation. By Property 29, g has finite variation function $|g|$. Let $u < z_0$. By Proposition 32, for every $z \in \mathbb{R}$ we have

$$\Delta_{uz}|g| = \Delta_{uz} g\,.$$

Since g vanishes outside the first quadrant around z_0, we have $\Delta_{uz} g = g(z)$; the variation function $|g|$ also vanishes outside the first quadrant around z_0; therefore $\Delta_{uz}|g| = |g|(z)$. It follows that $|g| = g$. ∎

The Jordan decomposition theorem for functions of one variable with finite variation has the following analog for functions of two variables.

34. Theorem. *Let $g : \mathbb{R}^2 \to \mathbb{R}$ be a real-valued function with finite variation, vanishing outside the first quadrant around a point $z_0 = (s_0, t_0)$. Then there are two functions $g_1, g_2 : \mathbb{R}^2 \to \mathbb{R}$, increasing and incrementally increasing, vanishing outside the first quadrant around z_0, such that*

$$g = g_1 - g_2\,.$$

Proof. By property 29, g has finite variation function $|g|$ which vanishes outside the first quadrant around z_0. We take $g_1 = |g|$ and $g_2 = |g| - g = g_1 - g$. Then $g = g_1 - g_2$ and g_2 vanishes outside the first quadrant around z_0.

By Property 22, $|g|$ is increasing and incrementally increasing. It remains to show that g_2 is increasing and incrementally increasing.

Let $z < z'$. By Proposition 30, we have
$$|g(z') - g(z)| \leq |g|(z') - |g|(z),$$
hence
$$g_2(z') - g_2(z) = |g|(z') - g(z') - (|g|(z) - g(z))$$
$$= (|g|(z') - |g|(z)) - (g(z') - g(z)) \geq 0,$$
consequently g_2 is increasing.

Then, by Property 23 we have
$$\Delta_{zz'} g_2 = \Delta_{zz'}(|g| - g) = \Delta_{zz'}|g| - \Delta_{zz'}g \geq 0,$$
hence g_2 is incrementally increasing ∎

J. Lateral limits

A function of one variable $g : \mathbb{R} \to E$ with finite variation has lateral limits at every point of $\mathbb{R}$.

For a function of two variables $g : \mathbb{R}^2 \to E$ with finite variation, the lateral limits do not necessarily exist. An additional condition will ensure the existence of lateral limits: finite variation of the partial functions of one variable $g(\cdot, t)$ and $g(s, \cdot)$. We shall prove the existence of the lateral limits under more general conditions, for a function $g : A^2 \to E$, where A is a dense subset of $\mathbb{R}$. In particular, we can have $A = \mathbb{R}$.

In this case, the increment $\Delta_R g$ is defined only for rectangles with all vertices in A^2. But we can define the variation $v(g, R)$ for any rectangle R in $\mathbb{R}^2$ bounded or unbounded, by the same equality:
$$v(g, R) = \sup \sum_i |\Delta_{R_i} g|,$$
where the supremum is taken over all partitions $P = (R_i)_{i \in I}$ consisting of rectangles R_i with vertices in $R \cap A^2$.

The variation function $|g|(z) = v(g, (-\infty, z])$ is defined for any $z \in \mathbb{R}^2$.

Properties 19 and 20 are still valid for $|g|$. If g has finite variation and if the partial function $g(\cdot, t_0)$ has finite variation for some $t_0 \in \mathbb{R}$, then $g(\cdot, t)$ has finite variation for any $t \in \mathbb{R}$. A similar property is valid for $g(s, \cdot)$. This will follow from the following lemma:

35. Lemma. *Let $g : A^2 \to E$ be a function, $s < s'$ and $t < t'$ in A. Then, for any $a < b$ in $\mathbb{R}$ we have*
$$v(g(\cdot, t'), [a, b]) \leq v(g(\cdot, t), [a, b]) + v(g, [a, b] \times [t, t'])$$

and
$$v(g(s',\cdot),[a,b]) \leq v(g(s,\cdot),[a,b]) + v(g,[s,s']\times[a,b]).$$

Proof. We shall prove only the first inequality. Let $\sigma : a \leq s_0 < s_1 < s_n \leq b$ be a division of $[a,b] \cap A$. Denote $R_i = (s_i, s_{i+1}) \times (t, t')$ for $0 \leq i < n$. Then
$$|g(s_{i+1}, t') - g(s_i, t')| \leq |\Delta_{R_i} g| + |g(s_{i+1}, t) - g(s_i, t)|.$$

Taking the sum we obtain

$$\sum_{0 \leq i < n} |g(s_{i+1}, t') - g(s_i, t')|$$
$$\leq \sum_{0 \leq i < n} |\Delta_{R_i} g| + \sum_{0 \leq i < n} |g(s_{i+1}, t) - g(s_i, t)|$$
$$\leq var(g, [a,b] \times [t, t']) + var(g(\cdot, t), [a,b]).$$

Taking the supremum for all divisions σ we get the desired inequality. ∎

36. Corollary. *Assume that $g : A^2 \to E$ has finite variation and that for some $s_0, t_0 \in A$, the partial functions $g(\cdot, t_0)$ and $g(s_0, \cdot)$ have finite variation. Then for every $s, t \in A$, the functions $g(\cdot, t)$ and $g(s, \cdot)$ have finite variation.*

Proof. If $t > t_0$ we apply the preceding lemma with t and t' replaced by t_0 and t; if $t < t_0$ we apply the lemma with t and t' replaced by t and t_0. ∎

The following theorem ensures the existence of lateral limits.

37. Theorem. *Let A be a set dense in $\mathbb{R}$ and $g : A^2 \to E$ a function. Assume that g and the partial functions $g(\cdot, t), g(s, \cdot)$ have finite variation for $s, t \in A$. Then g has lateral limits at every point $z \in \mathbb{R}^2$.*

If g has finite variation function $|g|$, then g has lateral limits at every point $z = (s, t)$ with s, or t, or both, equal to $-\infty$.

Proof. (cf. [L]). We shall only prove the existence of the right limit $g(s+, t+)$ when both s and t are finite. The existence of the other limits is proved similarly.

Let $z = (s, t) \in \mathbb{R}^2$ and let $z_n = (s_n, t_n)$ be a decreasing sequence from A^2 with $z_n \downarrow z$ and prove that $g(z_n)$ is a Cauchy sequence in E. We divide the proof into several steps.
a) If (s_n) and (t_n) are stationary, i.e., if there is an n_0 such that $s_n = s$ and $t_n = t$ for $n \geq n_0$, then $g(s_n, t_n) = g(s, t)$ for $n \geq n_0$; therefore $(g(z_n))$ is a Cauchy sequence.
b) Assume (s_n) is stationary and (t_n) is not stationary. By deleting a finite number of terms we can assume that $s_n = s$ for every $n \in \mathbb{N}$. It follows, in particular, that $s \in A$. Let $j \in \mathbb{N}$ and consider the division $t_j \leq t_{j-1} \leq \cdots \leq t_1$

of $[t_j, t_1]$. For the partial function $g(s, \cdot)$ we have

$$var(g(s, \cdot), [t, t_1]) \geq var(g(s, \cdot), [t_j, t_1])$$
$$\geq \sum_{1 \leq i < j} |g(s, t_{i+1}) - g(s, t_i)| = \sum_{1 \leq i < j} |g(s_{i+1}, t_{i+1}) - g(s_i, t_i)|,$$

therefore

$$\sum_{1 \leq i < \infty} |g(s_{i+1}, t_{i+1}) - g(s_i, t_i)| \leq var(g(s, \cdot), [t, t_1]) < \infty.$$

Let $\varepsilon > 0$. Since the above series is convergent, by the Cauchy criterion, there is an N such that for every $m > n \geq N$ we have

$$\sum_{n \leq i < m} |g(s_{i+1}, t_{i+1}) - g(s_i, t_i)| < \varepsilon.$$

Then, for $m > n \geq N$ we deduce

$$|g(s_m, t_m) - g(s_n, t_n)| = |\sum_{n \leq i < m} [g(s_{i+1}, t_{i+1}) - g(s_i, t_i)]| < \varepsilon,$$

therefore $(g(s_n, t_n))$ is a Cauchy sequence.

c) If (t_n) is stationary and (s_n) is not stationary, we deduce, by symmetry, that $(g(s_n, t_n))$ is a Cauchy sequence.

d) Assume now that neither (s_n) nor (t_n) is stationary. Let $z_0 = (s_0, t_0) \in A^2$ with $z_0 < z$. Let $j \in \mathbb{N}$ and consider the grid $\sigma \times \tau$ of the rectangle $(s_j, s_1] \times [t_j, t_1]$ with $\sigma : s_j \leq \cdots \leq s_2 \leq s_1$ and $\tau : t_j \leq \cdots \leq t_2 \leq t_1$.

Consider the rectangles $R_i = (s_{i+1}, s_i] \times (t_0, t_{i+1}]$ and $R'_i = (s_0, s_i] \times (t_{i+1}, t_i]$, for $1 \leq i < j$. These rectangles form a partition contained in the rectangle $[z_0, z_1] = [s_0, s_1] \times [t_0, t_1]$. Then

$$|\Delta_{R_i} g| = |g(s_i, t_{i+1}) - g(s_{i+1}, t_{i+1}) - g(s_i, t_0) + g(s_{i+1}, t_0)|$$
$$\geq |g(s_i, t_{i+1}) - g(s_{i+1}, t_{i+1})| - |g(s_i, t_0) - g(s_{i+1}, t_0)|,$$

hence

$$|g(s_i, t_{i+1}) - g(s_{i+1}, t_{i+1})| \leq |\Delta_{R_i} g| + |g(s_i, t_0) - g(s_{i+1}, t_0)|.$$

Similarly

$$|\Delta_{R'_i} g| = |g(s_i, t_i) - g(s_i, t_{i+1}) - g(s_0, t_i) + g(s_0, t_{i+1})|$$
$$\geq |g(s_i, t_i) - g(s_i, t_{i+1})| - |g(s_0, t_i) - g(s_0, t_{i+1})|,$$

hence

$$|g(s_i, t_i) - g(s_i, t_{i+1})| \leq |\Delta_{R'_i} g| + |g(s_0, t_i) - g(s_0, t_{i+1})|.$$

Then
$$|g(s_i,t_i) - g(s_{i+1},t_{i+1})| \le |g(s_i,t_i) - g(s_i,t_{i+1})| + |g(s_i,t_{i+1}) - g(s_{i+1},t_{i+1})|$$
$$\le |\Delta_{R_i} g| + |\Delta_{R'_i} g| + |g(s_i,t_0) - g(s_{i+1},t_0)| + |g(s_0,t_i) - g(s_0,t_{i+1})|.$$

It follows that
$$\sum_{1 \le i < j} |g(s_i,t_i) - g(s_{i+1},t_{i+1})| \le \sum_{1 \le i < j} (|\Delta_{R_i} g| + |\Delta_{R'_i} g|)$$
$$+ \sum_{1 \le i < j} |g(s_i,t_0) - g(s_{i+1},t_0)| + \sum_{1 \le i < j} |g(s_0,t_i) - g(s_0,t_{i+1})|$$
$$\le var(g,[z_0,z_1]) + v(g(s_0,\cdot),[t_0,t_1]) + v(g(\cdot,t_0),[s_0,s_1]),$$

therefore,
$$\sum_{1 \le i < \infty} |g(s_i,t_i) - g(s_{i+1},t_{i+1})|$$
$$\le var(g,[z_0,z_1]) + v(g(s_0,\cdot),[t_0,t_1]) + v(g(\cdot,t_0),[s_0,s_1]) < \infty.$$

Let $\varepsilon > 0$. Since the series is convergent, by the Cauchy criterion, there is an N such that for every $m > n \ge N$ we have
$$|g(s_m,t_m) - g(s_n,t_n)| = |\sum_{n \le i < m} [g(s_i,t_i) - g(s_{i+1},t_{i+1})]| < \varepsilon,$$

consequently $(g(z_n))$ is a Cauchy sequence.
e) We shall prove now that the limit $\lim g(z_n)$ is independent of the sequence (z_n). Let $z'_n \downarrow z$ and $z''_n \downarrow z$ with $z'_n, z''_n \in A^2$ and prove that $\lim g(z'_n) = \lim g(z''_n)$.
For this purpose we construct a sequence (z_n) with $z_n \in A^2$ and $z_n \downarrow z$ such that (z_{2n}) is a subsequence of (z'_n) and z_{2n+1} is a subsequence of (z''_n).
We take $z_1 = z''_1$; since $z'_n \downarrow z$, the rectangle $R_1 = [z,z_1]$ contains all but finitely many terms z'_n; we choose $z'_{n_1} \in R_1$ and denote $z_2 = z'_{n_1}$; since $z''_n \downarrow z$, we choose a term $z''_{n_2} \in R_2 = [z,z_2]$ and denote $z_3 = z''_{n_2}$, etc. Then $\lim g(z_{2n}) = \lim g(z'_n)$ and $\lim g(z_{2n+1}) = \lim g(z''_n)$. On the other hand $\lim g(z_{2n}) = \lim g(z_n) = \lim g(z_{2n+1})$, consequently $\lim g(z'_n) = \lim g(z''_n)$.
f) Let $a \in E$ be the common limit of all the sequences $g(z_n)$ with $z_n \in A^2$ and $z_n \downarrow z$ and prove that
$$\lim_{\substack{z' \to z \\ z' > z \\ z' \in A^2}} g(z') = a.$$

Assume the contrary: there is an $\varepsilon_0 > 0$ such that for any rectangle $R = [z,u]$ with $u > z$, there is a $z' \in R \cap A^2$ with $z' \ne z$ and $|g(z') - a| \ge \varepsilon_0$. We start with a decreasing sequence $z_n \downarrow z$ with $z_n \in A^2$ and $z_n > z$ and take $R_1 = [z,z_1]$; there is a $z'_1 \in R_1 \cap A^2$ with $z'_1 \ne z$ and $|g(z'_1) - a| \ge \varepsilon_0$; then

take $R_2 = [z, z'_1 \wedge z_2]$; there is a $z'_2 \in R_2 \cap A^2$ with $z'_2 \neq z$ and $|g(z'_2) - a| \geq \varepsilon_0$. By induction, we construct a decreasing sequence (z'_n) with $z'_n \neq z$, $z'_n \in A^2$, and $z'_{n+1} \leq z'_n \wedge z_{n+1}$, and with $|g(z'_n) - a| \geq \varepsilon_0$. Then $z'_n \downarrow z$; therefore $\lim g(z'_n) = a$ which contradicts the preceding inequalities $|g(z'_n) - a| \geq \varepsilon_0$ for all n. It follows that
$$\lim_{\substack{z' \downarrow z \\ z' \in A^2}} g(z') = a$$
and the theorem is proved. ∎

38. Theorem. *Let A be a set dense in $\mathbb{R}$ and $g : A^2 \to E$ a function. Assume that:*
(i) the function g and the partial functions $g(\cdot, t)$ and $g(s, \cdot)$ have finite variation for $s, t \in A$;
(ii) there is a subspace $Z \subset E^*$ norming for E, such that for each $x^* \in Z$, the function x^*g is partially right continuous with respect to one of the two variables.
Then g is right continuous and can be extended to a right continuous function $G : \mathbb{R}^2 \to E$ such that G and the partial functions $G(\cdot, t)$ and $G(s, \cdot)$ have finite variation for $s, t \in \mathbb{R}$, and such that for any rectangle R we have
$$var(G, R) = var(g, R).$$

Proof. a) Let $x^* \in Z$. To make a choice, assume x^*g is right continuous with respect to the first variable and prove that g is jointly right continuous on A^2.

Let $z_0 = (s_0, t_0) \in A^2$. By Theorem 37, the limit
$$\lim_{\substack{z \downarrow z_0 \\ z \in A^2}} g(z) = a$$
exists in E. In particular, taking $z = (s, t_0) \in A^2$, we get
$$\lim_{\substack{s \downarrow s_0 \\ s \in A}} g(s, t_0) = a.$$
Since z^*g is right continuous with respect to s, we have also
$$\lim_{\substack{s \downarrow s_0 \\ s \in A}} x^*g(s, t_0) = x^*g(s_0, t_0) = x^*g(z_0).$$
It follows that $x^*a = x^*g(z_0)$ for every $x^* \in Z$, hence $a = g(z_0)$; consequently g is right continuous at z_0.

b) Denote by $G : \mathbb{R}^2 \to E$ the function defined for every $z \in \mathbb{R}^2$ by
$$G(z) = \lim_{\substack{z' \downarrow z \\ z' \in A^2}} g(z).$$
By step a), for $z \in A^2$ we have $G(z) = g(z)$, hence G is an extension of g.

c) To prove that G is right continuous on $\mathbb{R}^2$, let $z \in \mathbb{R}^2$ and (z_n) a sequence from $\mathbb{R}^2$ with $z_n \to z$ and $z_n > z$, and prove that $\lim G(z_n) = G(z)$. For each n, there is a $z'_n \in A^2$ with $z_n \leq z'_n, |z_n - z'_n| < 1/n$ and $|G(z_n) - g(z'_n)| < 1/n$. Since $z_n \to z$, we have also $z'_n \to z$, therefore $G(z) = \lim g(z'_n)$. From

$$G(z_n) - G(z) = [G(z_n) - g(z'_n)] + [g(z'_n) - G(z)]$$

it follows that $\lim G(z_n) = G(z)$; hence G is right continuous.

d) To prove that G has finite variation, we note that, since G is right continuous, its variation can be computed by taking only grids with end points in A^2 (Proposition 13). But at the points of A^2, G and g coincide; therefore the variations of G and g are equal on any rectangle. By the same argument, the variations of the partial functions $G(\cdot, t)$ and $g(\cdot, t)$ are equal for $t \in A$, hence $G(\cdot, t)$ has finite variation for $t \in A$ and by Corollary 36 for any $t \in \mathbb{R}$. Similarly, $G(s, \cdot)$ has finite variation for any $s \in \mathbb{R}$. ∎

Taking $A = \mathbb{R}$ we obtain the following corollary.

39. Corollary. *Let $g : \mathbb{R}^2 \to E$ be a function and assume that*
(i) g and the partial functions $g(\cdot, t), g(s, \cdot)$ have finite variation;
(ii) there is a space $Z \subset E^*$ norming for E such that for every $z^* \in Z$, the function x^*g is partially right continuous with respect to one of the two variables.
Then g is right continuous on $\mathbb{R}^2$.

40. Theorem. *Let A be a set dense in $\mathbb{R}$ and $g : A^2 \to \mathbb{R}$ a real-valued function. Assume that g is increasing and partially right continuous with respect to one of the two variables.*

Then g can be extended to a right continuous, increasing function $G : \mathbb{R}^2 \to \mathbb{R}$.

If g is incrementally increasing, so is G.

Proof. If g is increasing, then the limit

$$\lim_{\substack{z' \downarrow z \\ z \in A^2}} g(z')$$

exists for every $z \subset \mathbb{R}^2$. Then we can use the proof of Theorem 38 to deduce the existence of the right continuous extension $G : \mathbb{R}^2 \to \mathbb{R}$, which, by right continuity, is increasing.

If g is incrementally increasing, then, again by right continuity, G is also incrementally increasing. ∎

41. Corollary. *If $g : \mathbb{R}^2 \to \mathbb{R}$ is increasing and partially right continuous with respect to one of the two variables, then g is right continuous.*

K. Measures associated to functions

Let $g : \mathbb{R}^2 \to E$ be a function. For any rectangle $R = (z, z']$ with $z \leq z'$ we set
$$m_g(R) = \Delta_R g.$$
We obtain a set function $m_g \colon \mathcal{P} \to E$ on the semiring $\mathcal{P}$ of rectangles of the form $(z, z']$. Denote by $\mathcal{R}$ the ring generated by $\mathcal{P}$. We have the following properties.

42. m_g *is finitely additive on* $\mathcal{P}$.

This follows from Proposition 5.

43. m_g *can be extended to a finitely additive measure* $m_g : \mathcal{R} \to E$.

This is a consequence of Proposition 7.2.

44. *If* $g' = g + C$, *with* C *a constant, then* $m_{g'} = m_g$.

The converse is not true, in general.

Example. Let $f : \mathbb{R} \to E$ be a nonconstant function and define $g : \mathbb{R}^2 \to E$ by $g(s,t) = f(s)$ for $(s,t) \in \mathbb{R}^2$. Then $m_g(R) = 0$ for any rectangle $R = (z, z']$ but g is not constant.

45. Proposition. m_g *has finite (resp. bounded) variation iff* g *has finite (resp. bounded) variation.*

For any rectangle $R \subset \mathbb{R}^2$ we have
$$var(g, R) \leq var(m_g, R) \leq var(g, \overline{R}).$$

Proof. Let $(R_i)_{i \in I}$ be a partition of rectangles with vertices in R. Then
$$\sum_i |\Delta_{R_i} g| = \sum_i |m_g(R_i)| \leq var(m_g, R),$$
therefore
$$var(g, R) \leq var(m_g, R).$$

Let now $(R_i)_{i \in I}$ be a partition contained in R. Then each rectangle R_i has vertices in $\overline{R}$. We have
$$\sum_i |m_g(R_i)| = \sum_i |\Delta_{R_i} g| \leq var(g, \overline{R})$$
hence
$$var(m_g, R) \leq var(g, \overline{R}). \qquad \blacksquare$$

46. Proposition. *Let* $R \subset \mathbb{R}^2$ *be a rectangle. We have the equality*
$$var(m_g, R) = var(g, R)$$

in each of the following cases:
a) $\inf R = -\infty$ or $\inf R \in R$;
b) g is right continuous.

Proof. We only have to prove the inequality

$$\mathrm{var}(m_g, R) \leq \mathrm{var}(g, R).$$

Assertion a) follows from the proof of Proposition 45, since in this case the rectangles R_i have vertices in R.

To prove assertion b) let $(R_i)_{1 \leq i \leq n}$ be a partition contained in R. Denote $R_i = (u_i, z_i]$. Let $\varepsilon > 0$. Since g is right continuous, for each i there is a point $v_i \in \mathbb{R}^2$ with $u_i < v_i < z_i$ such that

$$|\Delta_{u_i z_i} g| < |\Delta_{v_i z_i} g| + \varepsilon/n.$$

Then

$$\sum_i |m_g(R_i)| = \sum_i |\Delta_{u_i z_i} g| < \sum_i |\Delta_{v_i z_i} g| + \varepsilon \leq \mathrm{var}(g, R) + \varepsilon,$$

therefore

$$\mathrm{var}(m_g, R) \leq \mathrm{var}(g, R).$$

∎

47. If $g : \mathbb{R}^2 \to \mathbb{R}$ is a real-valued function, then $m_g \geq 0$ iff g is incrementally increasing.

48. Theorem. *If g has finite variation function $|g|$ and if g is right continuous, then*

$$|m_g| = m_{|g|}.$$

Proof. Let $R = (z, z']$ be a rectangle with $z = (s, t)$ and $z' = (s', t')$ and let $\varepsilon > 0$. There is a grid $\sigma \times \tau$ in R with

$$\sigma : s < s_0 < s_1 < \cdots < s_n = s' \text{ and } \tau : t < t_0 < t_1 < \cdots < t_m = t'$$

such that

$$\mathrm{var}(g, R) < \sum_{ij} |\Delta_{R_{ij}} g| + \varepsilon,$$

where $R_{ij} = (s_i, s_{i+1}] \times (t_j, t_{j+1}]$, for $0 \leq i < n$ and $0 \leq j < m$. Since g is right continuous, we have $\mathrm{var}(g, [z, z']) = \mathrm{var}(g, (z, z'])$. Then by Property 21 we have

$$m_{|g|}(R) = \Delta_R |g| = \mathrm{var}(g, [z, z']) = \mathrm{var}(g, (z, z'])$$
$$< \sum_{ij} |\Delta_{R_{ij}} g| + \varepsilon = \sum_{ij} |m_g(R_{ij})| + \varepsilon$$
$$\leq |m_g|(R) + \varepsilon;$$

therefore
$$m_{|g|}(R) \leq |m_g|(R).$$
Then
$$m_{|g|}(A) \leq |m_g|(A), \text{ for } A \in \mathcal{R}.$$
Conversely, let $A \in \mathcal{R}$. We have $A = \bigcup_{1 \leq i \leq n}(z_i, z_i']$ with $(z_i, z_i']$ mutually disjoint. By Property 23 we have

$$|m_g(A)| = |\sum_i m_g(z_i, z_i']| \leq \sum_i |m_g(z_i, z_i']|$$
$$= \sum_i |\Delta_{z_i z_i'} g| \leq \sum_i \Delta_{z_i z_i'} |g| = \sum_i m_{|g|}(z_i, z_i'] = m_{|g|}(A),$$

consequently,
$$|m_g|(A) = m_{|g|}(A), \text{ for } A \in \mathcal{R}.$$

∎

L. σ-additivity of the measure m_g

We shall prove now some theorems that ensure the σ-additivity of the measure m_g.

49. Theorem. *Let $g : \mathbb{R}^2 \to \mathbb{R}$ be a real-valued function, incrementally increasing and right continuous. Then the measure m_g is σ-additive on $\mathcal{R}$.*

Proof. By Property 43, m_g is finitely additive on $\mathcal{R}$. By Property 47 we have $m_g \geq 0$. It is enough to prove that m_g is σ-additive on the semiring $\mathcal{P}$. By Proposition 7.2 it will follow that m_g is σ-additive on $\mathcal{R}$.

Let $(z_n, z_n']$ be a sequence of disjoint rectangles with union a rectangle $(z, z']$ from $\mathcal{P}$ and prove that
$$m_g(z, z'] = \sum_{1 \leq n < \infty} m_g(z_n, z_n'].$$

We have first, for every n,
$$\sum_{1 \leq i \leq n} m_g(z_i, z_i'] = m_g(\bigcup_{1 \leq i \leq n}(z_i, z_i']) \leq m_g(z, z'],$$
hence
$$\sum_{1 \leq n < \infty} m_g(z_n, z_n'] \leq m_g(z, z'].$$

To prove the converse inequality let $\varepsilon > 0$. Set $z = (s, t)$ and $z' = (s', t')$. Since g is right continuous at $z = (s, t)$, it is right continuous in s at the point

(s, t') and right continuous in t at the point (s', t). Then we can find a point $a = (\alpha, \beta)$ such that $z < a < z'$ and

$$|g(z) - g(a)| < \varepsilon/3,$$
$$|g(s, t') - g(\alpha, t')| < \varepsilon/3,$$

and

$$|g(s', t) - g(s', \beta)| < \varepsilon/3.$$

Then

$$m_g(z, z'] - m_g(a, z'] = \Delta_{zz'} g - \Delta_{az'} g$$
$$= (g(z) - g(a)) - (g(s, t') - g(\alpha, t')) - (g(s', t) - g(s', \beta)) < \varepsilon.$$

For each n write $z_n = (s_n, t_n)$ and $z'_n = (s'_n, t'_n)$. Using the right continuity of g at $z'_n = (s'_n, t'_n)$, the right continuity in s at the point (s'_n, t_n), and the right continuity in t at the point (s_n, t'_n), we can find a point $a_n = (\alpha_n, \beta_n) > z'_n$ such that

$$|g(a_n) - g(z'_n)| < \varepsilon/(2^n \cdot 3),$$
$$|g(\alpha_n, t_n) - g(s'_n, t_n)| < \varepsilon/(2^n \cdot 3),$$

and

$$|g(s_n, \beta_n) - g(s_n, t'_n)| < \varepsilon/(2^n \cdot 3).$$

Then

$$m_g(z_n, a_n] - m_g(z_n, z'_n] = \Delta_{z_n a_n} g - \Delta_{z_n z'_n} g$$
$$= (g(a_n) - g(z'_n)) - (g(\alpha_n, t_n) - g(s'_n, t_n))$$
$$- (g(s_n, \beta_n) - g(s_n, t'_n)) < \varepsilon/2^n.$$

The compact rectangle $[a, z']$ is covered by the sequence $((z_n, a_n))$ of open rectangles; therefore there is a finite cover:

$$[a, z'] \subset \bigcup_{1 \leq i \leq N} (z_i, a_i).$$

Then

$$(a, z'] \subset \bigcup_{1 \leq i \leq N} (z_i, a_i],$$

therefore

$$m_g(a, z'] \leq m_g\Big(\bigcup_{1 \leq i \leq N} (z_i, a_i]\Big)$$
$$\leq \sum_{1 \leq i \leq N} m_g(z_i, a_i] \leq \sum_{1 \leq n < \infty} m_g(z_n, a_n].$$

It follows that

$$m_g(z, z'] \leq m_g(a, z'] + \varepsilon$$
$$\leq \sum_{1 \leq n < \infty} [m_g(z_n, z'_n] + \varepsilon/2^n] + \varepsilon = \sum_{1 \leq n < \infty} m_g(z_n, z'_n] + 2\varepsilon.$$

Since $\varepsilon > 0$ is arbitrary, we deduce that

$$m_g(z, z'] \leq \sum_{1 \leq n < \infty} m_g(z_n, z'_n].$$

∎

Denote by $\mathcal{D}$ the δ-ring of bounded Borel subsets of $\mathbb{R}^2$. The δ-ring $\mathcal{D}$ is generated by the ring $\mathcal{R}$.

50. Theorem. a) *Let $g : \mathbb{R}^2 \to E$ be a function. If m_g is σ-additive on $\mathcal{R}$ then g is incrementally right continuous.*
b) *Assume $g : \mathbb{R}^2 \to E$ is right continuous and has finite (resp. bounded) variation.*

Then m_g can be extended uniquely to a σ-additive measure $m : \mathcal{D} \to E$ (resp. $m : \mathcal{B}(\mathbb{R}^2) \to E$) with finite variation $|m|$ and

$$|m| = |m_g|, \quad \text{on } \mathcal{R}.$$

If g has finite variation function $|g|$, then

$$|m| = m_{|g|}, \quad \text{on } \mathcal{R}.$$

Proof. To prove assertion a) let $z'_n \downarrow z$ in $\mathbb{R}^2$. Then $(z, z'_n] \downarrow \phi$; since m_g is σ-additive we deduce that $m_g(z, z'_n] \to 0$, that is, $\Delta_{zz'_n} g \to 0$; consequently

$$\lim_{z' \downarrow z} \Delta_{zz'} g = 0,$$

that is, g is incrementally right continuous.

Assume now g is right continuous and has finite variation. From Proposition 45 we deduce that m_g has finite variation $|m_g|$. From Theorem 48, it follows that if g has finite variation function $|g|$, then $|m_g| = m_{|g|}$. Then we can apply the extension theorem 7.4 to prove assertion b). ∎

The measure m obtained from m_g is still denoted by m_g and is called the *Stieltjes measure* associated to g. If $|g|$ is finite, the extension of the measure $m_{|g|}$ from $\mathcal{R}$ to $\mathcal{B}(\mathbb{R}^2)$ is still denoted by $m_{|g|}$ and we have

$$|m_g| = m_{|g|}, \quad \text{on } \mathcal{B}(\mathbb{R}^2).$$

M. The Stieltjes integral

51. Assume $g : \mathbb{R}^2 \to E$ is right continuous and has finite variation and assume $E \subset L(F, G)$.

The space $L_F^1(m_g) = L_F^1(|m_g|)$ of Bochner $|m_g|$-integrable functions $f : \mathbb{R}^2 \to F$ is also denoted $L_F^1(dg)$ or $L_F^1(g)$.

For $f \in L_F^1(dg)$, the Stieltjes integral $\int f\,dg$ is defined by the equality

$$\int f\,dg = \int f\,dm_g.$$

If the variation function $|g|$ is finite, we have $L_F^1(dg) = L_F^1(d|g|)$ and

$$\left|\int f\,dg\right| \leq \int |f|d|g|, \text{ for } f \in L_F^1(dg).$$

For an $|m_g|$-measurable function f defined $|m_g|$-a.e. on $\mathbb{R}^2$ and with values in $[0, \infty]$, we define the Stieltjes integral $\int f d|g|$ by the equality

$$\int f d|g| = \int f d|m_g| \leq \infty$$

§30. PROCESSES WITH FINITE VARIATION

In this paragraph we associate to each process X with integrable variation a stochastic measure μ_X (Theorems 4 and 5) and prove that such a process is summable (Theorem 6) and its stochastic integral can be computed pathwise as a Stieltjes integral (Theorem 8).

A. Processes with integrable variation

In this paragraph we shall denote $\mathcal{M} = \mathcal{B}(\mathbb{R}^2) \times \mathcal{F}$. A process $X : \mathbb{R}^2 \times \Omega \to E$ is said to be *measurable* if it is $\mathcal{M}$-measurable. If X is a right continuous process such that X_z is $\mathcal{F}$-measurable for each $z \in \mathbb{R}^2$, then X is measurable.

1. Definition. *A process $X : \mathbb{R}^2 \times \Omega \to E$ is said to have finite (resp. bounded) variation, if for each $\omega \in \Omega$, the path $z \mapsto X_z(\omega)$ has finite variation on each bounded rectangle $(z, z']$ (resp. finite variation on $\mathbb{R}^2 \times \Omega$).*

If X vanishes outside the first quadrant, to say that X has finite variation means that $X(\omega)$ has finite variation on $(-\infty, z]$ for each $\omega \in \Omega$ and $z \in \mathbb{R}^2$.

2. Definition. *We say that a measurable process $X : \mathbb{R}^2 \times \Omega \to E$ has integrable variation if the total variation $|X|_\infty = \sup_z |X|_z$ is integrable.*
If X has integrable variation, then $X_z \in L^1_E$ for each $z \in \mathbb{R}^2$.

3. Definition. *A set $M \subset \mathbb{R}^2 \times \Omega$ is said to be evanescent if there is a negligible set $A \in \mathcal{F}$ such that $M \subset \mathbb{R}^2 \times A$.*
A σ-additive measure $\mu : \mathcal{M} \to E$ is called a stochastic measure if it vanishes on evanescent sets.

We shall extend for two-parameter processes the properties of one-parameter processes with integrable variation presented in §19.

B. The measure μ_X

We establish first the correspondence $X \mapsto \mu_X$ for incrementally increasing processes.

4. Theorem. *There is a 1-1 correspondence $X \longleftrightarrow \mu_X$ between the set of right continuous, measurable, integrable, incrementally increasing processes X on $\mathbb{R}^2 \times \Omega$ with $X_0 = 0$ and the set of positive, σ-additive, stochastic measures μ_X on $\mathcal{M}$, given by the equality*

$$\mu_X(M) = E(\int 1_M dX_z), \text{ for } M \in \mathcal{M}.$$

If ϕ is a real-valued, measurable process on $\mathbb{R}^2 \times \Omega$, we have

$$\phi \in L^1(\mu_X) \text{ iff } E(\int |\phi_z| dX_z) < \infty \, ;$$

in this case $E(\int \phi_z dX_z)$ is defined and

$$\int \phi d\mu_X = E(\int \phi_z dX_z).$$

Proof. The proof is similar to that of Theorem 19.7. Assume that $X : \mathbb{R}^2 \times \Omega \to \mathbb{R}$ is right continuous, measurable, integrable, and incrementally increasing. We prove first that for every $M \in \mathcal{M}$, the function $\omega \mapsto \int 1_M(z,\omega) dX_z(\omega)$ is integrable. This is true for sets $M = (z, z'] \times A$ with z, z' in $\mathbb{R}^2$ and $A \in \mathcal{F}$. In fact,

$$\int 1_M(z,\omega) dX_z(\omega) = 1_A(\omega)\Delta_{zz'}X(\omega)$$

and $1_A \Delta_{zz'} X \in L^1_E$. Then, by a monotone class argument, the assertion is true for every set $M \in \mathcal{M}$. We set

$$\mu_X(M) = E(\int 1_M dX_z), \text{ for } M \in \mathcal{M}.$$

It is easy to see that μ_X is σ-additive on $\mathcal{M}$ and vanishes on evanescent sets, hence μ_X is a stochastic measure.

Since X is incrementally increasing, the measure μ_X is positive.
For real-valued $\mathcal{M}$-step processes ϕ we have

$$\int \phi d\mu_X = E(\int \phi_z(\omega) dX_z(\omega)).$$

Assume now ϕ is a positive, measurable process and let (ϕ^n) be an increasing sequence of positive $\mathcal{M}$-step processes such that $\phi^n \uparrow \phi$. Then

$$\int \phi^n d\mu_X \nearrow \int \phi d\mu_X \leq +\infty.$$

For each $\omega \in \Omega$, the sequence $(\phi^n(\omega))$ of positive $\mathcal{B}(\mathbb{R}^2)$-step functions is increasing and $\phi^n(\omega) \uparrow \phi(\omega)$; therefore

$$\int \phi^n_z(\omega) dX_z(\omega) \nearrow \int \phi_z(\omega) dX_z(\omega) \leq +\infty,$$

consequently

$$E(\int \phi^n_z dX_z) \nearrow E(\int \phi_z dX_z) \leq +\infty.$$

Since for each n we have

$$\int \phi^n d\mu_X = E(\int \phi^n_z dX_z),$$

passing to limit we obtain

$$\int \phi d\mu_X = E(\int \phi_z dX_z) \leq \infty.$$

It follows that $\phi \in L^1(\mu_X)$ iff $E(\int \phi_z dX_z) < \infty$, and in this case we have

$$\int \phi d\mu_X = E(\int \phi_z dX_z).$$

If ϕ is a real-valued, measurable process, we apply the above reasoning to the positive part ϕ^+ and the negative part ϕ^- of ϕ. We deduce that $\phi \in L^1(\mu_X)$ iff $\phi^+, \phi^- \in L^1(\mu_X)$, iff $E(\int \phi_z^+ dX_z) < \infty$ and $E(\int \phi_z^- dX_z) < \infty$ iff $E(\int |\phi_z| dX_z) < \infty$, and in this case we have

$$\int \phi d\mu_X = \int \phi^+ d\mu_X - \int \phi^- d\mu_X = E(\int \phi_z dX_z).$$

Conversely, assume μ is a positive, finite, stochastic measure on $\mathcal{M}$ and prove that there is a right continuous, measurable, integrable, incrementally increasing process X such that $\mu = \mu_X$. For each $z \in \mathbb{R}^2$ and $A \in \mathcal{F}$ set

$$\mu^z(A) = \mu((-\infty, z] \times A).$$

Then $\mu^z : \mathcal{F} \to \mathbb{R}_+$ is a positive, σ-additive measure. Since μ is a stochastic measure, the measure μ^z vanishes on negligible sets $A \in \mathcal{F}$, hence $\mu^z \ll P$. By the Radon–Nikodym theorem, there is an $\mathcal{F}$-measurable, integrable, positive function Y_z such that

$$\mu^z(A) = \int_A Y_z dP, \text{ for } A \in \mathcal{F}.$$

If $z < z'$, then $\mu^z \leq \mu^{z'}$, hence $Y_z \leq Y_{z'}$, a.s. We have also $\Delta_{zz'} Y \omega) \geq 0$, a.s. In fact, for every set $A \in \mathcal{F}$ and every $z = (s,t) < z' = (s',t')$ we have

$$\begin{aligned}
\int_A \Delta_{zz'} Y dP &= \int_A (Y_z + Y_{z'} - Y_{(s,t')} - Y_{(s',t)}) dP \\
&= \mu^z(A) + \mu^{z'}(A) - \mu^{(s,t')}(A) - \mu^{(s',t)}(A) \\
&= \mu((z,z'] \times A) \geq 0.
\end{aligned}$$

Since $A \in \mathcal{F}$ is arbitrary, we deduce that $\Delta_{zz'} Y(\omega) \geq 0$, a.s. Let $N(z,z')$ be the negligible set such that for $\omega \notin N(z,z')$ we have

$$Y_z(\omega) \leq Y_{z'}(\omega)$$

and

$$\Delta_{zz'} Y(\omega) \geq 0.$$

Let N be the union of the sets $N(z, z')$ with z and z' rational. Then N is negligible, and for $\omega \notin N$ and for all rational points $z < z'$, we have

$$Y_z(\omega) \leq Y_{z'}(\omega)$$

and

$$\Delta_{zz'} Y(\omega) \geq 0.$$

Then, for $\omega \notin N$, $Y_\cdot(\omega)$ has right limit along rational points. For every $z \in \mathbb{R}^2$ set

$$X_z(\omega) = \lim_{u \downarrow z} Y_u(z), \; u \text{ rational, for } \omega \notin N,$$

and

$$X_z(\omega) = 0, \; \text{ for } \omega \in N.$$

Then X is right continuous and for each $z \in \mathbb{R}^2$, X_z is $\mathcal{F}$-measurable.

We prove now that for each $z \in \mathbb{R}^2$ we have $X_z = Y_z$, a.s.

Let $z \in \mathbb{R}^2$ and $A \in \mathcal{F}$. Let (u_n) be a decreasing sequence of rational points of $\mathbb{R}^2$ such that $u_n \downarrow z$. Then

$$\mu((-\infty, u_n] \times A) \to \mu((-\infty, z] \times A),$$

that is,

$$\int_A Y_{u_n} dP \to \int_A Y_z dP.$$

On the other hand, $Y_{u_n}(\omega) \mapsto X_z(\omega)$ for $\omega \notin N$, hence

$$\int_A Y_{u_n} dP \to \int_A X_z dP.$$

We deduce that

$$\int_A Y_z dP = \int_A X_z dP, \text{ for } A \in \mathcal{F},$$

hence $X_z = Y_z$, a.s. Since X_z is $\mathcal{F}$-measurable for each $z \in \mathbb{R}^2$ and X is right continuous, we deduce that X is measurable. It follows also that for $z < z'$ we have $\Delta_{z,z'} X = \Delta_{z,z'} Y$, a.s., hence $\Delta_{zz'} X \geq 0$, a.s., that is, outside a negligible set $N'(z, z')$. Since X is right continuous, we have $\Delta_{zz'} X(\omega) \geq 0$ for ω outside the union N' of the sets $N'(z, z')$ with z, z' rational. If we modify X setting $X_z(\omega) = 0$ for every $z \in \mathbb{R}^2$ and $\omega \in N$, then X is right continuous, measurable, increasing, incrementally increasing, and positive. Finally X_∞ is integrable. In fact,

$$\begin{aligned} E(X_\infty) &= E(\sup_n X_n) = \sup_n E(X_n) \\ &= \sup_n \int X_n dP = \sup_n \mu((-\infty, n] \times \Omega) \\ &= \mu(\mathbb{R}^2 \times \Omega) < \infty. \end{aligned}$$

For each $z \in \mathbb{R}^2$ and $A \in \mathcal{F}$ we have

$$\begin{aligned} \mu((-\infty, z] \times A) &= \mu^z(A) = \int_A X_z dP \\ &= E(\int 1_{(-\infty,z] \times A} dX_z) \\ &= \mu_X((-\infty, z] \times A). \end{aligned}$$

Since both μ and μ_X are σ-additive on $\mathcal{M}$, it follows that $\mu = \mu_X$. Finally, using Property 29.44, we have

$$\mu_X = \mu_{X-X_0},$$

therefore we can choose X such that $X_0 = 0$. ∎

We consider now the correspondence $X \mapsto \mu_X$ for vector-valued processes X with integrable variation.

5. Theorem. *Let $X : \mathbb{R}^2 \times \Omega \to E$ be a right continuous, measurable process with integrable variation $|X|$.*

There exists a stochastic measure $\mu_X : \mathcal{M} \to E$ with finite variation $|\mu_X|$ satisfying the following conditions:
a)

$$\mu_X(M) = E(\int 1_M dX_z), \text{ for } M \in \mathcal{M}$$

and

$$|\mu_X|(M) = E(\int 1_M d|X|_z), \text{ for } M \in \mathcal{M},$$

that is,

$$|\mu_X| = \mu_{|X|}.$$

b) *If $E \subset L(F, G)$ and $H : \mathbb{R}^2 \times \Omega \to F$ is a process, then $H \in L^1_F(\mu_X)$ iff H is μ_X-measurable and $E(\int |H_s| d|X|_s) < \infty$. In this case $E(\int H_s dX_s)$ is defined and*

$$\int H d\mu_X = E(\int H_s dX_s).$$

Proof. We note first that since $|X|_\infty \in L^1$, it follows that X_z and $|X|_z$ are integrable for every $z \in \mathbb{R}^2$.

We prove that for every $M \in \mathcal{M}$, the function $\omega \mapsto \int 1_M(z, \omega) dX_z(\omega)$ is integrable, the same way as in the proof of Theorem 19.3. Set

$$\mu_X(M) = E(\int 1_M dX_z), \text{ for } M \in \mathcal{M}.$$

Then $\mu_X : \mathcal{M} \to E$ is a σ-additive measure. In particular, since the variation $|X|$ is right continuous, measurable, incrementally increasing, and integrable, we associate to $|X|$ the positive measure $\mu_{|X|}$, by the equality

$$\mu_{|X|}(M) = E(\int 1_M d|X|_z), \text{ for } M \in \mathcal{M}.$$

Then, for every $M \in \mathcal{M}$ we have

$$|\mu_X(M)| = |E(\int 1_M dX_z)| \leq E(\int 1_M d|X|_z) = \mu_{|X|}(M),$$

therefore μ_X has finite variation $|\mu_X|$ satisfying $|\mu_X| \leq \mu_{|X|}$. We shall prove now that we have the equality $|\mu_X| = \mu_{|X|}$.

Since X is measurable, it is separably valued, hence we can assume E is separable. Then there is a separable space $U \subset E^*$ norming for E. Since $\mu_X \ll \mu_{|X|}$, by the Radon–Nikodym Theorem 2.34, there is a process $H : \mathbb{R}^2 \times \Omega \to U^*$ such that:

α) For every $u \in U$, the function $\langle H, u \rangle$ is $\mu_{|X|}$-integrable, $\mathcal{M}$-measurable, and

$$\langle \mu_X(M), u \rangle = \int_M \langle H, u \rangle d\mu_{|X|}, \text{ for } M \in \mathcal{M}.$$

β) $|H| \leq 1$, $\mu_{|X|}$-a.e.
γ) $|H|$ is $\mu_{|X|}$-integrable, $\mathcal{M}$-measurable, and

$$|\mu_X|(M) = \int_M |H| d\mu_{|X|}, \text{ for } M \in \mathcal{M}.$$

Taking $M = (-\infty, z] \times A$ with $A \in \mathcal{F}$, we get from theorem 4 that

$$\langle \mu_X(M), u \rangle = \int_M \langle H, u \rangle d\mu_{|X|} = E(1_A \int_{(-\infty,z]} \langle H_v, u \rangle d|X|_v).$$

It follows that

$$\langle X_z, u \rangle = \int_{(-\infty,z]} \langle H_v, u \rangle d|X|_v, \text{ a.s.}$$

outside a negligible set $N(u, z)$.

Let U_0 be a countable dense subset of U. The union N of the sets $N(u, z)$ with $u \in U_0$ and z rational is negligible and, by right continuity, we have

$$\langle X_z(\omega), u \rangle = \int_{(-\infty,z]} \langle H_v(\omega), u \rangle d|X|_v(\omega), \text{ for } z \in \mathbb{R}^2 \text{ and } \omega \notin N.$$

Let $\omega \notin N$. The function $X_{\cdot}(\omega) : \mathbb{R}^2 \to E$ is right continuous and has finite variation $|X(\omega)|_{\cdot} = |X|_{\cdot}(\omega)$. Let $m_{X(\omega)}$ be the Stieltjes measure corresponding to $X.(\omega)$. Then $m_{X(\omega)}$ has finite variation $|m_{X(\omega)}| = m_{|X|(\omega)}$ (Theorem 29.48). We can apply Theorem 2.34 for $m_{X(\omega)} \ll m_{|X|(\omega)}$ and deduce the existence of a function $G_\omega : \mathbb{R}^2 \to U^*$ such that:

α') For every $u \in U$, the function $\langle G_\omega, u \rangle$ is $m_{|X|(\omega)}$-integrable, $\mathcal{B}(\mathbb{R}_+)$-measurable, and

$$\langle m_{X(\omega)}(M), u \rangle = \int_M \langle G_\omega, u \rangle dm_{|X|(\omega)}, \text{ for } M \in \mathcal{B}(\mathbb{R}^2);$$

β') $|G_\omega|$ is $\mathcal{B}(\mathbb{R}^2)$-measurable and

$$|G_\omega| = 1, \, \mu_{|X|(\omega)}\text{-a.e.}$$

Taking $M = (-\infty, z]$ we deduce

$$\langle X_z(\omega), u \rangle = \int_{(-\infty, z]} \langle G_\omega(v), u \rangle d|X|_v(\omega).$$

From the two representations of $\langle X(\omega), u \rangle$ we obtain, for all $z \in \mathbb{R}^2$,

$$\int_{(-\infty, z]} \langle H_v(\omega), u \rangle d|X|_v(\omega) = \int_{(-\infty, z]} \langle G_\omega(v), u \rangle d|X|_v(\omega);$$

therefore there is a $\mu_{|X|(\omega)}$-negligible set $N'(\omega, u) \subset \mathbb{R}^2$ such that

$$\langle H_v(\omega), u \rangle = \langle G_\omega(v), u \rangle, \text{ for } v \notin N'(\omega, u).$$

The set $N'(\omega) = \bigcup \{N'(\omega, u) : u \in U_0\}$ is $\mu_{|X|(\omega)}$-negligible and for $v \notin N'(\omega)$ we have
$$H_v(\omega) = G_\omega(v).$$

Let $A = \{(v, \omega) : |H_v(\omega)| < 1\}$. Then $A \in \mathcal{M}$ and for each $\omega \in \Omega$ we have

$$A(\omega) = \{v : |H_v(\omega)| < 1\} \in \mathcal{B}(\mathbb{R}^2).$$

For $v \notin N'(\omega)$ we have $|H_v(\omega)| = |G_\omega(v)| = 1$, $\mu_{|X|(\omega)}$-a.e.; therefore $A(\omega)$ is $\mu_{|X|(\omega)}$-negligible; consequently

$$\mu_{|X|}(A) = E(\int 1_{A(\omega)}(v) d|X|_v(\omega)) = 0.$$

It follows that $|H_v(\omega)| = 1$, for $v \notin N'(\omega)$, hence, by β), we have, for $M \in \mathcal{M}$,

$$|\mu_X|(M) = \int_M |H| d\mu_{|X|} = E(\int 1_M(v, \omega)|H(\omega)| d|X|(\omega))$$
$$= E(\int 1_M(v, \omega) d|X|_v(\omega)) = \mu_{|X|}(M),$$

that is, $|\mu_X| = \mu_{|X|}$, and assertion a) is proved.

To prove assertion b), let $H : \mathbb{R}^2 \times \Omega \to F$ be a measurable process. Let (ϕ^n) be an increasing sequence of positive $\mathcal{M}$-step processes such that $\phi^n \uparrow |H|$. For each $\omega \in \Omega$, the sections $\phi^n(\omega)$ and $|H(\omega)|$ are Borel functions on $\mathbb{R}^2$ and $\phi^n(\omega) \uparrow |H(\omega)|$; therefore

$$\int \phi_z^n(\omega) d|X|_z(\omega) \nearrow \int |H_z(\omega)| d|X|_z(\omega) \le +\infty.$$

The mapping $\omega \mapsto \int \phi_z^n(\omega) d|X|_z(\omega)$ is $\mathcal{F}$-measurable for each n, hence the mapping $\omega \mapsto \int |H_z(\omega)| d|X|_z(\omega)$ is $\mathcal{F}$-measurable. It follows that

$$\int |H_z| d|X|_z \in L^1(P) \quad \text{iff} \quad E\left(\int |H_z| d|X|_z\right) < \infty.$$

Assume $E(\int |H_z| d|X|_z) < \infty$ and prove that: $H \in L_F^1(\mu_X)$, the integral $E(\int H_z dX_z)$ is defined, and we have the equality

$$\int H d\mu_X = E\left(\int (H_z dX_z)\right).$$

From the assumption $E(\int |H_z| d|X|_z) < \infty$ we deduce that there is a negligible set $N \subset \Omega$ such that for $\omega \notin N$ we have

$$\int |H_z(\omega)| d|X|_z(\omega) < \infty, \quad \text{a.s.}$$

This means that for $\omega \notin N$, the function $H_\cdot(\omega)$ is $m_{|X|(\omega)}$-integrable; since $m_{|X|(\omega)} = |m_{X(\omega)}|$ (Theorem 29.48), it follows that $H_\cdot(\omega)$ is $m_{X(\omega)}$-integrable, hence the integral $\int H_z(\omega) dX_z(\omega)$ is defined for $\omega \notin N$.

Let (H^n) be a sequence of F-valued, $\mathcal{M}$-step functions such that $H^n \to H$ and $|H^n| \le |H|$. Then, for $\omega \notin N$ we have $H^n(\omega) \to H(\omega)$ and $|H^n(\omega)| \le |H(\omega)|$. Since $H(\omega)$ is $m_{X(\omega)}$-integrable, we can apply Lebesgue's theorem and deduce that $H^n(\omega) \to H(\omega)$ in $L_F^1(m_{X(\omega)})$ and $|H^n(\omega)| \to |H(\omega)|$ in $L^1(m_{X(\omega)})$. It follows that

$$\int |H_z^n(\omega) - H_z^m(\omega)| d|X|_z(\omega) \to 0,$$

as $n, m \to \infty$, and

$$\int H_z^n(\omega) dX_z(\omega) \to \int H_z(\omega) dX_z(\omega), \text{ as } n \to \infty.$$

Since for each n the function $\omega \mapsto \int H_z^n(\omega) dX_z(\omega)$ is $\mathcal{F}$-measurable, it follows that the function $\omega \mapsto \int H_z(\omega) dX_z(\omega)$ defined for $\omega \notin N$ and extended with 0 outside N is $\mathcal{F}$-measurable and we have

$$\left|\int H_z(\omega) dX_z(\omega)\right| \le \int |H_z(\omega)| d|X|_z(\omega).$$

§30. PROCESSES WITH FINITE VARIATION 399

Since $E(\int |H_z(\omega)| d|X|_z(\omega)) < \infty$, by hypothesis, we can apply Lebesgue's Theorem and deduce that $\int H_z(\cdot) dX_z(\cdot)$ is integrable and

$$\int H_z^n(\cdot) dX_z(\cdot) \to \int H_z(\cdot) dX_z(\cdot), \text{ in } L_G^1(P);$$

therefore

$$E(\int H_z^n dX_z) \to E(\int H_z dX_z), \text{ in } G.$$

On the other hand, from

$$\int |H_z^n(\omega) - H_z^m(\omega)| d|X|_z(\omega) \le 2 \int |H_z(\omega)| d|X|_z(\omega)) < \infty,$$

we deduce, by Lebesgue's Theorem, that

$$E(\int |H_z^n - H_z^m| d|X|_z) \to 0, \text{ as } n, m \to \infty.$$

For each n and m we have, by the definition of $\mu_{|X|}$,

$$\int |H^n - H^m| d\mu_{|X|} = E(\int |H_z^n - H_z^m| d|X|_z),$$

hence (H^n) is a Cauchy sequence in $L_F^1(\mu_X)$. Since $H^n \to H$ pointwise, it follows that $H \in L_F^1(\mu_X)$ and $H^n \to H$ in $L_F^1(\mu_X)$; therefore

$$\int H^n d\mu_X \to \int H d\mu_X.$$

Since for each n we have

$$\int H^n d\mu_X = E(\int H_z^n dX_z),$$

and since, by the above,

$$E(\int H_z^n dX_z) \to E(\int H_z dX_z)$$

passing to limits in the preceding equality we get

$$\int H d\mu_X = E(\int H_z dX_z).$$

Conversely, assume $H \in L_F^1(\mu_X)$. We can assume that H is $\mathcal{M}$-measurable. We have $|H| \in L^1(\mu_X)$. Let (ϕ^n) be an increasing sequence of $\mathcal{M}$-step processes such that $\phi^n \uparrow |H|$. Then $\phi^n \to |H|$ in $L_F^1(\mu_{|X|})$, hence

$$\int \phi^n d\mu_{|X|} \to \int |H| d\mu_{|X|}.$$

We proved above that we have also

$$\int \phi_v^n d|X|_v(\omega) \nearrow \int |H_v(\omega)| d|X|_v(\omega) \le +\infty,$$

therefore

$$E(\int \phi_v^n d|X|_v(\omega)) \to E(\int |H_v(\omega)| d|X|_v(\omega)) \le +\infty.$$

Since, for each n we have

$$\int \phi^n d\mu_{|X|} = E(\int \phi_v^n d|X|_v)$$

we deduce that

$$\int |H| d\mu_{|X|} = E(\int |H_v(\omega)| d|X|_v(\omega)),$$

hence $E(\int |H_v(\omega)| d|X|_v(\omega)) < \infty$ and assertion b) is proved. ∎

C. Summability of processes with integrable variation

We are able to prove, in the next theorem, that the processes with integrable variation are summable. We prove also the relationship between I_X and μ_X.

6. Theorem. *Let $X : \mathbb{R}^2 \times \Omega \to E$ be a right continuous, cadlag, adapted process with integrable variation $|X|$. Then*
a) *The stochastic measure $I_X : \mathcal{R} \to L_E^1$ can be extended to a σ-additive measure $I_X : \mathcal{P} \to L_E^1$ with bounded variation $|I_X|$.*
b) *X is 1-summable relative to any embedding $E \subset L(F, G)$.*
c) *$\mu_X(B) = E(I_X(B))$, for $B \in \mathcal{P}$*
and

$$|I_X|(B) = |\mu_X|(B) = \mu_{|X|}(B) = E(I_{|X|}(B)), \text{ for } B \in \mathcal{P}.$$

The proof is the same as that of Theorem 19.13.

D. The stochastic integral as a Stieltjes integral

We state first a theorem asserting the density of the $\mathcal{R}$-step processes.

7. Theorem. *Let $X : \mathbb{R}^2 \times \Omega \to E \subset L(F, G)$ be a cadlag, adapted process with integrable variation $|X|$. Then*
a) *The set of measures $(I_X)_{F, L_G^\infty}$ is uniformly σ-additive.*
b) *We have*

$$L_F^1(\mathcal{P}, \mu_X) = L_F^1(\mathcal{P}, I_X) \subset L_{F, L_G^1}^1(X) = \mathcal{F}_{F, L_G^1}(X)$$

and the $\mathcal{R}$-step processes are dense in $L_{L_G^1}^2(X)$.

The proof is the same as that of Theorem 19.14.

The following theorem states the pathwise computation of the stochastic integral as a Stieltjes integral.

8. Theorem. *Let* $X : \mathbb{R}^2 \times \Omega \to E \subset L(F,G)$ *be a cadlag, adapted process, p-summable relative to* (F,G), *and with finite variation* $|X|$. *If* $H \in \mathcal{F}_{F,L_G^p}(X)$ *and if* $\int |H_z(\omega)| d|X|_z(\omega) < \infty$ *for every* $z \in \mathbb{R}^2$ *and* $\omega \in \Omega$, *then* $H \in \hat{L}_{F,L_G^p}(X)$ *and*

$$(H \cdot X)_z(\omega) = \int_{(-\infty, z]} H_z(\omega) dX_z(\omega), \ a.s.$$

The proof is the same as that of Theorem 19.16.

Chapter 10

Two-Parameter Processes with Finite Semivariation

We now extend, in this chapter, for two-parameter processes, the results of Chapter 5 concerning the one-parameter processes with finite semivariation.

§31. FUNCTIONS WITH FINITE SEMIVARIATION IN THE PLANE

In this paragraph we study the functions g with finite semivariation and show that if g is right continuous and with finite semivariation (rather than finite variation) and if c_0 does not belong to the codomain of g, then the associated measure m_g also has finite semivariation and is σ-additive. Using the integration theory of §5 we define the Stieltjes integral by the equality $\int f\,dg = \int f\,dm_g$.

A. Functions with finite semivariation

Let $g : \mathbb{R}^2 \to E \subset L(F, G)$ be a function.

1. Definition. *For any rectangle $R \subset \mathbb{R}^2$, bounded or not, containing or not containing its sides, the semivariation of g on R relative to the pair (F, G) is a number denoted by $sv_{F,G}(g, R)$ or $svar_{F,G}(g, R)$ and defined by*

$$svar_{F,G}(g, R) = \sup |\Sigma \Delta_{R_i} g\, x_i|_G\,,$$

where the supremum is taken for all partitions $P = (R_i)_{i \in I}$ of rectangles from R with vertices in R and all families $(x_i)_{i \in I}$ of elements from F with $|x_i| \leq 1$.

The semivariation function $\tilde{g}_{F,G}\colon \mathbb{R}^2 \to [0,\infty]$ is defined for every $z \in \mathbb{R}^2$ by
$$\tilde{g}_{F,G}(z) = svar_{F,G}(g,(-\infty,z]).$$

If the pair (F,G) is understood we shall write $sv(g,R)$ or $svar(g,R)$ instead of $svar_{F,G}(g,R)$, and $\tilde{g}$ instead of $\tilde{g}_{F,G}$.

We say that g has *finite semivariation relative to* (F,G) if $svar(g,R) < \infty$ for any bounded rectangle R.

We say that g has *bounded semivariation* relative to (F,G) if $svar(g,\mathbb{R}^2) < \infty$.

We say that g has *finite semivariation function* $\tilde{g}_{F,G}$ if $\tilde{g}_{F,G}(z) < \infty$ for every $z \in \mathbb{R}^2$.

We remark that the semivariation $svar_{F,G}(g,R)$ depends on the values of F and G only, but not on the norm of E; therefore it is the same whether the embedding $E \subset L(F,G)$ is an isometry or just continuous.

We list some properties of the semivariation.

2. For any rectangle R, we have
$$svar(g,R) = \sup_Q |\Sigma \Delta_{R_{ij}}(g) x_{ij}|,$$
where the supremum is taken for all the grids $Q = \sigma \times \tau = (R_{ij})$ contained in R and $x_{ij} \in F_1$.

3. If $R \subset R'$ are two rectangles, then
$$svar(g,R) \leq svar(g,R').$$

4. For any rectangle R we have
$$svar_{F,G}(g,R) \leq var(g,R).$$

5. Proposition. *Assume $E \subset L(F,\mathbb{R})$ isometrically. Then*
$$svar_{F,\mathbb{R}}(g,\mathcal{R}) = var(g,\mathcal{R}).$$

Proof. Let $P = (R_i)_{1 \leq i \leq n}$ be a partition with vertices in R and $\epsilon > 0$. For each i there is a point $x_i \epsilon F_1$ with $\langle x_i, \Delta_{R_i} g \rangle \geq 0$ and
$$|\Delta_{R_i} g| < \langle x_i, \Delta_{R_i} g \rangle + \frac{\epsilon}{n}.$$

Then
$$\begin{aligned}\Sigma_i |\Delta_{R_i} g| &< \Sigma_i \langle x_i, \Delta_{R_i} g \rangle + \epsilon \\ &\leq svar_{F,\mathbb{R}}(g,R) + \epsilon.\end{aligned}$$

We deduce that
$$var(g,R) \leq svar_{F,\mathbb{R}}(g,\mathbb{R}).$$

Using Property 4, we get the equality of the statement. ■

6. Corollary. *If $g : \mathbb{R}^2 \to \mathbb{R}$ is a real-valued function and if we consider $\mathbb{R} = L(\mathbb{R}, \mathbb{R})$, then the variation and the semivariation of g are equal.*

7. Proposition. *Let $g : \mathbb{R}^2 \to E \subset (F, G)$ be a function. For any rectangle R we have*
$$svar_{F,G}(g, R) \leq svar_{E^*, \mathbb{R}}(g, I) \leq var(g, I).$$
If the embedding $E \subset L(F, G)$ is an isometry then for any rectangle R we have
$$svar_{\mathbb{R}, E}(g, R) \leq svar_{F,G}(g, R).$$

The proof is similar to that of Proposition 4.12, concerning the semivariation of measures.

B. Semivariation and norming spaces

There is an alternative way of defining the semivariation, using a subspace $U \subset G^*$ which is norming for G. For each $u \in U$ we define the function $g_u \colon \mathbb{R}^2 \to F^*$ by the equality
$$\langle x, g_u(z) \rangle = \langle g(z)x, u \rangle, \text{ for } x \in F \text{ and } z \in \mathbb{R}^2.$$

In particular, considering the embedding $E = L(\mathbb{R}, E)$, for every $x^* \in E^*$ we have
$$g_{x^*}(z) = x^* g(z) = \langle g(z), x^* \rangle, \text{ for } z \in \mathbb{R}^2 .$$
Considering $g_u : \mathbb{R}^2 \to F^* = L(F, \mathbb{R})$, by Proposition 5 we have, for any rectangle R,
$$svar_{F, \mathbb{R}}(g_u, R) = var(g_u, R).$$

8. Proposition. *For every rectangle R we have*
$$svar_{F,G}(g, R) = \sup_{u \in U_1} var(g_u, R).$$
In particular, for any $z \in \mathbb{R}^2$ we have
$$svar_{F,G}(g, (-\infty, z]) = \sup_{u \in U_1} var(g_u, (-\infty, z]).$$

The proof is similar to that of Proposition 4.13.

We defined the semivariation function $\tilde{g}_{F,G}$ by
$$\tilde{g}_{F,G}(z) = svar_{F,G}(-\infty, z]),$$
for $z \in \mathbb{R}^2$.

From Properties 3–8 we deduce the corresponding properties for the semivariation function $\tilde{g}$.

3'. If $z \le z'$ in $\mathbb{R}^2$, then $\tilde{g}(z) \le \tilde{g}(z')$.
4'. $\tilde{g}(z) \le |g|(z)$, for any $z \in \mathbb{R}^2$.
5'. If $E \subset L(F, \mathbb{R})$ isometrically, then

$$\tilde{g}_{F,\mathbb{R}}(z) = |g|(z) \text{, for } z \in \mathbb{R}^2.$$

7'. $\tilde{g}_{F,G}(z) \le \tilde{g}_{E^*,\mathbb{R}}(z) = |g|(z)$, for $z \in \mathbb{R}^2$.
If the embedding $E \subset L(F, G)$ is an isometry, then

$$\tilde{g}_{\mathbb{R},E}(z) \le \tilde{g}_{F,G}(z), \text{ for } z \in \mathbb{R}^2.$$

8'. $\tilde{g}_{F,G}(z) = \sup_{u \in U_1} |g_u|(z)$, for $z \in \mathbb{R}^2$.
In particular, considering the embedding $E = L(\mathbb{R}, E)$ we have

$$\tilde{g}_{\mathbb{R},E}(z) = \sup_{x^* \in E_1^*} |x^* g|(z) \text{, for } z \in \mathbb{R}^2.$$

C. The measure associated to a function

Let $g : \mathbb{R}^2 \to E \subset L(F, G)$ be a function and consider the additive measure $m_g : \mathcal{R} \to E$ defined by

$$m_g(R) = \Delta_R g$$

for $R \in \mathcal{R}$.

9. For any $u \in G^*$ we have

$$(m_g)_u = m_{g_u}.$$

In particular, for the embedding $E = L(\mathbb{R}, E)$ and for any $x^* \in E^*$ we have

$$m_{x^* g} = x^* m_g.$$

The proof is straightforward.

10. Proposition. m_g has finite (resp. bounded) semivariation $(\tilde{m}_g)_{F,G}$ iff g has finite (resp. bounded) semivariation relative to (F, G).
For any rectangle $R \in \mathcal{R}$ we have

$$svar(g, R) \le svar(m_g, R) \le svar(g, \overline{R}).$$

Proof. For any $u \in G^*$, we have, by Proposition 29.45,

$$var(g_u, R) \le var(m_{g_u}, R) \le var(g_u, \overline{R}).$$

Taking the supremum for $|u| \le 1$ and using Proposition 8, Property 9, and Proposition 4.13, we get the desired inequalities. ∎

11. Proposition. Let $R \subset \mathbb{R}^2$ be a rectangle. We have the equality

$$svar(m_g, R) = svar(g, R)$$

in each of the following two cases:
a) $\inf R = -\infty$ or $\inf R \in R$;
b) g is right continuous.

Proof. By Property 29.46, in each of these two cases we have

$$var(m_{g_u}, R) = var(g_u, R),$$

for $u \in G^*$ and we take the supremum for $|u| \leq 1$. ∎

The following proposition states a property of $svar(g, R)$, but the proof involves the measure m_g.

12. Proposition. *Assume $U \subset G^*$ is a closed space, norming for G. Then*

a) $svar_{F,G}(g, R) < \infty$ *for any* $R \in \mathcal{R}$,
iff
b) $svar(g_u, R) < \infty$ *for any* $u \in U$ *and* $R \in \mathcal{R}$.

Proof. Assume b). By Proposition 29.45 we have

$$var(m_{g_u}, R) \leq var(g_u, \overline{R}) < \infty$$

for any $u \in G^*$ and $R \in \mathcal{R}$. Then, by Proposition 4.15 we deduce that

$$svar_{F,G}(m_{g_u}, R) < \infty$$

for $R \in \mathcal{R}$; consequently, by Proposition 10,

$$svar_{F,G}(g, R) < \infty,$$

for $R \in \mathcal{R}$, which is a).

Conversely, assume a). Then b) follows from Proposition 8, for $u \in U_1$ and then for any $u \in U$, since $g_{\alpha u} = \alpha g_u$. ∎

The following extension theorem of m_g is the analog of the Extension Theorem 29.50; but in case g has finite semivariation (rather than finite variation) we have to impose the condition $c_0 \not\subset E$.

Denote by $\mathcal{D}$ the δ-ring of the bounded Borel subsets of $\mathbb{R}^2$.

13. Theorem. *Assume $c_0 \not\subset E$ and let $g : \mathbb{R}^2 \to E$ be a function with finite (resp. bounded) semivariation relative to $(\mathbb{R}, E)$. Assume there is a space $U \subset E^*$ norming for E such that for every $x^* \in U$, the function x^*g is right continuous.*

Then the measure $m_g : \mathcal{R} \to E$ can be extended uniquely to a σ-additive measure $m : \mathcal{D} \to E$ (resp. $m : \mathcal{B}(\mathbb{R}^2) \to E$) with finite semivariation $\tilde{m}_{\mathbb{R}, E}$ on $\mathcal{D}$ (resp. on $\mathcal{B}(\mathbb{R}^2)$) and

$$\tilde{m}_{\mathbb{R}, E} = (\tilde{m}_g)_{\mathbb{R}, E}, \text{ on } \mathcal{R}.$$

If, in addition, $E \subset L(F,G)$ and if g has finite semivariation relative to (F,G), then m has finite semivariation $\tilde{m}_{F,G}$ on $\mathcal{D}$ (resp. on $\mathcal{B}(\mathbb{R}^2)$) and

$$\tilde{m}_{F,G} = (\tilde{m}_g)_{F,G} \ , \ on \ \mathcal{R}.$$

Proof. By Property 11, for any rectangle R we have

$$(\tilde{m}_g)_{\mathbb{R},E}(R) = svar_{\mathbb{R},E}(g, R).$$

Hence m_g has finite semivariation $(\tilde{m}_g)_{\mathbb{R},E}$ on $\mathcal{R}$; therefore m_g is locally bounded on $\mathcal{R}$. If g has bounded semivariation relative to $(\mathbb{R}, E)$, then m_g has bounded semivariation $(\tilde{m}_g)_{\mathbb{R},E}$ on $\mathcal{R}$, hence m_g is bounded on $\mathcal{R}$. For each $x^* \in U$, the real function x^*g is right continuous and has finite variation and $var(x^*g, R) \leq svar_{\mathbb{R},E}(g, R)$. By the Extension Theorem 29.50, the corresponding measure m_{x^*g} can be extended to a σ-additive measure on $\mathcal{D}$. In particular, m_{x^*g} is σ-additive on $\mathcal{R}$. By Property 9 we have $m_{x^*g} = x^*m_g$; therefore x^*m_g is σ-additive on $\mathcal{R}$, for every $x^* \in E^*$.

We can now apply the Extension Theorem 7.8 and deduce that m_g can be extended to a σ-additive measure $m : \mathcal{D} \to E$. If the variation function $\tilde{g}_{\mathbb{R},E}$ is bounded, then m_g is bounded and then m_g can be extended to a σ-additive measure $m : \mathcal{B}(\mathbb{R}^2) \to E$.

From Theorems 4.20 and 4.21, we deduce that m has finite semivariation $\tilde{m}_{\mathbb{R},E}$ and by Proposition 4.19 we have

$$\tilde{m}_{\mathbb{R},E} = (\tilde{m}_g)_{\mathbb{R},E} \ , \ on \ \mathcal{R}.$$

If g has finite semivariation $\tilde{g}_{F,G}$ then, by Property 11, m_g has finite semivariation $(\tilde{m}_g)_{F,G}$ on $\mathcal{R}$. Then by Corollary 7.6, m has finite semivariation $\tilde{m}_{F,G}$ on $\mathcal{D}$ or on $\mathcal{B}(\mathbb{R}^2)$ and we have

$$\tilde{m}_{F,G} = (\tilde{m}_g)_{F,G}, \ on \ \mathcal{R}.$$

∎

14. Remark. The measure m is still denoted by m_g and is called the *Stieltjes measure* induced by the function g with finite semivariation.

We emphasize that if $c_0 \subset E$, the theorem is not necessarily true, that is, a function g with finite semivariation and right continuous does not necessarily have a Stieltjes measure corresponding to it.

D. The Stieltjes integral for functions with finite semivariation in $\mathbb{R}^2$

15. Let $g : \mathbb{R}^2 \to E \subset L(F,G)$ be a *right continuous* function with finite semivariation relative to (F,G). Assume the corresponding measure $m_g : \mathcal{R} \to E$ can be extended to a σ-additive measure $m = m_g : \mathcal{D} \to E \subset L(F,G)$ with finite semivariation $\tilde{m}_{F,G}$, where $\mathcal{D}$ is the δ-ring of bounded Borel subsets of

$\mathbb{R}^2$. By Theorem 13, this can be done if $c_0 \not\subset E$ and if g has finite semivariation relative to $(\mathbb{R}, E)$; or if $c_0 \not\subset E$ and the embedding $E \subset L(F, G)$ is an isometry, since in this case $svar_{\mathbb{R},E}(g, R) \leq svar_{F,G}(g, R) < \infty$ for every bounded rectangle R.

We can then apply the integration theory of §5. Let $\mathcal{F}_{F,G}(m)$ be the space of Borel measurable functions $f : \mathbb{R}^2 \to F$ with

$$\tilde{m}_{F,G}(f) = \sup_{u \in U_1} \int |f| d|m_u| < \infty.$$

Then $\mathcal{F}_{F,G}(m)$ is a complete vector space for the seminorm $\tilde{m}_{F,G}(f)$ and we have

$$\mathcal{F}_{F,G}(m) \subset \bigcap_{u \in G_1^*} L_F^1(|m_u|).$$

For $f \in \mathcal{F}_{F,G}(m)$ we define the integral $\int f dm \in G^{**}$ satisfying

$$\langle \int f dm, u \rangle = \int f dm_u,$$

for $u \in G^*$ and

$$|\int f dm| \leq \tilde{m}_{F,G}(f).$$

For $f \in \mathcal{F}_{F,G}(m)$ we define the Stieltjes integral $\int f dg$ by the equality

$$\int f dg = \int f dm_g.$$

We denote further by $L_{F,G}^1(m)$, or $L_{F,G}^1(g)$, or $L_{F,G}^1(dg)$, the subspace of $\mathcal{F}_{F,G}(m)$ consisting of functions f with $\int f dg \in G$.

16. Assume g is right continuous and has finite variation. Then g has finite semivariation relative to any embedding $E \subset L(F, G)$ and

$$svar_{F,G}(g, R) \leq var(g, R).$$

The corresponding measure $m : \mathcal{D} \to E$ has finite variation $|m|$ and finite semivariation $\tilde{m}_{F,G}$ satisfying $\tilde{m}_{F,G} \leq |m|$. Then $L_F^1(|m|) \subset L_{F,G}^1(m)$ and

$$\tilde{m}_{F,G}(f) \leq \int |f| d|m|,$$

for $f \in L_F^1(|m|)$. For a function $f \in L_F^1(g)$, the Stieltjes integral $\int f dg$ is the same, whether we consider $f \in L_F^1(|m|)$ or $f \in L_{F,G}^1(m)$.

17. For any Borel functions $f : \mathbb{R}^2 \to F$ we shall write $\tilde{g}_{F,G}(f) = (\tilde{m}_g)_{F,G}(f)$. If the pair (F, G) is understood, we write $\tilde{g}(f)$ instead of $\tilde{g}_{F,G}(f)$. For a rectangle $R = (-\infty, z]$ we have then

$$\tilde{g}(1_R) = \tilde{g}(z).$$

In particular,

$$\tilde{g}(1_{\mathbb{R}^2}) = \tilde{g}(+\infty).$$

§32. PROCESSES WITH FINITE SEMIVARIATION IN THE PLANE

In this paragraph we show that, under the assumption $c_0 \not\subset E$ and $c_0 \not\subset G$, a process X with integrable semivariation is summable, the stochastic integral can be computed pathwise as a Stieltjes integral, and a measure μ_X can be associated to X, exactly as in the case of processes on the line, with integrable semivariation.

A. Processes with finite semivariation

Let $X : \mathbb{R}^2 \times \Omega \to E \subset L(F,G)$ be a process.

1. Definition. *We say that the process X has finite (resp. bounded) semivariation relative to (F,G) if, for every $\omega \in \Omega$, the path $z \mapsto X_z(\omega)$ has finite (resp. bounded) semivariation $\tilde{X}(\omega)$ relative to (F,G), where*

$$\tilde{X}_z(\omega) = svar_{F,G}(X_\cdot(\omega), (-\infty, z])$$

for $z \in \mathbb{R}^2$.

2. Definition. *We say that a measurable process X has p-integrable semivariation relative to (F,G) if $\tilde{X}_\infty \in L^p$.*

3. Remark. Let $X : \mathbb{R}^2 \times \Omega \to E \subset L(F,G)$ be a right continuous, measurable process with bounded semivariation relative to (F,G) and assume $c_0 \not\subset E$. Then, for each $\omega \in \Omega$, we denote by $m_{X(\omega)}$ the σ-additive measure $m_{X(\omega)} : \mathcal{B}(\mathbb{R}^2) \to E$ with finite semivariation $\tilde{m}_{X(\omega)}$ relative to (F,G), satisfying

$$m_{X(\omega)}(z, z'] = \Delta_{zz'} X(\omega), \text{ for } z < z' \text{ in } \mathbb{R}^2.$$

We have
$$m_{X(\omega)}(-\infty, z] = X_z(\omega).$$

For any measurable set M and $\omega \in \Omega$ we denote by $M(\omega)$ the section of M along ω. We have

$$\tilde{m}_{X(\omega)}(M(\omega)) = svar(X_\cdot(\omega), M(\omega))$$

in case $M(\omega)$ is a rectangle. In particular

$$\tilde{m}_{X(\omega)}(-\infty, z] = \tilde{X}_z(\omega).$$

For any measurable process $H : \mathbb{R}^2 \times \Omega \to F$ and any $\omega \in \Omega$, we denote by $H(\omega)$ the path $z \mapsto H_z(\omega)$ and we denote further

$$\tilde{X}(H(\omega)) = \tilde{m}_{X(\omega)}(H(\omega)) = \tilde{m}_{X(\omega)}(|H(\omega)|).$$

The following theorem ensures the measurability of $\tilde{X}(H(\omega))$.

4. Theorem. *Assume $c_0 \not\subset E$ and $c_0 \not\subset G$. Let $X : \mathbb{R}^2 \times \Omega \to E$ be a right continuous, measurable process with bounded semivariations $\tilde{X}_{\mathbb{R},E}$ and $\tilde{X}_{F,G}$ and let $H : \mathbb{R} \times \Omega \to F$ be a measurable process. Then:*
a) *The function $\omega \mapsto \tilde{X}_{F,G}(H(\omega)) \leq \infty$ is $\mathcal{F}$-measurable.*
a′) *If F or G is separable and $\phi : \mathbb{R}^2 \times \Omega \mapsto \mathbb{R}$ is a measurable process, then the function $\omega \mapsto \tilde{X}_{F,G}(\phi(\omega))$ is $\mathcal{F}$-measurable.*
b) *If $\tilde{X}_{F,G}(H(\omega)) < \infty$ for every $\omega \in \Omega$, then the function $\omega \mapsto \int H_z(\omega) dX_z(\omega)$ is $\mathcal{F}$-measurable and has values in G.*
c) *If $E(\tilde{X}_{F,G}(H(\omega))) < \infty$, then the function $\omega \mapsto \int H_z(\omega) dX_z(\omega)$ is integrable and has values in G.*

The proof is similar to that of Theorem 21.6.

B. The measure μ_X

We have the following analog of Theorem 30.5, for two-parameter processes with integrable semivariation:

5. Theorem. *Assume $c_0 \not\subset E$. Let $X : \mathbb{R}^2 \times \Omega \to E \subset L(F, G)$ be a right continuous, measurable process with integrable semivariation relative to $(\mathbb{R}, E)$. Then there is a σ-additive stochastic measure $\mu_X : \mathcal{M} \to E$ satisfying the following conditions:*
a) $\mu_X(M) = E(\int 1_M dX_z)$, *for $M \in \mathcal{M}$;*
b) *If $\phi : \mathbb{R}^2 \times \Omega \to \mathbb{R}$ is a measurable process with $E(\tilde{X}_{\mathbb{R},E}(\phi)) < \infty$, then $\phi \in \mathcal{F}_{\mathbb{R},E}(\mu_X)$. In this case the integral $E(\int \phi_z dX_z)$ is defined and we have*

$$\int \phi d\mu_X = E(\int \phi_z dX_z) \in E$$

and

$$(\tilde{\mu}_X)_{\mathbb{R},E}(\phi) \leq E(\tilde{X}_{\mathbb{R},E}(\phi(\omega))).$$

c) *Assume that X has integrable semivariation $\tilde{X}_{F,G}$. Then μ_X has finite semivariation relative to (F, G).*
If $H : \mathbb{R}^2 \times \Omega \to F$ is a measurable process with

$$E(\tilde{X}_{F,G}(H)) < \infty,$$

then

$$H \in \mathcal{F}_{F,G}(\mu_X).$$

In this case $E(\int H_z dX_z)$ is defined and we have

$$\int H d\mu_X = E(\int H_z dX_z) \in G$$

and

$$(\tilde{\mu}_X)_{F,G}(H) \leq E(\tilde{X}_{F,G}(H(\omega))).$$

The proof is the same as that of Theorem 21.8, replacing adequately the notations: the independent variable $s \in \mathbb{R}$ is replaced by $z \in \mathbb{R}^2$ and the variable $z \in G^*$ is replaced by $u \in G^*$.

C. Summability of processes with integrable semivariation

6. Theorem. *Assume $c_0 \not\subset E$ and $c_0 \not\subset G$. Let $X : \mathbb{R}^2 \times \Omega \to E \subset L(F,G)$ be a cadlag, adapted process with p-integrable semivariations $\tilde{X}_{\mathbb{R},E}$ and $\tilde{X}_{F,G}$. Then:*
a) *X is p-summable relative to (F,G).*
b) *$I_X(M)(\omega) = \int 1_M(z,\omega) dX_z(\omega)$, for $M \in \mathcal{P}$.*
c) *For every $M \in \mathcal{P}$ we have*

$$\mu_X(M) = E(I_X(M)).$$

d) *For every $M \in \mathcal{P}$ we have*

$$(\tilde{\mu}_X)_{F,G}(M) \leq (\tilde{I}_X)_{F,L_G^p}(M) \leq \|\tilde{X}_{F,G}(M(\omega))\|_p.$$

Let $H : \mathbb{R}^2 \times \Omega \to F$ be a predictable process such that the function $\omega \mapsto (\tilde{X}_{F,G})_\infty(H(\omega))$ belongs to L^p. Then:
a') *$H \in \mathcal{F}_{F,G}(\mu_X) \cap \mathcal{F}_{F,G}(X)$.*
b') *$\int H dI_X \in L_G^p$, $\int H_z(\omega) dX_z(\omega) \in G$, a.s.*
and

$$\left(\int H dI_X\right)(\omega) = \int H_z(\omega) dX_z(\omega), a.s.$$

b'') *We have $H \in L_{F,G}^1(X)$ iff the process $(\int_{(-\infty,z]} H_z(\omega) dX_z(\omega))_{z \in \mathbb{R}^2}$ has a cadlag modification. In this case we have*

$$(H \cdot X)_z(\omega) = \int_{(-\infty,z]} H_z(\omega) dX_z(\omega), a.s., \text{ for every } z \in \mathbb{R}^2.$$

c') *$\int H d\mu_X \in G$ and $\int H d\mu_X = E(\int H dI_X)$.*
d') *$(\tilde{\mu}_X)_{F,G}(H) \leq (\tilde{I}_X)_{F,L_G^p}(H) \leq \|\tilde{X}_{F,G}(H(\omega))\|_p.$*

The proof is similar to that of Theorem 21.12.

References

Ba.   D. Bakry, *Limites "quadrantales" des martingales*, Lecture Notes in Math. 863, Springer, Berlin Heidelberg, 1981.

Bar.   R. G. Bartle, *A general bilinear vector integral*, Studia Math. **15** (1956), 337–352.

B-D-S.   R. G. Bartle, N. Dunford, and J. Schwartz, *Weak compactness and vector measures*, Canad. J. Math. **7** (1955), 289–305.

B-P.   C. Bessaga and A. Pelczynski, *On bases and unconditional convergence of series in Banach spaces*, Studia Math. **5** (1974), 151–164.

Bi.   K. Bichteler, *Stochastic integration and L^p theory of semimartingales*, Annals of Prob. **9** (1981), 49–89.

Bo.   S. Bochner, *Integration von Funktionen, deren Werte die Elemente eines Vektorraumes sind*, Fund. Math. **20** (1933), 262–276.

Bou.   N. Bourbaki, *Intégration*, Hermann, Paris, 1952–1959, chap I–VI.

B-D.1.   J. K. Brooks and N. Dinculeanu, *Strong additivity, absolute continuity and compactness in spaces of measures*, J. Math. Analysis and Appl. **45** (1974), 156–175.

B-D.2.   J. K. Brooks and N. Dinculeanu, *Lebesgue-type spaces for vector integration, linear operators, weak completeness and weak compactness*, J. Math. Analysis and Appl. **54** (1976), 348–389.

B-D.3. J. K. Brooks and N. Dinculeanu, *Projections and regularity of abstract processes*, Stochastic Analysis and Appl. **5** (1987), 17–25.

B-D.4. J. K. Brooks and N. Dinculeanu, *Regularity and the Doob-Meyer decomposition of abstract quasimartingales*, Seminar on Stoch. Proc., Birkhäuser, 1988, 21–63.

B-D.5. J. K. Brooks and N. Dinculeanu, *Itô formula for stochastic processes in Banach spaces*, Conference on diffusion processes and related problems in Analysis, Birkhäuser, 1990, 349–397.

B-D.6. J. K. Brooks and N. Dinculeanu, *Stochastic integration in Banach spaces*, Seminar on Stoch. Proc., Birkhäuser, 1991, 27–115.

B-L. J. K. Brooks and P. W. Lewis, *Linear operators and vector measures*, Trans. AMS **144** (1974), 139–162.

C.1. R. Cairoli, *Une inégalité pour processus à indices multiples et ses applications*, Lecture Notes 124, Sem. Prob. IV, Springer, Berlin Heidelberg, 1970.

C.2. R. Cairoli, *Décomposition de processus à indices doubles*, Lecture Notes 191, Sem. Prob. V, Springer, Berlin Heidelberg, 1969–70.

C-W. R. Cairoli and J. Walsh, *Stochastic integrals in the plane*, Acta Math. **134** (1975), 111–183.

Cha. S. D. Chatterji, *Martingales of Banach-valued random variables*, Bull AMS **66** (1960), 395–398.

Cho. Nhansook Cho, *Weak convergence of stochastic integrals driven by martingale measures*, Stochastic Proc. and Their Appl. **59** (1995), 55–79.

D-M. C. Dellacherie and P. A. Meyer, *Probabilités et Potentiel*, Hermann, Paris, 1975–1980.

D-U. J. Diestel and J. Uhl Jr., *Vector measures*, AMS Mathematical Surveys **15**, Providence, RI, 1977.

D.1. N. Dinculeanu, *Vector Measures*, Pergamon Press, Oxford, 1967.

D.2. N. Dinculeanu, *Weak compactness and uniform convergence of operators in spaces of Bochner integrable functions*, J. Math. Analysis and Appl. **109** (1985), 372–387.

D.3. N. Dinculeanu, *Vector-valued stochastic processes I. Vector measures and vector valued stochastic processes with finite variation*, J. Theoretical Prob. **1** (1988), 149–169.

D.4. N. Dinculeanu, *Vector-valued stochastic processes V. Optional and predictable variation of stochastic measures and stochastic processes*, Proc AMS **104** (1988), 625–631.

D.5. N. Dinculeanu, *Intégrale Stochastique des processus à semivariation finie*, CRAS, Paris **326** (1998), 343–346.

D.6. N. Dinculeanu, *Stochastic processes with finite semivariation in Banach spaces and their stochastic integral*, Rendiconti del Circolo Mat. di Palermo **48** (1999), 365-400.

D.7. N. Dinculeanu, *Stochastic Integration for abstract, two parameter processes I. Stochastic Processes with finite semivariation*, Rendiconti del Circolo Mat. di Palermo **48** (1999).

D.8. N. Dinculeanu, *Stochastic integration for abstract, two parameter processes II. Square integrable martingales in Hilbert spaces*, Stochastic Analysis and Appl. (to appear).

D.9. N. Dinculeanu, *Stochastic integral of process measures in Banach spaces I. Process measure with integrable variation*, Rendiconti Accad. Naz. Sci., Memorie di Mat. e Appl. **22** (1998), 85-128.

D.10. N. Dinculeanu, *Stochastic integral of process measures in Banach spaces II. Process measures with integrable semivariation*, Rendiconti Accad. Naz. Sci., Memorie di Mat. e Appl. **22** (1998), 129-164.

D-F. N. Dinculeanu and C. Foiaş, *Sur la représentation intégrale de certaines operations linéaires IV*, Canad. J. Math. **13** (1961), 529–556.

D-K. N. Dinculeanu and I. Kluvanek, *On vector measures*, Proc. London Math. Soc. **17** (1967), 505–512.

Di-Mu. N. Dinculeanu and M. Muthiah, *Stochastic integral of process measures in Banach spaces III. Orthogonal martingale measures*, Rendiconti Accad. Naz. Sci., Memorie di Mat. e Appl. **23** (1999).

Di-U. N. Dinculeanu and J. J. Uhl Jr., *A unifying Radon–Nikodym theorem for vector measures*, J. Multivariate Analysis **3** (1973), 184–203.

Do.1. I. Dobrakov, *On integration in Banach spaces I*, Czech. Math. J. **20(95)** (1970), 511–536.

Do.2. I. Dobrakov, *On integration in Banach spaces II*, Czech. Math. J. **21(96)** (1970), 680–695.

Db. J. L. Doob, *Stochastic Processes*, Wiley, New York, 1953.

Dz. M. Dozzi, *On the Decomposition and integration of two parameter stochastic processes*, Lecture Notes 863, Springer, Berlin Heidelberg, 1980, 162–171.

D-P.   N. Dunford and B. J. Pettis, *Linear operations on summable functions*, Trans AMS **47** (1940), 323–392.

D-S.   N. Dunford and J. Schwartz, *Linear Operators. Part I. General Theory*, Wiley Interscience, New York, 1958.

E.   R. Elliot, *Stochastic Calculus and Applications*, Springer, New York, 1982.

Fi.   D. L. Fisk, *Quasimartingales*, Trans AMS **120** (1965), 369–389.

Fo.   H. Fölmer, *Quasimartingales à deux indices*, CRAS, Paris **288** (1979), 61–63.

G-P.   B. Gravereaux and J. Pellaumail, *Formule d'Itô des processus à valeurs dans des espaces de Banach*, Ann. Inst. H. Poincaré **10** (1974), 399–422.

Go.   M. Gowurin, *Über die Stieltjessche Integration abstrakter Funktionen*, Fund. Math. **27** (1936), 255–268.

Gr.   M. Green, *Planar Stochastic Integration Relative to Quasimartingales*, Real and Stochastic Analysis, CRC, Boca Raton, FL, (1997), 65–157.

H.   P. R. Halmos, *Measure Theory*, Van Nostrand, New York, 1950.

IT.   A. and C. Ionescu Tulcea, *Topics in the Theory of Lifting*, Springer, New York, 1969.

Ka.   G. Kallianpur, *Stochastic Filtering Theory*, Springer, New York, 1980.

K-M.   N. Karoui and S. Méléard, *Martingale measures and stochastic calculus*, Prob. Theory and Related Fields **84** (1990), 83–101.

Kl.1. I. Kluvanek, *On the theory of vector measures I*, Mat. Fyz. Časopis **11** (1961), 173–192.

Kl.2. I. Kluvanek, *On the theory of vector measures II*, Mat. Fyz. Časopis **11** (1961), 76–81.

Kl.3. I. Kluvanek, *Completion of vector measures*, Rev. Roumaine Math. Pures Appl. **12** (1967), 1483–1488.

Ku.   H. Kunita, *Stochastic integrals based on martingales taking their values in Hilbert spaces*, Nagoya Math. J. **38** (1970), 41–52.

K-W.   H. Kunita and S. Watanabe, *On square integrable martingales*, Nagoya Math. J. **30** (1967), 209–245.

Kus.1. A.U. Kussmaul, *Stochastic Integration and Generalized Martingales*, Pittman, London, 1977.

Kus.2. A.U. Kussmaul, *Regularität und Stochastische Integration von Semimartingalen mit Werten in einen Banach Raum*, Dissertation, Stuttgart, 1978.

Kw. S. Kwapien, *On Banach spaces containing c_0*, Studia Math. **52** (1974), 187–188.

L. Ch. Lindsey, *Two parameter stochastic processes with finite variation*, PhD Dissertation, University of Florida, 1988.

Me. M. Métivier, *Semimartingales*, de Gruyter, Berlin, 1982.

M-P.1. M. Métivier and J. Pellaumail, *Stochastic Integration*, Academic Press, New York, 1980.

M-P.2. M. Métivier and J. Pellaumail, *On Doleans–Follmer's measure for quasi-martingales*, Illinois J. Math. **17** (1975), 491–504.

M. P.A. Meyer, *Théorie élémentaire des processus à deux indices*, Lecture Notes 863, Springer, Berlin Heidelberg, 1981, pp.1–39.

N.1. J. Neveu, *Martingales à Temps Discret*, Masson, Paris, 1972.

N.2. J. Neveu, *Processus Aléatoires Gaussiens*, Les Presses de l'Univ. de Montreal, 1968.

Ni.1. O. M. Nikodym, *Sur la généralisation des intégrales de M. J. Radon*, Fund. Math. **15** (1930), 131–179.

Ni.2. O. M. Nikodym, *Sur les suites convergentes de fonctions parfaitement additives d'ensemble abstrait*, Monatshefte Math. Phys. **40** (1933), 427–432.

O. S. Orey, *F-processes*, Proc. 5th Berkeley Symposium on Math. Statistics and Prob. II, Vol I, Univ. of California Press, Berkeley, CA, 1965, 301–314.

P. J. Pellaumail, *Sur l'Intégrale Stochastique et la Décomposition de Doob–Meyer*, SMF, Asterisque **9** (1973).

Pe. B. J. Pettis, *On integration in vector spaces*, Trans AMS **44** (1938), 277–304.

Ph. R. S. Phillips, *On linear transformations*, Trans. AMS **48** (1940), 516–541.

Pra. M. Pratelli, *Intégration Stochastique et Géometrie des espaces de Banach*, Lecture Notes, Seminaire de Prob., Springer, Berlin Heidelberg, 1988.

Pro.   P. Protter, *Stochastic Integration and Differential Equations*, Springer, New York, 1990.

R.   J. Radon, *Über linear Funktional Transformationen und Funktional Gleichungen*, S-B Akad. Wiss. Wien **128** (1919), 1083–1121.

Rd.   E. Radu, *Mesures Stieltjes vectorielles sur $\mathbb{R}^n$*, Bull. Math. Soc. Sci. Math. Roumanie **9** (1965), 129–136.

KR.1.   K. M. Rao, *On decomposition theorems of Meyer*, Math. Scand. **24** (1969), 66-78.

KR.2.   K. M. Rao, *Quasimartingales*, Math. Scand. **24** (1969), 79–92.

MR.1.   M. M. Rao, *Decomposition of vector measures*, Proc. Nat. Acad. Sci. USA **15** (1964), 771–774

MR.2.   M. M. Rao, *Stochastic Processes and Integration*, Sijthoff and Noordhoff, Alphen aan den Rijn, The Netherlands, 1979.

MR.3.   M. M. Rao, *Foundations of Stochastic Analysis*, Academic Press, New York, 1981.

MR.4.   M. M. Rao, *Stochastic Integration: a unified approach*, CRAS, Paris **314** (1992), 629–633.

MR.5.   M. M. Rao, *Stochastic Processes: General Theory*, Kluwer Academic Pub., Dordrecht, the Nederlands, 1995.

S.   L. Stoica, *On two parameter semimartingales*, ZW **45** (1978), 257–258.

W.1.   J. Walsh, *Martingales with a Multidimensional Parameter and Stochastic Integrals in the Plane*, Cours de 3e Cycle, Paris VI, 1977.

W.2.   J. Walsh, *Convergence and regularity of multiparameter strong martingales*, ZW **46** (1979), 177–192.

W.3.   J. Walsh, *An Introduction to Stochastic Partial Differential Equations*, Lecture Notes 1180, Ecole d'Eté de Prob. de Saint Flour XIV, Springer, Berlin Heidelberg, 1985, 265–439.

Y.1.   M. Yor, *Sur les intégrales stochastiques à valeurs dans un espace de Banach*, CRAS Paris **277** (1973), 467–469.

Y.2.   M. Yor, *Sur les intégrales stochastiques à valeurs dans un espace de Banach*, Ann. Inst. H. Poincaré X (1974), 31–36.

Index of Notations

$(I_X)_{F,Z}$, 133
$(\mathcal{F}_s^1)_{s\in\mathbb{R}_+}$, 325
$(\mathcal{F}_s^1)_{s\in\mathbb{R}}$, 323
$(\mathcal{F}_t^2)_{t\in\mathbb{R}_+}$, 325
$(\mathcal{F}_t^2)_{t\in\mathbb{R}}$, 323
$(\mathcal{F}_t)_{t\in\mathbb{R}_+}$, 124
$(\mathcal{F}_z)_{z\in\mathbb{R}^2}$, 323
$(\tilde{I}_X)_{F,G}$, 125, 327
$(\tilde{I}_X)_{F,L_G^p}$, 133, 327
1_A, 3
$E(f, \mathcal{F})$45
$H \cdot X$, 138, 329
H^T, 125
H^{T-}, 125
H_T, 125
I, 133
I_A, 3
I_X, 125, 133, 327
$I_{F,Z}$, 133
$L^1(\mu)$, 13, 14
$L_F^1(\mu)$, 15
$L_F^1(\Sigma, \mu)$, 15
$L_F^1(dg)$, 208, 390
$L_F^1(g)$, 208, 390

$L_{F,G}^1(g)$, 251, 409
$L^\infty(\mu)$, 14
$L_F^\infty(\mu)$, 16
$L^p(\mu)$, 14
L_D^p, 124
$L_F^p(m)$, 26
$L_F^p(\mu)$, 15
$L_F^p(\Sigma, m)$, 26
$L_{F,G}^1(dg)$, 409
R_z, 321
$\mathcal{B}_D(\mathcal{D})$, 84
$\mathcal{C}_{\text{loc}}$, 3
$\mathcal{S}_F(\mathcal{R})$, 3
$\Delta g(t)$, 208
$\Delta_R(g)$, 322
$\Delta_R\, g$, 322
$\Delta_{(z,z']}\, g$, 322
$\Delta_{z,z'}\, g$, 322
$\Delta_{zz'} g$, 322
$\mathcal{F}_\infty^1$, 323
$\mathcal{F}_\infty^2$, 323
$\mathcal{F}_D(\mathcal{B}, \tilde{m})$, 84
$\mathcal{F}_D(\mathcal{B}, \tilde{m}_{F,G})$, 84
$\mathcal{F}_D(\mathcal{B}, m)$, 84

419

$\mathcal{F}_D(\mathcal{C}, \tilde{I}_{F,G})$, 133
$\mathcal{F}_D(\mathcal{C}, \tilde{m})$, 84
$\mathcal{F}_D(\tilde{I}_{F,G})$, 133
$\mathcal{F}_D(\tilde{I}_{F,L_G^p})$, 133
$\mathcal{F}_D(\tilde{m})$, 82
$\mathcal{F}_D(\tilde{m}_{F,G})$, 82
$\mathcal{F}_D(m)$, 82
$\mathcal{F}_D(m_{F,G})$, 82
$\mathcal{F}_F((I_X)_{F,G})$, 328
$\mathcal{F}_F((\tilde{I}_X)_{F,G})$, 134
$\mathcal{F}_F((\tilde{I}_X)_{F,L_G^p})$, 134
$\mathcal{F}_{F,L_G^p}(I_X)$, 134
$\mathcal{F}_T$, 124
$\mathcal{F}_{\mathbb{R}}(\mathcal{B}, \tilde{m}_{F,G})$, 84
$\mathcal{F}_{F,G}(I_X)$, 134, 328
$\mathcal{F}_{F,G}(X)$, 134, 328
$\mathcal{F}_{F,L_G^p}(X)$, 134
$\mathcal{F}_{T-}$, 124
$\mathcal{F}_{st}$, 325
$\mathcal{F}_\infty^2$, 323
$\mathcal{O}$, 124
$\mathcal{P}$, 124, 323
$[X]$, 297, 299
$[[X]]$, 297
$[[X]]^B$, 296
$\delta r(\mathcal{C})$, 3
$\int H dI_X$, 134, 328
$\int H dX$, 138, 329
$\int f d\mu$, 14, 15
$\int f dg$, 209, 251, 390, 409
$\int f dm$, 9, 27, 85
μ_X, 217, 220, 258, 391, 395, 411
$\overline{m}(A)$, 20
$\mathcal{R}$, 124, 323
$\sigma a(\mathcal{C})$, 3

$\sigma r(\mathcal{C})$, 3
$\Sigma_f(\mu)$, 13
$\tilde{I}_{F,G}$, 125, 133, 327
$\tilde{I}_{F,G}(H)$, 133, 327
$\tilde{I}_{F,L_G^p}$, 133, 327
$\tilde{g}_{F,G}$, 246, 404
$\tilde{m}$, 81
$\tilde{m}_{F,G}$, 63
$\tilde{m}_{F,G}(A)$, 10, 63
$\tilde{m}_{F,G}(f)$, 79
$\tilde{m}_{F,\mathbb{R}}$, 65
$\tilde{m}_{\mathbb{R},E}$, 65
φ_A, 3
$a(\mathcal{C})$, 3
$|g|$, 202, 374
g_u, 405
g_z, 245
gm, 28
m_g, 203, 385
m_z, 66
$m_{F,Z}$, 71
$r(\mathcal{C})$, 3
$sv_{F,G}(g, I)$, 244
$sv_{F,G}(g, R)$, 403
$svar_{F,G}(g, I)$, 244
$svar_{F,G}(g, R)$, 403
$svar_{F,G}(m, A)$, 63
$v(g, I)$, 200
$v(g, P)$, 366
$v(g, R)$, 366
$var(g, I)$, 200
$var(g, P)$, 200
$var(g, R)$, 366
$var(m, A)$, 20
$sv_{F,G}(g, R)$, 403

Index

Algebra, 2
Almost everywhere, 13, 78
Approximation
    by σ-step functions, 6
    by step functions, 3, 4

Bracket
    sharp, 193, 349
    square, 297

Cadlag modification, 125, 181, 344
Commutation condition, 323
Conditional expectation, 45, 46
Conditionally compact set, 60
Convergence
    in measure, 82
    in the mean, 14
    locally uniformly, 171
    theorems, 87–90, 134, 153, 156, 157, 331

Decomposition of local martingales, 285
δ-ring, 2
Distribution function, 120

Division, 200
Doob's inequality in $\mathbb{R}^2$, 344
Doob–Mayer decomposition, 193, 349
Dual projection of a measure
    optional, 273
    predictable, 273
Dual projection of a process
    optional, 274, 280
    predictable, 274, 280

Egorov theorem, 87
Evanescent set, 124, 391
Existence of dual projection, 278
Existence of left limits in L_E^p, 332
Extension of measures
    from a ring to a σ-ring, 112–116
    from a semiring to a ring, 109

Filtration
    in $\mathbb{R}$, 124
    in $\mathbb{R}^2$, 323
Finite measure property, 13
Function

422   INDEX

Z-weakly measurable, 8
Σ-measurable, 3, 4
Σ-step, 3
μ-measurable, 13, 26, 78
μ-negligible, 13, 26, 78
σ-step, 6
Bochner integrable, 15
cadlag, 324
characteristic, 3
increasing in $\mathbb{R}^2$, 363
increment of a, 322
incrementally increasing, 363
incrementally right continuous, 322
integrable, 14, 82
monotone, 363
right continuous, 322
semivariation, 244, 404
simple, 3
simply measurable, 7
totally measurable, 10
vanishing outside a quadrant, 376
variation, 201, 374
weak star measurable, 8
weakly-measurable, 8
with bounded semivariation, 244, 404
with bounded variation, 200, 367
with finite semivariation, 244, 404
with finite variation, 200, 367

Gaussian measure, 120
Gaussian random variable, 120
Gaussian space, 120
Grid, 364

Integral(s)
    Bochner, 15
    classical, 13
    indefinite, 28, 96
    of positive functions, 14
    of step functions, 9
    of stochastic processes, 134, 328
    pathwise stochastic, 239, 268
    Stieltjes, 209, 251, 390, 409
    stochastic, 138, 329
    uniformly σ-additive, 58
    uniformly absolutely continuous, 58
    with respect to gm, 32
    with respect to a measure with finite semivariation, 85
    with respect to a measure with finite variation, 27
Itô's Formula, 312

Jump(s)
    of a function, 208
    process of, 304

Kluvanek theorem, 49

Lateral limits of functions in $\mathbb{R}^2$, 379
Lebesgue theorem, 18, 19, 90, 157, 177, 331
Limits of the variation in $\mathbb{R}^2$, 369
Local martingale, 183, 281–283, 285
    decomposition of, 285

Martingale
    1–martingale, 344
    2–martingale, 344
    local, 183, 281–283, 285
    square integrable, 185, 349
    strong, 344
    summable, 182, 183
    two parameter, 343
    weak, 343
Measure
    σ-additive, 9, 49
    σ-finite, 13
    absolutely continuous, 36, 52
    additive, 9
    associated to a function, 203, 247, 385, 406

bounded, 23, 24, 67
control, 55, 56
defined by density, 28, 96
Doléans, 165
finitely additive, 9
Gaussian, 120
locally absolutely continuous, 52
locally bounded, 23, 24, 67
locally uniformly absolutely continuous, 53, 73
locally uniformly strongly additive, 102
optional, 216
predictable, 216
representing, 12
Stieltjes, 209, 251, 389, 408
stochastic, 216, 391
strongly additive, 102
weakly σ-additive, 57
with bounded semivariation, 63, 67
with bounded variation, 20
with finite semivariation, 10, 63
with finite variation, 20
Modification, 125

Nikodym theorem, 56
Norming space, 2, 66

Operator
Dominated, 12
weakly compact, 12
Order relation in $\mathbb{R}^2$, 321

Partition in $\mathbb{R}^2$, 364
Pettis theorem, 57
Predictable rectangle, 124
Process(es)
σ-elementary, 157, 174
adapted, 124, 324
cadlag, 124, 324
caglad, 159
elementary, 157

integrable, 138
locally integrable, 171
locally pathwise integrable, 239, 269
locally summable, 171
measurable, 211, 391
optional, 124
pathwise integrable, 235, 267
predictable, 124
raw, 211
semilocally summable, 242
summable, 126, 327
with finite semivariation, 254, 391, 411
with finite variation, 211, 391
with integrable semivariation, 254, 391, 411
with integrable variation, 211, 391
Projection of a process
optional, 215
predictable, 215

Quadratic variation
scalar, 297
tensor, 297
vector, 297
Quasimartingale, 165

Radon–Nikodym property, 16, 36
Radon–Nikodym theorem, 36
Refinement of a partition, 364
Riesz representation theorem, 11, 117

Seminorm $\tilde{m}_{F,G}$, 79
Semiring, 2
Semivariation
and norming spaces, 66, 245, 405
function, 246, 404
of functions, 244, 403
of measures, 63
of real measures, 65

Set
 conditionally σ'-compact, 59, 60
 negligible, 13, 77
 predictable, 124
σ-algebra, 2
 optional, 124
 predictable, 124, 323
σ-ring, 2
σ'-compact, 60
Stopping time, 124
 graph of, 124
 predictable, 124
Stopping time in $\mathbb{R}^2$, 323
Summability
 criterion, 168
 of processes with integrable semivariation, 262, 413
 of processes with integrable variation, 233, 400
 of square integrable martingales, 188, 353

of stopped processes, 149
of the stochastic integral, 161, 163, 340, 341

Taylor formula, 311
Topology of convergence in the mean, 14

Usual conditions, 124

Variation
 function, 201, 367
 of a function, 200, 366
 of a function corresponding to a partition, 366
 of a measure, 20
 of a real function, 378
 of a real measure, 24, 25
 process, 212
Vitali theorem, 18, 19, 88, 156, 331
Vitali–Hans–Saks theorem, 56

PURE AND APPLIED MATHEMATICS

A Wiley-Interscience Series of Texts, Monographs, and Tracts

Founded by RICHARD COURANT
Editors Emeriti: PETER HILTON and HARRY HOCHSTADT
Editors: MYRON B. ALLEN III, DAVID A. COX, PETER LAX,
    JOHN TOLAND

ADÁMEK, HERRLICH, and STRECKER—Abstract and Concrete Catetories
ADAMOWICZ and ZBIERSKI—Logic of Mathematics
AKIVIS and GOLDBERG—Conformal Differential Geometry and Its Generalizations
ALLEN and ISAACSON—Numerical Analysis for Applied Science
*ARTIN—Geometric Algebra
AUBIN—Applied Functional Analysis, Second Edition
AZIZOV and IOKHVIDOV—Linear Operators in Spaces with an Indefinite Metric
BERG—The Fourier-Analytic Proof of Quadratic Reciprocity
BERMAN, NEUMANN, and STERN—Nonnegative Matrices in Dynamic Systems
BOYARINTSEV—Methods of Solving Singular Systems of Ordinary Differential
    Equations
BURK—Lebesgue Measure and Integration: An Introduction
*CARTER—Finite Groups of Lie Type
CASTILLO, COBO, JUBETE and PRUNEDA—Orthogonal Sets and Polar Methods in
    Linear Algebra: Applications to Matrix Calculations, Systems of Equations,
    Inequalities, and Linear Programming
CHATELIN—Eigenvalues of Matrices
CLARK—Mathematical Bioeconomics: The Optimal Management of Renewable
    Resources, Second Edition
COX—Primes of the Form $x^2 + ny^2$: Fermat, Class Field Theory, and Complex
    Multiplication
*CURTIS and REINER—Representation Theory of Finite Groups and Associative Algebras
*CURTIS and REINER—Methods of Representation Theory: With Applications to Finite
    Groups and Orders, Volume I
CURTIS and REINER—Methods of Representation Theory: With Applications to Finite
    Groups and Orders, Volume II
DINCULEANU—Vector and Stochastic Integration in Banach Spaces
*DUNFORD and SCHWARTZ—Linear Operators
    Part 1—General Theory
    Part 2—Spectral Theory, Self Adjoint Operators in
        Hilbert Space
    Part 3—Spectral Operators
FOLLAND—Real Analysis: Modern Techniques and Their Applications
FRÖLICHER and KRIEGL—Linear Spaces and Differentiation Theory
GARDINER—Teichmüller Theory and Quadratic Differentials
GREENE and KRANTZ—Function Theory of One Complex Variable
*GRIFFITHS and HARRIS—Principles of Algebraic Geometry
GRILLET—Algebra
GROVE—Groups and Characters
GUSTAFSSON, KREISS and OLIGER—Time Dependent Problems and Difference
    Methods
HANNA and ROWLAND—Fourier Series, Transforms, and Boundary Value Problems,
    Second Edition

*HENRICI—Applied and Computational Complex Analysis
    Volume 1, Power Series—Integration—Conformal Mapping—Location of Zeros
    Volume 2, Special Functions—Integral Transforms—Asymptotics—Continued Fractions
    Volume 3, Discrete Fourier Analysis, Cauchy Integrals, Construction of Conformal Maps, Univalent Functions

*HILTON and WU—A Course in Modern Algebra
*HOCHSTADT—Integral Equations
JOST—Two-Dimensional Geometric Variational Procedures
*KOBAYASHI and NOMIZU—Foundations of Differential Geometry, Volume I
*KOBAYASHI and NOMIZU—Foundations of Differential Geometry, Volume II
LAX—Linear Algebra
LOGAN—An Introduction to Nonlinear Partial Differential Equations
McCONNELL and ROBSON—Noncommutative Noetherian Rings
NAYFEH—Perturbation Methods
NAYFEH and MOOK—Nonlinear Oscillations
PANDEY—The Hilbert Transform of Schwartz Distributions and Applications
PETKOV—Geometry of Reflecting Rays and Inverse Spectral Problems
*PRENTER—Splines and Variational Methods
RAO—Measure Theory and Integration
RASSIAS and SIMSA—Finite Sums Decompositions in Mathematical Analysis
RENELT—Elliptic Systems and Quasiconformal Mappings
RIVLIN—Chebyshev Polynomials: From Approximation Theory to Algebra and Number Theory, Second Edition
ROCKAFELLAR—Network Flows and Monotropic Optimization
ROITMAN—Introduction to Modern Set Theory
*RUDIN—Fourier Analysis on Groups
SENDOV—The Averaged Moduli of Smoothness: Applications in Numerical Methods and Approximations
SENDOV and POPOV—The Averaged Moduli of Smoothness
*SIEGEL—Topics in Complex Function Theory
    Volume 1—Elliptic Functions and Uniformization Theory
    Volume 2—Automorphic Functions and Abelian Integrals
    Volume 3—Abelian Functions and Modular Functions of Several Variables
SMITH and ROMANOWSKA—Post-Modern Algebra
STAKGOLD—Green's Functions and Boundary Value Problems, Second Editon
*STOKER—Differential Geometry
*STOKER—Nonlinear Vibrations in Mechanical and Electrical Systems
*STOKER—Water Waves: The Mathematical Theory with Applications
WESSELING—An Introduction to Multigrid Methods
†WHITHAM—Linear and Nonlinear Waves
†ZAUDERER—Partial Differential Equations of Applied Mathematics, Second Edition

*Now available in a lower priced paperback edition in the Wiley Classics Library.
†Now available in paperback.